Praise for *The Permaculture Handbook*

When Apollo 13 found itself hurtling through space on its way to the far side of the moon with a ruptured oxygen tank, a critical path was assembled that returned all souls safely back to Earth. In *The Permaculture Handbook*, Peter Bane has mapped the critical path to a safe landing for civilization in the 21st century. That path runs through the backyards of suburbia and across the rooftops and balconies of urban apartment houses. Like the Apollo mission, Bane has jury-rigged our carbon dioxide removal system by re-tasking other components — small scale horticulture, aquatic plants and foraged trash, for instance — to buy us breathing room. This is a must read.

— Albert Bates, author of *The Post Petroleum Survival Guide and Cookbook*, and *The Biochar Solution*.

*The Permaculture Handbook* offers practical examples from the author's many years of experience, coupled with information from permaculture practitioners from around the country. This makes the guidebook particularly useful to American readers. The chapters covering the integration of multifunctional species and breeds of domestic animals and wildlife are especially note-worthy. These chapters address the sometimes under-recognized value of animal products and services in the Permaculture literature. This book is a joy to read while thinking about applications to one's own endeavors.

— Donald Bixby, DVM, Retired Executive Director of American Livestock Breeds Conservancy

Peter Bane's characteristic generosity and mastery are on display in this long anticipated encyclopedia of permaculture knowledge and lore. *The Permaculture Handbook* offers a thorough treatment of the design principles and patterns that of necessity precede the positive, practical, and profitable solutions that are the hallmark of permaculture design. Bill Mollison famously taught that we can create the "Garden of Eatin'". Peter Bane shows us how.

— Claude William Genest, Former Deputy Leader of the Green Party of Canada. Producer, Host of the Emmy nominated PBS series "Regeneration: The Art of Sustainable Living"

As oil becomes more scarce and expensive, society must adapt by re-localizing food systems. *The Permaculture Handbook* offers us a set of essential conceptual and practical tools for doing this in a way that's intelligent, beautiful, and sustainable. Want to create a better world? Your backyard — and this book — are the places to start.

— Richard Heinberg, Senior Fellow, Post Carbon Institute, Author, *The End of Growth*

Here is an exquisite design manual for re-inhabiting our towns and cities, transforming them into flourishing and nourishing landscapes that provide food, fiber, energy and community. As you read this book, looking out from time to time at the paved, manicured tracts we've called home, you can almost see the world around you transforming — a fruit tree here, a greenhouse there, a hill of corn, beans and squash in the corner, a photovoltaic array that charges your electric bicycle nearby. We are not in an era of despair, of coming apart. We are in an era of creativity where strong seeds grow in the very cracks in the concrete of our old ways of living. I applaud Peter for the wisdom, intelligence and hard work he has invested in his detailed, beautiful map of the land of permaculture.

— Vicki Robin, author *Your Money or Your Life* and *Blessing the Hands that Feed Us* (Viking/Penguin 2013).

Peter Bane offers a powerful antidote to what we might call our "ephemeraculture"— the consumerist, oil-dependent, wasteful, and ruinous way of life produced by the marriage of industrialism and capitalism. In a voice at once practical and visionary, he tells how we can achieve security, practice self-reliance, and revive local economies by shifting to a regenerative form of agriculture, beginning in our own backyards. If you're hungry for hope, good food, neighborliness, conviviality, and survival skills, here is the book for you.

— Scott Russell Sanders, author of *A Conservationist Manifesto*.

Morally mature cultures live well with the Earth and with each other. Peter Bane's beautifully written book lays out a path to that maturity and wisdom. Future generations will thank him — and you if you read and use the wisdom he has gathered.

— Carolyn Raffensperger, Environmental lawyer, advocate for future generations, Executive Director of the Science and Environmental Health Network

The wealth of practical knowledge in *The Permaculture Handbook* is a welcome addition to North America's canon of books on permaculture design, and will serve other regions as well. Peter Bane's broad and well-tested experience in ecological living, expressed here in clear and engaging prose, will guide homeowners, gardeners, and small farmers toward designing and living in lushly productive and harmonious landscapes.

— Toby Hemenway, author of *Gaia's Garden: A Guide to Home-Scale Permaculture*

A book to carry us through to Eden, this keen, comprehensive design guide with its wisdom and instruction will be indispensable to any household aiming at flourishing, soil-based self-reliance. The practice is permaculture and Peter Bane is a true master.

— Stephanie Mills, author of *On Gandhi's Path* and *Epicurean Simplicity*

With a lifetime of Permaculture immersion under his belt, Peter Bane's contribution to domestic self-reliant security in this comprehensive handbook is nothing short of remarkable. This will empower thousands of seekers to move off the bleachers and into the game. He balances beautifully the right amount of intensive ecological information with delightful case studies and practical outworkings. A gem of a manual. Thank you, Peter.

— Joel Salatin, Author, Farmer

There are a lot of people out there who want to know how to begin making a better world, but don't know where to begin. I can't think of a better starting point for someone who wants to restore and transform their place than Peter Bane's *Permaculture Handbook*. Bane has taken the single most urgent question of our times "How shall we live?" and given a real and viable answer.

— Sharon Astyk, writer, farmer, teacher and author of *Making Home*

*The Permaculture Handbook* is worth reading and rereading not only for its depth of insight but also for its storytelling charm. Peter Bane's reflections on community-based efforts to realize permaculture's potential across entire landscapes helps us realize that this set of design principles are not vague abstractions, but something that we can smell, taste and see before our very eyes.

— Gary Paul Nabhan, ethnobotanist, writer, permaculture orchardkeeper, seed saver

# THE
# PERMACULTURE
# HANDBOOK

*by*

Peter Bane

new society
PUBLISHERS

Cover design by Diane McIntosh.
Main image: © iStock (Jim Schemel); Rooster: iStock (pastoor)

Printed in Canada. First printing May 2012

Paperback ISBN: 978-0-86571-666-7
Ebook ISBN 978-1-55092-485-5

Inquiries regarding requests to reprint all or part of *The Permaculture Handbook*
should be addressed to New Society Publishers at the address below.

To order directly from the publishers, please call toll-free
(North America) 1-800-567-6772, or order online at
www.newsociety.com

Any other inquiries can be directed by mail to:

New Society Publishers
P.O. Box 189, Gabriola Island, BC V0R 1X0, Canada
(250) 247-9737

New Society Publishers' mission is to publish books that contribute in fundamental ways to building an ecologically sustainable and just society, and to do so with the least possible impact on the environment, in a manner that
models this vision. We are committed to doing this not just through education, but through action. The interior
pages of our bound books are printed on Forest Stewardship Council®-registered acid-free paper that is **100%
post-consumer recycled** (100% old growth forest-free), processed chlorine free, and printed with vegetable-based,
low-VOC inks, with covers produced using FSC®-registered stock. New Society also works to reduce its carbon
footprint, and purchases carbon offsets based on an annual audit to ensure a carbon neutral footprint. For further
information, or to browse our full list of books and purchase securely, visit our website at: www.newsociety.com

LIBRARY AND ARCHIVES CANADA CATALOGUING IN PUBLICATION

Bane, Peter
The permaculture handbook / by Peter Bane.

Includes bibliographical references and index.
ISBN 978-0-86571-666-7

1. Permaculture — Handbooks, manuals, etc. 2. Permaculture — Case
studies. I. Title.

S494.5.P47B35 2012          631.5'8          C2012-902010-9

new society
PUBLISHERS
www.newsociety.com

MIX
Paper from
responsible sources
FSC® C011825

*To the memory of*
*Ivan Illich,*
*defender of our humanity*

# Contents

## PART III: OUTCOMES

# Acknowledgments

Many people have contributed to my permaculture education, particularly those colleagues who have taught, published or helped to promote courses and projects with me. Michael Gibson deserves a note of thanks for opening the door. My first teachers, Lea Harrison and Max Lindegger, were warm, accessible and inspiring. As mentors, role models and friends they helped me to feel at home in the world. Craig Elevitch and Ken Boche were nest mates: fledgling together we took the first steps in design. Chuck Marsh enabled me to begin teaching and continued to be a close friend, collaborator and companion for many years through numerous permaculture adventures. From him I learned a great deal about plants, horticulture and the landscape trade. From him I also caught a serious case of ecovillage fever that nearly did me in. Jerome Osentowski has been a fellow traveller, friend, mentor and source of tremendous inspiration for over 20 years. As students, co-teachers, neighbors, colleagues and mountain men, Lee Barnes and Andrew Goodheart Brown have enriched my life in more ways than I can say.

The most fully developed examples of garden farming in this book are the work of Jerome Osentowski, Lee Sturgis, Dave O'Neill and Ian Graham. I am grateful to them for inspiration, hospitality, challenge, persistence and for the glorious beauty and success they have achieved. I thank them for making their work and lives an example to the world.

Beverly and Carl Winge were godparents to permaculture in the US and to my young family. Beverly was also my first teacher of journalism. Guy Baldwin made possible my career as a publisher by wanting to end his own. It was through that opportunity that I learned much of what I have written here. Cynthia Edwards made connections for me that have endured and that changed the world. Michael Pilarski is a treasure to us all. George Sobol and Patsy Garrard have been allies and mentors from the moment we met. Their collegiality and support, like that of Tony Andersen, were important international bridges for permaculture and for me personally. Patricia Michael introduced me to bioregionalism and much more, and in that movement I was privileged to meet Gene and Joyce Marshall whose wisdom, compassion and humor have allowed me to feel that I belong to a tradition, however unorthodox. Dave Jacke remains an honored colleague, friend and teacher whose struggles have made my own journey easier. Trina Paulus always knew. Larry Patrick helped me to understand my calling. Andy Langford has been to me a teacher and a fount of creative thinking for many years — I don't expect we're through yet. Richard Griffith launched my international teaching career and introduced me to the Canadian permaculture community where I was fortunate to meet Stuart Hill, a great sage and teacher who has skillfully challenged all permaculture thinkers to go within and to consider the human dimensions of sustainability.

Albert Bates generously offered photographs, support and helpful criticism for this effort. More than anyone I know, his peripatetic explorations and catholic intellectual pursuits have brought the world of energy descent cultures closer to hand. Tim and Maddy Harland occupy a special place in my heart for they among few in the world share the daily struggles and joys that only permaculture publishers can understand. Arjuna da Silva and Paul Caron were constant companions, tender friends and bold antagonists on the frontiers of ecovillage life. My hat is off to them both for staying the course and for enriching my life immensely. Declan Kennedy, with true Celtic grace, has danced circles around the impossible, making it look easy. Toby Hemenway has broadened the popular knowledge and appeal of permaculture in North America, and even more importantly to me he generously contributed his knowledge and his talent to *Permaculture Activist* for many critical years. His clear thinking and ecological insights have been an inspiration.

Joseph Kennedy, Michael Smith and Catherine Wanek invited me graciously into the world of natural building. Gustavo Ramirez and Silvia Balado showed me the warm and lively culture of the pampas. Their work displays remarkable understanding of permaculture design, and from them I have learned a great deal. Diana Christian, fellow editor, ecovillage neighbor and now author, was a provocative interrogator and wicked confidante when life brought us together for several years. Her journalism continues to inspire me. Becky Elder displays an exuberant love for people and the living world that is as big as all outdoors. Her stalwart support and boundless enthusiasm no less than her buoyant company have brought me joy. Mark Krawczyk demonstrated the authenticity of the permaculture diploma at a time when few understood it in the US. He has been a delightful collaborator and steady correspondent for many years. Martin Crawford has cheerfully shown the value of persistence with tree crops, making a global example of his own work. I am grateful for his friendship. Darren Bender-Beauregard provided me a good background on the Guinea Hog and its care. He and his wife Espri have cordially opened their home and farm to my classes for a number of years, sharing their innovative garden farm systems with others. Scott Horton, editor, co-teacher, artist and advisor, held my hand through many difficult seasons, bringing his creative flair to the often mundane world of print journalism as he stretched the bounds of the permaculture community. John Stollmeyer brought me into one of the richest cultures I have ever had the pleasure of experiencing and where I felt improbably at home. His generosity enabled me to realize capacities I treasure to this day. Sandy Cruz has been a loyal friend and collaborator whose ability to find positive expressions for virtually every challenge in life seems both ordinary and astonishing. Teaching with Keith Morris helped me to sharpen my game.

My students, especially those who have sat through longer workshops and courses, have been a continous inspiration and a source of much valuable insight and growth. It is from and for them that I have written this book.

Though working colleagues have been the ground of my permaculture experience, none of it would have come about absent the indispensable creative endeavors of David Holmgren and Bill Mollison. We are all in their debt. I particularly want to thank David for the inspiration of his 2005 paper, "Retrofitting the Suburbs for Sustainability," which recapitulated in clear terms the germ of an idea long familiar in permaculture discussions, and which has become a central theme of this book.[1] David and his partner Su Dennett kindly put me in touch with Richard Telford who graciously made available his excellent iconic art to illustrate the permaculture principles and ethics in Chapters 3 and 4.

The words in this text are mine and its mistakes as well, but the graphics that accompany them have come from the hands and the vision of many others. In particular, I am grateful to photographers Patrick Brunner, Joshua Choate, Warren Kirilenko and Norm Shafer. They stand out from among the dozens, known and unknown, whose images grace these pages. Bill Hoover at Purdue University genially helped me to track down some obscure but valuable photos of intercropping. The North American Regional Climate Change Assessment Program (NARCCAP) kindly supplied the map of projected climate shift reproduced in Chapter 20. Despite a colossal legacy of destruction, the US federal government does no small amount of virtuous work. NARCCAP is funded by the National Science Foundation (NSF), the US Department of Energy (DoE), the National Oceanic and Atmospheric Administration (NOAA) and the US Environmental Protection Agency Office of Research and Development (EPA). The National Arbor Day Foundation cordially gave permission for me to use their climate zone hardiness maps. Much more importantly, however, they preserved access for the public to the data that underlie those maps when it was briefly released in 2004 and later retracted by the USDA. This was, of course, practical information routinely generated by government scientists about the northward migration of climate zones due to global warming that the G. W. Bush administration found politically inconvenient to acknowledge.

It would not have been possible for me to write this book or to have made an honest claim to being a garden farmer without the unstinting support of my life partner Keith Johnson, he of the ten green fingers and ten green toes. In no small way the vision we have forged together has propelled this book into the world.

A small but wonderful cast of players accompanied and made possible the work of writing *The Permaculture Handbook*. I want to thank Jonas Carpenter and Philip Shelton who did much of the heavy lifting at Renaissance Farm during the year when the manuscript was being born. They did so cheerfully and brought joy to our household when the work seemed endless. Rob Archangel aided me early on with research. Rhonda Baird read and critiqued the chapter on labor and even more importantly has cheerfully labored in the vineyard of permaculture design and teaching along with me for six years — taking on her shoulders great loads that eased my own burdens. Marie Fleming was the first to read the book in its entirety. Her creative suggestions contributed greatly to its panoply of illustrations, and her unfailingly positive response was deeply supportive at a time when I needed it the most. Jack Heimsoth, Abi Mustapha and Jami Scholl worked diligently and creatively to give spirited expression to concepts that were hard to photograph. Their drawings have given the book a life of its own. My daughter Liberty Bane and her husband Edward Carter played crucial roles in organizing the art and in beginning to produce some of the hundreds of photos that have been used here. I hope the book and its message may be as meaningful to them as their support has been to me.

A special word of thanks must go to Creighton Hofeditz who not only took scores of photographs but accomplished the seemingly impossible task of marshalling, scanning, formatting and presenting virtually the whole corpus of images reproduced herein, moving seamlessly between film negatives, digital photography, paper drawings and the bottomless well of the World Wide Web. He became an expert fisherman, trolling the interwebs and file cabinets for unlikely prey. Throughout the months just behind us he displayed by turns the delicacy of a snake handler, the charm of a stage actor, the discretion of a lawyer and the determination of a commando.

I am grateful to my editor Betsy Nuse, whose unfailingly kind and accommodating support made palatable and ultimately welcome the deep and potentially painful cuts to the original manuscript that have enabled it to be published. It is much stronger for her efforts. Heather Nicholas and EJ Hurst seduced me into New Society's tender trap and continue to shepherd the book's progress toward its audience. Sue Custance and Ingrid Witvoet, managing editors and producers of the work, showed grace under fire and remarkable patience. Thank you all.

— Peter Bane
January 2012

# Foreword

## by David Holmgren

Permaculture is a design system for sustainable living and land use first articulated by Bill Mollison and myself in Australia in the mid-1970s. Since then it has spread around the world stimulating creative household and community initiatives to reduce ecological footprint, increase resilience and relocalize economies. While the scope of permaculture applications ranges from aquaculture to design against disaster, from ecological building to local currencies, many people would understand permaculture as being a form of organic gardening.

The idea that gardening is the most sustainable form of agriculture and the basis for the relocalization of our economies, proposed in *Permaculture One* (1978), was reinforced in a short essay I wrote in 1991, *Gardening As Agriculture*. In that essay I asserted that gardening should be recognized as a serious and important form of agriculture that functions as an incubator for new farmers and farming methods.

Over the last three decades a small but growing number of pioneers informed by permaculture and related concepts have shown how this is possible. In recent years the grassroots explosion of interest in food gardening and farming is reshaping mainstream approaches to sustainability. This belated recognition is a hopeful sign that an abundant and resilient future is possible by redesign of food production and consumption.

With his chosen term "garden farming," long-time permaculture writer, publisher, teacher and practitioner Peter Bane crystallizes this concept for those new to permaculture as well as its seasoned practitioners seeking to extend their chosen way of life into a livelihood. In focusing on the productive transformation of our suburban and peri-urban allotments, Bane shows how these "problematic" landscapes could become the "solutions" in an energy descent world of ongoing climate change, expensive and unreliable energy and economic contraction.

In true permaculture style, this book combines empowering vision with grounded common sense, strategic thinking with nuts and bolts information.

Part One eloquently and simply conveys the principles and patterns behind this big-picture story. Bane's pattern language of garden farming, based on the classic pattern language of architect and town planner Christopher Alexander, is a major contribution to the ongoing evolution of permaculture design methods.

Part Two draws on the author's depth of experience as a permaculture teacher to explain the diverse components of permaculture design for creating a livelihood from garden farming in suburban and peri-urban landscapes. The metrics and rules of thumb necessary to make designs work are grounded in both the author's personal experience and his extensive observation of pioneering examples of garden farming.

The reality of garden farming is convincingly portrayed and further illustrated at intervals through the book with case studies informed by permaculture design principles and patterns.

Of all the permaculture books from Australia, America and around the world, this one most completely fills the big space between my own articulation of permaculture theory in *Permaculture: Principles and Pathways Beyond Sustainability* (2002) and my earlier intimate documentation of our own efforts towards garden farming in *Melliodora: A Case Study in Cool Climate Permaculture* (1995). This book is likely to become the classic design manual for those with the energy and enthusiasm to become the garden farmers of the future.

In the process, Peter Bane shows that, in hard times, the apparent ethical conflict between personal and household resilience on one hand and working for a better world on the other can be resolved applying permaculture ethics and design principles.

DAVID HOLMGREN is the co-originator of the permaculture concept. He is the author of *Permaculture Principles and Pathways Beyond Sustainability* (2002) and *Future Scenarios* (2009), among many other books and publications. A smallholder and garden farmer in the Australian state of Victoria, he has taught permaculture design in North America, Latin America, Europe and the Middle East for nearly two decades, and has played a critical role in advancing global understanding of the complex and multilayered challenges surrounding energy and resource use.

# Attitudes

# CHAPTER 1

# Garden Farming

New and old at the same time, garden farming has historical antecedents and contemporary examples all over the world. At its simplest it is no more than the efforts of people to provide for their own needs from their immediate surroundings, work that connects us directly to our Neolithic ancestors. Dressed up in the patriotic colors of Victory Gardens during World War II, garden farming grew 40% of America's produce.[1] Small farms are still the pattern in Japan, Korea, Taiwan, Poland, Slovenia and many other developed Asian and central European countries.

In the Soviet Union, where the state-run collective farms were notoriously inefficient and food shops were often empty or offered limited supplies, the Russian people learned to supplement their diet from produce grown at their *dachas*, small summer cottages just beyond the urban fringe. When the Soviet Union collapsed in 1991, these peri-urban garden farms were a lifeline that prevented mass hunger. People simply planted extra rows of potatoes and cabbages, and when they couldn't get fuel to drive out to the country, took the bus or rode a bike. Private allotments surrounding cottages today grow 50% of the country's vegetables, fruits and dairy on 7% of the land.[2]

These and other examples throughout this *Permaculture Handbook* will show that much of garden farming is about meeting household needs — what in permaculture is called self-reliance — a term that I distinguish from the more commonly used phrase, self-sufficiency. *Self-sufficiency* implies not needing any supplies from outside. Many 19th and early 20th century farms in Europe and the US were self-sufficient, buying only salt, tea or yard goods and other luxuries. *Self-reliance*, on the other hand, is about taking responsibility for one's own household needs as part of a resilient local economy. Trade and barter will be important components of a self-reliant economy. In the US, Amish communities produce a great deal of their own food, clothing, tools and household goods locally, but they are also involved in a great deal of trade. Where there is no local source for some items, they are purchased by mail order from other Amish producers or even "English" neighbors or commercial concerns.

## Building Self-Reliance

Self-reliance is an aim of the permaculture design system. At the household and community level it increases security and independence, thus *resilience*, or the ability to absorb shocks and disturbances and to recover quickly from them. Self-reliance reduces dependence on distant sources and suppliers and thus reduces the energy intensity of food and other essentials, shrinking our ecological footprint. By meeting most of our own needs, we minimize the damage and dependency caused by global trade, more easily regulate our consumption and conserve resources. We also bolster our own capacity for survival and prosperity as well as our ability to aid others around us. Self-reliance is not about isolation, nor is it a dogma; rather it describes a rational hierarchy of independence and interdependence from which we can make ethical decisions about what we consume…and what we produce.

My own permaculture teacher, Lea Harrison, told a story that has remained with me about self-reliance. When visting the flat lowlands of Nepal in the 1980s to teach permaculture, she saw — in an area called the Terai — many small farms with neat fields of millet, mustard, wheat and lentils, but there were few trees. As her hosts showed her around one village, she spotted a woodland across the valley which seemed to be of a very different character from the farms surrounding it, so she asked if they could take a look. It turned out to be the home compound of Mrs. Rai, a woman who ran a school for neighborhood children. Welcomed as a foreign dignitary, Lea was shown the grounds, which featured edible and economic species of every type; the children were pursuing their studies out-of-doors, under the tree canopy. Lea asked her Nepali host how she had come to this place, and why it was so different from all the farms around it. Mrs. Rai laughed and replied, "Well, you see, if I needed it, I planted it. And this is what happened," explaining that her method was to grow everything that she and her students would use for food, fiber and medicine. The result was a food forest of the sort that Lea had been teaching her own students how to plant and cultivate. As we have learned time and again, no one has a monopoly on good ideas.

We each begin our journey toward self-reliance with the most essential elements and those we can supply most easily, and we continue replacing things we consume with things we produce (or eliminating consumption of needless items altogether), until it ceases to be economically sensible or practical to do so. For example, our southern Indiana household is unlikely anytime soon to grow tea or lemons, or to forge our own wrenches or strike our own nails. We haven't turned off the water from the public system, but we use very little of it and should the need arise, we could supply our own for many months (or indefinitely) from roofwater caught and stored in tanks. We continue to shop in local markets, but we have food put by and we grow a lot of our own. Each year we increase the amount we grow and the amount we store, as well as the amount and variety of things we sell or trade. We are well started on the path to self-reliance, and I hope this book will encourage and empower you to set out on that path — and to make great strides along it.

Self-reliance is an important aim for any society that hopes to endure the challenges of the coming decades. When I visited Slovenia in 2005, I saw a small nation of small communities. The capital Ljubljana has the population size of Madison, Wisconsin, or Windsor, Ontario, while the whole nation is about as big as New Hampshire. It has a temperate climate somewhat similar to that of eastern Pennsylvania, stretching between a narrow coastline on the Adriatic Sea and the Julian Alps. Throughout the countryside and in small towns, almost every house seemed to have a garden. Up the hills behind the houses were rows of grapevines, which my friends told me enabled most households to make their own wine.

Slovenia is a prosperous nation of two million people, a former Yugoslav republic that has become a member of the European Union. When we visited, the roads, rail lines and other infrastructure of the country were in good shape, and the markets and shops seemed to be full of a wide variety of goods. There were many cars in use, most of them fairly new or in good condition. The food we ate was of good quality and served to us in generous amounts. The people looked trim and healthy. There was little evidence of environmental contamination or pollution. By any standard, Slovenians enjoy a good quality of life, but the nation also demonstrates a high level of self-reliance at the household level.

## Feeding the Cities

The idea that our towns and cities could feed themselves with a little help from the suburbs and the countryside nearby may strike most observers as preposterous, but, as this book attempts to show, the present state of agriculture and energy supply is so unstable that food security for town and country alike depends on our willingness to undertake self-reliance. There is no shortage of examples to inspire us. Hong Kong, one of the most densely settled territories on Earth, still managed during 2010 to grow a considerable amount of the poultry required by its inhabitants on land within the tiny enclave.[3] Havana today, as a result of changes in post-Soviet Cuban agriculture, has over 35,000 urban gardens producing about half the fresh food consumed in that city of two million, plus a market selling local produce in each one of its 2,000 neighborhoods.[4]

Tenochtitlan, a pre-Columbian city of 200,000 on the site of modern Mexico City, was substantially fed from intensively cultivated water gardens nearby. Historically, food from these gardens traveled to market by canoe.[5] Remnants of the gardens, called

Land is meticulously cared for in Slovenia, and domestic self-reliance is high: only single trees or small patches of forest are cut. Stacks of lumber can be seen curing throughout the countryside. Drying hay is roofed, and the flowers of elder and other perennials are harvested for home use and market. [Credit: Peter Bane]

Canals in the shallow lakes provide boat access to *chinampas* in Xochimilco, south of Mexico City. [Credit: Scott Horton]

*chinampas*, persist today and are grown on raised beds in the shallow lakes.

New York at the turn of the 19th century was fed by a complex and productive garden agriculture on Long Island and from small farms up the Hudson Valley and in New Jersey. The term *truck farming* comes from the practice of trucking produce from farm to city markets, a matter of a few hours travel by horse-drawn wagon.[6]

Nor is the practice of self-reliance remote from us in either time or space. Today in North America, the fastest growing forms of agriculture are small peri-urban farms of less than 20 acres growing vegetables for market.[7] Much of this movement is driven by a demand for local and organic food stemming from concerns about personal and ecosystem health. And quite a bit of it is off the official radar. North Americans are learning to be discerning food consumers, and farmers markets are burgeoning.[8] This also means that many more people are choosing to start farming than at anytime since World War II.[9]

## A New Sustenance

When the history of the 20th century is written two hundred years from now, one of the things I imagine writers will recognize as contributing the most to human evolution is the emergence of what is loosely termed *organic agriculture*. By that I do not mean the codification of certain practices under sanction of the US Department of Agriculture, but a broad movement that preceded and gave rise to that codification and which is likely to outlast that august but deeply corrupt agency. Rather, I point to forms of farming and gardening that generate soil fertility as well as crops for market and consumption — farming that can regenerate landscapes, ecosystems and even whole societies. Humans have practiced agriculture for 10,000 years without understanding why it works, and as a consequence most long-term agricultural practices and the civilizations built on them have collapsed from one or another failure to maintain ecological balance. Again and again, soils became exhausted, salted from irriga-

Faster than almost any other civilization, North Americans have destroyed their soils by plowing, poisoning and salting with fertilizer and irrigation.
[Credit: erjkprunczyk via Flickr]

tion, eroded because of plowing; the climate changed because too many trees had been cut or the population increased beyond the ability of farmers to provide food.

To understand the profound importance of regenerative agriculture, the kind of farming that builds natural capital, we need to see it not as a fringe or retrograde activity — "unable to feed the world," as conventional agronomists would claim — but as a heroic and undersung achievement in the face of overwhelming institutional neglect, cultural dissipation, economic monopolies and dire ecological challenges from chemical, nuclear and genetic pollution, climate change and an eroding resource base in the land and in society. Tracing the threads of progressive agriculture and ecology against the backdrop of the rise of a consolidated industrial food system will, I hope, show how the former carry a true populist mandate which we must now champion to reclaim the common wealth of the earth itself and to secure a future for ourselves and our descendants.

## Oil Turns the Tables

The end of World War II brought about a transformation of the US economy. The enormous productive capacities of wartime industry were shifted to provide housing, transportation, consumer goods and new machinery for agriculture. Chemical factories which had been turning out munitions and synthetics for the war switched to fertilizer and pesticide production, while the government through agricultural extension, centered at the research universities and bolstered by a triumphant chemical science, promoted the model of industrial farming. Forty-acres-and-a-mule was now out, and every John Doe farmer was urged to buy a John Deere tractor, a combine and a sprayer. The logic of Fordism gained the upper hand, and by the 1970s, Nixon's Secretary of Agriculture, Earl Butz, a brilliant but benighted agronomist from Purdue University, was announcing to American farmers the new national policy, "Get big or get out."[10]

Food growing shifted from regional economies to a national system of increasingly specialized producers and processors. The Interstate and Defense Highway System, initiated under Eisenhower in 1957 as a make-work guarantee against another depression, enabled trucking to displace rail delivery of many cargoes, but especially it enabled large California and Florida growers to exploit their long-season advantage to capture most of the nation's market for fruits and vegetables, leaving a hollowed-out agriculture through the Corn Belt hanging onto beef, hogs, corn, soybeans and some small grains and hay as the basis of a shrinking farm sector. First poultry processing, and later pork and beef production, were consolidated by a few large meat-packers, while the profitability of small, mixed farming was eviscerated. More and more farms went under the auctioneer's call, their impoverished and depressed former proprietors remaindered to fill low-end jobs in a booming industrial economy.

Even as local truck and dairy farms disappeared under suburban sprawl, organic gardening began to gain traction through the publishing and research efforts of J. I. Rodale and the stimulus of best-selling Connecticut author Ruth Stout, whose books on mulch gardening promoted a lazy approach.[11] Get spoiled hay from the farm down the road, she wrote: pile it on your garden and plant into it, and you'll never have to weed. Millions were inspired to try, even if the scientific basis was still obscure. Stout's work was important too in being an early urban critique of agricultural dogma, and an important call for differentiation between farming — with all its connotations of peasant drudgery — and gardening, which was blithe, sophisticated and smart. Stout was a *bon vivant* who enjoyed the company of New York's literati. She wanted her garden to flourish, but she didn't want it to rule her life. Home gardening and organic

Small-scale seed exchanges and wider networks are helping to maintain crop diversity in the face of corporate seed control and monoculture farming.
[Credit: Soren Holt]

methods got a further boost when Rachel Carson exposed the dangers of pesticides, vaunted by industry and government then and since as the guarantor of abundant food surpluses.[12]

The bigger-is-better thread of this story continued to unfold in ways that are increasingly familiar and deeply tragic. Industrial agriculture, striving for ever-greater economies of scale, grew hand-in-hand with food processing industries that consolidated many thousands of smaller producers into gigantic and increasingly multinational conglomerates. The Green Revolution, funded by the Rockefeller and Ford Foundations, exported the DNA of industrial agriculture to India, Mexico, the Philippines and Indonesia, major centers of both traditional farming and agricultural diversity.[13] Environmental contamination continued to increase, much of it driven by agrochemicals, pharmaceuticals and the industrial processing and packag-

ing of food for transport. These industries, whose roots run through the death camps of Nazi Germany and the laboratories of the nuclear and munitions complex of war and empire, eventually consolidated into a global oligopoly enclosing food, medicine and seeds, and vernacularly called Big Pharma. By 1970, the components of this cartel had already begun lobbying for patent protection for plant breeding to ensure the commercial viability of hybrid seeds; subsequently they would argue and connive for genetic manipulation of plants and animals to increase their control over the world's food supply.[14]

The popular response to this rising tide of industrial pollution and social sabotage erupted in 1970 with the environmental movement, but early legislative victories in the establishment of the Environmental Protection Agency, and the adoption of the Clean Water Act and Endangered Species Act among other landmark laws, had mixed results. US rivers and streams got cleaner, but coal-powered plants continued to evade smokestack regulation, so acid rain and mercury pollution spread further. The Reagan-Thatcher worldwide reaction against the progressive measures of the 1970s rolled out a program of corporate theft and upward concentration of wealth in the name of deregulation, "free" trade and other abominations of Milton Friedman's economic theology. With the installation of George W. Bush as Decider-in-Chief in the United States, a concerted rollback of progressive laws (such as the Bill of Rights) and a subverting of environmental regulation and oversight became the agenda of the national government in the US. Since 2008 we have seen the crash of this conceit, if not yet a cleanup of the wreckage or a thorough prosecution of the perpetrators.

It must be pointed out that none of this economic concentration would have been possible without a massive application of petroleum-based energy. Like fish in the sea, North Americans have swum in the ocean of cheap hydrocarbons for so long that we have almost no frame of reference for thinking about a world without them. Not everyone, however, has been so blinkered.

## Another Way

Mostly outside the lens of the mass media, the twinned thread of a resurgent popular sovereignty has gathered resources, asserting that another world is not only possible but necessary. Contesting complex and multi-dimensional issues in obscure global forums, the political activism of smallholders has been little understood by North Americans living in the cocoon of empire. As patent protection, intellectual property rights in plant breeding and later genetic engineering began to enclose the world's common heritage of crop diversity for private profit, seed-saving groups and networks sprang up around the world.

Indian nuclear physicist Vandana Shiva was an early and incisive voice against the destructive enclosure of agriculture. Her 1991 address to the International Permaculture Conference in Nepal was an important element in my own education about the politics of diversity.[15]

By the 1970s, the leisure pursuits of gardening, combined with concerns about food safety and a cultural wave of back-to-the-land reaction against industrial excess and soulless work led to an explosion of small-scale experiments in organic farming. *Mother Earth News, Harrowsmith Magazine* and *Acres U.S.A.* emerged as voices for this nascent eco-agriculture and chroniclers of the growth of renewable energies and other cultural and technical innovations supporting a new agrarianism. *Mother* and *Harrowsmith* aimed at new homesteaders, while *Acres* publisher Charles Walters marketed his papers to those trying to make a living in farming. *Acres* led the call for rural regeneration with a populist critique of big government agriculture policy and wed this to alternative scientific views of health taken in a variety of slices: through

soils, pasture plants, animal nutrition and of course into food, the human body and an environmental paradigm of medicine.[16]

Poet and Kentucky farmer Wendell Berry became a prophet of this new movement with the publication of his 1976 book, *The Unsettling of America.*[17] It was a call for a renewal of the agrarian roots of the country. Agronomist and Methodist lay preacher Wes Jackson entered the fray in 1980 with his collection of essays, *New Roots for Agriculture.* Jackson has continued and deepened his researches into a new ecological basis for agriculture at the Land Institute, arguably now Kansas's greatest cultural offering to the world.[18]

Farmers of the upper Midwest, reacting to the agricultural depression and dispossession of the early 1980s, banded together as the National Farmer's Organization to protest agricultural policy, but soon turned these political impulses into economic traction by founding the Coulee Region Organic Producers and Packers, a cooperative which has grown into the largest marketer of organic farm produce in the nation, doing business as "Organic Valley."™ Other marketing innovations helped the increasing number of small organic farmers who were experiencing productive success find outlets for their crops. In 1981, Robyn Van En of Indian Line Farm in Massachusetts helped pioneer in North America the Community Supported Agriculture (CSA) model of subscription farming that had originated in Switzerland and Japan in the late 1970s.[19] By linking food consumers directly with food producers, CSA reduced costs, spread the considerable risk of farming widely, helped small and new farmers get into business and flourish and built local economic connections that formed the basis of new communities of place. Over 12,500 CSAs are now operating through the US and more than 500 in Canada. Bob and Bonnie Gregson described this movement through their own experience on two acres outside Seattle in a small 1996 booklet.[20] They boasted of a comfortable middle-class income from half their land supplying city families with vegetables, small fruits and eggs eight months a year, with time in the winter for knitting, repairs, crafts and other leisurely or cash pursuits. They predicted that a torrent of others would follow in their wake, and indeed many tens of thousands have.

Not all the lines that now converge on garden farming began on tiny plots. The intensive grazing management insights of Frenchman Andre Voisin[21] were picked up by Alan Savory, who carried the grazing message wrapped in a broad systemic theory called "Holistic Management" to mostly larger farmers and ranchers in Australia, New Zealand and the western US. These ideas, which exploit the possibilities of movable electric fencing and portable watering systems to enable graziers to condense animals into compact areas of pasture and move them frequently, mimicked the impacts and outcomes of predator-herd interactions in nature that Voisin had written of and that Savory had seen in Africa. The resulting agricultural savannas were many times more productive than traditional pasture management and grew far healthier animals and meat than confinement feedlot operations. When entrepreneurs such as Joel Salatin, whose family had tried innovative farming methods in Venezuela in the 1950s before returning to Virginia in 1961, bred the direct-to-consumer marketing model with the productive power of rotational grazing, then backcrossed the offspring with a line of pure polyculture, a breakout champion was born.[22] Salatin has made the rounds of farm conferences across the continent and has leapt out of his self-described box as "a Christian, libertarian environmentalist" into avant garde circles through permaculture courses and even the eco-entrepreneurs' summit Bioneers conference in wealthy Marin County, California. More recently, perennial agricultural writer Gene Logsdon has described his own smaller-scale rotational grazing on a 32-acre garden farm in Ohio.[23] He predicted that millions could adopt his pasture practices on spreads as small as five acres.

## Descending the Energy Mountain

In the paragraphs above I have sketched the emergence of small-scale and regenerative farming as a viable alternative to the juggernaut of industrial agriculture and Big Pharma. The hope of the former is that it can survive and prosper through the continuing aggressive expansion and ultimate collapse of the latter. The growth of industrial agriculture and the emergence of global monopolies combining seed, food processing, chemical manufacture and pharmaceuticals track well with the increase in application of fossil energy to human purposes. How these immense aggregations of capital and power unwind cannot be entirely foreseen, though sudden collapse is a distinct possibility. That they have grown fat on an empire of oil, that their fortunes are tied to it and that the empire is now well into its final decades of decline is a central presumption of this book and a fact well enough established by others that I shall not argue it here.[24]

We are not going to eat food grown in test tubes, no matter how many science fiction novels have been published. We will continue to need to cultivate the soil of the Earth. But with changing and unstable climate everywhere, and in the wake of modernity's massive dislocation of traditions, we find ourselves in utterly novel and not entirely friendly territory. The familiar is no guide to the future, and in any case is being swept away before our very eyes.

The global monopolists are right about one thing: we do need a system. How else can seven billion people figure out how to live on a rapidly evolving (or devolving) planet? But we don't want a system that imposes one solution on everyone and everything. That's death. We need a system that offers us an unchanging reference point in a world of continual change—a system that is accessible to all human beings in all cultures and all places, and that empowers each of us to continue to unfold its mysteries. We call this empirical exploration and systematization of the self-organizing and phenomenal world *science*, and its growth is one of the great gifts of our plunge into the corrupting cesspit of fossil energy.

We cannot expect, however, to continue for long increasing our knowledge of the world by splitting atoms at great cost and peril, or by launching rockets into deep space, though these things are wondrous. The word science comes from a Latin root that means to split or cut, and most of the early and powerful discoveries of science came from cutting things apart into their pieces and discovering the relationship of those pieces to other pieces. The world, as a result, is pretty well blown up and is quite in danger of coming apart in great chunks if human beings don't quickly learn how to knit things together gracefully again. Fortunately, out of the ramifications of scientific exploration came the science of ecology which, unlike most fields of science until quite recently, began and continues as an integrative discipline (geography and anthropology are two similar fields that come to mind).

## A New Science of Holism

Like many branches of science, *ecology* is the study of relationships, but unlike most scientific disciplines, it is the study not of one element to another, but of all to all systematically. Ecology looks at living communities or assemblages of many species of plants, animals, microbes and the landscapes they inhabit to assess the quantity and quality of information and resources that flow within the system. It can identify system boundaries and can see where information is missing and when it is critically threatened (as, for example, when a key species goes extinct and an important function of the whole system

consequently fails). And like the novel discipline of systems analysis with which it shares a common intellectual structure, ecology is the study of feedback loops and their consequences. In other words, it looks at how living communities relate within and without.

Ecology operates with a novel kind of logic and solves problems differently than most disciplines. Ecological thinking is the fundamental tool for regenerative agriculture and for garden farming. It underpins permaculture design. The problems of ecosystems, or indeed of biology, like the problems of business or cities or politics or farming, are problems of *organized complexity*: that is, they have many elements, but not an infinite number, and all the elements within the system affect all the others. The problems of ecology are not like the problems science began to untangle in the 19th century when mass phenomena came into our awareness: elections in which millions voted, the organization of telephone exchanges or the marketing of consumer goods. These kinds of problems involve *unorganized complexity*, in which there are often literally millions of elements, none of which is in any dependent relation to the others, but which systems nonetheless display regular patterns of behavior and which can be studied and understood using tools such as statistical analysis.[25]

The problem-solving methods of ecology involve a kind of thinking we now call holistic. They require that we think about wholes and their relationships to other wholes. Each element has its own integrity and, if alive, is self-regulating as well, yet each is also related to others within the system, influenced by feedback.

*Holism* is the paradigm of the age we have recently entered. It erupted in the mass mind about four decades ago and has been accepted and integrated into the various intellectual disciplines and arenas of culture at different rates. In the monocultural paradigm of conventional agriculture, when a crop is

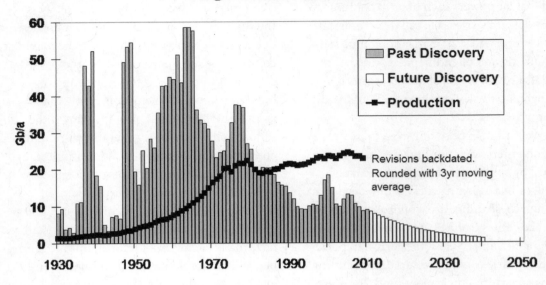

**THE GROWING GAP**
**Regular Conventional Oil**

Past Discovery
Future Discovery
Production

Revisions backdated. Rounded with 3yr moving average.

Worldwide oil consumption has outpaced discovery for over 40 years, reaching ratios of 5:1 in the last decade.

[Credit: Colin Campbell]

"attacked" by an insect "pest," the farmer calls in the cavalry to rout the bad bugs, just as in movie westerns. He (usually he) sprays a chemical to kill the insect. The crop is saved, the dollar cost of the chemical is justified by the much greater value of the crop preserved for sale at market, and great industries have arisen on these insights. Seen from holistic perspective, however, the chemical causes illness in workers at the factory, illness in farmers on the land, accumulates in the crops grown and thus reaches consumers invisibly, may drift to the neighbors, stay in the soil or run off in the waters, and it kills not only the pest organism, but 50 other kinds of insects, some of which were predators of the pest, a part of the biological information feedback of the ecosystem. The pest organism, which breeds by the millions and evolves rapidly, is now set back, but some of its population survive to breed greater resistance to the temporarily effective but permanently damaging chemical. The cycle repeats. From 1940 to 1984, crop losses to insects in the US rose from 7% to 13%.[26] The incidence rate of all cancers rose 1.1% per year from 1973 (the first year statistics were collected nationally in the US) to 1996, a massive increase. Approxi-

mately 400 cases of cancer occur every year per 100,000 persons, up from about 300 a generation ago.[27] Similarly, operating on the "bad-bug" theory of medicine, hospitals have become a major source of antibiotic-resistant pathogens.[28] But the small magic tricks of conventional medicine and agriculture are failing in the face of nature's much greater laws of balance. The "problems" that the monocultural mind sees as "pests" or "pathogens" are, in the holistic framework, forces or organisms trying to clean up or rebalance a failing system, indicators of ill health to be treated not by adding to the burden of death and morbidity, but by nurturing more and greater life.

## Permaculture Envisions a New Commons

It was the genius of Australians David Holmgren and Bill Mollison to seize on the critique of industrial civilization offered by the 1972 Club of Rome report, *The Limits to Growth* — which first brought systems analysis and computer modeling to the resource and energy flows underpinning the world economy — and marry it to practical applications of ecological theory.[29] They took seriously the book's predictions of resource and energy shortages by the early 21st century, saw all around them the wreckage of societies driven by centralized institutions and crafted a set of memes for tunneling out of the trap. The resulting analysis provided an intellectual foundation and direct guidance for many of the world's most promising cultural and agricultural experiments.

Holmgren and Mollison fused the subtitle of J. Russell Smith's 1929 book, *Tree Crops: A Permanent Agriculture*,[30] into a single word: Permaculture. Collaborating on a revision and expansion of David's undergraduate thesis in ecological design, bringing to bear Bill's long years of observing nature as a wildlife biologist, they authored and published *Permaculture I* in 1978, postulating that people could design synergistic assemblies of plants, animals and structures that served human needs but adhered to nature's logic.[31]

Permaculture was much influenced by the writings of H.T. Odum, who had illuminated the economic conundrum of modern society, dependent as it was on the continued extraction of fossil fuels.[32] *Permaculture I* and *Permaculture II* — authored by Mollison solo the following year — sketched out an ecological design system for smallholders that Permaculture's coauthors believed could be useful in all climates and cultures.[33] It drew on traditional knowledge, which Mollison as an indigenous Tasmanian fisherman, farmer and hunter/biologist had derived in considerable degree from his own family and local traditions as well as contact with aboriginal Australians. Permaculture connected many streams of the world's traditional knowledge with modern forms of science and urged ordinary people everywhere to continue that lineage of empirical investigation. The books were a prospectus for a worldwide distributed experiment in ecological subsistence agriculture for the post-industrial world.

That experiment is now over 30 years old, and I will argue that its fruits are abundant and that results have validated the original thesis well enough that we should expect it to meet the needs of a new generation of garden farmers whether they be former pastoralists settled into towns in Botswana or industrial workers made redundant by energy descent in Boston. In Chapters 3 and 4, I will introduce permaculture principles and ethics and their ecological underpinnings, which form the basis of the design system, while Chapters 5 and 6 examine permaculture practice and vision. But before we go there, we have to look at a largely unrecognized cultural phenomenon and answer the question posed by Chapter 2: who am I to farm?

# CHAPTER 2

# Who Am I to Farm?

Today only 0.3% of Americans and 2.2% of Canadians derive their primary income from farming.[1] This is the smallest proportion of the population devoted to farming in the history of either nation or in the history of the world. No other societies have made our basic connection to the earth and the garnering of sustenance such a marginal specialty. Are we, as economists and prophets of progress proclaim, more evolved and more efficient, freeing up labor from the drudgery of farming to perform more complex and rewarding tasks in industry or the creative professions? Or have we so lost ourselves in thrall to the logic of the machine, that we will sacrifice everything to it — the quality of our food, our health, the land, even our very souls?

The dynamic of the modern economy, by which large-scale production became dominant through the subsidy of fossil energy, has indeed made farming a marginal occupation at the bottom fringe of the system — a dirty and dangerous primary industry, akin to mining, logging or fishing. The vast prairie expanses of the United States and Canada have lent themselves to mechanized farming

so that only a few individuals are needed to manage holdings of thousands of acres.

Of course the statistics about farming as an occupation mask many ways in which the work of millions of people is hidden, so the "efficiency" and "progress" of our high-tech

Garden farmers are everyone and everywhere. [Credit: top row, l–r: USDAgov via Flickr, Steven Forrest, See-Ming Lee via Flickr; bottom row, l–r: ^W^ via Flickr, John Glavis, Peter Hellberg]

Most nutrient-rich food eaten in the US and Canada is picked by migrant workers, often undocumented.

[Credit: Bob Jagendorf]

As many as half of all meals today are eaten away from home.

[Credit: avlxyz "Alpha" via Flickr]

societies may be seen as an artifact of ideology as much as a sign of social evolution. More and more food is imported to North America from elsewhere in the world, where it is grown by Asian, African, Latin American, Caribbean or European farmers, usually on smaller farms and with more labor input. Even within our borders, the real food grown here — that is, the nutrient-dense food that sustains our health, such as fruits and vegetables — is picked and processed by an immigrant labor force of Mexicans, Jamaicans, Salvadorans, Haitians and other dispossessed farmers from the South. Even in our wealthy societies, we have many millions more farmers today than we acknowledge.

But what about most North Americans? Are we happy to be eating industrial food? Are we flourishing in our post-agricultural careers? Do we gladly forsake the countryside for city culture?

Certainly millions seem content or may never dream of asking these questions. But there is ample evidence that many of us have never completely relinquished our attachments to a more agrarian way of life. The US Frontier, and with it the opportunity for anyone to claim a piece of land from the government and homestead it, closed in 1890. Yet every wave of urbanization since World War I has been accompanied or followed by the resurgence of agrarian ideals. Thomas Jefferson's vision of the United States as a nation of yeoman farmers continues to echo down the ages. In the 1930s, M. G. Kains wrote a manual for erstwhile farmers, *Five Acres and Independence*. He introduced the book with a quotation from Henry Ford (which may be more than ironic) extolling the virtues of the land.[2] Already by 1935, the manic ups and downs of the capitalist business cycle were familiar enough that "return to the land" was a recurrent and well-recognized impulse in society.

Even after microbiology and engineering made cities less acutely unhealthy, industrial production, with coal as a primary fuel, made them dirty and often noisome places from which better-heeled residents sought relief at summer resorts and in "garden suburbs" where the amenities of a quasi-rural settlement could be combined with the convenience of swift rail transit to centers of commerce. Long before use-based zoning began to sort out industrial from residential sectors within the city, the dream of the suburbs had taken root.

An even more radical critique of industrial civilization arose from the lives and writing of Helen and Scott Nearing. Their 1954 book, *Living the Good Life*, and subsequent titles extolled the virtues of simple living close to the land. The Nearings not only turned away from the hubbub and clamor of city life but proposed an unconventional response to economics as well. After Scott, who was trained as an engineer and an economist, was blacklisted from academia following World War I for his socialist and antiwar views, the couple retired to the Vermont frontier, reduced their consumption of industrial goods and adopted a vegetarian diet based on homegrown food. They built their own house from local materials and disciplined themselves to divide their days equally between "bread labor" (work for sustenance), intellectual pursuits and socializing. Working six weeks a year in the late winter to make maple syrup and sugar afforded them enough cash income to pay taxes and even to travel. Not only did their forest farming and designed approach to living inspire a whole generation in the 1970s seeking a way back to the land, but they appear to have lived healthy, principled and successful lives without compromising their values. Scott lived to 100 years of age and ended his life by fasting in 1983, while Helen, a generation younger, survived him some 12 years in their second homestead on the Maine coast, surrounded by friends and admirers.[3]

While not agrarian by design, the post-World War II suburban boom appealed to the unrealized dreams of millions who left the countryside for war and better wages, but from whom the pull of a pastoral life had never entirely vanished. Men continued to enact, in mechanical and often neurotic ways, the rituals of making hay as they cut lawns into perfect green squares every weekend. Women organized ice cream socials and birthday parties like the collective celebrations of harvest that had ennobled the hard lives of their ancestors. Children were the real crop here. During the 1950s this patchwork of farm fields, forest remnants and village-scale neighborhoods, peopled by the children and grandchildren of factory workers, immigrants, ex-farmers and other groups newly enriched by the war economy, became the

dream landscape of the boomer generation, the largest in North American history. Small herds of children roamed this bucolic terrain, secure in the privilege their parents extracted as world conquerors until, of course, the next development took down a totemic patch of woods or replaced a mysterious meadow with a cul-de-sac of new houses. Perhaps it should come as no surprise that, as it reached adulthood, this age cohort sought meaning in nature amidst a world seemingly mad with the designs of human dominance, corporate conformity and mutually assured destruction from nuclear weapons.

Well into the 1970s when energy crises began to call into question the wisdom of a commuting way of life, suburbs continued to afford a new generation of children the same glimpses of a comfortable life embedded in nature. But the suburbs were changing too, as they grew to become the dominant habitat for North American societies.[4] City centers and their surrounding neighborhoods, under assault by highway builders, redlining and white flight born of racism, hollowed as their outer fringes spread.

The agrarian way of life found its greatest contemporary philosopher in Wendell Berry whose political views on farming, land use and culture reshaped debate in the US. If the Nearings had offered moral inspiration and

Cheap energy has enabled our cities to sprawl over once-productive farmland.
[Credit: Daquella Manera]

economic guidance, Berry's critique of urban civilization provided an intellectual foundation for the Vietnam-era pulse of "return to the land." Driven less by economics than it had been during the 1930s and more by cultural alienation from the turmoil of decaying cities and a general rejection of the values of industrial capitalism and war born of empire, this broad wave of hippie communes and organic homesteaders brought lifestyle issues into public consciousness. Vegetarianism and concern for wholesome food free of chemicals grew in direct proportion to the expansion of industrial agriculture with its emphasis on vast grain monocultures and the feedlot finishing of livestock.

Economic opportunities in the countryside continued to be constrained, however. The agrarian ideal struggled against industrial consolidation. The US economy began its long-term contraction about 1973 following the peak of national oil extraction. Farmers were squeezed by the relentless logic of the market — overproduction leading to large surpluses and low prices — while input costs rose with the inflationary price of oil, now set in international markets and no longer by the Texas Railroad Commission. A second oil shock and double-digit inflation piled on top of too much farm debt led to a severe depression in rural America in the early 1980s. Many farmers sold out. Not a few committed suicide.[5]

The traditional household pattern of life eroded as millions of women moved into the workforce in the 1970s and beyond, largely to compensate for falling incomes and inflating costs of living. While energy concerns and economic hardship during the 1970s put a temporary brake on expansion of the suburbs, military Keynesianism under Reagan combined with loose banking laws in the 1980s led to a glut of suburban housing and office developments occupying the new niches created by the federally funded interstate highway system. Flight from center cities, which had begun as a backlash against racial integration in the 1960s and 1970s, accelerated. A generation of sprawl had begun whose end we viewed in 2008 and 2009 as the so-called sub-prime mortgage crisis.

The depression of the 21st century, outwardly visible from 2008 onward, has been the occasion of much writing on the link between energy supply, settlement patterns and the shaky basis of the US economy. Social critic and geographer James Howard Kunstler has called the suburbs "the greatest misallocation of resources in the history of the world."[6] There can be little doubt that paving over much of the nation's best agricultural land and cutting old growth forests to frame shoddily built McMansions was a tragedy of epic proportions, but the question is not whom to hang but what can be done with it now? However disreputable its causes, the emptying out of many American cities and the spreading of the population over broad metropolitan regions marks a necessary and inevitable turn toward a state of lower social and technological complexity that will develop progressively as energy supplies decline.

The contraction of oil and other fossil fuel supplies must translate into a contraction of the economy and of industrial food production. We cannot expect to see a sustained increase in economic output ever again. Indeed, sustaining present levels of output may barely be possible with a full-scale national mobilization of resources to transform energy systems, transport and other infrastructure. This is, frankly, unlikely to be achieved. Many workers in the developed world will become permanently unemployed as farmers in the developing world have been in the past generation with the growth of global trade; food prices will rise with transport and energy costs. The stage is set for a new Agrarian Revolution, though whether this turns into a fulfillment of Jefferson's vision or a new feudalism depends on how we the people respond.

Inspired by permaculture, Charlie Headington planted his Greensboro, NC front yard with edibles and began changing his neighborhood.

[Credit: Chuck Marsh]

It turns out that landownership patterns matter a great deal, not only to the structure of society, but to the economy's ability to create wealth. The stars of the post–World War II economic boom were the East Asian Tigers: Japan, South Korea and Taiwan. Each of them either had land reform imposed upon it (in the case of Japan by the American occupying administration of General MacArthur) or adopted it early on in their rise to prosperity, and economists generally acknowledge that redistribution of land to many millions of farmers was essential in providing the broad-based access to wealth that sustained each nation's rise to the first ranks of the international economy.[7]

The epic "misallocation of resources" created in North American suburbia a fabric of many small land holdings packed close around our centers of population. In a clumsy, expensive and still incomplete way, we have marked out a pattern for a democratic yeomanry. Many potential garden farms are located on some very fine former farmland: northern New Jersey, northern Illinois, the south end of San Francisco Bay and the Lake Ontario lowlands. And even where the soil was not originally well developed, the land is usually flat to rolling. These territories have been supplied with extensive road and water networks, and both labor and a rich array of resources, biological and industrial, lie all around. The largish houses, especially those built after 1980, may be poorly configured at present, but they could accommodate the extended families and larger households that will be needed to grow food and manage land with lower-energy resources and technologies.

The emergence of garden farms is at hand. Under the pressure of necessity as unemployment rippled through the economy, millions of North Americans turned to gardening or expanded their gardens in 2009 as evidenced by a 40% increase in vegetable seed sales.[8] Urban homesteading is spawning its own literature as energy descent forces more and more households to adapt in place. With income constrained and energy and materials shortages looming, the only resources capable of filling the gaps in livelihood are imagination, information and knowledge, in particular a deeper understanding of the material cycles and energy flows of nature. For that understanding, we turn in the next chapter to permaculture, a language derived from the patterns of the world around us.

# CHAPTER 3

# Gardening the Planet

Permaculture is a language of design for the creation of post-industrial human habitats: homes, neighborhoods, towns and a productive countryside. Resting squarely on the science of thermodynamics, particularly on the way the laws of energy manifest through living communities or ecosystems, permaculture is itself a design for cultural transformation. It is a self-replicating set of memes set loose to infect cultures wherever industrial economies have disrupted traditional land use — which is to say, almost everywhere.

In the early 1970s, Mollison and Holmgren set out, along with many others, to rally the world to a course of change, and had such efforts across society not encountered massive and organized pushback by elite interests, we might today be well advanced on the path of transformation.[1] Alas, we are 30 years behind the need. The job before us today is much the same as the one we faced in 1972 when *The Limits to Growth* first warned us of the impending world crisis: to create the resources, energy flows and cultures that will succeed oil-based societies.

Permaculture is an extension of the empirical practices of science and of indigenous knowledge that preceded it. Our efforts, both in teaching and in demonstrating permanent culture, should be guided by Care of the Earth, Care of People and Fair Distribution of Surplus. These permaculture ethics are often shortened to the mnemonic phrase "EarthCare, PeopleCare, FairShare and LimitsAware."

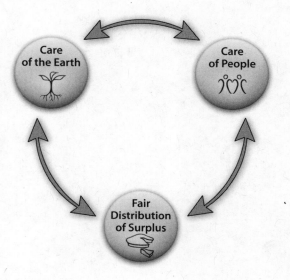

The Permaculture
Ethics
[Credit: Icons—
Richard Telford]

Permaculture provides people who have been cut off from their own traditions, land bases and even from basic contact with nature with the means to restore a healthy and productive relationship to the natural world around them. One basis of that relationship is *ecology*, or informed observation of the living world; the other is *design* — a positive, creative response to our own needs and the logic of natural systems. Permaculture is thus a system for taking responsiblity for our lives at the most fundamental level, that of energy. We eat, drink, pass wastes and shelter ourselves; we live together in villages, towns and cities; we move around the landscape. The way we do those things presently is destructive of the natural world, making it harder and harder for us to continue as a species. It needn't be.

Permaculture's genius as a social movement lies in its grassroots means of propagation: Each One Teach One. The science and design systems on which permaculture rests are not trivial — they are not possible to convey adequately in a few words — but they are accessible to people from all walks of life and all levels of education. The initial training takes about two weeks and is offered all over the world in many formats, including by online correspondence. My own students have ranged in age from 8 years to 86 and

have included professionals, artists, people in government service, students, householders, college professors, doctors, farmers and ranchers, grandmothers and fathers, the unemployed. Not a few of them have gone on to become accomplished teachers in their own right.

You do not have to train in permaculture design to practice it in your own life, but you can give yourself a strong leg up in your understanding of ecological design if you do. Plus, you will gain immediate broadband access to a group of friends and colleagues who can support you to continue in this process of action learning, as well as a worldwide network of helpful researchers.

## Ecosystem Insights

We look first at some of the fundamental insights permaculture draws from ecology, because to understand how to create successful human-centered ecosystems, or permacultures, we need to understand how natural systems work.

### 1. Ecosystems Have Open Boundaries

All ecosystems, whether they be forests, grasslands, swamps or suburban neighborhoods, exist in a matrix of other forces. The most important of these are the energies of the sun which drive all metabolic processes. No ecosystem is isolated and separate. Even islands with their clearly defined edges are affected by oceanic storms, the migration of marine mammals, wind-borne bacteria and, of course like all other parts of our planet, by waste products of industrial civilization: nuclear fallout, heavy metals, coal and dust settling out of the atmosphere and the impact of chlorofluorocarbons (CFCs) and other gases on the ozone layer.

### 2. Ecosystems Are Dynamic

Because spores, seeds, microbes, pollen, insects, birds and animals from next door or from around the world may drift, fly, swim

Landscape Catchment,
Pattern #1
(See Chapter 6)
[Credit: Jami Scholl]

or walk into ecosystems and set up shop, the composition of living communities is always changing. And, while healthy ecosystems maintain a dynamic balance of species and resist large-scale changes, natural and human forces can throw off that balance and make for sudden — or gradual — shifts. Earthquake, fire, flood and windstorm may disturb soils and vegetation, may remake the landforms, may deposit fertility or sweep it away; traffic, hunting, too much harvest, tillage or the introduction of a pathogen or alien species by people may erode soils, degrade biomass or throw off predator/prey relationships, causing the ecosystem to reorganize itself in a new direction. This can happen within the span of a few years or decades.

The American Chestnut (*Castanea dentata*) was the dominant species in eastern US forests for centuries, but between 1904 and 1938, most of several hundred million trees of that species died or were cut, all within the lifetime of the oldest humans alive today.[2] Though the loss of the chestnut was an enormous economic and cultural tragedy, the forest retained its basic composition and function as other species already present moved toward dominance. On the slopes of Mt. Saint Helens in Oregon, dramatic regeneration has occurred since the volcano exploded in 1981. Most of the species returning to life there are natives. Indeed, large-scale disturbances may present the acid test for adaptability.

### 3. The Larger the System, the More Stable It Will Be

Small patches of woods are easily colonized by exotic species, which find niches on the edge. Vast plantations with low levels of diversity are equally vulnerable to disruption by a single new pathogen. On the other hand, large, undivided forest tracts provide fewer opportunities for new species to become established. The largest living system we know is the Earth itself, which maintains remarkably stable levels of atmospheric gases, surface temperatures and overall levels of precipitation for very long periods. This phenomenon is called *homeostasis*, the tendency for organisms and ecosystems to remain stable.

### 4. Energy for Life on Earth Comes from the Sun

While small amounts of energy enter the Earth system (Gaia) from the planet's core and from the burning of fossil fuels and nuclear reactions set off by human beings, the vast bulk of energy used by life comes from *sunlight*, converted by plants into carbon bonds and stored as sugars, starches, cellulose, lignins, amino acids, fats and other related compounds. Once fossil fuels are exhausted, only the widely distributed and diffuse energy from the Sun, the Earth's core and the plants and animals that grow between them will be available to meet our needs. Though human-released energies are modest compared to solar energy, energy capture by the atmosphere induced by the human burning of fossil fuels and the destruction of forests and soils has dramatically shifted the solar energy equation so that the planet is warming and may continue to do so with potentially catastrophic consequences.

Shelter in the Sun, Pattern #12 (See Chapter 6)

[Credit: Jami Scholl]

## 5. Everything Eats Something Else

Even plants eat sunlight, a relatively harmless activity, and one on which all animals, including ourselves, are utterly dependent. In living communities, feeding relationships form webs of direct and indirect effects. For example, when wolves and lions are hunted out — as they have been across most parts of the lower 48 states of the US — deer populations proliferate, causing increased browsing pressure on understory herbs in the forest.

In the worst cases, the plant population may be set back so much that a permanent collapse of the browsing animal population may result. That's called *overshoot*, and it means that a population has exceeded the carrying capacity of its environment. The expression of overshoot does not appear immediately, but only after a lag, because most environments have some measure of resiliency. The food source will attempt to grow and reproduce in response to being grazed, browsed or predated, but if intense feeding pressure persists, the plants or animals being eaten will be unable to grow and reproduce before being hit again. On a pasture, this may mean that the sheep eat closer and closer to the growing tips of the grasses until they literally eat down to the roots, or even tear them out.

All populations oscillate within some limited range. Typically, a population may as much as double without immediately exhausting its food supply. However, numbers must soon fall thereafter, or the resource base may be permanently diminished. Sometimes the resource base may shrink due to natural causes. In his book *Collapse*, Jared Diamond describes the death of the Norse society in Greenland in the 14th century. Cut off from overseas supplies by colder climate and spreading pack ice and unwilling to shift their cultural behaviors to emulate the well-adapted Inuit peoples who fished from the rich ocean waters surrounding them, the Norse persisted in their pattern of animal husbandry and grain agriculture until their crops failed from the cold. The tragic end came as the last survivors ate their young livestock, foreclosing the possibility of regeneration.[3]

Humanity has been in overshoot on planet Earth since about 1989.[4] We are eating, hunting, fishing, killing, culling, harvesting and displacing more than the earth is yielding sustainably. Each year, there is less produced and we go further into ecological deficit. The energy that permits us to maintain any facade of normalcy comes entirely from fossil fuels. While some regions and countries maintain *ecological footprints* (total impacts on the natural world) within the scope of their natural resource bases, most of those areas are considered "poor" or "underdeveloped." The planet as a whole cannot support the human numbers now present at the levels of consumption pursued by industrial societies. Overshoot has been demonstrated in numerous ecological studies. But in defiance of both ethics and sound science, we are now running an uncontrolled experiment involving humanity and the biosphere. It is likely to prove that humans are not an exception to the laws of nature.

## 6. All Eating Results in Some Wastes

This is a basic law of energy in biology. It's also a good-news fact because each waste product is a potential food for some other being. We can see this as a form of sharing. When the lion kills a zebra, it eats the choice parts and leaves the rest. The jackals and hyenas feed next, then vultures, then ants, flies and other insects, then fungi, bacteria and other soil organisms. Of course, the lion also defecates, and dung beetles get some of that action. Each of these creatures has a niche (a food resource and a habitat), and each of them makes wastes which in turn become food for other organisms.

Plants do it too, much to the benefit of all other forms of life: they eat sunlight, breathe in carbon dioxide and exhale oxygen and water vapor. They give off a certain amount of heat, produce sugars and starches, pump these down to their roots and release some to fungi and soil microbes, which in turn help the plants take up minerals in the soil. The very nature of eating supports increasing diversity of life on Earth.

### 7. Energy Flows Through Ecosystems

Energy for life comes primarily from the Sun — a constant stream modulated by the seasons of growth and our planet's angular orientation to its star. But of the Sun's energy that reaches the Earth, about half is reflected off its surface or scattered by layers of the atmosphere; plants and microbes capture only a tiny fraction of the rest. Nevertheless, the energy reaching the Earth is more than 5,000 times the amount of energy used in the industrial economy, while the energy captured through photosynthesis is about five times the energy generated by fossil fuels.[5] Gaia stores this incoming energy in the oceans, in forests and soils as carbon (biomass and humus) and in the increasing diversity of life forms (proliferating genetic information). Because the Sun keeps on shining, we are in no danger of running out of primary solar energy anytime soon. We can, however, exhaust the stored solar energy on Earth, both locally and globally, and this is what matters to life. Stored solar energy in the form of carbon and diverse life forms is the battery that powers regeneration on our planet. Exploiting and depleting stored energy is the main way human beings have advanced the project of civilization over the last 10,000 years — chiefly by cutting forests and tilling soils. We have been running down our batteries, and as a consequence the current of Life is weakening.

### 8. Materials that Make Up All Living Bodies Cycle

Most of plant and animal biomass (97%) consists of elements that are gases in the atmosphere: carbon, oxygen, hydrogen and nitrogen. At and just above the surface of the earth, those gases are always and everywhere available, therefore they are not limiting to the life and growth of organisms. The other 3% are minerals which are stored as salts and other compounds in the soil, the oceans and the earth's crust. In healthy ecosystems, minerals move up and down, in and out of living and dead bodies, but remain in local circulation. Only rarely are any lost through leaching, erosion, fire or migration, and these are usually offset by gains from deposition or the breakdown of bedrock by tree roots. But harvest without returning residues and wastes to the ecosystem can degrade the mineral content of soils and biomass. This has happened to most of our agricultural lands, with the result that plants and animals (including humans) who live from those lands are weakened, diseased or may be incapable of reproducing because the minerals in their soils, diets and bodies have been depleted. Minerals are often a limiting factor in the health of ecosystems, especially in the parts of nature most impacted by humans.

### 9. Information Flow Is the Chief Resource in Ecosystems

*Information* arises in ecosystems as feedback and is stored genetically (and among settled peoples, culturally). The loss of information can be more disruptive to life than the loss of biomass or even of minerals. A pollinator, a fungal associate or a seed-dispersing animal can be a limiting factor which, if removed, can result in the failure of another species to thrive or reproduce. Recent studies have shown that clearcutting of Douglas Fir forests in the Pacific Northwest may result in a failure of the trees to regenerate naturally, even where soils have not been seriously eroded and seeds are present. It appears that germination of the fir seedlings depends on a fungus infecting the seed, and that the spores of this fungus are transported in the gut flora of squirrels living in the trees. When the trees are clearcut, squirrel habitat vanishes, the squirrels decamp and the reservoirs of the critical fungus in the system leave with them.[6]

Most healthy ecosystems have multiple pathways by which energy can circulate through food webs, so the loss of any single element doesn't throw the system off. We call this *redundancy* or resiliency. By analogy, if the kitchen is out of ham, redundancy means that you can still get a meal if you are willing to eat a cheese sandwich or a chicken salad. But simplified systems (such as our agriculture) are easily disrupted by a change of information because they lack redundancy. In 1971, most of the US corn crop failed due to a rust organism, causing billions of dollars in losses. This was only possible because the information content of most of the farm fields was very low, with only a single variety of a single species present. The picture is not very different today. Conventional agriculture, with its emphasis on monocultures, is literally "biologically stupid."

Forested Ridges,
Pattern #6
(See Chapter 6)
[Credit: Jami Scholl]

### 10. Species Composition and Ecosystem Architecture Change Over Time

Disturbance triggers *succession*, though it doesn't guarantee it. Small, short-lived plants (weeds or opportunists) show up after any disturbance of soil. They grow quickly in the abundant sunshine, gather soil minerals, make lots of seed which blows away and then die, helping to cover the once-bare soil. In doing so, they provide a niche for larger and longer-lived plants (pioneers) to establish in their midst. These grow taller, shading and cooling the soil, holding more moisture; they often fix nitrogen or otherwise mobilize soil minerals and by producing small fruits and nuts attract small mammals, birds and reptiles to disperse their seeds. These small animals concentrate nutrients through their manures (nitrates, phosphates and trace minerals) to further enrich the soil; they also bring in seeds of other plants not already present. The pioneer shrubs and small trees break the wind, keep large animals off the recovering soil and provide a shaded nursery for long-lived tall trees to sprout and grow. Eventually, these canopy-bearing trees will shade out most of the smaller plants and establish a forest that makes the rain, modifies the regional climate, builds huge amounts of carbon in the soils and biomass and spends most of its energy maintaining a stable environment.

Most human agriculture consists of cultivating weeds in simple systems. Farmers who plow and suburban lawn mowers both work against succession. That effort takes an enormous amount of energy and work.

### 11. In Nature, Cooperation Is the Rule, Competition the Exception

Individuals within species compete for similar food resources, but within the larger community species are most often in relationships of *cooperation*. The lion, disemboweling the zebra, may snarl at the jackals to keep them away until it is finished eating, but then gives over most of its prey to smaller animals which eat in their turn. When two species require the same niche, it is far more likely that one will adapt its behavior to eliminate competition than that either will go extinct. This may

happen by one shifting its feeding schedule, finding a new food source or altering its territory or nesting behavior. Unlike many perverse cultural attitudes in human society, including the example of the Greenland Norse above, nature enthusiastically demonstrates that it is better to be different than to be dead.

The emergence of diversity, with its sources in the inefficiency of eating and in the adaptability of organisms eager to survive, is the big story in ecology, something that ought to inspire in us wonder and reverence, especially as life at the level of individual organisms is so transient and often difficult.

Permaculture has extracted many salient insights from ecology and by the use of memorable phrases and lively stories has made them into a functional system of design — a way of making culture from a new model. The aim is to meet the needs of people everywhere by organizing the energy flows and material cycles of the local environment into patterns that mimic the structure and function of natural systems. One of the key strategies we use is to enrich those systems with information that may not be locally available, and to do so in a way that is regenerative, empowering people to adopt, improvise and evolve their own living environments. Education is built in.

The permaculture system of design is not limited in its application to agriculture. The term itself has also come to be understood as "permanent culture," a paradoxical notion that nevertheless conveys the aim of an enduring adaptation to the natural world.

Besides the design of productive landscapes — which of course include human dwellings and other structures — permaculture has been applied to the creation of "invisible structures" such as businesses, community currencies, labor exchanges, ecovillages, urban neighborhoods and educational bodies. There is a permaculture credit union in the United States,[7] an international permaculture-inspired university operating worldwide[8] and many examples of village and town-scale communities strongly shaped by permaculture principles.[9]

Wildland Foraging, Pattern #10 (See Chapter 6) [Credit: Jami Scholl]

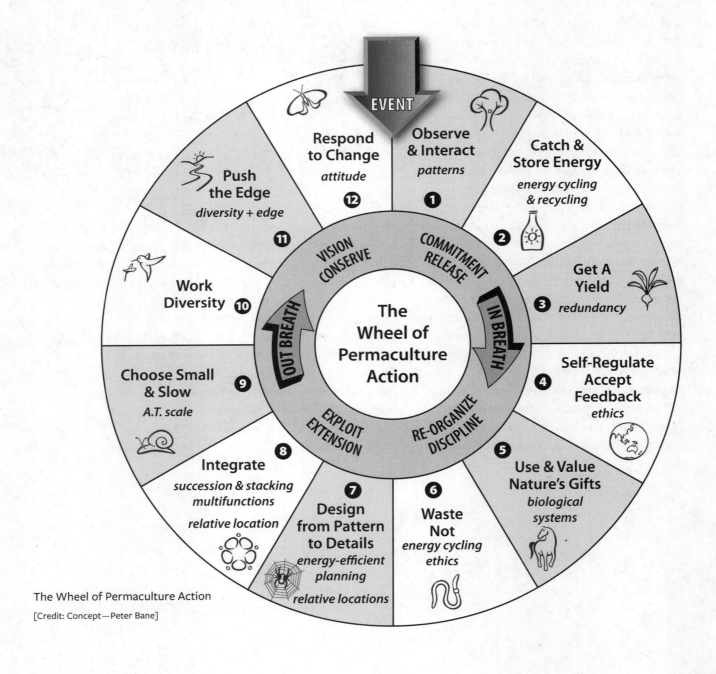

The Wheel of Permaculture Action

[Credit: Concept—Peter Bane]

# CHAPTER 4

# Permaculture Principles

In 2002, David Holmgren, permaculture's younger and quieter co-author, reentered the global permaculture conversation decisively with the publication of a new book.[1] While this book had much to say about the evolution of the permaculture system of design and permaculture's impact on the culture of *energy descent* — a phrase of David's coinage — it most decisively established a template of working principles that has gradually taken precedence over the hodgepodge of good ideas, aphorisms and guides for the designer that had gone by the name principles up to that time. David's increasingly visible teaching and writing work, disseminated through his world travels, his books and over the Web, has persuaded many second- and third-generation teachers as well as newer students to bring this conceptual lens into focus through practice.

I teach my students this template of 12 principles, and I have reproduced them here with the attractive icons designed by Richard Telford. I have ordered them in their formal sequence, but show them as part of a repeating cycle, or spiral pattern, that helps reveal their deeper connections and meanings. We learn not only from words but from shapes, and permaculture design depends on an integration of these two basics ways of knowing.

## Understanding the Patterns of Nature

Permaculture practice involves the disciplined application of ethics and of principles which, with mastery, become so familiar that we can think of and apply them all simultaneously. However, we can learn the system more easily if we recognize that each cycle of design or encounter with a landscape or a human project begins with an event — a tree falling in the forest, the act of moving into a new house, a street widening, a school closure — and from this event our sequence of principles unfolds dynamically as well as logically. The 12 principles that follow draw on the insights of ecology and from cultural wisdom about human capacities and needs.

### 1. Observe and Interact

Gathering information is the beginning of any design, design the beginning of any responsible action. We smell the wind and taste the earth, we monitor our feelings, take the pulse of opinion, use instruments, read maps, consult public records. We notice indicators, anomalies, tracks, flows and remnants. We collect stories and ask questions of neighbors, old-timers, professionals. We notice what is present and what is absent, analyze what has happened to shape the terrain.

This first principle and the two that follow it form the first of two phases of ecological principles or adaptations to the world around us. In human terms, the first three principles comprise a phase of *Commitment*. In ecological terms, any event (or disturbance) *releases* energy that has been bound up in old structures and patterns. When energy is freed from storage, the first response of any ecosystem is a heightened alertness — something new has occurred. Perhaps it means food, perhaps danger. Turn over a clod of earth in your garden and walk away: the birds will be all over it in short order, looking for worms. A car crash never fails to collect dozens of onlookers.

Whether we react to an event or proactively design for change, the beginning of any creative or regenerative process requires "protracted and thoughtful" observation. Mollison has contrasted this with "protracted and thoughtless labor," which he argues is the pattern of our materialist industrial culture. Permaculture, increasingly like contemporary science and differing from classical science, admits that our observations are interactions: they change both subject and system. There is no such thing as a detached observer. And so an ethical stance is required of us. Our presence in the world makes a difference.

Observation is greatly enhanced by the recognition of *pattern*, which is a kind of language of form that enables us to see trends and influences from the past as they recur. Steady and repeated observation, even at great or infrequent intervals, will reveal patterns. Thus, the wisdom of our culture is embodied in elders, in teaching stories and in other forms of the long view. This first principle is linked to "the application of pattern to design," a useful guideline for the designer.

### 2. Catch and Store Energy

Whenever energy is released, ecosystems mobilize elements to capture this valuable resource. Though in the strictest sense almost all stored energy on Earth begins as carbon bonds formed by plants through photosynthesis, in practical terms there are many forms of stored energy, including the *embedded energy* of manufactured things, infrastructures and human cultures. *Social capital* (law, family bonds and corporate structures) and the education and health of people are forms of stored or embedded energy. We can draw on this to generate new activity.

In order to take a yield or harvest surplus from any living system, which is the aim of agriculture and of permaculture as well, there must be an abundance of life. The many energy transactions this entails in turn depend on a reservoir of stored energy. Older understandings of permaculture principles spoke of "cycling and recycling energy," but without the explicit injunction to capture and store energy, we are not going deep enough into ecosystem processes to be fully responsible and may rely too much on already overburdened natural systems.

Working against our education as a species, we have a cultural memory and an enduring image of the Earth as a vast space and of ourselves as tiny, almost insignificant creatures scratching out our existence on its surface. We recapitulate this cultural memory as each generation grows from impotent infancy toward independence and power, and ultimately to old age and death. The notion that humans could harm, let alone exhaust, the Earth has been unimaginable until the past few generations.

We are now in a period of great release of stored energy from the deep geologic past enabled by the human exploitation of fossil fuels. Until the business model of global capitalism — with its enormous infrastructure of extracted capital, its training centers in business schools and its tight integration with major governments — either exhausts itself, experiences a massive intellectual and religious upheaval or is put to death by a rebellion of popular dissent against its negative consequences, we will see continued worldwide struggle over stored energy and resources, land and culture.

If we are to continue some form of civilized life, we must rapidly reorganize our economic activities and life patterns to capture solar energy, not just for current use, but for long-term storage and for the regeneration of natural communities. Our job from here on is to plant forests, build soil, store water and save seed (and other forms of life). With all due respect to the current generation of carbon sequestration warriors, the importance of these common resources is not adequately measured by the market, so we cannot expect the market economy to help us much in pursuit of these necessary aims. They can be met, but we will have to look outside the market. This is why permaculture puts attention on the household economy as the location where these activities can most easily be established and from which they can be propagated.

These four elements — forests (biomass), soils (humus), water and seed (genetic diversity) — are the great building blocks of ecosystem abundance. Together with a judicious conservation of the wastes and salvage of industrial society — including its learnings, some of its technologies and parts of its infrastructure — we can hope to build a prosperous and sufficient future for ourselves, our children and their descendants. Garden farming is one of the most promising ways we can do this.

### 3. Get a Yield (or Harvest)

Our efforts to work with nature must bear fruit that can provide for our needs.

Humans belong in the system, certainly in designed systems or permacultures, and thus have legitimate claims on the yields of those managed ecosystems. The principle of *harvest* expresses those claims. A system that cannot provide for its cultivators is bankrupt in the short term, however productive it might be in some imagined long term. *Yield* is the expression of surplus in ecosystems. Individual elements (plants, animals or structures) may be said to have yields. Chickens, for example, lay eggs, give off heat, produce feathers, drop manures and breed surplus roosters which can be eaten. But yields always arise at the culmination of a cycle of energy capture and transformation. Grasses produce seed, a potential yield, only after they have germinated, rooted, grown and photosynthesized over a period of several months. We may be interested in the grain, but there are also straw, root biomass and a host of grain-eating animals, birds, insects, rusts, fungi and other organisms that represent potential yields from the same cycle. We must stretch observation and imagination to recognize the other yields of this and similar growth cycles, and we should see which of these are surplus (and thus can be harvested) and which must be left to feed the next cycle. As building material, straw becomes shelter for animals or for people. As animal bedding, enriched with manures, it can be returned to the fields as a source of fertility. The stover (stubble and root mass), if left standing over winter, will harvest snow and rain, protect the land from some erosion and feed soil organisms as it breaks down. Hunters take a yield from fowl that glean the grain fields: quail, pigeons, turkeys and pheasant.

Each time energy cycles or transforms, another resource is created and another yield may be possible. Our job is to imagine the sequence of yields, both from individual elements and from the system as a whole. Then we can design relationships and organize energies and technologies to realize those yields.

To continue our analogy, grain after harvest may remain as seed for the next year's crop, or may become cereal, bread, animal feed, mash and thence beer, whiskey or transport fuel. It may be traded for other goods or sold for money. Each of these is a potential energy pathway. Which will have the greatest yields? Which will generate the most cycles? The animal feed may become meat, wool, manure or traction. The mash may become alcohol, then substrate for mushrooms, then animal feed or fish food, then meat. The alcohol may become inebriation, gifts, medicine (by itself or with botanical additions) or tractor fuel. The bread may become food, which turns into labor and loyalty and the ability to plan and design for another day.

By harvesting, we turn the wheel of energy transformation. In doing so, we hasten the onset of abundance. Of course, skill and experience are required to achieve optimal results — the ripe fruit is more satisfying than the unripe — but we should not neglect harvest whatever our level of knowledge and experience. Indeed, knowledge and experience are one of the first yields of any system or of any cycle of energy. No matter what we do or in what state of development our system may be, there are many possible ways to find and take a yield that will enhance and not disrupt its growth. The ritual of harvesting first fruits is common to many religious traditions and a recognition across cultures of the beginning of energy cycling and of the significance of harvest.[2]

### 4. Self-Regulate and Accept Feedback

Just as harvest represents positive feedback, or the energy of encouragement, this principle expresses its complement: negative feedback, or restraint, an equally important if sometimes less popular concept.

No system can grow forever; it is always bounded by resource and energy limits. Recognizing these is crucial to health and balance. This principle begins the second ecological phase in our cycle of principles, one characterized by *Reorganization* and, in human terms, by *Discipline*. Think of this principle and of the two that follow it as helping us determine the functional boundaries of our system.

Mature ecosystems are characterized by efficient use of resources, conservation of materials and high levels of internal communication. In human terms this is parallel to self-knowledge. Maturity is much about good judgment and appropriate restraint of the exuberances of life and growth. Self-regulation keeps us out of trouble. It may be well to harvest your laundry greywater by diverting it into the landscape instead of putting it down the sewer, but if your drainpipe empties into your neighbor's flower bed, you're going to hear about it. Good design requires us to think about downstream impacts and other effects that in the mainstream economy are sometimes called *externalities*, and to incorporate these into the feeding and resource cycles of our own ecosystems.

The conventional culture encourages us to imagine that there is a place called *away* where unwanted things go, but in reality away is simply out of sight or invisible in plain sight. We must change our behavior to act as though we all live downstream, because we do. This principle insists that we "deal with our shit," because 1) we don't want to slip, fall down and wind up with it all over our faces and 2) it's actually a source of wealth which we ignore to our detriment.

When we accept feedback — even brusque, misguided or unbalanced feedback — we deepen our ability to navigate the waters of an unknown and complex future. The most important feedback humanity is being given today is reflected in the breakdown of natural systems. If we fail to accept this feedback the message will only get louder.

### 5. Use and Value Nature's Gifts

The current economy recognizes nature primarily as a resource base and a dumping ground for wastes. Permaculture also recognizes the necessity of living from the yields of nature, but this principle expands our awareness of the many *services* that nature provides as well as the goods we are accustomed to taking from it. Understanding the value of nature's services enables us to protect the goose in order to continue collecting golden eggs. Soils are not merely support for plants but are the primary site where wastes are transformed into new life. Wind may cool us and our buildings; it can also generate energy, aerate ponds to maintain aquatic health, reduce fungal diseases in fruit orchards and distribute seed and pollen.

Many of the least appreciated and therefore most abused environments are also the most important ecologically. Wetlands, for example, have been massively disturbed, drained, plowed, paved or otherwise destroyed over the past 150 years in North America. Yet wetlands were once a source of enormous fertility and valuable buffers in the hydrologic cycle. Shorn of its once vast wetlands, the Mississippi River has repeatedly shown its tremendous capacity to flood and damage towns, cities and farms along its banks. Low places, where water collects and life flourishes, are also sites where wastes are renewed. We can draw on the model of swamps and wetlands to create our own constructed systems for purifying water.

The symbol of the horse, associated with this principle, reminds us that animals once provided important measures of work in the human economy — and will again as fossil energy wanes. Horses have been used as much to pull plows as for any other purpose, but many animals can perform tillage or substitute for human labor. The concept of a *chicken tractor* (a mobile cage in which the birds can be moved strategically to scratch up spent garden beds) is no longer novel, but it continues to be a rich tool for managing fertility that has many applications on the garden farm. Goats can clear brushy, thorny and viney edges of unwanted plants. With the right management, pigs can build small earthworks such

as terraces and swales; they can also root out brambles, trees and shrubs, seal ponds, excavate stones out of soil and have famously been used to hunt truffles, a flavorful mushroom that grows unseen underground.

Noticing services, which are freely given, helps us understand the many functions that each element of landscape can provide. Almost everything has many potential functions, and the success of small-scale systems depends on observant and artful design to use them. Seeing services as well as products as yields of an element or of a system strengthens our capacity to make connections that heal and enliven the world around us.

### 6. Make No Waste

Another way to understand this principle is by the formulation "Waste = Food." In nature nothing is wasted. And at the same time, all resources are ultimately degraded into forms with less embedded energy. That energy is lost as heat radiated into the environment and ultimately into deep space. Structures invariably break down, whether they are the ephemeral walls of cells, the delicate tissues of leaves, the bones of animals or the very rocks of the Earth's crust, but none of the parts are lost (save a few molecules of hydrogen and helium that may float away from the outer reaches of the atmosphere). The leftovers are remade in a miraculous dance that has endured for four billion years.

We depend on healthy populations of soil organisms to capture and reorganize mineral nutrients and all the complex molecules of organic life into new raw materials for plant growth. And we depend on plants to take up these organic nutrients and grow leaves, stalks, flowers, seeds and roots that we and our livestock can eat in turn. Close links between the yields, surpluses or wastes of each element — and the needs of another — mark a well-organized system.

Resources that cannot be digested by a system appear to be wastes. Usually this is a lack of anything local to feed on them — an opportunity for imagination and design. Mollison has observed that slugs in a mulched garden are not a failure of the cultivation system, but merely reflect "a deficit of ducks."

With the exception of radioactive materials (which we should forswear producing or distributing), everything on the Earth can be and ultimately is consumed by some other life form. Phosphorus is a particularly important mineral present in all cells and used for all biological energy transformations. It is in such demand by living organisms that every atom of phosphorus not bound up in the Earth's crust is cycled an average of 46 times per year.[3]

David Holmgren has pointed out the relevance of this principle to the evolution of mature ecosystems, including those of the coming period of energy descent. Established ecosystems, having achieved a successful form and adaptation, seek to hold on to the resources that support them, spending proportionally less effort on growth. In old forests, large amounts of carbon accumulate in soils and biomass, phytochemical diversity proliferates and physical layering thickens from subsoil depths to canopy. Human economies can mature as well, and it's a reasonable prediction that much of the next phase of deindustrialization will involve both more restructuring and more maintenance of those systems that are supportive of life. It would appear that we have entered a future in North America which will see little new building, a good bit of demolition and remodeling and — where the resources are still available — determined preservation. The Cuban economy has lumbered along on the fringes of the industrial world without an automotive industry to speak of by repairing and maintaining a modest fleet of pre-revolutionary US vehicles from the 1940s and 1950s. Such will be the fate of all motoring economies eventually.

In our time of throwaway excesses, maintenance has earned a bad name, mainly because the skills, materials and support structures for repair activities in our culture have in themselves degraded over the past two generations. Which of us has not at some time or another run into the impossibility of repairing an appliance or a consumer device because spare parts were no longer available or it had not been designed for disassembly? Timely maintenance, it turns out, usually requires less effort than wholesale repair or replacement — thus the proverb "A stitch in time saves nine" — which is why we often nurse along an old car, piece of furniture or pair of shoes until it can no longer be fixed.

Ordering the movement of resources down the cascade of function is one of the most important sets of skills that a good garden farmer or post-industrial citizen needs. Helena Norberg-Hodge, in her fine monograph on the people and culture of Ladakh, describes thrift by an example that has echoed vividly in my mind since I first read it some 12 years ago. Living high on the western Tibetan plateau in a severe environment, Ladakhis carefully placed scraps of clothing too thin to be used as patches or quilt stuffing to line the bottom and sides of irrigation canals, thereby reducing erosion. In a world without fossil fuels, our health and happiness may come to depend on just such a keen sense of value.[4]

These first six environmental principles form a long in-breath of preparation and adaptation to the environment. The six that follow allow us to express our intentions, expand the power and scope of the ecosystems we inhabit and streamline their function for the long haul. The next three principles mark a phase of *Exploitation* of the resources that have been carefully built up in the preceding half of the cycle with the aim of *extending* the system to which we have heretofore apprenticed ourselves.

### 7. Design from Pattern to Details

Awareness of patterns sharpens our perception of the forces in flow around us. The meander of a river reduces its speed and therefore the erosive power of the moving water. The branching of a tree enables it to spread its leaves over the greatest surface with the least energy devoted to a support structure.

This principle reminds us that *top-down thinking* precedes and informs *bottom-up action* as a formula for successful work. Permaculture applies a fundamental pattern to organizing living systems for people called zone-and-sector. The next chapter will explain how this template offers guidance to mediate between natural and cultivated ecosystems and to place all the elements of the landscape in the most productive, secure and energy-efficient spatial relations.

Natural patterns are widely applicable and have their genesis in the very nature of matter itself. However, more complex, specific and human-oriented patterns can also be observed — for example, in the optimum shape and size of public spaces or the relationship of settlements to their surrounding topography. Successful buildings follow well-established cultural and architectural patterns for reasons that have been carefully articulated by Christopher Alexander and his colleagues.[5] Because they engage the right side of the brain which perceives form and spatial relations, patterns access our organic or body intelligence. They are inextricably rooted in form and thus are grounded in the body's experience of the world. Indeed, these kinds of patterns may be intrinsic to culture itself, and those that are most deeply rooted in human anatomy, physiology and instinct cross cultural boundaries altogether.

The dynamic aspect of this principle also reminds us that design, the creative act of shaping the world around us, is an iterative process, best done in a series of thoughtful stages where each new layer of the work grows out of the previous ones. Of these stages, the first is therefore the least dependent, but even at the beginning of any human endeavor, we are well-advised to study carefully the signature of nature's energies from both recent and deep time on the landscapes and peoples we would influence.

### 8. Integrate, Don't Segregate.

This principle goes to the heart of permaculture in speaking of connection and relationship. Buckminster Fuller expressed it as the adage "unity is at minimum two." The mind may separate elements, species or categories of things, but in the physical world they exist together. Our design work therefore must consider things as components of integrated systems.

Permaculture seeks to understand both the context and dynamic of our human communities and the forces that shape our inhabited landscapes as a way of harmonizing the natural and cultural processes. We place elements near each other where this provides functional benefit. And for our systems we select elements that can serve multiple functions, therefore enhancing economic yield and efficiency.

For example, the building in which I am writing was formerly a garage. Its long side faces the driveway slightly south of east. From May to August of the first three years we lived here, most days began with a tremendous buildup of heat on that side of the building. Asphalt and concrete absorbed and reradiated solar gain to a fault. The interior became uncomfortable, air conditioning costs were high and the outdoor space was scorching and inhospitable. We began adapting by hanging large bedsheets off the eaves in front of the big picture windows that had replaced the original garage doors. Smaller drapes hung in front of two smaller windows. Then we added insulating blinds to the inside of all the windows. Both measures helped but weren't adequate, and the bedsheets and small drapes were a nuisance. They billowed, sometimes ripped loose, flipped up and stuck on the roof in high winds and had to be put up and taken down each year. In 2009, we built

This grape-shaded awning integrates forces in conflict: seasonally variable sunlight, rain, foot traffic and entrance transitions.

[Credit: Peter Bane]

a small overhang with a clear polycarbonate roof over the sidewalk that fronted the building all along its east side. Planting beds just beyond the sidewalk already had low plants established, and to this mix we added grapes, morning glories and groundnuts, all vines.

Standing dry under the awning during a rainstorm is not only invigorating but tremendously informative. The trellis supports deciduous vines which provide shade and active transpiration to make the building more comfortable in summer, but which can be removed as the onset of cool weather makes the sun welcome again. The plants also create a certain amount of privacy and give us food and beauty during a season when many visitors come onto the property. The groundnuts are legumes that will fertilize the grapes and other plants in the beds. We use the dry space thus created along the building for crafts, food drying and temporary dry storage. The awning roof waters the plants without letting moisture get on the grapes, thus limiting fungal problems. And even the asphalt driveway and concrete sidewalk contribute beneficially by helping to ripen the

fruit and dry the herbs. This previously ugly, uncomfortable and useless space has become supremely valuable through the addition of a few elements that have integrated the forces at play.

The study and formulation of patterns appropriate to a design prepares the way for an integration of elements placed within those patterns. In the example above, the patterns "Positive Outdoor Space" and "Courtyards Which Live" taken from Alexander et al. informed our thinking about the space between our two buildings, which is now becoming a series of Outdoor Rooms — yet another pattern. There is even an echo of the pattern Arcades. The seasonal alternation between sunshine-that-warms-us and sunshine-that-bakes-us was a pattern that cried out for resolution; it matches the deciduous habit of certain plants.

By seeing the multiple functions that each element may serve and designing those to match the needs of other elements in the system, we can achieve the integration that permits small-scale systems to thrive and outperform larger, disjointed operations.

Since small scale offers many other benefits, integration is the key principle we must observe to pack small systems with enough life and vigor to support us.

### 9. Choose Small and Slow Solutions

Choosing to work with small, slow technologies and systems may seem paradoxical in the face of daunting social change. Shouldn't we hurry up and get ready? Well yes, civilizational decline and economic contraction should engender in us a kind of urgency. It need not provoke haste. Most people are still sleepwalking toward the future, so it can seem that we must awaken them in a hurry. But haste and the waste it makes are the hallmarks of our energy-abundant culture and the cause of much of our present distress.

Despite our great belief in material progress, despite the manifold accomplishments of high-powered technology coupled with fossil fuels, human beings remain curious and fallible monkeys. We have good eyesight and a keen feel for the main chance, but just because we can see things doesn't mean we understand them. There are limits to our capacities. With the loss of our *energy slaves*,[6] we will become more and more aware of those limits.

This may be as much a gain as a loss, however. Much of the distress of modern life comes from moving and acting in strange and distant environments, from the loss of the familiar and from speed. Air travel is clearly disorienting and does violence to the body's circadian rhythms (jet lag), but even prolonged periods of travel at the speed of cars, trains or buses can be debilitating. Multitasking may bring a certain exhilaration, but a constant diet of it ultimately exhausts and fragments the mind's capacity for attention, inducing stress disorders. An engineer texting on his cellphone brought about the worst US commuter train crash in 36 years in southern California in 2008.[7] Speed kills, and so does divided attention.

Technologies that require large amounts of energy are inherently *entropic*, that is, they are wasteful and dissipative. Usually they have many complicated parts or are produced and made effective by complicated processes. Therefore, they are prone to fail catastrophically and are hard to repair. They also leave a large trail of "externalities" or unintended consequences. Absent cheap fossil energy, they are very costly and thus available only to a few, therefore they are unjust and undemocratic.[8] Millions of Americans depend on nuclear fission to make their toast and brush their teeth each morning, leaving to their descendants a ghastly legacy for the most banal of purposes.

Small and slow mark the processes of nature and particularly of life. Photosynthesis captures trillions of megawatt hours of energy every year but takes place one cell and one plant at a time. Forests add about 1% of biomass each year, though very young plants may exceed that rate for a while. Small and slow systems and technologies are easily managed by human beings. They afford us essential opportunities to learn, the lifelong capacity for which may be our greatest strength.

Small and slow means local, human-scale, intimate and familiar. It means steady progress and setbacks that do not ruin us. It means appropriate technology: tools that help us but do not enslave us. A new forest can be grown in a backyard, beginning with thousands of seeds in planting beds or pots. And its care requires no more than a few minutes a day. Planting that forest in the ground will come in time, but that too can be done by a few hands working steadily a little at a time. Hand labor and simple machines built the pyramids of Egypt and the great ceremonial mounds of North and Central America. By 3,000 years ago, hand methods and patience had domesticated all the plants and animals we use to feed and clothe ourselves today.

Most North Americans, if they survive the upheaval of the next three decades, will find themselves intimately involved in organic processes of cultivating land, living in communities where transport is expensive, local or nonexistent and applying hand labor to a multitude of processes that today are mechanized. We are grievously unprepared. For a time, we may enjoy the small luxury of learning how to get work done through biological systems, renewable energies and good design without having to depend upon the outcomes for our very lives. We should use this interval wisely.

### 10. Cultivate Diversity

The last quadrant of the wheel of permaculture action is characterized ecologically by *Conservation* of the resources and energies that have been built up through the cycle. The human orientation to this phase of refinement is one of *Vision*, reflecting the need to see deeply into the essential nature of the world in order to conserve its life.

It is estimated that humans have used 100,000 plants for food, medicine, fiber, dye, fuel and weaponry, yet today only 20 plant species make up 90% of the human diet.[9] Even among the few hundred species that are commercially cultivated, there has been a considerable loss of varieties over the past century, reducing the functional diversity of our farm fields. Healthy ecosystems support diverse populations of organisms, and as we have seen already, diversity is one of the most important energy storages in any ecosystem and a good measure of the level of available information.

In our cultivated systems we should always include some native plants for the value of the connections they make to local pollinators, soil organisms and beneficial insects. More than this, we should spread our bets on many crops to ensure that we will eat, whatever conditions of weather or pests may come our way. In a tiny lot where I lived in the mountains of North Carolina, we grew over 150 species of plants in less than 2,000 square feet. We used virtually all of them in some part of the household economy. It's fair to say that we weren't trying very hard, and that most of our time then was taken up with building and community affairs. The garden was a part-time endeavor that nevertheless provided us with all the fresh fruit and salad we needed from May through November. About an hour away from that little garden, on 1½ acres outside the nearby town of Burnsville, Joe Hollis cultivates over 1,000 species of plants which provide him his living as an herbal apothecary. These examples may be a little uncommon, but they point at the scope of effective small-scale systems. Most gardeners are doing very well if they raise 30–50 types of fruits and vegetables and a handful of herbs, but 300–500 species would be typical in a working permaculture. The average Cherokee woman at the time of European contact knew and used approximately 800 species of plants for food, fiber and medicine.[10]

We should cultivate diversity as a basic approach to exploring the limits of the system in which we live. Some crops will show themselves to be better suited than others to heavy yields, and these four, ten or two dozen will become our mainstays, while other species may be used in small amounts (for system support) or may occupy temporary niches. By pushing the limits of what is climatically possible where we live, we can assist species from neighboring regions to move as they will need to in response to changing climate. As I have written elsewhere, we already know that climate zones in North America are marching poleward at twice the maximum natural dissemination rate of native tree species.[11] Only with human support can we hope to preserve the diversity of our native forests.

Whatever upheavals come about from a destabilized climate, we must carry into the future as many potentially economic and useful species as we can. Alan Kapuler has written eloquently of the concept of *deep diversity*, by which he refers to the inclusion of plants from every known taxonomic order in gardens whenever possible.[12] The transmission and reproduction of domesticated plants and knowledge about them is possibly our most valuable heritage; it is one we have an absolute moral obligation to pass along to future generations undiminished. The scientists at the Vavilov Institute in Leningrad took this moral injunction so seriously that during the German siege of the city in World War II, many of them died of starvation rather than diminish the treasure in their charge: a wealth of food seeds collected from around the world.[13] Collection and testing of regionally adapted species is one of the first tasks of any permaculture designer working in a new area.

We cannot depend on the institutions of government, let alone commercial seed merchants, to maintain diversity for us. Worldwide, ⅔ of seed production is in the hands of fewer than 10 giant multinational corporations.[14] Their agenda is profit, not preservation, and they are all deeply implicated in chemical agriculture, pharmaceutical medicine and manipulation of the public purse. Indeed, many of them are involved in preventing scientific research about the diabolical products of their own laboratories. For these and many other reasons, the emergence of home-scale seed saving and local and regional networks of seed exchange is imperative. However marginal these enterprises may be economically, they are of central importance to the survival of human cultures.

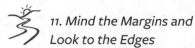

### 11. Mind the Margins and Look to the Edges

Permaculture students have been repeating to each other for years, "the edge is where the action is," a mantra of wisdom about ecosystem dynamics. Edges, or ecotones in landscape, blend the resources of two or more environments — forest and meadow, for example — and thus present a richer field in which life may proliferate. Sensitive indicator species such as amphibians often occupy the edge between major types of environments.

The ecosystems with the greatest amount of edge — estuaries, swamps and forests — capture the most solar energy as biomass.

Our cultivated systems should be rich in productive edges where designed connections can enhance the sum of yields. Rare species may only occupy deep woods or require large, unbroken territories over which to roam, and these terrains where they exist should not be disturbed by human activities. But our agroecosystems, our townscapes and our managed forests are already fragmented into a complex mosaic of uses. Designing to enhance the productivity of these edges can let us depend more on environments already disturbed and dominated by humans, leaving the rest of nature to demonstrate its own evolution.

Permaculture itself, both as an intellectual effort and as a global movement, has arisen at the margins of society and has occupied disputed territory at the edges between disciplines such as agriculture, forestry, architecture and economic development. Its willingness to cross boundaries has won it few admirers among academics keen to defend their expert claims to turf, and it has almost entirely escaped the attention of governments, but the same transgressive attitudes have won it a following among the many disaffected segments of society who see the handwriting of global resource shortages on the wall and wonder who in charge is paying any attention. As John Michael Greer has pointed out, in a time of epochal change, cultural upheaval and complete uncertainty about the future, *dissensus* (the agreement to disagree) is the safest strategy for society as a whole to pursue.[15] The successful adaptations, necessary acts of preservation and renewal and most useful insights into science, technology, culture and education are likely to come from completely unexpected corners. Who, in the age of dinosaurs, would have bet on a bunch of shrew-like creatures scurrying around in the duff to come out on top? And while the meek may or may not be the sole inheritors of the

Earth, at the moment, the odd stand a better chance than the average.

 ### 12. Cultivate Vision and Respond to Change

In this last phase of the cycle of response to disturbance, encompassing this and the previous two principles, our role is to guide the systems we have repaired and expanded on a stable course through a process of *refinement*. We have been deepening our awareness of the ecosystems around us and enhancing their internal connections. The culmination of the cycle described by the 12 permaculture principles is a practice of inculcating *vision* based on rich understandings and a sensitivity to the dynamic of change. We cannot know what forces may next disrupt the landscapes and communities we inhabit, but if we have done our work, we will have the reserves of energy, capacity and insight to respond appropriately when those changes come.

Only by immersing ourselves in the world of nature, recognizing our complete interdependence with it and accepting our role as active cultivators, observers and celebrants of its richness can we hope to ensure the survival of our own species, of our humanity and of our most treasured cultural values. Ethics anchor the power of vision. So a commitment to care for the Earth and for people and to share surpluses fairly for the benefit of all life is our touchstone in a time when change may be the only constant.

We must keep our focus on the enduring values of biological diversity, the immense stabilizing power of forests, the preciousness of clean water and the alchemical potential of the soil as we walk our path. Future generations will care not a fig for our success in the stock markets, or the number of presidential elections we allowed to be stolen or the millions of doodads we shipped around the world, nor will they look kindly on the problems we have created and not solved in the form of nuclear waste, genetic pollution and extinction of species. They will care very much about inheriting a habitable world well-stocked with the elements, skills, tools and systems for regeneration.

# CHAPTER 5

# Learning the Language of Design

*Culture* is our adaptation to the world around us. And design is the conscious process of making culture. *Design*, as Holmgren has written, is also a new kind of literacy that builds on existing forms of reading, writing and numeracy. It specifically entails fluency in the realm of form, process and context. For design to be alive and responsive to the historical demand of energy descent, it must be ecologically informed. The shaping of productive environments for people that work within nature's limits will be the central project of the next era of human history.

## Observation Is the Beginning

The key to any design process is observation, complemented by other forms of data gathering. This means getting to know your home place. If you have lived in the same bioregion for many years, you may already have a sense of the range of normal climate and may even perceive how that is changing under conditions of climate shift. If you are new to the area, it's even more important that you take some time to gather basic impressions. Walk the land in all seasons and at different times of day and night. See it in the rain. Watch where the water runs and see how storms buffet the landscape. Pay special attention to the extremes of sun and shade, of temperature and moisture, of growth and decay.

### Microenvironments

Walk the land and notice different microclimates. It's especially fruitful to walk about late on a sunny afternoon in autumn or winter when the temperature begins to fall rapidly. In a landscape with any relief, pockets of cold air will form and flow downhill, and the atmosphere will stratify to reveal warmer and cooler layers. These microenvironments can be critical to the long-term success of tree crops and to the comfort of homes and animal housing. If you have a chance to see the land under a late winter or early spring snowfall, these same warm and cool microclimates can be graphically obvious as the snow begins to melt. Take note of warm and cool slopes, different plant communities and any animal tracks and paths you see. Keep an eye out for wet spots in the soil, rock outcrops and any

41

anomalies of landform that may indicate previous human activity.

Use all of your senses, and let your feelings of comfort and discomfort, danger, anxiety or lightheartedness rise to the surface. Your subconscious can register significant influences that may not be observable by the five ordinary senses. The body is capable of detecting invisible influences, including flowing underground water — this is how dowsers can locate suitable spots for drilling wells — but also landscapes or buildings in which great violence has occurred. Not everyone will respond in the same way to these subtle influences and psychic footprints, but if your intuition tugs at you, it may be offering you a lead that needs to be further investigated. One way to gather these quieter inputs is to sit for a bit on the land and let yourself begin to daydream or nod off. On the edge of the dream state, many non-verbal influences can come forward that we would otherwise not allow into consciousness. In addition to signals of past activity on the land, your dream consciousness can offer you visions and symbols that may help you formulate a picture of the whole system and its larger meaning. This is especially important when you are starting fresh with a piece of land, even if that is some place you already live but are ready to see with new eyes.

You may focus your direct observations by returning time and again to a selected spot in the landscape. Choose it by how it feels to you; it should be comfortable and offer a range of views and environments. Allow yourself to develop a practice of sitting quietly in that spot for at least 20 minutes at a time. Try to do so daily or at least several times a week. When you first arrive, the birds and the animals will go silent and still. After 20 minutes of your quiet presence, however, they will resume their ordinary business; then you can really see the life of the place.

## Accessing Written Records

You will need to collect other information, much of which is in the public domain, about the property where you live or where you intend to live. The Natural Resource Conservation Service has offices in every county in the United States. These provide local soils maps.[1] Increasingly, soils data has been digitized. Though the government of Canada has emphasized soils mapping for agricultural districts — meaning that large tracts of the nation's soils are unmapped — the many maps that have been created are now readily available by Web download.[2] If you live in an urbanized area, online map sources may offer detailed aerial views of the property and the neighborhood. These can be extremely helpful and are scalable. Notice what may be nearby, especially industrial infrastructure such as railroads, factories or quarries and important natural features such as large wooded areas, lakes or significant hills or mountains.

These features especially may affect your local winds, making them different from the prevailing winds of the region. The US National Oceanic and Atmospheric Administration and Environment Canada keep records of weather across the continent.[3] Regional

GoogleEarth map of an urban area shows a high level of detail.

[Credit: ©2011 Google; Imagery © GeoEye, DigitalGlobalSanborn, US Geological Survey, USDA Farm Service Agency; Map data ©2011 Google]

weather history provides a start. You should plan to augment that information by recording the high and low temperatures, precipitation and significant biological phenomena at your own site. Keep a daily log of these data. Local airports are a source of information on wind direction and intensity.

County government, usually the office of the recorder of deeds or the tax assessor, will have records of properties in your neighborhood. It's worth studying these to see the distribution of landownership and to learn the current valuations. These records may also contain detailed information about buildings in the area.

## Reading the Landscape

The data you gather become a picture with meaning only after analysis. This is how you begin to make connections. Ultimately, you are going to tell a story about what has happened in your place, and you will begin adding to that story as you implement your own design. If you live across the street from a school, you can deduce that there will be regular patterns of traffic congestion as buses and parents arrive to drop off and pick up their children. If you are looking at a sloping piece of land, it's reasonable to assume (and you should check on the ground) that the soils at the bottom of the slope will be deeper and richer than those at the top. If a property has a old barn, it's likely that soils around it will be compacted, but also enriched with manures and littered with iron and other human detritus. In the rural mountains of North Carolina where I lived for many years, I learned that piles of stone in the woods indicate that the area had once been tilled: the stones had been piled up by farmers long gone. The time of abandonment can be determined by the age of the trees and species composition of the forest. In southern Indiana where I live now, the pioneer tree species on old farm fields are Eastern red cedar, sassafras and often persim-

mon. Their presence tells a story of social and economic change.

Plants give us clues to the underlying soils and patterns of past land use that help us analyze our observations. We call these familiar and recurring associations *indicators*. For example, common plantain (*Plantago major*) is a Eurasian species that earned the name "white man's foot" because it appeared in the wake of settlement. It grows well in compacted soils and can usually be found flourishing between wagon ruts, in driveways or along old road tracks. Boxwoods, vinca, daffodils or day lilies are clear signs that a house was once nearby, as they are all cultivated ornamentals that persist long after humans have departed and buildings have collapsed. In many parts of the the US, anomalous cedars may suggest a graveyard. Reeds, cattails, horsetails or sedges indicate a high water table or periodic inundation of the soil.

Whatever data you have gathered you must sift through and begin to develop into a story of the place. How have people lived here in the past? When did they come and how long have they been gone? How have the plant and animal communities changed over time? Is the environment healthy or degraded? Is there pollution or contamination and of what types? What are the special qualities of the place? How does it change its face over the course of the seasons? From which directions might danger come? Design aims to answer these and many other questions and to resolve the conflicts and problems presented by the site and the needs of its present and future inhabitants.

## The Use of Maps

All maps are useful, but the map is not the territory. Better than a map is the process of mapping itself, which you should undertake. You can use your eyes and your body to help you create a graphical representation of your landscape.

Stand on the property and a make a sketch on a piece of paper showing rough shapes and the location of important landmarks. Be sure to include any buildings that will remain, roads and driveways, special trees and anything like wells, cisterns or gates that will influence the use of the land. You can shade in vegetation in patches: here's the lawn, this was a garden, that area is covered in brambles,

the woods begin there. Now pace off some distances and jot them down on your sketch map. You can check your figures and refine the map with time, but any mapping process begins with something like this. Today, thanks to the power of satellite photography and the Internet, you may be able to print an aerial photo of your landscape and, with some paced figures from your sketch map, estab-

## MEASURING WITH THE BODY

Learn to use your body as a measuring tool and you will always have measuring instruments with you, no matter where you go.

The principal body measurements of use are four: hand span, arm span, arm reach and stride. With these four you can measure - by yourself - any object from the size of a softball up to a football field or even larger. The only exceptions are objects much taller than yourself, such as trees or tall buildings. For those, some simple tools are needed.

Get a friend to help you and measure each other's dimensions. Be sure that you write the numbers, as they will not necessarily be memorable.

**Hand span** is the distance between the outside of your thumb and the outside of your little finger when those digits are stretched as far apart from each other as possible. The number can be anywhere from about 6 inches up to more than 10 inches for very large hands. With a series of hand spans you can quickly size windows, shelf dimensions, counter heights, rabbit cages, tree trunk diameters or the size of cartons or furniture.

**Arm span** is the distance between your middle fingertips when your arms are stretched out wide to either side of your body. This unit is useful for measuring room sizes and smaller buildings, primarily in the horizontal dimension. You can quickly measure off lengths of rope or twine by stretching them between your

hands. Arm spans are most accurate when you can measure along a fixed surface or a line with which your hands remain in contact. When you try to measure a distance through empty space, it's easy to introduce error, since there's nothing to hold onto after each measurement, but the method is still reasonably accurate over short distances.

**Arm reach** is the distance between the ground or the floor and the tip of your outstretched middle finger when your arm is fully extended overhead. Your dominant arm is probably slightly longer, so use that when measuring. The reach is primarily useful for vertical measurements of buildings, rooms or smaller trees and shrubs. If you can reach 7 feet, then by holding a stick of 3-6 feet, you can measure dimensions up to 13 feet. This will reach the roofline or at least the gutter of most one-story buildings, and will often reach the windowsill height of second-story rooms. By standing on a ladder and holding a stick, you can reach 18-20 feet. Most younger fruit trees can be measured or estimated with this unit.

**Stride** is the distance covered when you take a step. To be consistent when making multiple strides, measure from the back of the heel. To measure your stride, which varies slightly as you walk, lay out a course of 100 feet. Of course, this can also be done in meters, 50 or 100. Mark the ends with sticks, bricks, colored flags or chalk on the sidewalk, then pace this distance back lish a scale and fill in details not visible from Earth orbit.

*Exclusion by Overlays*

Ian McHarg, a professor of landscape architecture at the University of Pennsylvania and the author of *Design with Nature*,[4] developed a method of design in the 1960s that involved making a series of overlays on translucent paper or transparent film that could be laid down over a base map. The base map would show all existing and permanent features of the landscape, including property boundaries, buildings, water bodies, forests and any cultural features such as roads, power lines or cemeteries. The overlays would separately map soils, water drainage, solar aspects and slopes, wind and noise patterns and any

---

and forth at least twice in each direction. Try to lay the walking course out on level ground because your body will shorten your stride when you walk a descending grade and lengthen it as you move up a slope. Average the results of your four pacings - this should eliminate any error due to uneven ground. You can divide the distance by the average figure to get an average stride, but actually, the stride is rarely as accurate over a short distance as your arm reach. It is more useful to know that you take 44 strides to cover 100 feet, for example, than what the length of one imaginary stride may be. The stride is mainly useful for measuring ground over distances of 20 feet or more. By striding the length and breadth of a garden, a yard, a lot or a field, you can quickly obtain its area, and from this you can estimate the need for mulch or seed, or the carrying capacity of foraging chickens.

### Measuring Tall Trees and Building Heights

To measure very tall objects, get a stick of 20-40 inches. Any straight piece of light wood will do - a meter stick, a broom handle or a piece of tree branch. It doesn't need to be calibrated. You will need a friend (or two) to help you. Hold the stick vertically at arm's length in front of you and stand at a distance from the object you are measuring so that when you look over the top of your fist clenched at the base (or in the middle) of the stick you will see the base of the tree or ground level of the building. At the same time

and from the same place, you should be able to look over the very top of the stick to see the very top of the tree or building. You are making an imaginary triangle between your eye and the top and bottom of the measured object and you are aligning the top of your fist and the top of the stick with two of the sides of that triangle. Adjust your stance, distance from the object and grip on the stick until all these points line up appropriately. Now, rotate your arm so that the stick turns 90° to align with the ground at the base of the tree. It is always easiest to rotate the hand toward the center of the body, so if you are right-handed, you'll typically rotate your right hand counter-clockwise. You have just laid the tree on the ground (virtually)! Ask your friend to stand in a line with the tree perpendicular to the line that your body makes with the tree, and have him or her move out from the side of the tree until reaching the point that aligns with your line of sight past the top of your now pivoted stick. Your friend is, in effect, standing at the top of the "fallen" tree. You have laid the imaginary triangle on the ground. Mark the place where your friend stands, and measure from there back to the center of the base of the tree at ground level. This is the height of the tree, which you have just neatly calculated without climbing a single branch. The same procedure works for buildings, cliffs, light poles or any tall, otherwise immovable object.

number of other environmental influences. By progressively narrowing the filter for potential land uses, the McHarg method offers a kind of design by exclusion. If some use is not contraindicated by environmental factors, it can be considered. While this form of analysis is very complex and time-consuming and may not be appropriate for a small property, it yields powerful insights about the suitabilities of land to a very fine degree of resolution.

The most important uses of mapping are to preserve information that may otherwise be lost from memory and to reveal relationships that might not otherwise be obvious. Mapping can also reveal changes in the landscape as it develops over time, thus helping us avoid unnecessary problems. Dave Jacke and Eric Toensmeier, in *Edible Forest Gardens,* outline a method of mapping that is useful for permaculture design.[5]

I love maps and have been around them all my life; my father was a surveyor. But like the proverbial doctor's children or the carpenter's house, the map of our small property has never gotten a lot of professional attention. It exists in an unfinished state today, six years

An unfinished map of the author's suburban farm
[Credit: Creighton Hofeditz]

after we moved in and long after we have planted many trees, planned several structures and built some of them.

Our need for mapping, however, is growing as the plantings and the placement of structures become more dense and complex with each season. The things we needed most to know and that a map from a previous owner could have helped tell us were underground. Only some of them were known to or were findable by the utility companies. The location of our septic field, for example, was purely a matter of surmise. We never talked to the previous owner, who lived out of state; the home's builder was long dead, and the realtor knew nothing about it. Even the inspector we paid to evaluate the house failed to find it. We discovered the leach field when we began to dig for water lines. So, while mapping has many important uses and is a valuable tool, you may be forced to forego it. And, you can do good design on a small scale without it — for a while.

That said, I recommend that you get comfortable with sketching, mapping and other visual means of communicating. And I urge you to document your work on maps as you develop your property. Most especially record the identity of plantings and the presence of water lines and other underground infrastructure that would not be visible to a later observer.

## Developing Vision

Some people are visually oriented and can imagine the shape of things to come. Most people, I have discovered, are not. Nevertheless it is possible for anyone to develop a *vision* of what they want their home landscape to be like. A helpful way to start is by listing all the things you want or need to have around you. Alongside that make another list of the qualities you would like the landscape to exhibit: dappled light, birdsong, bodies of water, aromatic scents, the sound of children (or not), colors in the vegetation, details of buildings, distance to neighbors, direction of the light, a sense of mystery. Be as specific and inclusive as you can. Think of environments you have liked or in which you have felt comfortable. What qualities did they embody that evoked those feelings of resonance?

Another way to access vision is to identify and list all the *names* of the property you would design, beginning with the aboriginal names of the territories that encompass it. Look at maps, old and new. Are there ranges of hills or mountains nearby? Toward what stream does your property drain? Are there vernacular features of the landscape around you: cuestas, bluffs, buttes, arroyos, prairies, dunes? What geologic and biotic processes have shaped the landscape? Gather up these clues in your investigations. What features have natives, settlers, travelers and scientists noted before you, and how have these been named? All these names are clues to the identity of your land. They are a part of the vision you must develop in order to see the land clearly and help it to manifest its highest possibilities.

If you already live on the property you are designing, it's necessary to begin by "cleaning the slate" so that you can imagine the best outcome without too much prejudice from existing patterns. This is not always easy to do, but go outside, walk away from the place and try to empty your mind. Approach the property, if possible from a direction you don't always take, or from a high point where you can see much of it as you draw near. Using your hands and arms, imagine sweeping away the existing infrastructure to see the landform beneath it. Focus on the shape of the land and try to see the energies of wind, water, sun and plants that have molded it. Imagine it under the downpour of thousands of years of rainstorms. See its place in the valley of which it is a part — grasses or woods rolling off to the horizon. Imagine all the plants that have lived and died on it, the animals that have crossed it, the worms and pillbugs that have churned

up its soil. If you live in a part of the continent where glaciers once flowed, imagine the ice. If your region was once at the bottom of the ocean, visualize the land being gradually lifted up out of the water; see what it was like as the shoreline retreated and the first grasses sprouted on the dunes. If people have lived on the property for hundreds of years, imagine all their comings and goings, all the energies used for cooking and staying warm; perhaps there were births or deaths there.

Ask yourself how how you will contact the essence of the place and give it new clothing. What would you eliminate of things already there, what would you keep and what would you add to make it more alive?

## Turning Problems into Solutions

Now that you have begun by identifying and observing the landscape, reading its clues, studying maps and documents and polling your own opinions, needs and desires and those of your household, you must distill these into a design *aim* or set of aims. Typical aims for a household economy based on permaculture principles would be the provision of sustenance from the land and the immediate community — what I have called self-reliance. Other aims might include comfortable and functional spaces for living or a rich, diverse and productive community of plants and animals. Some people will want to create home employment that engages the whole family; others will emphasize privacy and serenity or an environment that promotes creativity and playfulness. Some will seek stronger connections with neighbors and to facilitate social gatherings and cooperation.

Low-cost, visual harmony and beauty, a changing tapestry of color and scent throughout the year — all these are possible aims, and you can think of many more that are specific and appropriate to your situation.

It's important to choose the principal aims and to make a clear statement incorporating them that describes the feeling you have for

# SITE EVALUATION QUESTIONNAIRE/CHECKLIST

This can be used as a questionnaire with clients or as a checklist for your own needs and resources.

Date: _____

Name: _____

Email/ Phone/Fax/Cellphone/URL/Website: _____

Mailing Address/Property Address: _____

Number of acres or hectares or square footage (+/-): _____

Number of people living onsite, relationships, names, ages:
_____

Occupations/Interests/Hobbies: _____

Lifestyle(s) (e.g., home-based, commuting, long-distance travel, work-intensive, children, retired): _____

Eating patterns: _____

Disabilities or limitations: _____

Ability to invest in the site (e.g., time on and off land, labor, budget): _____

Onsite resources (e.g., water, wind, light, biomass, buildings, infrastructure): _____

Footprint (in total square feet) of existing buildings (give details): _____

Security of land tenure (e.g., mineral/water rights, deed restrictions, zoning, easements, leases): _____

Elevation: _____

Relief (highest point/lowest point): _____

Average annual precipitation: _____

Frost dates (average first/last, more detail if known): _____

Temperature (average annual minimum/maximum): _____
_____

Potential catastrophes (e.g., fire, flood, frost, lava, hurricanes, tornadoes, hail, rebellion, development): _____

List of plans, maps, drawings available: _____

History of land (e.g., logged, cropped, inhabited, pastured, sprayed, mined): _____

Known problems: _____
_____

Level of food self-reliance desired: _____

Degree of household privacy/sociability desired: _____

Desires and dreams, wants, needs: _____

Short- and long-range priorities: _____

Access (existing roads, driveway, farm roads, rights-of-way, pavement type): _____

Aspect(s) (N, S, E, W and slope gradients): _____

Water (springs, ponds, tanks, surface water, public system, size and volume of watershed, flood levels, drainage patterns): _____

Vegetation (flora, percentage tree cover and stocking density, exotics present, ground covers, poisonous species, rare): ____

Habitat (e.g., fauna, native and introduced, pests): _____
_____

Features (e.g., rock outcrops, waterfalls, caves, swimming holes, wind or waterpower sites, views): _____
_____

Soils (sand, silt, clay, loam, organic content, mineral deficiencies): _____

Erosion (historic or ongoing): _____

Available utilities/services (e.g., power, phone, sewerage, gas, photoelectric, hydro, shops, schools, public transit, hospitals, fire departments, waste disposal/recycling): _____

Local resources (e.g., people, organizations, schools, government agencies, sawmill, farmers market, free plants or seeds, biomass sources, sand, gravel, stone, timber, mulch, processing waste, manure): _____
_____

Off-site influences (pollution, noise, hazard, distance to neighbors, environmental conditions, social or legal constraints): _____
_____

Additional Comments: _____
_____
_____
_____

your objectives. To use our own experience in suburban Bloomington, Indiana, as an example, when we moved here in 2006, my partner and I sought the freedom to create a beautiful and exciting demonstration of permaculture principles that would uplift and inspire our community and galvanize positive change in the neighborhood. We wanted to create a productive landscape that would provide us abundant and nourishing food in all seasons, that had comfortable and functional spaces for working and living both indoors and out and that would enable us to shrink our carbon footprint as close as possible to zero within five years while we earned our livelihood at home. (See a mind-map of the Farm Vision on page 56.)

Such a set of aims is both tangible and spacious enough to help us find the right solutions. However, it says nothing about what plants we would grow (except that they would demonstrate permaculture and help feed us). It doesn't say when or whether or which animals we might tend. It doesn't commit us to building a barn (we may) or to keeping our shed (only a little longer…) nor does it direct us to sell things here, off-site or not at all. Those are all decisions that come further down the design path. They involve *strategies*, which bring the aims into being over time, and the techniques, elements and resources needed to embody the design.

Once you have clear aims, once you have made a statement of your intentions, then the gap between reality and ambition comes into focus. At that point different design methods become useful.

## Pattern Languages[6]

As we articulate our aims and goals, we are connecting to cultural and landscape patterns that have shaped our world and our psyche. We are reaching into a language of patterns that is part collective unconscious and part visible infrastructure of our cultural setting. Pattern languages for cultural design

are human, and each place speaks its own dialect. At the same time, all inhabited places on Earth share some common elements of a universal pattern language. As designers, it is most helpful if we can acknowledge the many deep layers of this pattern language so that we can select the right patterns from it to shape the statements of place we wish to make.

Students of pattern language have flourished in many areas, creating pattern languages for ecological bioregions,[7] for progressive social change[8] and for forest gardens.[9] These embrace both the built environment and invisible structures of relationship.

My vision of garden farming arises in the context of a suburban and peri-urban

Many visions of landscape are possible. What will yours be?
[Credit: (clockwise from top left) Figure A: Jon Roberts, Nelson Minar, Lalitree Darnielle, Piush Dahal; Figure B: Denna Jones, David Owen, Creighton Hofeditz]

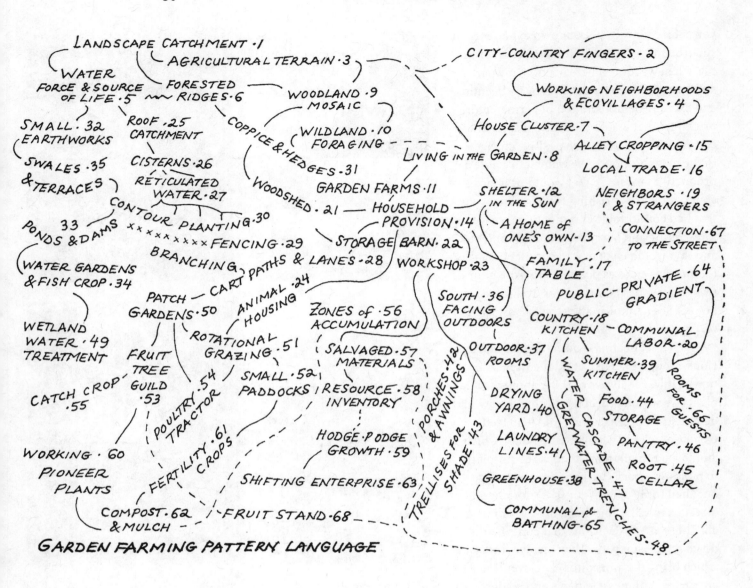

**GARDEN FARMING PATTERN LANGUAGE**

A Mind Map of the Language and Some of its Internal Connections

[Credit: Peter Bane]

landscape, so I have composed a pattern language appropriate for that kind of environment. Although much denigrated as a nowhere place, the area surrounding our cities and towns once played a vital role in provisioning urban populations and in transferring urban wealth to rural areas; it will again. At the end of this pattern language section, I will detail two methods of analysis and placement. Taken together with the pattern language and layered over a sound analysis of observed and gathered data, these methods should enable you to identify and achieve important synergies and economies between your needs and the capacities of the land.

## Garden Farming Pattern Language and How to Use It

Think of this language as an inventory of possibilities from which to shape your home landscape. The patterns are numbered: those at the top of the list are the largest in scope, while those at the bottom evince the smallest scale. From this selection you'll begin to see the elements and some of the relationships that need to be developed to manifest your vision in a living landscape.

A listing of the patterns follows, while a more complete articulation of the patterns may be found in Chapter 6. Each of the patterns has many potential functional connections, some of which are suggested in the commentaries that accompany them.

*Patterns 1–10 describe the ecological and social ground of garden farming.*
1. Landscape Catchment
2. City-Country Fingers
3. Agricultural Terrain
4. Working Neighborhoods and Ecovillages
5. Water: Source and Force of Life
6. Forested Ridges
7. House Cluster
8. Living in the Garden
9. Woodland Mosaic
10. Wildland Foraging

*Patterns 11–20 describe the energetic and economic organization of the garden farm itself.*
11. Garden Farms
12. Shelter in the Sun
13. A Home of One's Own
14. Household Provision
15. Alley Cropping
16. Local Trade
17. Family Table
18. Country Kitchen
19. Neighbors and Strangers
20. Communal Labor

*Patterns 21–24 describe major building elements.*
21. Woodshed
22. Storage Barn
23. Workshop
24. Animal Housing

*Patterns 25–27 describe the emergence of the water system.*
25. Roof Catchment
26. Cisterns
27. Reticulated Water

*Patterns 28–32 describe the shaping of the land into major subdivisions.*
28. Branching Cart Paths and Lanes
29. Fencing
30. Contour Planting
31. Coppice and Hedges
32. Small Earthworks

*Patterns 33–35 describe the movement and use of water through the system and its effects.*
33. Ponds and Dams
34. Water Gardens and Fish Crop
35. Swales and Terraces

*Patterns 36–43 describe solar influences and outdoor living spaces.*
36. South-Facing Outdoors
37. Outdoor Rooms
38. Greenhouse
39. Summer Kitchen
40. Drying Yard
41. Laundry Lines
42. Porches and Awnings
43. Trellises for Shade

*Patterns 44–46 describe food handling and reserves.*
44. Food Storage
45. Root Cellar
46. Pantry

*Patterns 47–49 describe the use of water and its return to the landscape.*
47. Water Cascade
48. Greywater Trenches
49. Wetland Water Cells

*Patterns 50–55 describe major and minor cultivation systems.*
50. Patch Gardens
51. Rotational Grazing
52. Small Paddocks
53. Fruit Tree Guild
54. Poultry Tractor
55. Catch Crop

*Patterns 56–63 describe the harvesting of resources from the environment.*
56. Zones of Accumulation
57. Salvaged Materials

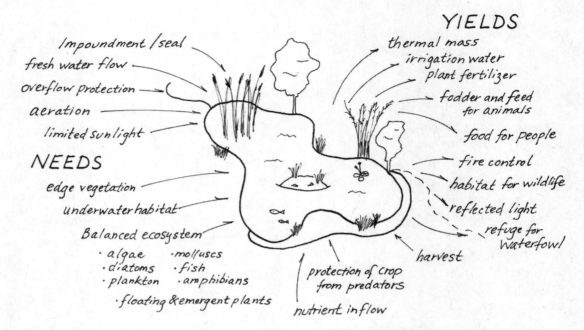

**YIELDS**

- thermal mass
- irrigation water
- plant fertilizer
- fodder and feed for animals
- food for people
- fire control
- habitat for wildlife
- reflected light
- refuge for waterfowl

Impoundment/seal
fresh water flow
overflow protection
aeration
limited sunlight

**NEEDS**

edge vegetation
underwater habitat
Balanced ecosystem
- algae        · molluscs
- diatoms    · fish
- plankton   · amphibians
- floating & emergent plants

harvest

protection of crop from predators

nutrient inflow

The Needs and Yields of a Fish Pond

[Credit: Peter Bane]

*Patterns 63–68 describe exchange functions of the garden farm.*

Permaculture has borrowed many tools and concepts in applying design to ecosystems. It shares many methodologies with other design disciplines, but two of its methods stand out as primary and distinctive, one logical and the other spatial. The first of these is *needs-and-yields analysis*, sometimes called niche analysis, based on the logic of ecosystems, which are communities of interacting species and landforms. The second method is *zone-and-sector analysis*, which uses spatial methods and functional connections to organize the elements of a garden, farm, household, neighborhood, business or town.

## Needs-and-Yields Analysis

A needs-and-yields analysis is not unlike writing descriptions of all the workplaces and jobs to be done when creating a business. By determining all the needs and yields of the system, we can figure out what elements must be present to support each other — and what connections or relationships between them will result in the greatest satisfactions. If our organization has jobs that are not done by workers within it, then we, as the boss, have to do that work. If our chickens have no access to soil in which to deposit their manure, we'll have to shovel it, whereas they would prefer to drop and scratch it into the soil themselves. If we arrange for the chicken's hunting grounds and its toilet to be in our garden beds at certain times, then we can avoid both work and pollution.

Take, for example, the needs of a fish pond. The fish pond needs a source of water and containment for it. The pond must be protected from pollution and excessive sedimentation. It also needs good aeration at all seasons and times of day. The fish need a source of food and other species to consume their wastes, which means they need other aquatic organisms as well as insects, and nutrient to grow plankton. The pond may need protection from summer heat. The fish need protection from predators, whether cats, herons or raccoons.

The yields of a fish pond are nutrient-rich water (for irrigating and fertilizing plants), fish of course, insect control (the fish and tadpoles may eat them), reflected light, humidity, sediment and other nutrient wastes from the bottom, moist soil habitat (around the pond), thermal mass, fire protection and control, recreation if the pond is big enough, protection from predators (it's a barrier to many creatures), contemplation, habitat for various kinds of wildlife and waterfowl, water for domestic livestock (including bees) and a limited amount of waste processing.

With the needs and yields identified, we can begin to design connections between the fish pond and the landscape in which it is located. Some of the yields of the pond are only available when it's in certain areas or near other uses. Its microclimatic influence, for example, is small and probably doesn't extend much beyond half its radius. The best location to use the pond's tempering influence, therefore, would be on the shore, or even better, on an island or peninsula surrounded by water. And its value as a barrier, whether against fire or intruders, would depend on its shape, length and location. Some of the pond's yields such as sediment, if not used, become pollutants. Even fish, if not harvested selectively, may become ecologically unbalanced, with too many small individuals or not enough prey organisms.

The aim of permaculture design is to meet the needs of the system and all its elements, including its human inhabitants, primarily from within the system itself, thus reducing costs and increasing resilience. By identifying needs and yields, we are analyzing the niches or opportunities within the ecosystem, and by matching the needs of one element with the yields of others, we may achieve a self-regulating, self-reliant system that requires only modest intervention from us.

## Analyzing Zones

Zone-and-sector analysis reveals the impact of habitation and the influences of the environment on a site. It has two components that work in opposite directions. A third, quieter element of the pattern is elevations, or using gravity to advantage. Zones describe a progression of territories surrounding the center of a system. On a homestead or farm, that center is typically the house where people dwell. Sectors correspond to environmental influences that come from beyond the site, and they always have a direction.

Zones measure impact of the dwellers on the land and the intensity of their interventions in the system; zones are numbered from 0–5. As the numbers get larger, the size of territory increases while human impacts and management decrease.

The analysis of elements for placement within the zone pattern begins by determining the management requirements of each plant, animal or structure. A henhouse, for example, needs our attention twice a day — when we let the birds out and put them to bed, and not a few times at night, as we may be called to chase off predators. To make upwards of 730 trips a year as efficiently as possible, we should keep the distance between our back door and the chicken coop short. So proper placement of the henhouse is somewhere within or at most on the edge of zone 1, perhaps some 50 feet from the back door, rarely more. Imagine the distance you would be willing to run, in a hurry, at night, in the rain, not quite awake, to chase off a coon or interrupt a weasel. Conversely, walnut trees need a little attention when planted, some water and mulch in the first year or two, maybe a little weeding at the same time, and then perhaps only once-a-year attention for a decade until they come into bearing, and then no more than 10 days out of each year to collect nuts or prune limbs for the next 80 or 100 years until the tree is harvested for timber. Walnuts, which have antagonistic interactions with many other crops, are well-suited to zone 4.

There are no dogmas about the location of elements. A dwarf or cordon apple (on wires) might be right outside the back door of the house, an espalier apple against the very wall of the house itself (zones 1 and 0 respectively) while a semi-dwarf apple in a forest garden might be in zone 2, surrounded by companion shrubs and perennial herbs, perhaps 100 feet away. An orchard of standard apples could be a main cash enterprise in zone 3, and an untended seedling apple providing extra animal forage in pasture might be in zone 4. In Ohio and elsewhere in the footsteps of Johnny Appleseed, apples are found in untended hedgerows growing wild — arguably a part of zone 5 on many rural properties, or even along railroad corridors in urban areas, also potentially zone 5. The numbers correspond to the intensity of relationship and interaction, and to distance from the organizing center.

Scale also changes on a gradient, increasing from inner to outer zones. This is true for plants and animals, water storages, tree systems and fertility management. Fertility in zones 1 and 2 will be managed by hand-weeding, mulching, coppice of shrubs and small animal rotations, while in larger zones, rotational grazing by larger animals, cover crops and tree coppice will be the primary sources. Buildings, however, are kept near the center (with few exceptions), and the largest may be the dwelling itself.

In dense urban settlements, the outer zones are almost always shared, public or common land and are most often discontinuous with the inner zones, e.g., community gardens, street trees, edible parks and boulevards and utility or rail corridors. Even zone 2 may overlap among households as described in the pattern Alley Cropping (see Pattern #15, Chapter 6).

There is also an economic correspondence to the zones. Dwellings without connection to landscape are highly dependent on collective systems of sustenance. In today's world, that typically means distant resources. Large properties obviously have more economic

## Permaculture Zones and their Elements

|  | Elements | Economic function |
|---|---|---|
| Zone 0 House | house and attached structures, other building interiors, pantry | consumption, processing, storage |
| Zone 1 Garden | intensive garden beds, laundry and drying yard, woodshed, garden, tool storage, small greenhouse, piped water, outdoor rooms, summer kitchen, rabbits, poultry, children's play; animals needing special care; (z.1 or z.2) root cellar, cisterns, sauna | self-reliance, household provision |
| Zone 2 Orchard | productive fruit trees and shrubs, piped water, small ponds, poultry forage, compost piles, greywater treatment, dairy barn (at the edge of z.3), workshop, storage barn, mulch crops, nursery crops, living fences, resource inventory, tank aquaculture | barter, local trade |
| Zone 3 Fields | staple crops, larger fish ponds, field shelters for geese or turkeys on range, hogs, larger greenhouses, hedgerows, alley cropping | cash cropping for market |
| Zone 4 Pasture/ Woodlot | pastures, woodlot, larger ponds, windmills, windbreaks, permanent fences, sheep, cattle, goats, horses, llamas, nut groves, unselected fruits, silvopasture | investment, life cycle returns |
| Zone 5 Wild Areas | forest, coastlines, unmanaged prairies and wetlands; ridgelines and other high places | insurance against catastrophe, genetic reserves, selective foraging |

amplitude, while households on smaller properties share common resources as a base for their cash income and investment strategies. All human productive landscapes are and must be embedded in a matrix of wild lands on which we are ultimately and utterly dependent for our deepest layers of economic and physical security. The human economy is always dependent on the natural world.

*Analyzing Sectors*

The second half of the zone-and-sector pattern describes the influence of sun, wind, migrating wildlife, noise, pollution, genetic drift, views and all manner of forces that originate outside the system. Each of these sectors is primarily identified by its direction. It's not possible to place elements properly in zones without also analyzing the impact of sectors on them.

The sun in temperate North America is always to be found centered on the southern horizon. In winter it moves from the southeast to the southwest, in summer from the northeast to the northwest, inscribing ever-increasing arcs higher and higher in the sky as the season progresses from winter to summer solstice before reversing, and always passing through the southern sky at midday. The direction and angle of the sun are important factors in designing buildings, perennial crop systems, outdoor spaces and in evaluating the microclimatic conditions of different plots of sloping land. Similarly, winds come from prevailing directions, while storm winds may vary more or less predictably. Winter and summer winds differ in their prevalence and force, and so we would take account of these sectors to design windbreaks, corridors for ventilation of orchards or to channel cooling breezes toward a house or energy toward a wind turbine. Wildlife moves along observable corridors. Seeds and pollen are dispersed by wind and gravity among other vectors, and thus we can take advantage of these forces

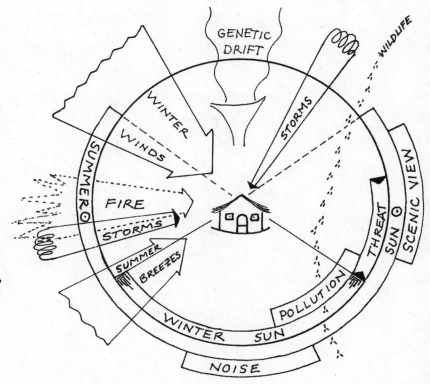

A Sector Diagram
[Credit: Peter Bane]

when planning restoration projects. The neighbor's driveway might be a good reason to put up a screen or a barrier fence, while a busy highway might warrant a thick hedge or embankment to suppress noise and hide the view.

By analyzing the different sectors influencing a site, we can determine how best to focus beneficial energies such as winter sun, summer breeze or a view of the mountains, while scattering or deflecting hostile energies such as storm winds, pollution or fire. Sectors, along with zones, enable us to place every element of the system in the best possible location for beneficial function.

## Putting It All Together

A design begins with a vision, which comes from within us and also from within the landscape. The process of making that vision a reality requires that we observe the site for a prolonged period, gather information from maps and other printed sources and analyze our findings. From the analysis, informed by our intuition and feelings, we craft a story,

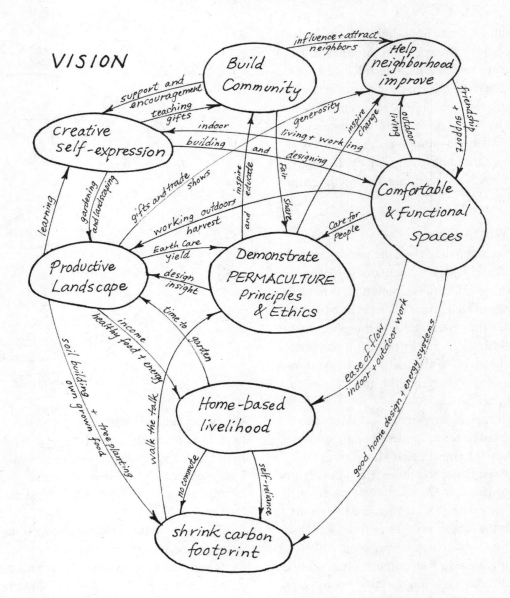

**A Bubble Diagram of the Farm Vision**

[Credit: Peter Bane]

incorporating past and present, that begins to shape the vision of what may come to be in that place.

The environment will present limiting factors and challenges which we hope will be revealed by our investigations and analysis. The vision will reveal the scope of the design, and from this big picture we must attempt to identify all the components. At this stage, pattern language can be of great help, since it presents us with many well-formed words in a cultural language. We can borrow these patterns, starting with the largest that matches the breadth of the vision, then arrange them according to the logic of the language and of the landscape, and further refine their relationships with each other through a needs-and-yields analysis. Then we begin to locate these elements by zones and sectors.

Design is not a linear process, but iterative. That is, it moves by cycles of learning and integration. We take a step, gather information or feedback from the landscape and from people who are involved; we come to a new understanding, and then we take the next step. However, design does have a dynamic: it must begin with the largest patterns and move toward smaller and smaller levels of detail.

# A Garden Farming Pattern Language

I offer this language as an aid to designing permaculture systems on urban and suburban properties and for the creation of garden farms at whatever distance from city centers.

Each pattern consists of a problem statement and a solution, with each *problem* being a common conflict or demand in the terrain of garden farming that needs to be resolved, and each *solution* being a directive about how to solve it by placing and shaping landscape, building and social elements. While the patterns describe common settings, these always exist in the context of larger wholes, and each pattern in turn will be filled out by smaller patterns or wholes.

For those readers who, like myself, want to integrate permaculture design with the cultural legacy transmitted by the Pattern Language project, the language subset in this chapter branches at two points from the original book.[1]

1. Following the opening sequence of patterns that describe the regional landscape: The Distribution of Towns, City-Country Fingers, Agricultural Valleys, Lace of Country Streets, Country Towns and The Countryside

2. After the third sequence which describes the character of towns and village-scale communities and which begins with Community of 7000, Subculture Boundary, Identifiable Neighborhood and Neighborhood Boundary.

We can, for example, expect to see town-sized communities of 7,000 emerge as new centers amidst the suburban sprawl now dominated by automobiles. The Transition (Towns) movement, emerging worldwide, is homing in, albeit somewhat unconsciously, on the power of this scale of settlement.[2] As transportation slows and resources contract away from a global economy, more traditional market towns will nurture the renaissance of local economies. Within these trading territories, village-scale neighborhoods will emerge as natural social and economic communities.

## 1. Landscape Catchment

Uplifted landscape catchments are the source of potential energy that enables ecosystems to differentiate and to sustain complex life forms. The erosion of mountains provides the elements to build soil, while the cascade of water from peak to valley creates endless niches for life to inhabit. The newer the uplift, the richer the mineral content of soils. Uplands are the source of new material, of rain, of pure water. Soils there are least developed, most quickly eroded. Conversely, lowlands are zones of accumulation and deposition that hold richer soils, larger reserves of carbon and also carry the greatest load of contaminants.

Though nutrient flows downhill by gravity, it is often returned from sink to source by life processes. Anadromous fish such as salmon and shad swim upstream to spawn and die and feed enormous food webs high in the catchment, returning oceanic minerals to deep continental interiors. Goats, which evolved as mountain dwellers, sleep in their native habitats above the tree line to minimize predation but browse at lower elevations, thus moving nutrients uphill. Human cultures too, have evolved means of balancing the downhill flow of nutrients. *Transhumance* is the migration of herds and herders into high mountain pastures in summer to harvest forage (and to return manures), while coastal dwellers gather in seaweeds to fertilize their farm fields.

Our obligation is to slow the descent of water through catchments; this augments the growth and cycling of life forms while reducing the loss of minerals and soil by gravity. To this end, keep ridges and riparian zones covered in trees, create terraces to retain soil and restore wetlands and meanders along all permanent streams. Maintain the health of coastal and riverine fisheries, especially of those ocean-going fish that spawn upstream.

To complete this pattern, direct agricultural development to appropriate landforms (3-Agricultural Terrain), plant trees in uplands (6-Forested Ridges), store water high and release it harmlessly (5-Water: Source and Force of Life). Create check dams, terraces and swales to reduce runoff (32-Small Earthworks) and hold water in Ponds and Dams (33). Plant and manage productive woodlands (31-Coppice and Hedges), move animals through grasslands quickly and intensely (51-Rotational Grazing) and encourage the economic use of these ecological reserves to ensure their preservation (10-Wildland Foraging). Observe, harvest and cycle resources from Zones of Accumulation (56).

## 2. City-Country Fingers

The present smear of urbanization in our city regions wastes land and dilutes culture. As energy contraction deepens, economic rationalization will intensify settlement and services along transit corridors that follow upland roads, rivers and coastlines. In between these city fingers reaching out from the urban core, suburban areas will aggregate into town centers surrounded by working farm landscapes comprised of village-scale neighborhoods of 1,000 acres or so.

North American metropolitan regions contain an average of two acres per person overall (more than enough land to provide food and fuel for their inhabitants) while cities proper average five persons per acre — not a very dense standard compared to Manhattan or older European and Asian

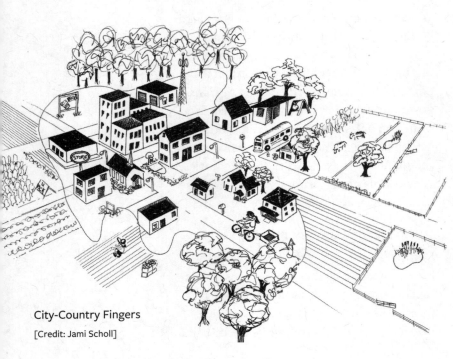

City-Country Fingers

[Credit: Jami Scholl]

cities, but in many ways more appropriate for a low-energy future.[3] To help this pattern emerge, encourage transit-oriented development along main routes and rail lines, give some measure of land use authority and civic infrastructure to village-sized territories (4-Working Neighborhoods and Ecovillages) and support agrarian development in suburbs and towns between the urban fingers (11-Garden Farms). Connections between city and country can be enhanced by creating bicycle boulevards — safe thoroughfares for slow traffic — that extend transit lines and main city roads into country areas. Place farmers markets at the edge of the fingers to stimulate exchange (16-Local Trade). In both city and country, encourage the planting and use of edible landscapes to increase food security (8-Living in the Garden).

## 3. Agricultural Terrain

Not all land is well-suited for cultivation. In the 20th century the best land for agriculture near cities was also seen as the best land for building, leading to enormous development conflicts and the submerging of much first-class farmland under pavement and housing. In the urban devolution of the 21st century, the confluence of city markets with suburban homes and lawns presents one of our greatest opportunities. The classic understanding of McHarg, Alexander and others — that valleys should be reserved for agriculture and the slopes above them for settlement — was not wrong, but now must be viewed through a finer lens.

Soils indeed wash down into valleys, making them richer, but so does cold air. Hill slopes, if moderate and offering good solar aspects, are ideal for housing and tree crops because they enjoy superior microclimates, warmer than valley floors. The choice locations are near the point of slope inflection from convex to concave (called *key points* or when connected, *a keyline*) wherever 60 feet of relief exists. Therefore, place farm settlements to optimize productivity of the land, while minimizing energy costs of shelter and access to markets. On broad uplands locate homes to the poleward side of plateaus, backed by woodland. Use the upper part of north-facing slopes for some tree crops, while reserving the flatter areas to the sunny side for more intensive cultivation, water storage and solar collection. On moderate sun-facing slopes locate housing at keyline elevations to benefit from thermal belts, with trees above and gardens and ponds below. Locate lowland settlements out of valley bottoms but just to the high side (and poleward if possible) of open ground, backed into woods and always out of the flood plain.

To complete this pattern, create and preserve woodlands and plant farm shelterbelts among the settlements (9-Woodland Mosaic), place houses together in clusters of 6–12 homes (7-House Cluster) oriented to the sun (12-Shelter in the Sun). Allow each homestead ½ acre or more of ground for cultivation (11-Garden Farms) and create level ground and water storage with Small Earthworks (32). Use poleward slopes for trees (31-Coppice and Hedges) and buffer lowlands with water (33-Ponds and Dams).

## 4. Working Neighborhoods and Ecovillages

People have a need for personal connection to the land and to community where they live. For reasons of biology, the appropriate human scale for regenerative and self-reliant settlement averages about 500 persons across cultures.[4] Other group sizes are significant for other purposes: work groups of 3–7, extended families or classes of up to 40, business and military units of 150,[5] but the village range from 250–1,000 spans from the low end of genetic health to the threshold of social pathology. It is in these aggregations that humans have lived for most our time in the Holocene.

Investigations formalized since 1990 but with antecedents from the early 20th

Working
Neighborhoods
and Ecovillages
[Credit: Jami Scholl]

century have attempted to clarify what this ancient pattern may mean for post-industrial humans.[6] A settlement providing access to all the elements of daily life — food production and consumption, housing, employment, sociability, basic healthcare and education, self-governance in most local matters and spiritual practice — has the best chance of meeting human needs at the lowest ecological cost while maximizing democratic well-being.

To be economically effective and regenerative, these newly conceived communities must become both ecological and self-reliant. None of this will happen overnight; the effort must be understood as a multi-generational project. Therefore, identify and begin to form neighborhoods and village-scale communities within and beyond the devolving metropolitan regions. They should have land bases of from 300 to several thousand acres with an eccentric nucleus in the direction of the greater urban core. Help each village or neighborhood achieve a clear boundary in natural features of landscape — water, woods, bluffs, ridges — or infrastructure (main roads, industrial zones). Accentuate the neighborhood entrances with gates and signs, be these pillars, trees, sculpture or buildings close by the road. Ensure that each village has a

gathering place — ideally the conjunction of an outdoor plaza with a meeting hall — and support the growth of self-reliance and self-government by investing the communities with some authority over local land use and responsibility for common resources: woods, parks, gardens, streets and waters.

The long-term success of ecovillages and neighborhoods as human communities will depend on the emergence of new forms of economic cooperation among their residents, coupled with a durable and functional landscape unique to each, including likely a school and some place of common spiritual practice or civic encounter. To aid these ends, plant edible landscapes throughout (8-Living in the Garden), organize the housing into small groups (7-House Cluster), provide market spaces (16-Local Trade) and design both casual and formal meetings between residents to bolster social capital (19-Neighbors and Strangers, 20-Communal Labor, 67-Connection to the Street). Especially foster the development of local land care (11-Garden Farms).

## 5. Water: Source and Force of Life

Water is essential to life and to the distribution of fertility, but water is heavy and as it moves it can be very destructive. Therefore, catch and store rain and runoff water in high structures. Spread and cycle it to enhance the growth and decay of biomass. Ensure that its release is clean and harmless.

Begin catching water from the sky by covering steep ridges and slopes with trees (6-Forested Ridges) and let every roof play its part (25-Roof Catchment, 26-Cisterns). Reshape catchments subtly to retain runoff (32-Small Earthworks) and create surface water storages (33-Ponds and Dams) to support life and yield (34-Water Gardens and Fish Crop). In the household, implement conservation measures (47-Water Cascade) and treat effluent to recapture nutrient and restore purity (49-Wetland Water Cells, 48-Greywater Trenches).

## 6. Forested Ridges

Narrow ridges and steep slopes in hilly or mountainous country are subject to severe erosion, but their elevation is a primary resource of potential energy for living systems. Wherever moisture and temperature permit, nature covers these harsh landscapes with trees. We should do the same. Therefore, reforest and maintain permanent tree cover on steep uplands and all slopes of more than 20%. These forests make rain, hold soil and begin the local cycles of water. Avoid building homes on narrow ridges and steep slopes.

To complete this pattern, let new and old ridge forests run into and help form a broader landscape cover (9-Woodland Mosaic). Encourage the sensitive use of these forests (10-Wildland Foraging) to ensure they are valued and enhanced by the community. Retain soils and manage runoff with earthen structures (32-Small Earthworks). Push the succession of new forests and woodland by planting and slashing supportive shrubs, herbs and small trees (60-Working Pioneer Plants). And let the choice of species be weighted to natives, but emphasize economically valuable and, wherever possible, edible plants (8-Living in the Garden).

## 7. House Cluster

Most of us meet our social and common cultural needs amongst a circle of 30–40 people.[7] Studies have shown that most people limit interactions with neighbors to the houses immediately adjacent and also directly across alleys or streets where traffic is light. A cluster of 6–12 houses near each other can provide the social and physical resources for effective mutual support in times of hardship. Common interests are likely to emerge, and petty conflicts wane as real economic pressures mount. Therefore, place houses in new developments to foster sociability and ready access. In older neighborhoods identify natural house clusters within limits of about 300 feet and create connecting paths, gates and alleys to increase contact. Arrange mailboxes and driveways to facilitate chance meetings. Provide some land and resources: streets, sidewalks and paths, community gardens, boulevards with edible landscapes, wooded hedgerows, alley crops — for which neighbors may exercise cooperative care. Keep through traffic and parking to one side or the outside of a house cluster, or work to narrow streets and slow traffic.

Nestle these clusters in the landscape according to the guidance of Agricultural Terrain (3) in order to conserve the productive capacities of the land and provide optimal microclimates. And whenever possible connect them to Working Neighborhoods and Ecovillages (4) so that their larger communal needs can be met. To fill out the House Cluster pattern, let productive trees form privacy screens between and around houses (31-Coppice and Hedges), line driveways and alleys with edible perennials (15-Alley Cropping), organize occasions for work parties and other celebrations (20-Communal Labor), make reciprocity and mutual protection a hallmark of neighbor relations (19-Neighbors and Strangers) and ensure that every house in the cluster is approachable (67-Connection to the Street).

## 8. Living in the Garden

Humans claim our place in nature through the garden. Our urban settlements are now for the most part desolate and eroding landscapes that produce no food. Culture in these places is inherently groundless and shallow where it survives at all. Food security and a collective sense of well-being will both be served by the planting and tending of edible landscapes everywhere people live and gather. These create useful work, shape humane spaces, heal soil and soul at the same time and bolster physical and economic prosperity.

Therefore, plant productive trees, shrubs and other perennials along streets, in parks and around public buildings. Encourage grazing and harvesting for home use. Invite neighborhoods to take stewardship of public plantings. Give away edible perennials to support the spread of small orchards and forest gardens. Fruit gathering and gleaning is an entry-level opportunity for young entrepreneurs to build social capital. Have some patience, sponsor city food fairs and get local chefs involved.

To manifest this pattern more fully, plant and maintain a fabric of many useful and edible perennials over the broader region (9-Woodland Mosaic), encourage respectful use of the wild and public landscapes (10-Wildland Foraging) and foster development of small land care enterprises (11-Garden Farms). Let all roadways and parking lots be bordered with fruits or nuts (15-Alley Cropping) and nurture the growth of soils and ecosystems by the regenerative power of trees (31-Coppice and Hedges). Surround those crop trees with supporting plants (53-Fruit Tree Guild). Also, maintain grasslands, lawns and meadows with animals whenever practical (51-Rotational Grazing).

**Woodland Mosaic**
[Credit: Jami Scholl]

Help spread the psychological benefit of this pattern by making many spaces for public and private gathering out-of-doors (37-Outdoor Rooms). Ensure soil fertility by vigorous cycling of biomass (60-Working Pioneer Plants, 61-Fertility Crops, 62-Mulch and Compost).

## 9. Woodland Mosaic

Humans evolved from forest-dwelling primates to become successful hunters of the African savannas. We still seek to occupy a niche at the edge of these two ecosystems. Thus we open clearings in wooded regions to plant crops and surround our homes with lawns. In grassland terrain we shelter our houses with trees. The most food-productive landscapes will take the form of a woodland mosaic with small sunny garden clearings, paddocks, working hedges and modest patches of productive trees. Aim for 40–70% tree cover in city regions. Let trees cover steeper slopes, protect riparian zones, shade pavement, provide shelterbelts for croplands and dot pastures to form agricultural savannas. Street trees and edible public parks help to form this pattern even in dense city districts.

A healthy form of this pattern will entail sensitive public use of wild zones (10-Wildland Foraging) and a strong anchor of cultivation (11-Garden Farms). Within settlements, small tree elements help to extend the mosaic (31-Coppice and Hedges, 53-Fruit Tree Guild). Be sure that tree planting and growth does not impede solar access for housing (12-Shelter in the Sun). Use the resulting increment of wood trimmings and thinnings for fuel (21-Woodshed) and fertility (62-Mulch and Compost).

## 10. Wildland Foraging

Our Pleistocene ancestors and all agricultural and pastoral peoples since have practiced wildland foraging. In the most rugged environments, foraging is the only sound adaptive strategy for survival. Ethics of care and balance emerged of necessity from foraging cultures. In the hyper-civilized terrain of modern cities, we fulfill our foraging instincts in ways that are often dysfunctional: crime and shopping. While mass populations today can't rely solely on foraging, wild foods with their intense flavors and superb nutrition provide a necessary tonic and supplement to gardening and agriculture. The model of foraging gives us a picture of shifting mosaics of abundance in the world around us.

Zones of forest and water storage are needed to protect vulnerable terrain and to provide ecological services, and humans need a nurturing relationship with these landscapes as well as with more domesticated ones. Hunting, fishing and the gathering of herbs, berries, mushrooms, dye materials and fiber plants provide a vital release for the deepest part of our beings. Therefore let every public landscape be open to foraging, teach respectful use of the wild in public schools and establish seasonal limits to protect the health of wild plant populations just as we now limit hunting.

To create this pattern, establish foraging in public places as an expected behavior (8-Living in the Garden), use more livestock amongst settlements because pastures provide an important buffer between wild and cultivated lands where livestock and game overlap (51-Rotational Grazing). Respect the enterprise of small-scale cultivators (11-Garden Farms) whose landscapes also nurture wildlife and diversity. Understand that these cultivated hotspots will need protection from

wild animals (29-Fencing). Let some of that protection take the form of living fences that may be shared (31-Coppice and Hedges).

## 11. Garden Farms

Large-scale agriculture has been destructive of the land and of rural culture, while industrial food has made modern people increasingly sick. Nor can this continue in a crowded world with declining energy. Gardening produces two to seven times the yield per unit of land of conventional farming. It gains that productivity from more eyes, hands and feet on each acre of ground.

Much of this new agriculture will emerge around cities. To cope with a 70% reduction in fossil fuels use over the next generation, we need millions of new farms, each with a working household to care for it. Urban regions in North America have organized water supplies and ready access to markets. Garden farms will range in size from less than half an acre to over 20 acres. Most will be managed by owner occupiers, some will be tended by renters. Some will expand into vacant lots with or without permission, and not a few may be sharecropped on neighboring yards. A changed economy will provide the conditions for increasing household size, and new arrangements for extended family, apprentices, live-in help and cooperative enterprises will provide the labor needed to achieve high productivity on small parcels. Therefore, remove legal barriers to urban, suburban and

Garden Farms
[Credit: Jami Scholl]

peri-urban farming. Enable smallholders to keep livestock and to sell produce directly to consumers from home; increase public market space in all districts.

The garden farms economy rests on a solid basis of self-reliance (14-Household Provision), requires the support of a larger labor force (17-Family Table, 20-Communal Labor), will be most easily organized by people in their own homes (13-A Home of One's Own) and needs a vital commerce in the immediate area (16-Local Trade). Agriculture depends on a good supply of water, which on a small scale can be met largely by Roof Catchment (25). The garden farm will also need an efficient access pattern (28-Branching Cart Paths and Lanes). As it grows, this sector will attract tremendous interest that must be filtered to create value (19-Neighbors and Strangers). Marketing will be the most immediate challenge to success, so encourage on-farm produce sales (68-Fruit Stand).

## 12. Shelter in the Sun

In cold and subtropical climates, homes should be oriented with the long side toward the sun to permit passive collection of solar energy in cool seasons. The mid-elevation of moderate sun-facing slopes offers the choicest microclimate for shelter. Much built housing ignores these rules of ecological design, but east- or west-wing additions, added south wall glazing and greenhouses can enhance energy capture by older buildings. Therefore, place and remodel homes for solar gain, to best exploit the advantages of landform and to allow cold air to drain away. Use overhangs, awnings and insulative window coverings to deflect heat in summer. Place adequate thermal mass where it can absorb direct radiation, enclose it in a well-insulated envelope and further protect the building, by trellises and screening shrubs, against cold winds and summer heat.

To develop this pattern, support home ownership (13-A Home of One's Own) with the aim to eliminate conflicting incentives between ownership and occupation. Orient buildings to place useful outdoor space in the sun (36-South-Facing Outdoors) and develop outdoor extensions of the living space on porches, decks, patios and balconies (37-Outdoor Rooms). Add a solar growing space to aid energy capture and food production and for sheltering cool-season activities (38-Greenhouse). Shield the building from excessive heat in summer (42-Porches and Awnings, 43-Trellises for Shade) and buffer the cold sides of the house with shrubbery and dense vegetation (31-Coppice and Hedges). Always collect runoff from the roof for irrigation, stock watering and washing (25-Roof Catchment).

## 13. A Home of One's Own

The need to remodel much current housing for energy efficiency and increased occupancy as well as for new economic activities favors owner occupancy as a more effective model of management than control by either public authorities or absentee landlords. Where small parcels of land accompany housing, as in North American suburbs, this form of settlement has good potential to foster democratic community and provide a bulwark against the reemergence of feudalism.

Work to secure home ownership for the great majority of households using collective action, political pressure, the pooling of resources through credit unions and mutual lending syndicates and squatting where necessary. Landlords and public officials have some role to play in providing rental housing and shelter to transient populations, but let those be limited to multi-unit buildings and the occasional second house near the owner's primary residence. Discourage conversion of all but the very largest homes into apartments, but support cohousing in its many forms — common wall townhomes and cluster housing among them. Permit unrelated individuals to occupy houses as co-owners and retain or extend homestead exemptions from property taxes.

To make this pattern durable and effective, educate for the aim of self-reliance (14-Household Provision) and expand legal and social tolerance for large households of all types (17-Family Table). Make a socially expansive and economically useful kitchen the center of the household (18-Country Kitchen), with a well-developed shading of public space into private from the street to the innermost realms of the home (64-Public-Private Gradient). Give the household a place to make and repair things (23-Workshop), amplify its useful living space by surrounding it with Outdoor Rooms (37) and give it a strong point of interaction with the public (67-Connection to the Street). Let the strengthened household population and economy provide a solid ground for extending agrarian values to the surrounding community (19-Neighbors and Strangers).

## 14. Household Provision

Self-reliance and the home economy will regain prominence as energy descent unfolds. Social insurance in the mid-21st century will consist of access to land, both private and common. Three-car garages will become home businesses and workshops, and we'll find new need for the pantries and cellars of our great-grandparents. Therefore, revise the model of the household to center on gainful work at home: food production, processing and storage. This will entail shelter for more helpers of all ages, but especially for multi-adult families. Also, utility spaces and structures that have been simplified, reduced or eliminated will have to be recreated: pantries, root cellars, drying yards, woodsheds, workshops, barns and animal shelters. Food and fuel storage is the point of beginning. The next steps involve creating more working spaces, both indoors and out.

Central to the completion of this pattern are a large working and socializing space around food (18-Country Kitchen). Almost as important are a series of ancillary spaces for storing and processing food, fuel and other useful materials (21-Woodshed, 40-Drying Yard, 22-Storage Barn, 45-Root Cellar, 46-Pantry). Support for year-round food production comes in the form of a Greenhouse (38). Protein and fertility needs will be supported by Animal Housing (24), while the making and repair of tools, jigs, cabinetry and other useful items needs a Workshop (23). All of this capacity supports an adaptive, flexible economy that will have many expressions (63-Shifting Enterprise).

## 15. Alley Cropping

This pattern derives from the agroforestry practice of planting rows of trees separated by alleys for arable crops. The need for more trees is global and immense, but space in our urban regions is constrained and sun-loving annuals remain an important part of diet. Rows of crop trees can frame accessible lanes, planting beds or linear paddocks. The trees take time to mature, leaving many seasons when competition for light between the canopy and the smaller plants in the alleys is minimal. Trees make more complete use of

Household Provision
[Credit: Jami Scholl]

soil moisture, pull up subsoil minerals, improve drainage, protect soil and can provide mulch, fuel and fodder.

Therefore, make rows of productive trees the backbone of the garden farm, and use all the space around and between them (the alleys) for compatible and mutually supportive species. Fit the trees in along driveways and fences, in parking lots, among planting beds and in pastures.

Design the alley crops to follow Branching Cart Paths and Lanes (28), mix nitrogen-fixers and other Working Pioneer Plants (60) amongst the trees and manage the woody perennials by cut-and-come-again practices (31-Coppice and Hedges). Plant along contours where possible to retain water (30-Contour Planting). Animals can be grazed among the trees (51-Rotational Grazing, 52-Small Paddocks), or space can be used for vegetables or grains (50-Patch Gardens). Even where the trees will eventually close canopy, the ground between them can provide ample opportunity for short-term yields (55-Catch Crop). Maintain soil fertility by all these means (rotations, animals, coppice, pioneers) as well as by Mulch and Compost (62).

## 16. Local Trade

Future economic security depends on the restoration and enrichment of the land and the production of useful goods and services made from local materials or exploiting local advantages. To remain habitable, every region must balance population with the sustainable yield of local resources, especially food and fuel. A robust and diverse agriculture of smallholders ensures both a strong supply of these primary products and a demand for tools, household wares and a wide range of basic manufactured goods. While innovation and items of unique provenance will continue to induce long-distance trade, supplies of food, energy and building materials must be developed in all locales. Local trade presently asserts the values of freshness, character and community solidarity, but it will regain advantages of price and reliability as energy costs rise and more producers enter the market.

To this end, support local growers and providers whenever possible. Establish public markets for produce, farm goods and local wares and handicrafts. Sanction home sales and roadside stands, and exempt from sales taxes not only all foods and produce, but all sales directly from individual or home producers to final consumers.

The elements of this pattern call for flexible production (63-Shifting Enterprise), the ability to make as well as grow things (23-Workshop, 11-Garden Farms) and a regular flow of connection between the household economy and its surrounding society (19-Neighbors and Strangers). Larger households will have the resources for selling as well as producing value (17-Family Table). Much

Local Trade
[Credit: Jami Scholl]

produce will be sold directly from the farm (68-Fruit Stand).

## 17. Family Table

Energy descent means more hands will be needed at home, yet population imperatives argue for a falling birth rate. Parents and children brought back together out of hardship is one model for a more economic household size, but new forms of relationship are also possible: housing or homesteading cooperatives, grandparents and grandchildren, a teacher with apprentices, a widow with student lodgers, young couples with one or two children amongst them, even boardinghouses with work sometimes exchanged for meals. The economics of group living are powerful, and the psychological benefits of multilateral intimacy are undersold in our current television culture. Challenges center on autonomy and the division of labor. A wealth of wisdom is available from social experiments with intentional community.[8]

A durable model is the extended family, where life cycle differences beget strength through the satisfaction of complementary needs. A measure of age diversity, and particularly the presence of elders, helps greatly to ameliorate conflict and submerge egos beneath the waters of common life. If the middle-aged, who are society's empowered and who are often sandwiched between school- and post-school-aged children and aging parents, could reach out in both directions — not even necessarily to their own children and parents — small miracles might occur.

Therefore, expand household numbers and share the economic gains justly by inter-generational wealth transfer, lowered living expenses and mutual aid. Change codes that restrict the building of in-law suites, develop backyard cottages and studios above (or in) the garage, as well as walk-out basement apartments — or add solar wings to old houses in order to increase the capacity of well-placed households. Expand kitchen and bath facilities and add entrances. Make clear arrangements about work and money, and treat all adults as fully autonomous. Aim to combine youth and middle age, so that the unavoidable additions of young children and the dependent aged remain in balance with the strength of working adults. Whatever the makeup of the extended household, its heart beats around the family table. Make this the center of the garden farm, the crucible from which a new culture can be born.

To serve the extended family, make the center of the home a large working social space (18-Country Kitchen), and put it at the center of the Public-Private Gradient (64) that begins at the street and ends in the most intimate realms of the house. Even here, create rituals of sharing to deepen the bonds of intimacy (65-Communal Bathing). Make regular occasions for collective effort (20-Communal Labor), invite in others selectively (19-Neighbors and Strangers) and provide space for temporary and transitional residents and guests (66-Rooms for Guests) so that the household can grow and change as needed.

## 18. Country Kitchen

The center of any home is the kitchen, and like the farm kitchens of yore, garden farms need big ones to accommodate 4–12 residents and guests, food preparation and much of the coming and going of a working household. Meetings over meals help sort out work and social life. Food storage needs are significant, and preparation of more meals from scratch ingredients requires ample counter and table space. As a new locus of social life, the home kitchen needs to grow to welcome neighbors and visitors, make space for play and generally become the hub of group living.

Therefore, make a space of nearly 20 by 30 feet at the heart of the house (this can include adjacent dining or sitting rooms in an open plan). Create a pantry nearby. Ensure good visual and vocal connection throughout, and provide multiple niches, seats or alcoves for

Country Kitchen
[Credit: Jami Scholl]

work, play, reading, crafts and household repairs. The cooking area needs at least 12 feet of counter space, lots of cabinets and shelves and a triangular pattern between a double sink, stove and refrigerator. Use overhead space for pots, pans and dishes and have several tables in the space for gatherings of various sizes — at least one big enough for the whole household and guests.

To complete this pattern, keep the kitchen at the midpoint of the Public-Private Gradient (64). This will almost certainly put it adjacent to some sort of Outdoor Room (37). Ensure that the household has plenty of food put by (44-Food Storage) using appropriate spaces (45-Root Cellar, 46-Pantry). Provide an outdoor cooking and food preserving space for use in hot weather (39-Summer Kitchen), and direct used sink and vegetable washwater through the household (47-Water Cascade) before it is reclaimed for soil nutrient (48-Greywater Trenches, 49-Wetland Water Cells).

## 19. Neighbors and Strangers

In gathering the energy to transform landscapes and culture, garden farmers have to draw on labor and goodwill from outside the household. New farms must knit themselves into the fabric of a neighborhood where their activities may be anomalous.

To make this pattern function well, form a well-articulated gradient between the front gate or most public area of the garden farm and your private spaces, and know just how far people are welcome into it (64-Public-Private Gradient). Offer a variety of simple goods for exchange that represent your farm fairly but that sell for only a few dollars and require little attention on your part (68-Fruit Stand). Develop your gate, your signs and your schedule to make it easy for visitors to come by, but hard for them to intrude (67-Connection to the Street). Make sure you have extra housing for guests, including guest workers, that doesn't compromise your own privacy (66-Rooms for Guests). Invite interested persons of goodwill to participate in periodic work parties where you provide some food and some direction so that everyone can learn a little and have a good time (20-Communal Labor).

## 20. Communal Labor

Food is everybody's business. In traditional cultures, seeding, harvest, food preservation and the slaughter of large animals brought the community together in work. So too did barn raisings and other large construction projects. In the neighborhoods of 21st century towns and suburbs, professional building, commuting to jobs, commercial shopping and divergent lifestyles have supplanted ties to land and subsistence lifeways. The garden farmer, swimming upstream against this troubled current, must weave together threads of collective memory, practical need and the excitement of novelty to recreate a fabric of communal labor. Therefore, develop networks of mutual aid to transform large projects into rituals of play and meaning, with neighbors when possible, but also with colleagues, volunteers, customers and guests. Let work be balanced with food, learning and stories. Make sure there is a role for everyone from 5 to 95 and that there are enough hands to make the job fun.

This pattern can more easily take form if there are open spaces on the farm that can accommodate group labor while containing the collective energy, for example, a Drying Yard (40). Informal but comfortable spaces for group gathering and eating are also important (37-Outdoor Rooms, 18-Country Kitchen). Make hot weather food preservation a special occasion and give it good support (39-Summer Kitchen). Keep a good selection of tools right at hand (23-Workshop). Have a sacrifice area of pasture or a widened driveway entrance that can double as extra parking (67-Connection to the Street). And keep a spare room, barn loft, extra tent, fold-out couch or cot handy for overnight and longer-term guests (66-Rooms for Guests). Let everyone who visits know there is work to be done (14-Household Provision).

## 21. Woodshed

Though design for passive solar gain is the most economical and ecological source for space heating, in few North American regions is it likely to be fully adequate for all needs. Therefore, to move the household toward a smaller carbon footprint and locally sourced energy, make all responsible improvements to house design for solar capture and storage, including insulation, glazing and mass, and then add biomass heating.

Even where sunshine can provide a substantial portion of home heat, most older houses are poorly configured to capture it, and fully effective retrofits may be out of reach. Wood, though by no means available everywhere, is the most widely distributed renewable fuel in cold regions. Useful amounts of it can be grown in backyards, hedges and street tree plantings, and more can be harvested from the wastestream of construction debris, used pallets and urban forestry. Newer wood stoves achieve 65–70% thermal efficiencies, and the best technologies (masonry heaters, rocket stoves and similar exhaust-harvesting devices) can extract 90%

or more. Wood provides services as it grows (soil building, habitat) that coal, gas or oil can never offer. And, wood is carbon neutral when supplied from sustained yield forestry and wastes.

Optimum energy yield from wood requires that it be air- and sun-dried for 18–24 months before burning; this means covered storage. Adopt conserving behaviors, install efficient stoves and build covered sheds adequate to hold two winters' fuelwood. Place the shed near the loading door for each building it must serve, but also provide it cart or truck access. Leave a yard no less than 15 × 20 feet near the woodshed where logs can be dumped, cut and split — and carts and trucks loaded and emptied without congestion. Let the woodshed double as a visual screen, as part of a fence or as a windbreak. Place it to receive several hours of direct sunshine a day. Capture rainwater from the roof to irrigate nearby gardens or feed a pond. Let the rafters hold surplus lumber, pipes and other resource inventory (58).

Fill out this pattern by harvesting your own firewood from productive multipurpose trees and shrubs (31-Coppice and Hedges). Keep an eye out for windfalls of dead or downed timber in your neighborhood or scrap wood you may come across (57-Salvaged Materials). Have a number of areas on the farm convenient to the street where a load of unprocessed firewood can sit

Woodshed

[Credit: Jami Scholl]

harmlessly for a few months (56-Zones of Accumulation). Connect these by good pathways to the woodsheds (28-Branching Cart Paths and Lanes). Regularly add marginal wood culled from your piles of lumber and salvaged burnables (58-Resource Inventory). And always connect significant building roofs to a water collection system (25-Roof Catchment).

## 22. Storage Barn

While garden farms have smaller needs than the mixed farms of the late 19th and early 20th centuries, some cheap, spacious shelter will still be required for a host of purposes: animal housing, tool and material storage, crop processing and temporary guest and worker housing. A classic dairy barn would have been placed about 100 feet from the house and on the edge of larger pastures and fields. Its multiple levels allowed for integrated feed storage and animal shelter. In narrow Appalachian valleys, the barn was often placed at the edge of fields tucked into the woods, for it needed neither solar gain nor flat land on both sides. Farmers everywhere have noticed that cutting the barn into a bank for vehicle entry on two levels is a boon.

A storage barn can be as simple as a rented semitrailer parked in a convenient location, but a two-story structure with good ground level access would be superior. An existing

Workshop

[Credit: Jami Scholl]

garage may be converted by ousting one or more of its automotive inhabitants. An old carriage house or purpose built structure with a small apartment, office or studio above approaches an ideal.

Build or rededicate a large multipurpose space within 100 feet of the main house as a storage barn. Collect rainwater from its roof, provide vehicle access on at least one side and make it easy to reach all parts of the building. If possible, orient the structure for solar gain so that it can be partially used as a workshop or apartment. Give it deep eaves to enhance the utility of the building exterior.

Complete this pattern by making part of the barn interior into a workshop and tool room (23-Workshop), provide temporary or longer-term housing in an upper section (66-Rooms for Guests) and shelter animals on the ground level, either inside or in lean-to sheds (24-Animal Housing). The barn may provide the best location for an attached Greenhouse (38), but with or without it, keep flat ground open on the sunny side of the building (36-South-Facing Outdoors).

## 23. Workshop

No self-reliant household economy can survive for long without the capacity to build and repair furniture, implements, small structures, tools and basic machinery. The weekend tinkerer in a garage or basement can still be found in most settled neighborhoods. Even men who've scarcely lifted a hammer in a decade may own extensive tool collections inherited from fathers and grandfathers. This cultural legacy runs deep and deserves a new set of clothes. As soon as you can afford it, build or convert a space for sheltered work with tools. The garage may have to serve. The minimum is a utility room, mudroom, porch or large vestibule that can be outfitted with cupboards and hooks for tool and material storage combined with a covered patio or adjacent paved driveway where projects can be tended during fair weather conditions.

The workshop should have a minimum long dimension of 20 feet, a level smooth stone or concrete floor, wide doors, good natural light and, if fully enclosed, a small stove for intermittent heat in winter. A three-sided structure with a deep awning open to the south would offer much of this for three seasons. Electric power is helpful; any water supply should be frost protected. Combine the workshop with a storage barn or guest rooms in a single structure. You will eventually want walls or a lockable tool closet.

To make the workshop complete, front it onto an open, paved space with good circulation (37-Outdoor Rooms, 40-Drying Yard). Make sure that it has a deep overhang to support the transition between inside and outside work areas (42-Porches and Awnings). Let the workshop support many pieces of the household economy (63-Shifting Enterprise).

## 24. Animal Housing

Small, cheap and designed for function, animal cages, pens, coops and sheds may be built from recycled and scrap material. Many a one-room settler's cabin or camper-trailer has ended its life as a chicken coop. Each structure must fully meet the shelter needs of animals throughout their daily and life cycles and should afford ready access to whatever range or forage the animals use.

Small livestock especially need protection from a host of wily predators: snakes, weasels, rats, skunks, raccoons, possums, dogs, cats and larger varmints. A good roof, access to water, ease of cleaning out manures and tight enclosure provide the basics. Allow four square feet each for chickens, about five for rabbits and 15–25 for goats, depending on size and number. Consider also the vertical possibilities: rabbits over chickens, chickens over ducks, hogs or poultry over a pond or, in the manner of the Swiss, the house over the herd. Provide a little extra secure space to store feed. Plant nitrogen-loving species like comfrey, bamboo or large grasses downslope to mop up manure runoff. Direct roofwater to stock watering tanks and troughs, and place animal shelters within easy walking distance of the dwelling (30–100 feet) but not upwind. Surround the animal housing with hedges or trees for shelter and fodder.

The connections from this pattern lead to mobile shelters (54-Poultry Tractor) and other forms of intensive grazing management (51-Rotational Grazing). Animals can be used to create and turn wastes into fertilizer (62-Mulch and Compost) and, with the right timing, can apply this directly to garden soil between crops (50-Patch Gardens). Lay out access to forage for quick, intense impact (52-Small Paddocks). Ensure that your livestock stays where you want it (29-Fencing).

## 25. Roof Catchment

Runoff water is the only significant untapped source of fresh water for meeting a thirsty world's future needs, and every roof is also a water collection device. Only gutters, downspouts, some simple plumbing and storage are needed to complete the picture. One inch of rain on a square foot of roof yields ⅝ gallon of water. The runoff from a 1,000-square-foot roof in Los Angeles is 8,000 gallons per year, in New Orleans 40,000 gallons, in Boston 27,000 gallons and in Toronto, 20,000 gallons. A family of four can live on 40 gallons of water per day while using an efficient clothes washing machine, water conserving plumbing fixtures (including a 1.6-gallon flush toilet) and a few simple changes of habit. That's about 14,000 gallons per year.[9] Gardening and livestock, of course, require additional supply.

Since concentrated runoff from roofs can be damaging to soils and buildings unless its release is controlled, direct roofwater from every structure into some useful collecting body and ensure that overflow from each of these cascades or soaks into a subsequent and larger storage, ultimately the soil.

**Cistern**
[Credit: Jami Scholl]

Complete the catchment pattern with plastic, metal or ferrocement tanks (26-Cisterns). Connect these with plumbing to convenient use points around the property and in the buildings (27-Reticulated Water). Take the overflow from tanks into larger storages (32-Small Earthworks).

## 26. Cisterns

Having clean water stored in good quantity onsite and available to flow by gravity where needed is the cornerstone of home security.

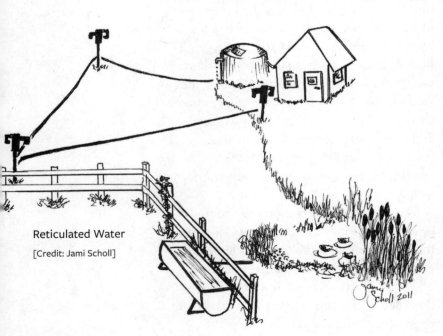

**Reticulated Water**
[Credit: Jami Scholl]

Among the earliest constructed waterworks were aqueducts and cisterns, for the concept of concentrating runoff and storing it for future use was well understood by the ancients. Cisterns are common in Australia, but North Americans may only have ever encountered an old and failing one, so may draw the wrong conclusions about their utility.

Cisterns are the batteries of a home water system. Pipes and channels for distribution are essential to make use of the resource once it is collected (27-Reticulated Water). Each cistern will have an overflow, and this must be directed safely to larger water storages, ultimately to soil, so arrange this cascade for maximum beneficial function (33-Ponds and Dams, 34-Water Gardens and Fish Crop, 35-Swales and Terraces).

## 27. Reticulated Water

The widespread movement of water through pipes is one of the great unsung achievements of the modern world, for water is heavy and tedious to carry in any quantity. Even more, it is indispensable. Therefore, enable water to be delivered with minimal effort to any zone where crops will be grown, animals housed or ranged or construction done. Locate hydrants or taps at protected spots near all main path intersections so that a 50- or 75-foot hose will reach from them to all likely use points. In semi-arid regions, irrigation lines should be installed in a similar pattern to all growing beds and perennial crops, except that the smallest flexible lines need not be buried deeply. If possible, supply the house and buildings with water stored high enough that it will flow by gravity, especially if you depend on a pump. Have ready access to 2,000 stored gallons of clean water for each adult.

To complete this pattern, first determine the access ways throughout the farm (28-Branching Cart Paths and Lanes), then let water lines follow them. It may be possible to install hydrants at the boundaries between main paddocks so that stock-watering tanks can be filled in each (51-Rotational Grazing). Annual crops of smaller plants must have access to about one inch of water each week during the growing season (50-Patch Gardens), but don't neglect perennials: allow a drip emitter for each main tree or shrub with its supporting plants (53-Fruit Tree Guild) or give the guild a good soaking by hose each week. Conserve water by mulching all perennials and garden beds (62-Mulch and Compost).

## 28. Branching Cart Paths and Lanes

A garden farm needs a network of paths and lanes connecting major use areas to aid harvest, construction and distribution of nutrient. Any area without vehicle access is of limited economic value. But because compaction makes paths into sacrifice zones, with little plant growth and limited soil development, paths should be permanent and their area kept to a minimum by good design.

After determining the placement of major buildings, connect them by lanes wide enough for a truck to pass from the public road to near the center of activity in the system. Branch smaller paths from these lanes, and connect the ends of paths to form loops through the core of the working zones. Lead garden and orchard paths off these loops so all areas are within a few feet of a path or lane. Where possible, lay paths on contour and orient them to harvest runoff for garden irrigation. Install water bars on sloping paths to prevent erosion, but make sure that wheelbarrows or carts can roll over them. Surface main lanes with stone, brick or gravel, smaller paths with old carpet or scrap wood and the tiniest garden paths with wood chip or coarse mulch. Encourage prostrate herbs in the pathways and along edges, but suppress grasses.

Path systems should take advantage of level ground wherever possible to harvest water (30-Contour Planting), with paths laid into the rear of terraces (35-Swales and Terraces). Harvest topsoil, decomposed mulch

and worm castings out of pathways to build up garden beds (50-Patch Gardens). Install stakes or bumpers at major path junctions to keep hoses in the paths and off the beds (27-Reticulated Water). Where paths run through grazing areas, make sure to provide alternate routes so that the grazing animals and the movement of loads along pathways do not conflict (52-Small Paddocks).

Branching Cart
Paths and Lanes
[Credit: Jami Scholl]

## 29. Fencing

The fragmented ecosystems of city regions provide nearly ideal habitat for deer and many other pest wildlife. To limit crop losses and to contain and manage your livestock, you will need working dogs, fencing or both. Dogs are regenerative and flexible if well-trained.

Fencing
[Credit: Abi Mustapha]

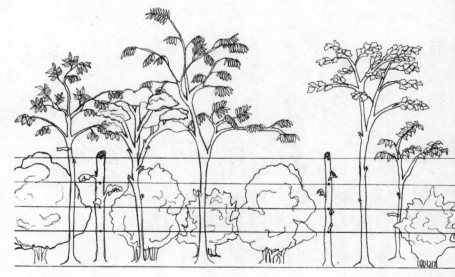

If you raise animals for food you will have meat scraps and offal with which to feed your guardians. If you live close by neighbors, your dogs may disturb them, for however well-behaved the animals may be, they will sometimes have to bark at night against potential intruders. Other farmers may understand this, but not everyone will be sympathetic. If you have five acres or more, dogs make sense. With less land, consider fencing.

Fences, while needing occasional maintenance, don't need to be fed. Deer can leap over 10-foot vertical barriers if they have a running start and a clear field of vision, but with good design a 7- to 8-foot barrier will keep them out. You can grow a living fence or effective hedge in 5–7 years, but will need metal or electric fencing as a transition. Establish a perimeter fence as soon as you can afford it.

A compact half-acre will cost about $1,000 to surround in welded wire; long, thin or irregularly shaped properties will be more costly to fence. Plant densely branching, useful shrubs and trees just inside the perimeter at 1½- to 2-foot spacing to grow a living fence. You may manage livestock rotations with mobile cages, water tanks or electric fencing, but on larger properties some internal fencing may also be needed, especially for cattle.

For internal fences, let lines follow major land use divisions, and place gates in corners of paddocks near intersecting fence lines. Fence lines must run in straight segments with all bends and corners reinforced. So lay out fencing early and make a rational design that minimizes bends and corners while adhering to useful divides and access ways.

Grow trees along permanent fences as reinforcement and replacement for posts, and manage the trees for mulch and fodder (31-Coppice and Hedges). Keep gates in the corners of main land subdivisions near each other to facilitate movement of livestock (51-Rotational Grazing). Subdivide main pastures with movable electric fencing to sort stock into Small Paddocks (52).

## 30. Contour Planting

All landscape has some slope, and cropping involves both traffic and soil disturbance. These create the conditions for erosion. Since erosion is destructive to soil fertility and can't be altogether prevented, it must be curtailed, bounded and the conditions for its easy reversal set in place everywhere. Organizing the landscape according to contour is the chief means of response. Align planting beds, tree lines and paddocks on contour, with paths and lanes to the uphill side to enable runoff to pool and infiltrate productive soil. Make terrace breaks above the paths where necessary. Dig out the paths and add the resulting topsoil to the beds. Use the paths for mulching of coarse material.

Contour planting helps fulfill the Alley Cropping (15) pattern. The use of this pattern is further amplified by creating Swales and Terraces (35). Trees and other perennials help to reinforce the terraces, but must be managed to support the growth of main crops (31-Coppice and Hedges). Contours help to define grazing areas for lowest impact (52-Small Paddocks). Contours also form the essential structure of gardens and should harvest water for their support (50-Patch

**Contour Planting**

[Credit: Abi Mustapha]

Gardens). Lay out beds to uniform widths as much as possible to facilitate the use of mobile poultry pens (54-Poultry Tractor).

## 31. Coppice and Hedges

Trees are the backbone of the living world, but tall trees are difficult to harvest. In the domestic ecosystem only a few trees can be tall without limiting the growth of other valuable crops; most must remain small. Virtually all the angiosperms (flowering trees — in North America mostly deciduous hardwoods) will regrow from stump sprouts if cut while relatively young. This can be repeated almost indefinitely. This is even more so with shrubs, which are adapted to being browsed by animals and damaged from the falling limbs and trunks of larger trees. Orchard trees are stunted by grafting and regularly pruned to keep them to manageable shapes and sizes. The ancient practices of *coppice* (the cutting of trees at multiyear intervals), *pollard* (similar to coppice but with cuts at head height to prevent animals browsing the regrowth) and of laying hedges (partially felling selected species to manage regrowth as fences and field divisions) should become standard tools of the garden farmer.[10]

Therefore, use trees and shrubs as living fences, as mulch and fertility crops, for fruit, mast, poles, fuel, fodder, edible leaves, nuts and to create windbreaks and trellises. Prune them at intervals of two months to seven years, depending on how fast they grow and what you need from them. Manage the regrowth by weaving, pleaching, browsing or chopping to obtain the full suite of yields and ecological services that trees can give.

Let this pattern shape much of the woody element of the garden farm, but especially to create living fences (29-Fencing), silvopastures (52-Small Paddocks) and forest gardens (53-Fruit Tree Guild). Coppice employs trees to pump nutrients into the system while keeping them small enough to be useful (61-Fertility Crops, 62-Mulch and Compost).

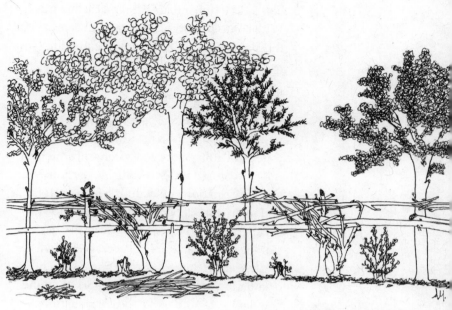

Coppice and Hedges
[Credit: Abi Mustapha]

## 32. Small Earthworks

While mechanical agriculture attempts to flatten and simplify landform, garden farming thrives on the productive edges created by small changes in elevation and surface shape. In drylands, pitting of the soil surface and sunken gardens both serve to collect and hold valuable moisture, seed and nutrient. In sloping landscapes, terraces are essential to retain runoff and provide working ground. Swales everywhere support groundwater recharge and nurture downslope plantings. Check dams restore catchments to health.

Small earthworks are within the grasp of any gardener. They need little engineering, can be quickly built, easily repaired and give both immediate and long-lasting results. Therefore, learn and practice the management of water with spades and mattocks, shovels and rakes to divert, spread, soak and hold back runoff for the greater growth of plants and animals.

Manifest this pattern by creating surface water storages (33-Ponds and Dams) and planting beds in sloping terrain (35-Swales and Terraces). Raised beds for Patch Gardens (50) and trenches for water purification and nutrient capture (48-Greywater Trenches) both add to the impact of small earthworks. With contour in mind, virtually all aspects of this pattern form Zones of Accumulation (56) as a basis for resource harvest.

## 33. Ponds and Dams

Ensuring an adequate supply of water is prerequisite to success in farming. Any body of water will enliven the ecosystems of which it is a part. Stored water is valuable for domestic supply, irrigation, aquaculture, fire control and microclimate. Earthen dams provide the cheapest and largest form of surface water storage practical in a small acreage.

Therefore, at the outset of any design or occupation of land, identify all possible pond and dam sites, and make the construction of water storages a primary task. Use dams to carry roads and paths. Shape the pond with ⅓ of its area in shallows less than three foot deep; include islands and peninsulas for ample edge. Seal ponds and dams using local clay whenever possible. Revegetate or mulch all disturbed earth immediately after shaping. Provide each pond an upstream silt trap, a spillway for safe release of floodwater, a drain if over 20,000 gallons and thick edge vegetation, but keep all trees off any dam. Consult

an engineer or experienced equipment operator on impoundments over ¼ acre. Avoid damming any permanent stream.

Complete this pattern by directing the runoff from roads into swales and the overflow of swales into ponds (35-Swales and Terraces). Add the surplus from roof runoff as well (25-Roof Catchment). Use the pond shallows for aquaculture (34-Water Gardens and Fish Crops). Place Greenhouses (38) to the north side of ponds or surround them by ponds to reflect light and moderate temperatures. Use the resulting microclimates to aid the growth of tree crops (53-Fruit Tree Guild). Create crenellated edges on all ponds as optimal environments for gardening (34-Water Gardens and Fish Crop).

## 34. Water Gardens and Fish Crop

Water gardens — which imitate swamps, estuaries and shallow lakes — yield more calories per square foot with less labor than any other form of gardening. Design small and shallow bodies of water with crenellated edges and low peninsulas to optimize access by foot or by boat. Plant the land with vegetables, surround the water with fruit trees, ensure a supply of nutrients to the pond (perhaps from animal manures) and raise fish and other organisms in the water. Make both the water channels and the peninsulas narrow and the water no more than five feet deep. Protect the land edges with adapted vegetation (willows, grasses). Periodically dredge the channels onto the beds to renew fertility. Use weirs and nets to control and harvest fish.

Complete the pattern of water gardens and fish crop by reinforcing the banks with water-loving trees and shrubs that can be cut for fertility, fuel and materials (31-Coppice and Hedges). Shelter some animals near or over the pond for the benefit of their manure fertilizing the crops (24-Animal Housing). Plant the land peninsulas and pond perimeters with vegetable crops (50-Patch Gardens) that will benefit from the subirrigation, reflected light

Ponds and Dams

[Credit: Jami Scholl]

and temperate microclimate. Place trellises from one bank to another over the water to increase growing area and to protect fish and water from predators and too much solar heat (43-Trellises for Shade). And use this pattern to surround and buffer the climate for Greenhouses (38).

## 35. Swales and Terraces

Sloping land requires small levelling structures to make it suitable for intensive cultivation. At the same time, in semi-arid landscapes, water-concentrating earthworks can make possible the growth of gardens and trees that could not otherwise be established. Use terraces to stabilize hill slopes and dig swales to store intermittent runoff in the soil. Space swales at vertical intervals of six feet or horizontal intervals of 60 feet, whichever is less, and otherwise wherever runoff is substantial. Stabilize terrace banks with stone or trees and swale banks and berms with trees, shrubs, deep-rooted perennial herbs and mulch. Pitch terraces toward the hill slightly, make swales broad and shallow and give both a planned overflow point reinforced with stone and plants for controlled release of floodwater.

Terrace benches can support either arable crops or grazing (50-Patch Gardens, 52-Small Paddocks). Their edges can be a place to introduce productive trees (53-Fruit Tree Guild). Design runoff to spill slowly from one terrace to the next, losing as little elevation as possible each time (47-Water Cascade). Place pathways to the rear of terraces (28-Branching Cart Paths and Lanes) and let swales develop into terraces by accretion of soils and organic matter (56-Zones of Accumulation).

## 36. South-Facing Outdoors

Land in the shadow of buildings, walls or evergreen trees is only of limited agricultural use. Cold ground can advantageously hold moisture and retard the blooming of trees and shrubs and may suit the cultivation of mushrooms, but in general, polar-aspected ground is best used for unheated storage, woodland and protective shrubbery (around buildings). Even roads do poorly if shaded in winter: ice forms easily and snow melts slowly. Extra moisture causes pavements to fail faster. Psychologically, humans are uncomfortable crossing a band of shade to reach sunny ground, so a poorly placed building or plant can negatively impact a large area. Therefore, orient buildings, pavements, courtyards, work areas and growing zones to take in the sun. Use multipurpose deciduous plants to provide shade in warm seasons.

This pattern helps to fulfill the needs of energy-efficient solar housing (12-Shelter in the Sun). Take advantage of the solar orientation of land to place a special growing zone (38-Greenhouse) or work area (40-Drying Yard). Give every contained outdoor space good access to the sun (37-Outdoor Rooms), sheltering those that need some shade under trellises and awnings (43-Trellises for Shade, 42-Porches and Awnings). Use north-aspected slopes and spaces for cold storage (45-Root Cellar).

## 37. Outdoor Rooms

We can never afford to enclose all the space we need to use. Every house in a temperate or subtropical climate must be able to expand its functions outward in warm weather and contract inward in cold. Porches, patios, balconies, decks, courtyards, gazebos, pavilions and even wooded glades and groves can usefully serve as places for gathering, eating or working during fair weather. These can be roofed with open sides, sheltered by vegetation or contained by building walls, fences or trees while remaining wide open to the sky.

Connect every house and working building to an outdoor room on at least one and preferably more sides. Make the transition between indoors and outdoors as easy as possible, using wide doors, steps, porches, awnings and trellises. Enhance the productivity of small spaces at the core of the system

Outdoor Rooms

[Credit: Jami Scholl]

by organizing them into outdoor rooms using multifunctional hedges, walls, fences and buildings. Especially valuable are those outdoor rooms lying between and connecting buildings.

To provide comfortable summer seating or working space in outdoor rooms, extend roofs and awnings out from buildings (42-Porches and Awnings) and grow vines on lightweight structures overhead (43-Trellises for Shade). A most important outdoor room is the canning or Summer Kitchen (39), though it may be partly or wholly enclosed with a roof and screened walls. One of the farm's larger outdoor rooms should be open to the sun and to breezes so that it may be used for processing material (40-Drying Yard).

## 38. Greenhouse

Year-round food means some form of special protection for plants in all but the mildest climates. Modern greenhouses employ lightweight plastic covers to reduce static roof loads. Sophisticated air circulation systems can transfer surplus daytime heat underground to warm soils and plants through the night, reducing or eliminating the need for backup heat. Whether free-standing or attached to a house, barn or workshop, the greenhouse is an essential adjunct to the gar-

den, providing an early start to spring crops, a long harvest in fall and space for fresh greens in winter. In summer it can ripen subtropical fruits in places with cool nights or be used for drying herbs. Sometimes a workout room or a laundry yard on rainy days, the greenhouse plays many roles.

Essential design requires good connections to the outdoors for ecological balance and pest control, plus adequate thermal mass, ventilation and well-made doors. Perennials add another dimension: try a peach, grapevines, a fig, a lemon or a loquat if these are marginal where you live. Place the greenhouse where it gets good sun in all cool seasons and near a body of surface water to reflect light and to moderate temperatures. Give the greenhouse plenty of vents, and in climate zones 5 and colder actively store heat in the soil of the beds. Glaze it with UV-resistant plastic sheeting or polycarbonate, both of which pass better light for plants than glass. Establish permanent beds and perhaps a pond or fish tanks inside, and use perennial herbs, shrubs and trees to give the greenhouse ecosystem resistance to pests. Use the structure for many purposes in all seasons.

Back up solar heating with the body heat of animals either within or adjacent to the greenhouse (24-Animal Housing). Similarly, the bath is a natural enhancement which offers extra heat and moisture and benefits from a verdant setting (65-Communal Bathing). Whether in tanks within or a pond without, aquaculture matches the needs and yields of a greenhouse superbly (34-Fish Crop and Water Gardens). The multifunctions of a greenhouse make it a pivotal support for the farm economy's many faces (63-Shifting Enterprise).

## 39. Summer Kitchen

In temperate climates the late summer ripens a surplus of fruits and vegetables conveniently ahead of the cold seasons of scarcity to come. But canning that harvest — the most reliable means of preservation — puts an onerous burden on the home kitchen. Keeping heat and extra moisture out of the house in August is a top priority for comfort and the longevity of the building.

The shift to outdoor eating in warm weather can be part of the logic of a summer or outdoor kitchen. The use of efficient wood-fueled rocket stoves needs the open air, and so do solar ovens. The shade of a nearby tree can help keep temperatures comfortable, or a trellis or patio umbrella may serve. The ideal setup would enclose all this with screens, but fly strips and traps can minimize flying pests at low cost, and the advantage of an ad hoc facility is that it can be set up on the patio or deck in an otherwise prime location that for most of the year can revert to preferred uses: seating, threshing or outdoor crafts.

Locate the outdoor kitchen in a shaded level area with a clean surface to make working safe. Place it within 50 feet of the main kitchen and pantry, so that transferring supplies is efficient, and specialized needs can still be met indoors (washing jars and pans in hot water, for example). Provide running water, two or more burners, and at least 12 and up to 20 feet of working counter space (two or three long tables) that can be easily cleaned. Keep animals and young children out from underfoot.

This pattern helps make possible the main objective of Food Storage (44). The summer kitchen is an ideal venue for a pergola, temporary awning or canopy (43-Trellises for Shade, 42-Porches and Awnings). Lead the sink drain out to Greywater Trenches (48) or dump your wastewater buckets onto nearby gardens (47-Water Cascade).

## 40. Drying Yard

Our future economy depends a great deal on the use of solar energy for process heat. Many traditional cultures recognized the value of an open, level pavement near the home. Such spaces were and still are used for threshing

Drying Yard
[Credit: Abi Mustapha]

grain, drying fruits, fermenting coffee beans and many other post-harvest operations. The suburban home has its equivalents: the patio and the driveway. When not covered by cars or hosting a basketball contest or picnic supper, these zones sport yard sales, dry out tents and tarps and provide the site for impromptu lawn mower or bicycle repairs. They will be perfectly useful adjuncts to the farm economy, which if they didn't exist would need to be created.

Make a broad, level, paved outdoor room near the house which can serve as a drying yard. The pavement may be stone, brick or existing concrete or asphalt, but need be no more than packed earth with a topping of cow manure, sand or pea gravel to keep it from becoming muddy or dusty. Grade the pavement to drain away from buildings. Ensure that the yard has good sunshine much of the day.

Solar drying of laundry may benefit from this open space, though a grass surface is often preferable (41-Laundry Lines). The drying yard will be more valuable if it is has small outdoor rooms on its perimeter (42-Porches and Awnings, 23-Workshop) where tools, containers and work in progress can be organized and kept dry or shaded.

Laundry Lines
[Credit: Abi Mustapha]

### 41. Laundry Lines

Homeowners association rules against this most basic solar technology are the least defensible of collective regulations, just waiting to be defied. Provide your household with upwards of 100 feet of lines. Run the lines east-west if you can, make them permanent and put them over turf if possible: it keeps the clothes cleaner when they fall to the ground. The best locations get all day sun and good breezes and are no more than 20 or 30 feet from laundry machines or the washing area.

Empty your laundry discharge into Greywater Trenches (48) nearby. Run the rabbit tractor over the yard (24-Animal Housing) or graze the sheep there when laundry isn't hanging. Double the laundry lines up with the wood yard (21-Woodshed) — the two uses are seasonally complementary and both benefit from sunlight.

### 42. Porches and Awnings

The porch performs a primal function of reception and outlook for the house. It is an outdoor room of the highest order — literally a part of the house where news is exchanged with the world and guests are received. Awnings protect doors and windows from sun and rain, but even more importantly, they make all entrance transitions and openings more graceful.

Give the main front and rear entrances of every house a porch, with seating if possible, but at minimum, a place to take off boots, hang hats or stash umbrellas. Give the porch an awning roof if not full enclosure, and let it look out on life. Make sure that the larger of the two porches is on a cool side of the house (east or north), and for every other door or solar window, provide an awning or overhang to mitigate summer heat and rainfall. Gutter the porches and awnings where pathways pass under their edges to avoid drenching people, and collect that runoff in a barrel or in buckets (25-Roof Catchment).

Porches and awnings can incorporate or be extended by trellises and vines to create deciduous shade (43-Trellises for Shade). In rainy climates or anywhere in a pinch, a linear awning or large porch can support Laundry Lines (41) or drying herbs (44-Food Storage).

### 43. Trellises for Shade

The great contrast in sun angles between warm and cool seasons in temperate latitudes requires some combination of well-placed overhangs and deciduous vegetation to keep buildings cool. Plants transpire moisture, enhancing the cooling effect of their shade, which makes them superior to simple awnings. To get effective deciduous shade from a tree in summer requires that it be very close to the building it is to protect, or that it be very tall, or both. The combination is problematic, as all trees throw leaves into gutters, branches onto roofs and sometimes fall with devastating consequences. A trellis with deciduous vines offers a better solution and fruit as well.

Wherever summer shade is needed in the same place that winter sun is welcome, create trellises and train deciduous vines over them to shelter building windows, walls and walkways. Choose fruiting, flowering and fragrant plants (tomatoes, melons, cucumbers, grapes, kiwis, clematis, morning glory, hops). Run the vines on wires held up by skeletal structures. Place clear awning roofs over walkway trellises along buildings. Grow other trellises entirely over outdoor rooms. Let awnings drain onto planting beds, so these structures become self-watering. Use annuals while perennials are getting established.

Select complementary species for trellises (53-Fruit Tree Guild) including some Fertility Crops (61). Make these linear plantings into small garden beds (50-Patch Gardens) and supply them with soil-improving amendments (62-Mulch and Compost) to keep them healthy.

## 44. Food Storage

Like all previous cultures that depended on agriculture, our households and communities must learn to store food as a primary responsibility. Having at all times a reserve of 6–12 months' food supply is traditionally the minimum condition for precluding starvation from crop failure. At the household level, this should include a wide variety of energy-, protein- and nutrient-dense foods in quantities of 1,000–2,000 pounds per person. These must be stored in forms that are stable against oxidation and protected against vermin, heat and other adversity for up to two years. This task will inevitably require much time, as well as specialized tools, structures and skills — and will take many different forms.

Drying by solar energy preserves a very high level of nutrient in fruits and vegetables, but may require supplemental heat in humid climates. Freezing preserves nutrients and appearance, but is energy-intensive and vulnerable to power outages. Canning compromises vitamin C and color, but secures stable storage for years and is easily achieved at home. Smoking and curing of meats and fish has often been a fall and winter activity because slaughter and butchering work are more safely and economically accomplished in cold weather: fire is needed and meat spoils less quickly in the cold. Grain and pulse crops primarily need protection from rodents and insects. Milk is traditionally preserved at the home scale and shipped economically as cheese. Cellaring of roots, tubers, fruits and bulky vegetables such as cabbage can reliably bridge between fall harvest and early summer abundance. Pickling, salting, storage in oil, in ashes or in the ground have all been practiced successfully, and, of course, no culture has ever overlooked fermentation, both for its preservative and nutritional benefits as well as for its often delightful by-products. Storing seeds is the capstone of these processes for it enables regeneration of the garden itself.

So begin now, wherever you live, to put food by. Acquire jars, crocks, baskets, buckets, barrels, pots, urns, vats and tanks suitable for food storage even before you plant a seed. Store food grown by others. Buy in bulk at farmers markets; glean fruit and nuts from street trees or neighbors. Learn with friends and family to dry, can and freeze food. The work of storing food involves art as well as science — it's complex, tedious and rewarding — a perfect arena for making culture. Surround your work with pleasing and familiar rituals. Celebrate the harvest, the fire, the first fruits and the good day's labor — they are the affirmation of life itself.

As you can, create appropriate larger structures for food preservation: shelves, cabinets, dryers, pits and clamps, caves, grain bins, silos and smokehouses as well as the obvious (46-Pantry, 45-Root Cellar, 22-Storage Barn). Place and build them for protection against vermin and wildlife, fire and flood. The cooking aspects of food preservation often want special shelter (39-Summer Kitchen), while the work is ideal for many hands (20-Communal Labor).

## 45. Root Cellar

Frost-free cold storage is valuable at all times and easily achieved using earth shelter during half the year. Below the local frost line, earth temperature remains close to 55°F year-round. With proper ventilation, a root cellar can hold even cooler temperatures from mid-autumn well into spring — just the season when bulk, moist storage of fruits and vegetables is needed.

Placement should be against a north basement wall or on a north-aspected slope, or dug into ground permanently shaded by a building or evergreen trees or shrubs (31-Coppice and Hedges). A cellar can also be sheltered partly in and partly below ground by a large water tank (26-Cisterns), and this may be important where the water table is high. Insulation is still important, even when the cellar is earth-sheltered. Low-power fans and thermostatic controls allow cooler night air to be drawn in automatically whenever thermal differentials favor air exchange. Entrances in particular should be shielded against direct sunlight; doors should be heavily insulated.

Access is important, so basement root cellars work better when a direct walk-in entrance to the basement can streamline loading and unloading. The root cellar should be no more than 100 feet from the kitchen door. Much closer is better. The root cellar may double as a storm shelter for properties without a basement.

## 46. Pantry

Houses built before 1920 usually had a pantry. These were mostly eliminated by builders as commercial distribution of groceries became widespread. They need to be recreated in all households. Proximity to the kitchen, protection from excess heat and ample wall space for shelving are primary criteria — a basement will do, though same-floor access is preferable. A porch, a sewing room, a child's former bedroom, a guest room, a large closet, even a back or basement stairwell may serve. Generally, the space need not be heated, though should never freeze, and windows should be curtained or shuttered against light most of the time.

This pattern helps to fulfill the pattern Food Storage (44). Therefore, build or convert a room or rooms to provide 20–25 linear feet

Root Cellar and Pantry

[Credit: Abi Mustapha]

of shelf space (10 inches deep) per household member for canned and packaged food storage. Locate this indoors near the kitchen where light, heat and access can be easily controlled, and stock it with a balanced array of preserved foods, herbs and cooking supplies.

## 47. Water Cascade

Water has many duties both inside the house and through the landscape. Drinking, cooking and the rinsing of dishes need the highest-quality water, but dishes can be washed in water of much lower quality. The use of basins not only limits the amount of water used for washing, but allows its reuse. Water for bathing should be clear and free of gross pollutants: rainwater or even well water with heavy mineral content should be more than adequate. Upstairs bathwater might regularly be drained into the washing machine for a load of laundry. Used dishwater suffices to flush toilets.

In the landscape, clean runoff from roofs and other surfaces should be caught and held high. After basic filtration, it is suitable for most domestic purposes from washing to stock watering and irrigation. Greywater discharge from laundry and bathing can readily be applied to trees or garden beds by subirrigation, though caution is needed in arid landscapes to limit the buildup of salts. Runoff entering ponds and waterways should always pass through soil, vegetation or mulch to intercept silt and recapture nutrients.

Design all water flows to capture elevational potential, heat energy and nutrient and to use and reuse water at its highest potential. When it is polluted and low in the system, direct water through biotic assemblies of soil and plants, fungi and microbes to purify it for release into the ground or surface water. The purification of used domestic water needs a soil or water medium in which filtration and active aerobic breakdown can occur (48-Greywater Trenches, 49-Wetland Water Cells).

## 48. Greywater Trenches

Water from laundry and bathing is easily and safely disposed of in landscape where its heat energy, nutrient load and irrigating value are benefits worth capturing. A shallow trench filled with wood chips, either above or in the ground, suffices to absorb the nutrient, filter the water and at the same time creates ideal conditions for converting coarse mulch into compost and worm castings, a high-grade fertilizer.

Let the trench run close to contour. Line the trenches so the medium can be changed periodically, and to prevent tree roots completely filling in these rich pockets of life. Give the lining a drain at the lower end so the filtered effluent can escape, either to swales in the garden below or directly to groundwater. Sufficient volume is needed to contain the maximum flow expected in ½ hour (usually from a bathtub emptying or a load of laundry — figure 50 gallons or less, about 6.5 cubic feet); a 12-foot trench 16 inches wide by 12 inches deep should be ample. A perforated pipe can be run through the trench lengthwise to ensure even distribution, or simply allow the greywater to enter at the upper end and percolate. Check the chips every year and replace them when they have mostly turned black and crumbly.

If releasing greywater to trees or woodland, move the discharge point regularly or divide the flow so you don't waterlog the soil at any point. A well-mulched Fruit Tree Guild (53) should absorb up to 30 gallons spread over 60 square feet in a week without any problems. A cold-weather or cold-climate variant would be to place the trench in a Greenhouse (38).

## 49. Wetland Water Cells

Swamps and marshes are the kidneys of the landscape; they purify water. We can model these wetland ecosystems to the same end. Wetlands slow the flow of water, causing it to drop its sediment load. Into these often

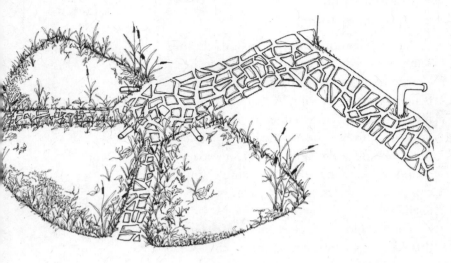

Wetland Water Cells

[Credit: Abi Mustapha]

anaerobic shallows, macrophytic plants (those with hollow stems) root; by pumping oxygen to their submerged tissues, they sustain colonies of aerobic bacteria (the sort that breakdown organic matter). It is these microbes that digest *E. coli* and other pathogens, while the plants absorb nutrients dissolved in the water.

Artificial wetlands need to be isolated from surrounding soil by impervious liners so that water treatment can proceed completely before effluent is released. Both black and greywater can be purified in this way, for a household of four or a town of 4,000. Also, some preliminary settling and skimming may be needed, especially for black water. A typical septic tank or similar device can be used upstream of the cells. The wetland cells themselves are filled with gravel or with layers of gravel, sand and soil to a depth that will not freeze, with the inlet above the outlet and both below the surface of the solid matter filling each cell. Plants grow, supported by the gravel-sand-soil medium with their roots in the flowing subsurface water. Two parallel cells (with plumbing to switch between them at four-day intervals) can both be drained into a downstream *finishing cell*. About 10 cubic yards per cell are needed for a household of four. Regular harvest of the plant biomass for use as fodder and mulch ensures the recycling of nutrients back to the food system.

Put treated effluent into an aquaculture system (44-Fish Crop and Water Gardens) or irrigate crops with it (50-Patch Gardens). In very cold climates build a Greenhouse (38) over the treatment cells. Keep trees out of the cells. Ensure that water flows by gravity throughout.

## 50. Patch Gardens

Placing the right cohort of plants on each small patch of ground and introducing animal impacts for selective harvest and fertilization promise optimal use of land at the garden scale. Linear elements — trellises, fences and building edges — remain part of the picture, but cease to dominate. Contour management of water and soil, branching paths and the clustering and stacking of vertical elements for best economy of light, moisture and nutrient introduce a new patterning akin to the mosaic of rich natural systems.

Diverse architecture and species polyculture among plants confuse pests, reducing losses without poisons. A perennial backbone to the garden and top-down soil building based on coppice and mulch eliminate most justification for tillage and, without the need for tillage, long straight lines hold only a mental appeal.

Patch gardens increase edge effect, enhancing nutrient exchange and other forms of synergy while limiting soil compaction from foot traffic in the growing areas. They may, but need not, be tiny. Grain gardens and small paddocks may cover a compact ¼ acre, though they may just as likely extend along contour in strips with complementary plants or trees for fertility.

Work from the beginning to design an orderly layout for buildings, pathways, water storages and woodland elements that takes best advantage of sun, slope and microclimate. Let garden beds and animal forage zones emerge organically as patches amidst and around these larger elements. Organize diverse plantings to exploit the many niches thus created. Keep good records of your annual and seasonal changes and rotate up the small plants and annual crops to maintain soil and garden health.

Let many of the gardens center around trees and form vertical patches (53-Fruit Tree Guild). Size them to be cleaned up by a Poultry Tractor (54). Cycle fertility into larger areas using animals (51-Rotational Grazing) and green manures (61-Fertility Crops). Use every niche of unoccupied space to grow something — even weeds (55-Catch Crop) — and let nothing go to waste (62-Mulch and Compost).

## 51. Rotational Grazing

The wild ancestors of our domesticated livestock, like many other prey animals, responded acutely to predators by herding, flocking and moving frequently. The cultural mimic of natural herd and flock behavior involves confining animals at high density (up to 40,000 pounds per acre) for very short periods (12–48 hours) before moving them. Electric fencing and portable watering systems have made this practical. For sheep, guard animals such as donkeys and llamas reduce losses to dogs and coyotes. All animals need access to water, but geese will follow their watering basins to new ground at your direction. Smaller poultry require some form of shelter against raptors during the day and small carnivores at night — a mobile coop is often used.

These systems have been adapted with good success to cattle, sheep, hogs, chickens, geese, ducks, turkeys and even rabbits, and can build organic matter in soils at two tons per acre per year. Each breed has its own needs, but the principles of management are similar. Let animals forage to their benefit, herd them along promptly and rest the land behind them. Animal polycultures (chickens after cattle, geese with pigs) offer synergies that reduce risks and disease. Swards become more diverse over time, but forbs (broad-leaved forages) and native grasses may be introduced to hasten succession by seeding just ahead of the herd.

In view of the immense numbers of bison once found on the North American Plains and of the vast herds of wild ruminants on the African savannas today, there may be no theoretical limit to the number and types of animals that may productively be ranged in this manner. And while mob stocking may play no role on the garden farm (beyond the occasional poultry flock of a few hundred birds), even a single mobile coop of $5 \times 3$ feet with five 3-pound birds represents more than 40,000 pounds per acre of grazing impact.

Keep your livestock tight, and move them often. Organize the grazing space to rotate animals through other cultivation systems (gardens, tree crops, arable ground) for fertility and pest control. But keep the herds and flocks out of waterways except for dedicated ponds for waterfowl. Instead, bring water to the animals. Use permanent fencing for perimeters and major land divisions, but portion your paddocks and move your animals with lightweight electric fencing or portable shelters. Heavy animals may need to be taken off pasture in winter, but British work suggests this may not be necessary where dense- and deep-rooted grasses can be established and forage stockpiled.[11]

The natural fulfillment of this pattern requires many subdivisions of land (52-Small Paddocks). For a few smaller animals, mobile coops are very flexible and satisfactory (54-Poultry Tractor). Use perennial borders around main fields and let these be forage species (31-Coppice and Hedges).

## 52. Small Paddocks

Tight stocking means small paddocks, sized to maximize the impact of the herd or flock. In order to adequately rest the sward, there must also be enough paddocks to move the stock every day or two for a month. Large animals such as cattle and horses will be few on

Small Paddocks
[Credit: Abi Mustapha]

garden farms. More often dairy will be from goats, and poultry or rabbits the primary source of meat. Close attention to the needs of the animals, the condition of the forage and the state of the fences will mark the successful herder.

Lay out paddocks along contour alleys between rows of potential fodder trees and shrubs, or clean up spent garden crops with stock penned briefly among the bolting lettuce, overripe radishes or corn stubble. Rotate grain gardens or bean patches in and out of pasture. Let the hogs stir up the orchard ground to root out Bermuda grass and other unwanted volunteers. Browse goats on weedy woodland edges. A paddock should afford one or at most two days forage. On a garden scale, active seeding of paddocks or patch gardens for poultry is practical. See that gates, paths and fence openings be organized to simplify moving stock from one paddock to another. In whatever forage situation you contrive, be sure that you provide water.

By using confined animal grazing, weeds can be turned into manure (61-Fertility Crops). Likewise, a day or two of grazing can clean up crop wastes, prepare ground for new seed and thus afford extra yield (55-Catch Crop).

The smallest paddocks are those temporarily covered by a mobile pen or coop (54-Poultry Tractor).

## 53. Fruit Tree Guild

To use trees properly, we must understand that they always grow in communities. In nature, few plants grow in pure stands: crop trees should always have smaller plants around them. The orchard arrangement of grass under trees profits only the makers of mowing machines. Fruits, most nuts and other productive trees are blossoming plants while grasses are not. Woody plants thrive in fungally dominated soils, grasses in bacterially dominated ones. Young trees are adversely affected by grasses, which produce allelopathic chemicals to suppress tree growth. The tree roots feed mostly in the same soil levels that are crowded with thick mats of grass roots vying for the same moisture and nutrient.

A far better arrangement supports crop trees with guilds — cooperative assemblies or polycultures — of flowering species, especially nitrogen fixers, dynamic accumulators, aromatic herbs, pest-repellent plants and nurse crops. Berries, particularly currants and gooseberries, fruit well in partial shade. Mints form an excellent ground cover. Alliums and marigolds ward off pests, just as daffodil bulbs repel voles underground. There is even room for sun-loving annuals in the early years, and later to the sunny side of the trees. Small-flowered plants of the carrot and cabbage families are excellent food sources for beneficial wasps and other pest predators.

Use woody mulch around trees, grow smaller companion shrubs, herbs and vines that will dynamically accumulate soil minerals. Plant tree guilds with light levels and ultimate sizes in mind. You can crowd things at the outset, but you must be prepared to prune, coppice and even transplant or remove things as the larger elements mature. Plant trees in guilds of useful species at every level from the canopy to below the ground. Don't neglect vines, fungi and habitat for beneficial insects, including pollinators.

The woody elements of a fruit tree guild respond well to cut-and-come again management (31-Coppice and Hedges). Open space around a fruit tree is prime terrain for vegetables while the tree is small (55-Catch Crops). Birds can easily be ranged through the guild to clean up windfalls and control pests (54-Poultry Tractor). Some large-leaved plants like comfrey, rhubarb, burdock, horseradish and arums make fine and sturdy sources of mulch (61-Fertility Crops). Dig wide holes for the perennials, sprinkle in some minerals and spread their roots, but put any manure or mulch only on the surface after planting (62-Mulch and Compost), not in the hole itself.

## 54. Poultry Tractor

The mobility of animals is their most profound characteristic and a great advantage to the farmer, for it allows livestock to gather their own feed. But left to roam at will, domestic livestock can quickly damage plant systems and even buildings. The solution to this conundrum is to confine stock in cages, pens and small paddocks that can themselves be moved or reconfigured. A nonconventional but now well-explored model exists in the *poultry tractor*, a small, mobile coop that holds 3, 10 or 50 birds but can easily be moved by one person. Tractors enable small flocks to till the ground while gathering much of their own food doing instinctive behaviors (in the case of chickens, pecking for insects and grubs, scratching and manuring) that are dysfunctional in the artificial environment of fixed confinement.

Tractors must protect their inhabitants from predators, while supplying water, supplemental feed and a little shelter from rain — and — for laying hens and ducks — a nest box from which eggs may be collected. The coops should be sized and shaped to move easily over whatever ground is to be grazed

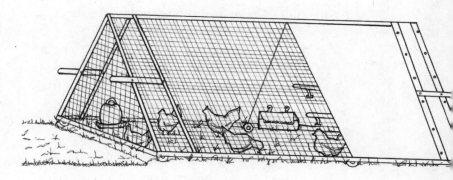

Poultry Tractor

[Credit: Abi Mustapha]

or tilled — a garden bed, for example. Wheels, runners or handles are helpful. If a canopy, roof or tree cover is provided to ward off raptors, more flexible confinement with electric netting can be used with greater numbers or with heavier, flight-averse birds.

Give poultry good access to fresh greens and live protein. Allow two to four square feet per bird. Move tractors when the ground has been well cleared of weeds and crop residues, or if on pasture, before the birds expose the soil. To keep predators off-balance, never leave a tractor in one place for more than two days. Allow ground to have rest or a new crop for at least a month between impacts.

Design larger, fixed coops to connect easily to the mobile fleet (24-Animal Housing). Move in the birds to capture crop wastes or to cycle green manures into mulch and animal manure (55-Catch Crop, 61-Fertility Crops).

## 55. Catch Crop

With annual vegetables, young tree crops and even mature woodland, areas of unused sunlight abound between plants, crops and seasons. Since soil fertility and total system yield depend entirely on plant growth and decay, unused sunlight on soil is a wasted resource. Nature responds with weeds — the smart gardener with a short-term, fast-growing crop.

During the early spring, many woodland ephemerals sprout and complete as much as ⅔ of their annual photosynthesis in a few weeks before the overstory leafs out. The same pattern can be applied to annual vegetables or even berries under and around tree crops.

It takes young fruits several years to bear, and they aren't very large for the first five years.

Therefore, keep all soil growing some crop at all times, preferably two or more crops at different layers, sizes and states of maturity. Seed or transplant the next crop as soon as or, better, before the maturing one is harvested. Apply this pattern in the Greenhouse (38) as well as out-of-doors in Patch Gardens (50). Use legumes and other Working Pioneer Plants (60) to build biomass in the early stages of system establishment. Even weeds can be a catch crop, returning accumulated minerals to poor soil (61-Fertility Crops).

## 56. Zones of Accumulation

This pattern takes many forms: cities are zones of accumulated material extracted from far and wide, fashioned into buildings and artifacts, regularly discarded. The culture of dumpster diving in wealthy cities attests to the lucrative nature of waste. With irrational cultural values rampant, cities discard food, organic matter, packaging, furniture, metal of all sorts, slightly or scarcely damaged consumer goods, surpluses for which there is no market and even whole buildings, neighborhoods and industries. In the most impoverished rural landscapes, windborne plastic trash may come to rest along fence lines where its role

as a mulch may be crucial to the germination of bird-dropped seeds and therefore to the regeneration of forests.

The adept garden farmer must develop an eye and a nose for where wastes collect and where they may be captured as resources, especially biomass, but also useful scrap lumber, furniture, metal and "spare parts." Entrepreneurs from landscapers to compost makers have organized trade in recyclable wastes and can often be paid to enrich themselves on the unrecognized treasure of others.

Closer to home, system components should be designed to collect runoff, soil, nutrient and food resources. The farm itself will inevitably develop zones of accumulation where waste resources can be stockpiled, and if these are not anticipated and designed, their sudden emergence may be problematic. Place them in a flow of collection, delivery, processing and distribution near main paths but not on them. Some materials will sit longer than others, but everything must eventually move or transform. Put to more needy areas any materials that are toxic or so degraded as to have no further constructive use in the near term, or which require industrial reprocessing. Organic matter poses no problems except in scale or handling, though occasionally bulk biomass may be a vector for unwanted seeds or pest organisms. Expansive vines must be thoroughly dried before laying them back on soil, or they will root and grow with a vengeance.

Collect high-grade wastes in the form of Salvaged Materials (57) and organize them, along with biomass, into a substantial and orderly Resource Inventory (58). Because the stream of these materials is erratic and low-cost projects may depend on them, expect Hodgepodge Growth (59). Catching carbon from the atmosphere and minerals from the soil makes all Fertility Crops (61) one kind of Zone of Accumulation. More broadly, let all your soils be covered, so that they all become ZAs (62-Mulch and Compost). Be sure

Zones of Accumulation

[Credit: Abi Mustapha]

that your entrance zones will accommodate temporary windfalls (67-Connection to the Street).

## 57. Salvaged Materials

The decay of industrial civilization is well underway and generates ample castoffs, some of which can be helpful to the restructuring of society and the repair of landscapes. Ultimately, only living communities and organisms are regenerative, but selectively slowing the cascade of entropy can be a source of useful salvage. A variety of sizes and types of material are more valuable than a huge quantity of any one thing, though do not hastily reject a bonanza. Dimensional lumber in larger pieces is always a good find, as are bricks, blocks and stone in small pieces. Intact sheets of wood, metal, screen and glass have many uses. Containers, especially if suitable for food or garden use, warrant collecting. Compound materials are difficult to recycle, so pass them by unless they meet a specific need and are in good shape. As a guide to collecting, experience with reuse and skills in repair are far more reliable than imagination. Some treasures are unlikely, and most require an abandonment of conventional thinking: venetian blinds, for example, are a disgusting window treatment, but make durable, reusable plant labels when cut into five-inch lengths. To everything, however, there is a limit.

Accept or seek out only those waste materials and goods for which you have storage space and that can be reused, repaired or easily deconstructed into useful components. The ability to reuse odd material is a valuable support for economic activities that must respond to changing conditions (63-Shifting Enterprise).

## 58. Resource Inventory

The purpose of collecting salvage is, at least for the near future, to avoid otherwise needless trips away from home to the supply depots of commerce. In 20 years, a resource

Salvaged Materials
[Credit: Jack Heimsoth]

inventory may be necessary simply to operate machines, of which there will then be many still serviceable. For the time being, a supply of useful material enables many small repair, retrofit or adaptive projects to be done in odd bits of time with little cost. For some otherwise useful gadgets or artifacts, repair parts must be improvised in any case, as they were either not designed for repair, or came from makers long gone out of business.

What makes a pile of scrap an inventory, beyond the tongue-in-cheek reframing, is its level of organization and ease of access. As the household creates deliberate energy storages (woodshed, barn, guest rooms, cellars) and acquires the muscular capacity to transform materials (workshop, drying yards, animals), it will inevitably gain new niches for ordering surplus materials along with the strength to process and put them to use.

Make every utility space indoors and out an orderly depot of useful resources: hardware, building and landscape materials, containers, biomass, lumber, extra cloth, clothing and rags. Purge these spaces and collections periodically of low-grade items. Collect useful materials as they appear during your foraging circuits (16-Local Trade), bring them into Zones of Accumulation (56) according to weight, size and nature and sort these zones at least twice a year in spring and fall, or when they reach capacity. Organize materials so everything can be accessed independently; make racks and shelves rather than stacks. Use bins, jars and cartons to aggregate similar small items and label them. Do not depend on tarps for dry storage beyond a few weeks.

An inventory of resources supports the here-and-there, now-and-then expansion of the household economy (59-Hodgepodge Growth). Perhaps the most frequent turnover of resources will be those for building soil (62-Mulch and Compost), which may only be stockpiled for a few days or at most weeks before being spread on the garden. Any business venture will require certain specialized items, but if several can draw on the same tools and resources, you will have an advantage (63-Shifting Enterprise).

## 59. Hodgepodge Growth

In order to take advantage of the unplanned devolution of economic and social complexity and to work successfully with the rhythms of nature, it is sometimes required to move quickly and often necessary to wait. Projects will stutter along as labor, weather, budget, materials and insight permit. This is not a bad thing. Slow progress affords time for learning from one's mistakes, improvements to design and unexpected economies. The result can be a superior structure or plan at lower cost. It can even mean disaster averted.

A garden farm involves a complete reordering of society and economy and will take time to achieve. Therefore, be flexible with all new endeavors, especially building and landscape projects. Design well in advance so that you know your needs and you understand the whole system of which the present piece is a part. Complete work in phases so as

not to lose ground, e.g., pour or fill the footer as soon as it's excavated so the trench doesn't collapse; roof a structure once the frame is up to protect it from weather. But do not dismay if things follow their own course, take longer than planned or if other projects intervene. Buildings and gardens take years to develop. A durable, energy-efficient water system can scarcely be completed in a single year. Understand and celebrate each small step toward your goal.

As your system develops, take every opportunity to add fertility (61-Fertility Crops). Becoming comfortable with uncertainty makes you more economically fit as opportunities arise and fade over time (63-Shifting Enterprise).

## 60. Working Pioneer Plants

Transforming landscapes and society requires hardy and self-reliant individuals, groups and species. In the process of ecological succession, pioneers are those mid-sized plants, often shrubs with small fruit, that can get going in unpromising terrain and make it much more amenable to life. They attract birds and small animals, who in turn bring in manures and seeds of other useful plants, increasing diversity and complexity. Pioneers often fix their own nitrogen or mobilize phosphorus and other nutrients. They are able to cover the ground, exclude large animals and accelerate soil repair by building biomass. They bear browsing and coppice well.

The role of pioneers is to prepare the way for taller, canopy-bearing species that will maintain the restored ecosystem. Pioneers are fertility pumps and are the main drivers of succession. Therefore, pack your system with many small- to mid-sized trees, shrubs and large herbs, especially those known to build or attract fertility, then work them hard: slash, browse, coppice and mulch the pioneers as much as they will bear to build up soil, feed animals and stimulate soil climaxes.

In warm temperate to subtropical cli-

Working Pioneer
Plants
[Credit: Abi Mustapha]

mates, pioneers are often tree and shrub legumes which can provide an essential nurse crop with their open canopies. They need to be cut frequently (31-Coppice and Hedges). In cooler climates, the legumes are more often ground covers and are used in crop rotations or as green manures (61-Fertility Crops). The essential methods with all these materials are to chop-and-drop, graze-and-manure or slash-and-mulch (62-Mulch and Compost).

## 61. Fertility Crops

Living soil is built by the continuous growth and decay of plants. Animals can accelerate this process by eating and excreting the plant biomass, making its nutrients more rapidly available to soil microbes, which in turn pass them back to plants and store some of them in humus. Fungi also play a role in soil fertility by decomposing woody matter.

Any deep-rooted plant is a good candidate for soil building. Traditional agriculture developed and continues to use forage grasses, legumes and forbs to produce large volumes of hay and pasture growth on which animals can be grazed. Plants can also be slashed or mown and the resulting material used as a mulch or fodder.

Many legumes are known to harbor at their roots bacteria that fix nitrogen, while other plants have fungal associates that also mobilize nutrients. Nitrogen-fixing plants, when grazed or cut, release some of their ac-cumulated nitrogen into the soil, making it available to surrounding plants. Other weedy species and herbs dynamically accumulate, or draw up mineral nutrients selectively into their tissues. As they die and decompose, these pass into the topsoil for use by other plants.

Jeavons estimated that 70% of cultivated land should be in fertility crops, but this assumes only herbaceous or weed-level planting to make compost and no use of animals.[12] Organic farmers practice rotating cash crops with cover or fertility crops, thus using as much as half the land to maintain fertility. The greater complexity that results from adding trees and shrubs, perennial herbs, fungi and animals can increase yields without a loss of fertility.

On a smallholding, at least 30% of the landscape should be growing woody perennials and other permanent fertility crops. These can include fruit and nut-bearing species. Pasturage might cover another 20% with some grazing and foraging under trees and in fields and gardens to glean and turn under spent crops. Another ⅓ of the land might be occupied by staple and vegetable crops with 10% or 15% in aquaculture. Each of these should have one or several fertility components, whether coppice, manure, legumes or fish waste. The more these systems can be integrated, the greater will be their sustained yield.

By the use of Alley Cropping (15), trees and shrubs can provide much of the fertility for arable crops without diminishing yields due to shading. Large biomass producers such as forage grasses, large herbs and tap-rooted perennials should be slashed or grazed to yield mulch and manure (60-Working Pioneer Plants, 62-Mulch and Compost).

## 62. Mulch and Compost

Soil is best built where undisturbed. It comes alive from the remains of biomass eaten by organisms. Animals can consume palatable plant material, including leaves and young shoots of trees and shrubs. Older, woody material must be shredded by soil fungi. Toxic, thorny or otherwise unpalatable plants must be digested by detritivores. Compost is a second possible pathway for the cycling of biomass. The process of combining green and brown or nitrogenous and carbonaceous materials in the right ratio (between 1:20 and 1:30) with ample air and moisture leads to hot and rapid breakdown by bacteria. The resultant compost is dark, rich in organic compounds, about ⅓ the weight of the original material and relatively free of seeds and pathogens (so that manures, including humanure, can be rendered harmless). Making compost is, however, a labor-intensive process, and also wasteful of carbon, nitrogen and metabolic energy. Compost, for all of its many virtues, cannot be the primary source of fertility on farm — it is best reserved for growing seedlings, nurturing transplants as part of a potting mix or top-dressing newly planted perennials.

The third main pathway for decomposition is actually the primary way that nature returns organic matter to soil: through mulch. Gravity carries everything to earth sooner or later, and organic matter breaks down in the presence of soil microbes and oxygen, which is ubiquitous near the surface of the soil. The presence of a deep litter layer ensures a healthy soil biota, and long cycles without disturbance foster the development of rich fungal networks to recycle woody biomass.

Therefore, build soil primarily by using mulch. Chop-and-drop (62-Fertility Crops) in all zones. Rotate animals through orchards, fields and gardens periodically to stimulate soil development with manures and to harvest wastes and control insect pests (51-Rotational Grazing, 54-Poultry Tractors). Channel noxious, seedy and coarse plant materials into hot compost along with household wastes and biomass of unknown origin to produce substrate for seedlings and other special uses.

## 63. Shifting Enterprise

Garden farmers are close to their customers. They also work at the edge of cultural change, or, as business strategists term it, in emerging markets. An enormous variable is weather, and another challenge is the collective change that occurs in the farm ecosystem over time. All these forces combine to demand flexibility, creativity and responsiveness. The dynamic of succession opens new opportunities as fruits, meats, honey and other products of a maturing system become available in quantity. The same dynamic may close other doors as the demands of annual cropping seem less warranted or competition takes profit out of formerly lucrative niche markets. Specializations based on unique microclimates, local demand for ethnic or values-based food or skillful pro-

Shifting Enterprise

[Credit: Jami Scholl]

cessing may emerge to anchor the home economy for a time. Niche products carrying a significant premium — baby salad greens, fresh herbs, flowers and other crafts for instance — will have the most variable markets. More durable yields will come from a good variety of vegetables and ordinary food of high popularity — tomatoes, eggs, apples, squash, clean meat, baked goods, jam and fruit butters, salad — especially in off-seasons. Perennials, animals and multiple crops through the year will spread the risk of weather.

The most reliable approach to steady income rests on two legs: information and diversity. You must constantly monitor real costs in time, land, seed and inputs against yield, both by volume and by market price. You must learn what you are good at, what your customers will buy and what the land will give. You also need a diverse array of crops: no fewer than four nor more than a dozen, to spread the risk of failure while avoiding unmanageable complexity. You can grow 60–100 varieties of fruits and vegetables for home use or barter — and probably should — but don't try to make money on more than a dozen. The labor demands and seasonality of your crops and your markets must be complementary and spread throughout the year.

Keep in close touch with your customers by maintaining an active and healthy public edge (64-Public-Private Gradient). Start small, and try new crops one at a time (59-Hodgepodge Growth). Use the convenience of steady trade at your front door to test odds and ends (68-Fruit Stand).

## 64. Public-Private Gradient

Working at home inevitably imposes challenges to the sanctity of family life and the integrity of work. One has to schedule time off and also private days when work may happen but visitors and callers are excluded. Otherwise, the public and the market reach into every corner and crevice.

Public-Private Gradient

[Credit: Jami Scholl]

The property itself must have a physical point of reception (67-Connection to the Street) that is obvious and regulated — a prominent gate is helpful, as are clear parking and signage to orient visitors, customers and new guests. This is most powerfully achieved by shaping spaces with plantings and landscape elements that direct the flow of traffic intuitively. Just as important as the physical gate are clear boundaries regarding phone and digital communication, schedules, seasons and practices surrounding onsite sales. The sooner you establish the limits of your accessibility to the public, the sooner you'll be in control of your time. If Saturday is always a market day, make sure that Sunday is spent with family and Monday is quiet too. If you have to take calls, get an answering machine, make clear you'll return calls at a certain hour or day and stick to your word. Don't carry your cellphone around the farm. Limit visitors to certain days of the week and weeks of the year, or hold paid (or free) tours or field days by appointment only.

To support this pattern, establish a clear public-private gradient across all fronts. Make the gradient long, so that the public end is clearly public, and the private end most secluded. Don't mix them up. Be responsive, courteous and entertaining when you agree to be, and turn people away politely at all other times. You'll be glad you did, and they will adjust.

## 65. Communal Bathing

Hot tubs have become widespread in North America; jacuzzis are a regular fixture of public lodging and high-end homes. Wet baths are wonderful, but require more sophisticated and expensive containment and protection for the plumbing, and thus have less flexibility for use in very cold weather when they might be most welcome. The Finnish sauna, now gone round the world, has its origins in the rustic north woods where dry heat, small spaces and steam offered the cleansing intimacy of the bath with no more water than might be melted from a pot of snow. Small sauna shacks were traditionally the first shelter built by Finnish families, in which they both lived and bathed until more ample quarters could be built. To get naked and sweat with others, often in the dark, is, as the Finns, the Russians and many native Americans well understand, to be reborn; it may be the minimum gesture of civilization that a people living in the wild must make.

The sauna room, like the back-deck hot tub, is very often detached from the house and invites bathers to be reborn into the open air. For a household that works together, communal bathing affirms that relaxation and renewal are not solely a private affair, and that the cycle of group life can regularly return to its root. Care for the body makes the boundary around Family Table (17) appropriately porous.

Place the communal bath in a well-protected location toward the private end of the household gradient. Combine its heat functions with food drying, a Greenhouse (38) or a utility space such as a Workshop (23). Let it be detached or directly opening to the outside of a building and near a pond for dipping if possible (33-Ponds and Dams). Provide a roofed area for clothing to hang and benches for changing. Surround the bath with clean decking or pavement that can be swept and is free of hazards to the feet. Design the stove or water heater to run on wood fuel, so that this most primal of survival functions can be met from raw elements of the world at hand.

## 66. Rooms for Guests

The need to expand the household seasonally—for brief periods or over the longterm—provides ample reason to create comfortable, multifunctional guest lodging. This can be a fold-out couch in the living room, a spare bedroom doubling as a pantry, a straw-filled loft in the barn or a guest cottage sometimes rented for income.

Establish the hospitality function in the household repertoire as early as possible, and learn how to make guests comfortable while keeping them from taking over your space and your daily rhythm. Put guests to work and make them a part of household chores and food preparation. Lodging will evolve as the system matures and more interior spaces are finished or refurbished, but you may need to say no to visitors from time to time. You will also learn, and will perhaps be able

Communal Bathing

[Credit: Abi Mustapha]

to make this systematic, when you can most benefit from extra hands.

Guest rooms ideally fall somewhere in the middle of the Public-Private Gradient (64). Visitors may enter the system to share work (20-Communal Labor), participating in special efforts and perhaps taking meals out-of-doors (37-Outdoor Rooms). Or, they may be invited to share the Family Table (17). Longer-term visitors, interns, WWOOFers, family and friends may be invited to mix more deeply into household rhythms (65-Communal Bathing).

Connection to
The Street
[Credit: Abi Mustapha]

## 67. Connection to the Street

The garden farming household belongs to a larger culture of earth care; its self-reliant ethic acknowledges interdependence with neighbors and the surrounding community. The farm too will inevitably have a market to which it belongs, and if the implications of local food revival in open and democratic societies are to be taken seriously, customers will want to know where their food came from. Therefore, design the connection to the street to shape reception.

The front gate or driveway entrance should be self-evident, with parking for several vehicles close by. Direct the flow of arriving traffic by landscape indicators more than signs. Water bars across the driveway can serve the dual function of harvesting runoff and slowing vehicles. A place for easy turning around is valuable on all driveways longer than 50 feet — this may be as simple as a small spur or wide parking apron. Establish your identity with an attractive and permanent sign. Keep the mailbox a little away from the entrance in a location convenient to the postal carrier.

The actual entrance should be clear and so should the point beyond which strangers ought not venture (64-Public-Private Gradient). A good way to anchor the public end of this gradient is with a small market (68-Fruit Stand).

## 68. Fruit Stand

You may decide never to sell directly from the farm, but it offers many built-in advantages — no transport costs and flexible scheduling chief among them. Set the terms of sale for a roadside stand to match the level of traffic. On a quiet back street, your neighbors are the likeliest to stop; there, self-service can work to your advantage. Be sure to provide shade cover, ice for perishable items, easy pricing and some small change in a secure box or can with clear instructions. If you live on a busy road or highway, a safe pull-off is essential to attracting business. Your farthest signs must also give motorists at least 20 seconds warning (at 50 mph, this is ⅓ mile, at 35 mph, over 1,000 feet). The signs must be bold and direct: "Fresh Fruit, 300 Yards." Use color and a simple graphic style. Leave your name off the sign unless you have a regional reputation that will draw custom.

Keep the stand on your own land and best within sight of the house or a location where someone is likely to be working. A capable youngster or retired elder of the household can tend a busy stand if a backup is available. You are better advised to offer a good selection on a few days than dribs and drabs every day. Fruits, juices, honey and tomatoes in season are always a draw. A few unusual items among the staples will keep your customers coming back.

# PART II

# Elements

# Renaissance Farm, Bloomington, Indiana, USA

[*Color Insert #1 includes photos of this property.*]

**LOCATION:** Flat ground in a 50-year-old suburb with next-door neighbors on four sides and about 40 houses within a ¼ mile. Unglaciated limestone uplands, elevation 820 feet, climate zone 6b trending toward 7a with 44 inches precipitation per year and a tendency toward dry periods from July to November. Tornadoes frequent the area: a low-intensity twister struck the city in the spring of 2011 causing extensive damage to trees. Spring flooding is common but does not threaten the farm infrastructure.

**DISTANCE TO URBAN CENTER:** Two miles by state highway.

**ECOSYSTEM:** Mixed hardwood forest with beech-maple co–dominants, now widely fragmented with many exotics. Heavy brown clay soils bearing evidence of farming and habitation. Heavy deer browsing pressure plus raccoon, woodchuck, skunk, possum, rabbit, squirrel and chipmunk. A good population of songbirds and even some raptors plus migratory geese, cranes and others. Small steady and seasonal streams drain deeply dissected terrain with about 100–200 feet of local relief.

**SIZE OF PROPERTY:** 0.7 acres, approximately 30,000 square feet.

**OPERATORS:** Peter Bane and Keith Johnson, assisted by seasonal interns and student volunteers.

**ESTABLISHED:** 2006.

**PRODUCTS SOLD:** Bedding and nursery plants, seed, salad, herbs, some vegetables.

**SERVICES OFFERED:** Publishing, consulting, teaching and apprenticeships.

**MARKETS:** Neighbors and farmstand visitors, local farmers market.

Our aim at Renaissance Farm was to establish a permaculture demonstration and achieve a good measure of household self-reliance, emphasizing perennials. My partner and I purchased this suburban property with two small ranch-style houses and a decrepit shed in March, 2006 after living for many years in the North Carolina mountains. We are both permaculture teachers and landscape designers and had been together for eight years when we moved to Indiana. Our connection to the community began in 2003 when we were asked to give a permaculture course for students at Indiana University. That has continued and has provided us with a wide window into many strata of this diverse small city.

When we arrived in the early spring of 2006, the yard was covered in a mix of lawn and remnant pasture grasses with two very large silver maples and another eight trees between 20 and 30 years of age on the periphery, mostly bird- or wind-planted: hackberry, ash, slippery elm, sugar maple. Some were dead or dying; one large mulberry was bearing heavily. Brambles and young pioneer trees covered a small neglected patch in one corner of the lot. Although the property had once been farmland, there was no evidence of gardening here nor at any of the other 53 houses in the mile-long neighborhood. Even ornamentals were scarce. A single woman neighbor tended some fruit trees in her yard.

The houses we bought were in poor but serviceable condition, and their repair and refurbishing, which is ongoing, took most of three years to reach a stable condition. Appliances, roofs, heating systems, windows and doors, interior and exterior trim, plumbing, wiring, carpets and floor finishes, chimneys, porches, insulation, interior walls, cabinets and more were replaced, repaired, moved or removed, installed or upgraded.

Nearly constant remodeling took lots of time and whatever money we could spare; we used most of our income to pay down the debts on the house because we were concerned about the unstable condition of the economy. Using a patchwork of loans from family (about 30%), a slow transfer of stranded equity in a house and lot in the intentional community where we had been living (about 20%) and a rolling string of too-good-to-be-true, no- and low-interest credit card advances from the big bad banks (the remaining 50%), we succeeded in financing our investment and all the improvements with a minimal outlay of interest, making the last loan payment in September 2008, just as US and global financial systems hit the wall.

The consequence of all this restoration and race to security was a slow development of the garden farm. Keith, who has gardened organically for 35 years, immediately recontoured the front yard into raised beds and dug drainage ditches to divert runoff from the building foundations. We transplanted fruiting shrubs and small trees that we salvaged from our gardens in North Carolina, began collecting mulch for the new beds and planted out salad crops as well as flowers and favorite perennials. First-year yields were modest, but the front-yard garden attracted generally positive attention from the neighbors, all of whom had a bird's-eye view of developments as they slowed down into the curve fronting the property on the only road that led into the neighborhood.

After triage on the houses stabilized their envelopes and appliances in the first year, we turned more of our attention to the farm in year two. Selective removal and major thinning of the trees on the lot yielded over four cords of fuelwood that were stored in a new woodshed and later burnt to heat both houses using newly installed wood stoves. Also in 2007, we cleared a space in the weedy neglected corner of the lot farthest from the street and built a 10,000-gallon ferrocement cistern above ground. After completing the tank in November, we installed underground plumbing to connect it to hydrants across the property, providing the basis for a permanent irrigation system. That same month we also erected a 20×48 foot-high tunnel greenhouse with metal ribs and plastic cover that enabled us to take salad crops, herbs and hardy greens through the winter. More contoured raised bed gardens took shape, while we continued to import wood chips, compost, straw and manure to bolster soil fertility.

With the completion of the water tank and lines, most of the major earthworking lay behind us, but surface-shaping activities continue today. A small pond that started as a volunteer project during a permaculture work bee was eventually expanded, lined and stocked with goldfish. In the summer of 2010, we added four small ponds in front of the greenhouse for microclimate, water treatment and aquaculture, tripling our surface water storage.

During the third year (2008), garden surpluses enabled Keith to take spring and early summer produce to the local farmers market two miles distant, while neighbors stopped in frequently to buy salad; most of the garden yield was still being consumed at home. That spring new metal roofs and gutters on both houses enabled us to connect the catchment system to the water tank and begin using stored rainwater on gardens as well as for laundry and one toilet. From a high of 8,000 gallons during the first year, we cut mains water use below 1,000 gallons a month, none of it for the garden. We also undertook a planting of 23 major fruit trees on semi-dwarf rootstocks: 10 apples, 5 pears,

2 plums, 3 peaches and 3 cherries. Thornless blackberries transplanted from Carolina, as well as black raspberries, currants, gooseberries and the large mulberry provided spring fruit crops for a couple of months, while the largest of several fig trees gave a handful of fruits that autumn.

Soil development still held back yields, but we enjoyed fall lettuce crops as well as spring surpluses in the third year. There were some potatoes, peas, peppers and squash to store. A generous crop of tomatoes ripened indoors all the way to the end of December. Onions and garlic lasted into January, and there were sporadic harvests of cabbage for slaw and sauerkraut. The greenhouse began supplying a few cold season harvests where in its first winter the plants had only survived but not yielded. A major top dressing of wood chips on all the outdoor beds helped create better soil conditions going into year four.

The weather in 2009 was wetter and cooler in spring and summer than in previous years, which suppressed tomato yields, but the berry harvest increased and salad continued to flourish. Keith took bedding plants to the farmers market in the spring. Roadside visits increased, and neighbors continued to stop by for vegetables. Potato crops expanded, and peppers, surviving partly under cloches through a cold, wet October, put on a late burst of fruit during an unseasonably warm November. Onions and garlic met household needs for the remainder of the 2010 winter, and the average bulb size was up from the year before, indicating better soil fertility. The fourth growing season also saw the first yields from tree crops with a handful of peaches, crab apples and Asian pears, a couple of quarts of figs and surprisingly good first yields from a 10-year-old che, or Chinese mulberry, most of which were probably nabbed by a skunk or possum.

At the instigation of an intern who wanted to learn beekeeping and who assembled one hive and installed a nucleus colony, bees arrived late in the spring and hung on with supplemental syrup well into winter, promising better pollina-tion and higher fruit yields in 2010, not to mention the chance of some local honey. Though these froze in January, we acquired another package of bees and took them through the 2010 season with a good honey harvest. The bees again died over winter in 2011—cause unknown (high losses among area beekeepers)—but we persisted and had two hives alive going into spring 2012.

Year four (2009) was also the year of the fence, when garden yields had grown to the point of attracting significant predation from deer, a herd of which loitered in the neighborhood, smoking cigarettes and comparing notes about their favorite organic vegetable snacks (Renaissance Farm being the best salad bar for several miles around). We planted 200 saplings of small-to-medium sized trees and shrubs in May and June on the three sides of the lot not already fenced. These went in at 18-inch spacing, but it became immediately apparent that they were years away from being effective as a hedge, so we turned to industrial methods and strung up a six-foot welded wire perimeter, extending its reach by another two feet using sticks and electric fence wire (sans current). After some of the herd tested the new fence by crashing over it a few times, they seemed to get the message that it wasn't worth breaking a leg for a little chard or some beans. Simple gates at both driveways and where paths connected the property with each of the neighboring yards completed the exclosure.

We erected a second woodshed in the spring, and by year's end foraging in our forested neighborhood had nearly filled it. Two winters' worth of heating fuel was aging under cover. Fully dried wood and increasingly effective thermal insulation and solar heat storage strategies in the buildings brought down fuel use to 2.4 cords for the 2,000 interior square feet in the winter of 2009–2010. As I wrote this chapter, we were tracking on a comparable rate of fuel use for the winter that followed, about 0.6 cord per 1,000 heating degree days (average daily temperature below 65°F times number of days).

Also in the mix for 2009 was the extension of an aboveground root cellar attached to the north and northeast side of the cylindrical cistern. A partial donut-shaped cave with gently sloping roof executed in ferrocement got spray foam insulation on the outside in mid-winter and then a carpet/plastic sandwich membrane and soil substrate for a living roof as the summer of 2010 ripened. Other big projects for the fifth year (2010) included a large covered front porch that became a useful venue for seed processing and which supported a solar array that meets all of our electric needs on an annual basis. More tree trimming opened the solar window wider and combined with skillful foraging enabled us to fill both woodsheds with close to four years of fuelwood.

Plans were afoot in the late fall for a major expansion of outbuildings in 2011 and beyond to include a composting toilet, barn, workshop, guest quarters, a second cistern, a sauna, animal housing and a potting shed. (We erected the barn in September of that year and began the cistern.) Harvesting from neighborhood trees and the farmers market in fall 2010, we put by a year's worth of apples and squash, and began using the root cellar as a walk-in cooler. We started making sauerkraut and kimchi in the late fall of 2010 and have kept this up, greatly enriching our diet and health.

How does all this add up? And what lessons can we learn from this relatively young system?

Self-reliance and food storage are both increasing. Soils are improving. The growing season is now year-round. Facilities are adequate for a complex and functional mixed farm and still improving. We are debt-free, firewood-rich and approaching carbon-neutral. We have both excluded our main garden pests (deer) and increased connectivity with our neighbors. The woody elements of the system are in place. Biomass growth is accelerating but still insufficient to need. Diversity is high, and we are refining varieties and timing. Insect damage is modest, but small mammal pests remain troublesome. The niches for livestock are not filled. Ground covers are insufficient, and weed pressure remains high. All nutrients are being recycled onsite. Waste going to landfill is minimal. Specialties are emerging. Farm income is low, and labor remains a challenge.

A net cultivated area of just over 25,000 square feet hasn't been fully planted, but offers the three-year prospect of adequate fruit and vegetable production for household use and surpluses over the long term. Some crops are already adequate to need (onions, garlic, salad), though only herbs, seed and bedding plants are annually in surplus. Salad is now available throughout the winter and is abundant in spring and fall, when we sell some. Bulk and staple food production is still at a low level (about 20% of need based on potatoes, sweet potatoes, kohlrabi, beets, carrots, popcorn, tomatoes and some dried beans). Soil fertility and tilth are improving slowly from minerals, mulch and cropping; adding animals would boost the rate of fertility gain, as will increasing amounts of woody mulch from tree and shrub prunings. The site is still importing biomass though we can already foresee the day when pruning and thinning will provide enough mulch. The barn construction turned a corner on investment in building—most of that is behind us. Significant smaller infrastructure projects remain ahead, but many of the elements of a self-reliant food system are in place (food and water storage, winter production, permanent garden beds, perennial fruit and coppice crops). Animal housing will permit us to add rabbits and poultry, which should provide a lot of our meat and eggs. We presently share in surpluses from a neighboring flock of chickens.

Progress has been rapid by objective standards as most visitors comment pointedly. From the inside, it hasn't seemed fast enough. The step-wise strategy of expanding the gardens as fertility and time permitted was matched to a very busy schedule that, besides renovation and new construction, also supported full-time home-based employment for one of us and about half-time employment for the other (in publishing, teaching and design). Limits on the farm expansion have also come from our choice of property. Being middle-aged when we started here (and coming with a certain reaction to primitive living from too long in the woods), we chose to buy low-cost existing housing. We also had to accommodate a well-established home business that couldn't be disrupted. We got the liability of compromised design and worn-out infrastructure that went along with it (the houses are oriented north-south, the wrong way for solar gain). Two buildings are harder to heat than one of an equivalent size would be (but there's lots of edge and great outdoor spaces around and between them). There was no garage, so for storage we've rented a 28-foot semi-trailer and parked it on our driveway; it will go when the barn is finished. We've had to use our living room for a workshop more than a few times.

Labor remains a limiting factor. In response, we have hosted interns for three of the past four years to help with a variety of homestead tasks. Through experiment and persistence, an agreeable cultural routine has emerged around work, food and learning. We find that we have been able to expand our household cooperatively by trading food, lodging space, tool use, knowledge and guidance for high-quality support in building up the farm. We are training ourselves and our helpers in a wide array of skills for self-reliance, from making months' worth of sauerkraut and apple pie, to maintaining solar panels through the winter, to identifying songbirds and making furniture from local woods.

Marketable quantities of honey, apples, perhaps seasonal eggs and, of course, high-value crops of salad, herbs, seeds and bedding plants can be expected to continue or to emerge within the next three to five years. Use of the greenhouse to produce winter vegetables may prove economic. A logical development toward value-added production suggests fermented vegetables, cider, salsa and meads, though marketing may prove more trouble than these are worth. To the extent possible, the approach of selling to neighbors and in the local farmers market (two miles distant) may be augmented by direct sales to restaurants or to event caterers and organizers in a community that already strongly supports local food.

Renaissance Farm has become part of a distinct movement within its eclectic neighborhood toward a more organic way of life. The neighborhood has a handful of rentals, not many children and few residents under 40. Most of the homes, some up to 5,000 square feet, have only one or two occupants, or even part-year residents. Expensive lots remain unsold, but a growing number of studios and workshops are visible too, and some signs of wood heating. With our assistance, the next-door neighbors have acquired a small flock of chickens and expanded it to 30 birds; they planted a garden for the first time in 2009 and have continued it. The single woman fruit enthusiast planted a vegetable garden, has gotten bees and built a chicken coop. Her neighbors have begun raising vegetables. The neighborhood includes at least one hovel without running water, a couple of multi-generation families (parents and adult children) and, most promisingly, a grass-based livestock farm of some 60 acres. This larger farm, with 9 or 10 head of cattle moved by electric fencing, is clearly still a hobby enterprise—it had no animals until two years ago and is fallow again this winter—but it makes an enormous space for agriculture at the very center of this suburban pretense. Though a long way from being a village, this palimpsest of American extravagance is also a holdout from an earlier and more sanely tempered era. It's a good place to be homesteading.

# CHAPTER 7

# Land—Scales and Strategies

Everyone needs access to land—for walking, for playing, for contemplating the beauty of nature—especially in cities. But nature is not a pet, and providing land as a support for health and spirit is a remnant of a far deeper and more complex relationship of humans to land. The most important bond between us and the earth is that we eat what it provides. We developed the habit of walking and running in order to pursue game, to roam in search of roots and berries. We contemplated the infinite complexity of land and sky around us in the process of stalking animals, netting fish and searching for ripe seeds. But for many modern people, food comes from the grocery store, not from farm fields, forests, streams and meadows. This conceit masks a vast, complicated and energy-wasteful reality: hundreds of millions of acres are mostly unseen and uncared for. And our necessary connection with the earth serves largely immaterial purposes.

For more than 30 years permaculture has directed its efforts toward helping ordinary people everywhere create local resources from the land, including food. Not only is this a way that people can meet their own needs, but it represents the best possibility for improving care of the land. Indeed, not only do people need the land, but the land, in order to flourish, needs people. Biologists and anthropologists have documented a strong correlation between high levels of biodiversity and high levels of cultural interaction between people and the land.[1]

Many people in North America have little or no access to land at the same time

Vast swaths of once densely urbanized land like this Detroit neighborhood are opening up for new possibilities.

[Credit: Robert Monaghan]

that much land in our metropolitan regions is wasted or derelict. Even much farmland is poorly used and could produce more food for people if it were cultivated differently. Methods of cultivation that yield more crop per acre and that can heal ecosystems and regenerate soil fertility — which has been so badly damaged by industrial agriculture — will almost certainly require different forms of land tenure and more people working and living on the land.

Garden farmers pioneering vacuous suburban landscapes offer a model of how this revolution in land use might begin. Suburbia is a cultural context dominated by diverse ownership of small parcels of land. Markets are close, infrastructure is already well-developed, and much of this land was,

in the recent past, used to grow food. It will again. A new vision is needed.

Many people in the modern world live in urban apartment blocks without private land around them. Consequently, permaculturists have developed strategies for guerrilla and communal gardening of public spaces, as well as for the use of vertical surfaces both in and around buildings to enhance food security and to green the city. Some of these strategies are adaptable to the garden farm. Foraging is practiced by subsistence farmers the world around. Urbanites do it too.

And everyone who can do so should grow a little food. Serviceberries from the bank parking lot, sprouts on your kitchen shelf, mushrooms in your basement or cherry tomatoes in pots on your balcony may greatly enhance your health and buoy your spirits, but they will not provide enough calories for an adequate diet. More land is needed. How much?

Living in the Garden,
Pattern #8 (See Chapter 6)

[Credit: Jami Scholl]

## The Standard US Diet

People in the US now eat an average of 2,000 pounds of food per person per year. This provides a daily intake of 3,800 calories per person, some 40% above the level recommended for healthy, active adults (2,700 calories).[2] The not surprising result is that most people in the US are overweight and many are obese. The current US diet consists of (by weight, not by calories) about 30% milk and dairy products, 20% vegetables, 14% fruits, 13% meat, 10% cereals, 7% sugar, 4.5% fat and about 2% nuts, legumes and other foods. The numbers do not add to 100% because of rounding, nor are they fully indicative of the dietary impact of each of the different food groups. Fats, for instance, carry about 20 times the calories per pound of vegetables. The same diet measured by calories shows that people in the US get 25% of their calories from fats, 22% from cereals, 15% from meat, 14% from sugars, 10.5% from dairy products, 6% from vegetables, 5% from fruits and 2.5% from nuts and legumes.[3]

Needs vary from season to season and with age, gender and other health factors. Physically active people — farmers and gardeners among them — generally need more food than office workers. Nutritional needs also vary with activity levels, climate and the nutrient density of foods making up the diet. Food system analysts have estimated that a healthy level of food intake for people in temperate countries (who must contend with the extra nutritional demands of cold weather) is about 1,000,000 calories per year, of which at least 10% should come from animal foods.[4] This is about 2,700 calories per person per day, less than ¾ of what people in the US now eat. A healthier diet would certainly provide less food than the current standard, and it would emphasize different foods, just as nutritional counsel to the public has advised for years: more fruits and vegetables and less meat, fat and sugar.

A third way of looking at food intake is nutrient density measured by portions of vitamins and minerals available. The nutrient content of food has declined since the advent of industrial agriculture due to the destruction of soil fertility wrought by tillage and chemicals. We are eating more calories in part to obtain the real nutrition that our bodies demand, delivered in smaller and smaller amounts as soil health and nutrient density decline. A detailed numeric accounting of vitamin and mineral levels in various foods, however, will not contribute much to our assessment of garden farming and land use, especially as nutrition can vary widely depending on soil conditions and cultural practices, so I refer readers to published sources.[5] Suffice it to say that pound for pound, fruits and vegetables deliver much higher levels of these essential nutrients than any other food groups, though nuts, organ meats, whole grains and legumes — and to some extent dairy products — contain significant amounts of particular and needed nutrients. This is promising for any attempt

to take responsibility for our diet and for food production. Fruits and vegetables can be easily grown at the garden scale, and some of the other nutrient-dense foods can also be raised in small areas.

Most fats and common sugars contain very little nutrition, though they provide almost 40% of the calories in the US diet. A case can be made for including some honey, maple or sorghum syrup, butter and coconut, olive, flax and hemp oils in the diet, as these contain minerals or essential fatty acids of importance to health. A healthier, trimmer diet — and one on which we should base our plan for a new food system and a new economy — would eliminate most of the sugars and some of the fats now consumed, reduce the amount of meat and dairy products and increase somewhat the amount of fruits, vegetables, eggs, cereals, legumes and nuts we eat.

A fourth and necessary way of examining food intake is to measure the amount of land required to grow the food. This gets to the heart of the issue, because land is becoming a limiting factor for the survival and health of humanity. The 2,000 pounds of food eaten per person per year in the US come from three acres of crop and pasture land.[6] In the aggregate, it takes some 900 million productive acres of farm and ranch land to feed the population of the United States. The country has this much agricultural land, but doesn't use it all for food production today. US cropland under cultivation at present totals somewhat more than 300 million acres, while pasture and rangeland totals about 600 million acres.[7] Many millions of acres of grazing, fallow and conservation land are not part of this total, but the US is still a net food importer.

Most of the crop acreage in the US is planted to four commodities: corn, soybeans, wheat and hay. These are converted by the industrial system into sugar, animal feed, ethanol, paints and inks, food supplements and hundreds of feedstocks for industry. Very little human food — mainly bread and cereals —

comes directly from these four crops. About 3% of crop acreage, some 9.3 million acres, grows fruits, vegetables, legumes and nuts. Significant "phantom" acreage in other countries also contributes these nutrient-dense foods in the US diet.

The broad picture is one of very few farmers raising large quantities of commodity crops, mostly in monocultures, with the labor-intensive and nutrient-dense components of the food system offshore or managed with immigrant labor. Food-miles are high, energy costs are unsupportable and there is very little resilience in the system. Even basic food reserves in government storage, once thought to be vast, are now almost nonexistent.[8] This problem presents an opportunity to achieve greater land efficiency and a more resilient food system by developing millions of garden farms that eliminate most of the food-miles in the present system, increase access to fresh and healthy foods, increase the total food supply and put greater value on food and its production.

Vegetables and fruits contain a high proportion of water, so yields per acre are high. Vegetables grown under conventional mono-cropping systems in North America yield an average of 9 tons per acre. Fruits yield 10 tons per acre. Grains range from a high of 3.5 tons per acre for rice to 0.7 tons per acre for rye. Oilseeds are generally less productive per acre (from 1.5 tons for peanuts down to 0.4 tons per acre for mustard seed) because of the energy intensity required by plants to synthesize fats. Tree nuts are also high in fats and yield about 0.5 ton per acre.[9]

Food yields of meat and milk are more difficult to measure, in part because milk is a fluid product not directly comparable to other foods and its output varies dramatically by breed. However, grass farmers in well-watered regions may be able to generate about 500 pounds (0.25 tons) of beef per acre per year. Production on western rangelands is much lower. Smaller animals are generally more productive as they are more efficient at converting feed into meat than cattle, but may not utilize pasture as well as ruminants. Poultry or hogs may yield up to twice as much meat per pound of feed as cattle, but much depends on the systems of cultivation and husbandry.[10]

At the other end of the spectrum from industrial agriculture lies the work of Alan Chadwick and John Jeavons.[11] Biointensive gardening has been more systematically studied and measured than most other forms of garden-scale agriculture, so it's a useful measure of what may be possible. Jeavons reports that it requires 4,000 square feet (0.09 acres) to grow a balanced plant-based diet of vegetables, fruits and legumes for one person for a year. Jeavons uses 70% of that 4,000 square feet to raise compost crops. These plants sustain soil fertility under intensive cropping conditions. In other words, only 1,200 square feet per person per year are actually needed to grow the food that gets to the table, but another 2,800 square feet are needed to grow the fertility that makes that crop land highly productive.

Thus we have a 30-fold range of land efficiency between the industrial farm system (with its high inputs of energy and capital and high outputs per hour of labor) at the low end and the biointensive garden model (requiring high inputs of labor and yielding high outputs of food per square foot) at the high end. Many more farmers would be needed to raise our food from the biointensive garden system, but much less land — and almost no capital or external energy would be used. Somewhere in between these two poles, I hope to describe a system that can become a net energy producer. It will also show greater ecological stability and diversity, while providing a modest, healthy, omnivorous diet and net gains in soil fertility from season to season. This is the hope of a regenerative or permanent agriculture, and the imperative for our survival.

## A New Notional Diet

We still have to match land area to food needs. If one million calories are needed per person per year, how might they be apportioned and what land would be needed to grow that food using regenerative garden farming methods?

### A Possible Diet for One Person for One Year

| Category | Weight (lbs) | Calories | Land Area (sq ft) |
|---|---|---|---|
| Milk and Cheese | 450 | 115,000 | 4300 |
| Vegetables | 350 | 75,000 | 1750 |
| Fruits | 300 | 72,000 | 600 |
| Cereals | 150 | 248,000 | 1500 |
| Potatoes | 150 | 52,500 | 450 |
| Meat | 90 | 80,000 | 2100 |
| Legumes | 80 | 125,000 | 1600 |
| Eggs | 50 | 30,000 | 550 |
| Fats | 30 | 120,000 | 700 |
| Nuts | 30 | 78,000 | 300 |
| Fish | 15 | 10,000 | 600 |
| Honey | 15 | 22,500 | — |
| **Totals** | 1,710 | 1,028,000 | 14,450 |

Credit: Peter Bane

My purpose is to propose a diet that would be acceptable to most North Americans for its variety, level of fats, protein and bulk and that could be grown and foraged by small-holder methods in suburban front and back yards and "empty lot" pastures with a modest amount of cooperation among neighbors and some local trade.

Quantifying food production in this manner is both necessary for assessing the prospects of self-reliance and provocative because it begs a host of questions about methods and assumptions. I will address a few of these here.

Cereals obviously pack a big calorie punch per pound and per acre. This is why they have formed the basis of diet for most of humanity over the past 8,000 years. Easy to store but hard to process, grains provide a good amount of protein but inadequate amino acid balance for complete nutrition. Agricultural peoples eating primarily grain

show higher levels of degenerative disease and poorer teeth and bones than foraging cultures.[12] Cereals in the diet are best mixed with legumes, which complement their amino acid profile, or with some milk, meat, eggs or fish. Grain and legumes combine well in the field too, either side by side or in rotations; the latter provide nitrogen for the the former. Seeds or nuts can also be helpful in providing complete dietary protein. Non-tillage systems are less well developed for grains and legumes than for other crops, but have been demonstrated in the work of Masanobu Fukuoka.[13] This diet envisions some use of corn, which is the most productive of the grains per acre and which grows well over most of the USA and southern Canada, but could also include oats, wheat, buckwheat, sorghum, millet or even rice where the climate permits.

The meat allotment is imagined to include rabbit, poultry and some lamb, pork or goat. Where land is more expansive or grazing shared, a little beef might be possible. Rabbits provide a high-protein, low-fat meat and are easy to raise in cages and to process. They are the preferred meat where land is scarce

Animal Housing, Pattern #24 (See Chapter 6)

[Credit: Jami Scholl]

and would be fed from grasses, legume tops, weeds, waste vegetables and coppice trimmings. Poultry would feed on system wastes, small amounts of seed and controlled rotations through garden beds. Home-raised meat would likely be supplemented by game.

Milk is not digested well by everyone, but when consumed raw is well-accepted by most, and it is one of the most productive ways of obtaining protein and fat from low-grade plant matter. Dairy animals, here envisioned to include goats and sheep, can feed on woody browse as well as grass and herbaceous forages, turning plant wastes into high-quality food. In colder climates, ruminants are especially important. Smaller breeds would fit well into the garden farm.

The traditional Irish diet of milk and potatoes was not a result of poor imagination, but of oppression and the need to raise maximum nutrition from minimum space. It turns out that calorie yields of potato nearly match those of grain and are more reliable in a cool, wet climate. The crop can also be stored in ground, out of sight and safe from seizure or taxation. When matched with milk protein, potatoes provide a well-balanced staple. In hotter climates, sweet potatoes are an attractive and nutritious alternative. Home processing and storage of potatoes is in many ways easier than threshing and grinding grains, despite their bulk. Cooking is also less energy-intensive.

Fats in the diet above are imagined to be largely animal in origin, probably lard or goose fat, both the basis of long life and good health in traditional diets. These animals can be fed on garden, farm, woodland and roadside wastes. Pigs are omnivorous; they were traditionally fattened from woodland foraging, spoiled milk and whey, along with surplus and windfallen fruits. Geese feed on grasses and forbs.

The fish component might be expanded to replace some of the meat, especially in wet landscapes. But unless exotic species are employed, productivity will likely be less than for rabbits. Molluscs and crustaceans are well worth exploring.

Honey production requires virtually no dedicated space, thus the lack of a figure for area.

The figure of 14,500 square feet per person, or about ⅓ of an acre, provides enough land for a complete diet. It envisions closed-loop fertility after an initial period of system buildup, with recycling of all manures, including humanure. Some aspects of this diet would be more easily produced on larger acreage (beef, milk, lamb, pork, nut crops), so cooperation between growers may optimize yields and diversity. Even at the garden scale, some specialization is likely to bring rewards.

If these methods and assumptions were adopted across the United States, the nation's population could be fed from ⅓ of the current crop acreage — about 100,000,000 acres, or 150,000 square miles (an area the size of California or 5% of the lower 48 states). The Canadian population could be fed from roughly 20,000 square miles (less than the available farmland in southern Ontario). In the nature of garden farming, the hurdles to market entry are small, as are the risks. Land not presently growing food can be mobilized to help meet our needs. So the strategies implied by these ideas are well-suited to creating a safety net against the predictable failures of industrial agriculture without disrupting present systems.

I hope to show how the synergies from permaculture design make this level of land productivity possible. This requires intensifying our use of imagination and intelligence and of the land itself, but not of energy or capital.

## Scales of Production

Despite a certain levelling, not all suburban or urban lots are the same size. Garden farms will emerge at several different scales. I think we can classify these into five orders.

### The Tiny City Lot

Many city homes in older districts were built on lots of 6,000–8,000 square feet. A nominal standard that emerged in the postwar era for most urban residential districts outside New York and a few dense older cities in the East and Midwest was ¼ acre, or about 11,000 square feet. There are many millions of such homes. I take this to be the upper end of the smallest scale of garden farms and a suitable home for smaller households. Properties of this size are obviously limited in their choice of cropping systems and the amount that they can produce, so the emphasis has to go on foods that provide the most nutrition per square foot, and which thus make the most impact on diet for reasons of freshness, freedom from contamination and scarcity. This can certainly include animals such as bees, chickens and ducks, pigeons, quail, rabbits or guinea pigs. Grain production would be for seed or at most a small amount of popcorn. Cash income would come mostly from adding value (e.g., berry jams and jellies or botanicals), through sale of seed, herbs, nursery plants or breeding animals or perhaps from flowers. For more suggestions, see Chapter 13, Setting Plant Priorities.

### Multi-Lot or Suburban Plots (Microfarms)

The next scale up represents garden farms of ¼ to ¾ of an acre. Later suburban developments somewhat farther from city centers, upscale homes from an earlier era not yet subdivided, older exurban properties now surrounded by new development or combined urban lots may aggregate this much land. Properties on this scale can achieve a significant fraction of self-provision with some surplus for local trade, barter and supply to neighbors. Selected high-value crops might be grown for income. Some medium-sized animals such as sheep, goats, geese or pigs could be kept successfully. Tree and shrub crops might be extensive and yield surpluses for trade or processing. Productive aquaculture

is possible. A system of this scale wants a full-time farmer during the season. There would be some periods when two to three people are needed, which implies a household with two or more adults, and perhaps with older children capable of helping out, or even with seasonal help in the form of interns or occasional hired labor.

### McMansion Redux or Green Island Plots

The middle of five scales ranges from ¾ acre up to 2.5 acres. This might be a large lot in a sprawled outer suburb, or half a dozen contiguous or nearby backyards closer to town. A farm of this size could support either a commercial intensive mixed vegetable operation capable of yielding a full household income ($40,000+) or a mixed farm provisioning a household of 4–8 people. Cash cropping should be part of the farm plan with a significant tree crop or livestock component. A small milk cow or several goats would not be unwarranted. An acre of berries could provide much of the cash income. A larger, multi-adult household would be needed to provide enough labor, or a plan should be devised for hiring seasonal help.

### Mini-Farm

Remnant properties from older eras not yet subdivided may offer opportunities in the range of 2.5 — 8 acres. Mini-farms of this size may be found just a few miles out from

House Cluster, Pattern #7 (See Chapter 6)

[Credit: Jami Scholl]

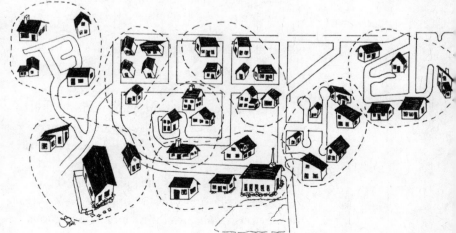

smaller cities as well. There are also possibilities for leased acreage on this scale from larger farmers or public lands dedicated to agriculture, such as school or other campus landscapes. A farm of this size lends itself to nursery crops, rotational grazing of livestock polycultures, egg and meat sales, fruit crops from a vineyard or cider orchard, a significant element of agroforestry or aquaculture (if the water resource is available) as well as all the elements of the smaller scales. Energy self-reliance should be possible, but full-time work may be needed from two to four people, and no single crop is likely to be adequate for economic support. Some small machines may be warranted. A household of 8–12 could be provisioned.

### Green Acres

Generally on the fringe of the metropolitan region, these largest of garden farms blend into the more familiar picture of older mixed farming, but with new thinking in evidence. From 8–25 acres, broadscale strategies begin to apply. Extensive aquaculture in dams should be economical. Rotational grazing of larger animals is possible. Specialty woodland crops or coppice could be developed from an existing forest or over time. A property of this size could be developed as suburban cohousing with a resident population providing a captive market for mixed farm products while channeling income from multiple households to support artisanal farming. Cheese or wine-making may be profitable at this scale.

## Strategies for Access to Land

For the smallest growers, whose market volume will be low and who will grow most things for self-provision, a location within biking distance (2–4 miles) of an existing or likely farmers market or in the midst of a residential section where neighbors can absorb all the surplus makes the most sense. There's a huge range of properties offering ¼–½ acre. Obviously, those seeking a larger contiguous plot will have more restricted choices, but a cursory look about most metro areas will reveal scores or hundreds of good potential microfarms.

Many future garden farmers already own the farm — just look out at your back or front yard. My low-equity, low-cost strategy has been to find a rundown house with still-functioning infrastructure (roof not leaking much, foundation solid, electrical system and plumbing intact, kitchen serviceable, toilets operative). Don't buy very much house — just what you need. Look for modest houses on larger lots — these will generally be older. A garage or shed is a welcome feature; a basement is less useful unless it has a direct entry door from the outside. If you live in a dicey or changing neighborhood but have an OK house, it may be possible to find a vacant lot across the street or down the block, or a house with more space to cultivate. Look for blocks with neighbors who will stay put, who have kids or who garden. This is very place specific, but clues may be flower beds, unusual paint schemes, statuary, well-occupied front porches. Good taste is not the point — care and concern for the public sphere is.

In US Rust Belt cities like Detroit, Pittsburgh, Youngstown, Cleveland and Knoxville, one can sometimes acquire several contiguous acres and a serviceable house for less than $25,000. Quite a bit of urban homesteading is going on in these places. Detroit alone has 50 square miles (32,000 acres) of vacant land or lots on which the house should be pulled down. That's 8,000 mini-farms and food for over 100,000 people. Many blocks have only a single house still standing occupied: that's a future farm landscape just waiting to be re-envisioned. Smaller cities generally have lower density to begin with, which translates into larger lots and more open land.

Jurisdiction may affect suitability. Cities vary in their regulatory regimes, but generally are more restrictive than unincorporated areas nearby under county or township rule. This is especially true regarding animals. Times are changing, and pressure will be put on public officials to accommodate urban farmers, but it's never fun to be the pioneer who has to fight political battles to keep chickens or to grow veggies in the front yard.

## What Can You Afford?

If you want to farm for a living, you'll need a plot in the Green Island/McMansion category (close to two acres) though you can make a partial living on less. Your likeliest scenario for a good income would be marketing vegetables, fruits, eggs and honey through subscription and farm stand customers or in bustling local farmers markets. Three-quarters of an acre should be enough to supply a 40-share subscription scheme for 20 weeks with extra produce for farm stand or market sale, depending a little on your climate. Over time and with experience this could translate into an adequate but below-median income, say $28,000 or more. Forty-five to fifty thousand dollars is likely the top of what two people working together in a metropolitan market could make on two acres, an area that can be managed with minimal equipment and only spot hired labor. The challenge is to find a property that you can pay for from a modest living. The classic loan evaluation formula suggests that you can afford to finance 2.5 years income, so on $28,000 income, the farm could pay off a $70,000 mortgage. Add in your down payment of $10–15,000, and that indicates what you can afford. Anything above that figure implies prior savings or outside income.

In New York City, Washington or San Francisco, even $50,000 is barely a middle-class income for two people, but if you are willing to look at less pricey communities, small acreage should produce enough income to pay a mortgage and a living. It might not be enough for high-priced healthcare and would certainly not put three kids through private

A Home of One's Own, Pattern #13 (See Chapter 6)
[Credit: Jami Scholl]

college. It helps to have some savings before you start, but that needn't be much.

Generally, you will want to buy property for no more than $100,000. Anything beyond is almost certainly speculative value or based on too large a house; over the long term the cost will not be a good investment for farming. Dave O'Neill and Lee Sturgis of Radical Roots Farm (see Case Study D) bought 4.7 acres of farmland for $61,000. They borrowed money to buy the land, which had no structures and few trees on it. They moved onto it with a camper-trailer. The farm operation cleared enough money to pay off the land, build a barn/apartment (into which they moved from the trailer) and later a permanent home (into which they moved from the apartment). Eight years into their operation, they are out of debt, well-housed, have most of their infrastructure in place and are wondering how much they want to work.

There are obviously many formulas for success. If you own land and it's paid for or your mortgage is affordable, you're ahead. Conceivably a property might have some commercial or rental value and acreage for farming to boot, so that you could afford more property than the farm itself would support. If you have some savings, you may be able to find a well-located property bigger than you need and sell off part of it. In a depressed area you may be able to acquire several contiguous or nearby lots in a single block. However if you intend to raise food primarily for your household and perhaps to trade a bit with neighbors, or if your motivations are largely to increase your food security or relieve the strain on family budgets, then working with the land you already have, perhaps sharecropping with neighbors or even renting an empty lot in the neighborhood could meet your needs.

### What about Sharing?

The most common form of land sharing in farming is rental, either on a cash or share-cropping basis, and these options exist in the urban and suburban environment as well. Young and energetic would-be farmers have been soliciting the use of their neighbors' yards in western US cities such as Eugene, Oregon, and Oakland, California, for several years, putting together urban CSAs from nearby lots by offering "grazing rights" or outright crop shares to the usually older landowners, who are in most cases unable to garden any longer or may be too busy. The advantages and disadvantages of such arrangements are fairly obvious: it takes very little money to put together a working farm using this strategy, but clear agreements are needed and making permanent improvements or even significant changes (pruning trees, perennial plantings, irrigation) may be out of reach. Cropping will likely be restricted to annuals and short-lived perennials such as strawberries or cane fruit. Parcels may be fragmented, adding some burdens to organization and resource management (where to put the compost pile and the tools?). Fencing may be inadequate, and all infrastructure supports, however temporary, will likely be awkward: a shed on skids or wheels, temporary trellising. Nonetheless, this approach requires mainly chutzpah and a little charm plus good communication skills and a willingness to innovate, while offering the possibility to begin farming with little risk. Renting a vacant lot might offer slightly more flexibility but less security, with the added burden of developing a water supply and warding off predators or thieves.

If you have limited means, or more social than financial capital, you might want to consider buying property together with others with the clear intention to farm it, either cooperatively or by dividing the land, functionally or legally. Sharing landownership works well if the relationship between the parties is clear and harmonious. Choose your partners well: consider their fitness for the job at hand, their emotional and financial stability and their long-term intentions and liabilities. Make clear agreements and write them out, being sure to include provisions for dissolving the agreement and a mechanism for any party to withdraw financially. Limit the number of partners to as many as you might share a house with — trying to get agreement among more than about five people is rather challenging, though not impossible. Ensure that each party has a measure of autonomy under the agreement, including independent access to whatever area of the land is theirs to work. If improvements are to be financed among the partners, be sure to include a financial mechanism such as annual payments or labor obligations in your written agreements. Empower any of the partners to pay the taxes or other legal obligations of the property on behalf of the whole group and provide a clear mechanism for balancing equity if this should ensue. For example, the real estate taxes come due and Kevin is broke and can't come up with his share. Sue and Jerry pay the taxes, and until Kevin pays them back, they have a lien against his share of the land. If Kevin falls into drink or loses interest in farming or simply can't manage his money and keeps failing to pay, eventually Sue and Jerry will have to force him out. It's better to anticipate this possibility and provide a way to achieve it without violence or endless frustration. There may still be friction and heartache, but at least there will be a resolution that doesn't put everyone out of business.

### Reclaiming the Commons

A third method of sharing land is to use common land and resources. This used to be the norm in European and early American societies, where grazing land, woodlands and even crop lands were shared resources regulated for the well-being of all members of the community. But enclosure (or privatization) of the commons began some 700 years ago and has proceeded so far that many people have no concept of *the commons*. This makes managing the remaining commons especially difficult as communication about the resource is important to conserving it.

Most people would acknowledge that the atmosphere is a global commons. We can identify other common resources such as our language, the airwaves, open-pollinated heritage seeds and plants, works of literature that have passed out of copyright, public rights-of-way and parks. These things belong to all people. State-managed lands such as public forests and grazing lands are common resources, but sometimes these are leased to individuals, effectively taking them out of the commons. Resources pass out of private ownership when they are discarded: trash is a kind of informal commons.

Neglected land is somewhat similar. The most obvious form this takes is roadside and median strips, so-called tree lawns or parking verges along streets, which fall into the public right-of-way but are usually tended by the adjacent property owner. In many areas these may become weedy, indicating a lack of care and attention. I have foraged a lot of fruit from untended trees in public and private spaces. Formerly owner-occupied properties are turned into rentals while their remnant plantings may continue bearing for decades. In college towns such housing is often occupied by students who value the lower rents of older buildings but may never pick the plums.

In economically depressed areas, houses, lots and even larger tracts may become derelict. People squat in abandoned buildings, trash is dumped on empty lots and into ravines. The public interest is served when local people take stewardship of these abused landscapes. Witness the proliferation of pro bono trash collection along US highways in recent years. Short of this, the edges of most properties are a kind of commons where minor incursions and short-term use are ignored or go unnoticed. One might effectively graze along road edges with one's goats or other herd animals, provided they were kept moving and the numbers weren't large. Harvesting fuelwood is similar. These biomass elements are mostly a maintenance headache for land managers and are thus often allowed to accumulate for months or years without any effective harvest. While contacting the landowner and asking permission is usually preferable, it's sometimes neither practical nor strictly necessary. In many cases, ownership is absentee.

The mortgage crisis is also an opportunity for a new suburban vision to take hold.
[Credit: Jeff Turner]

As the mortgage crisis deepens and neighborhoods erode, opportunities may arise for remaining property owners to band together to cast a safety net around vacant houses by gardening the lawns. Whether this takes the form of a maintenance-for-ground-rent agreement with the bank or mortgage holder (if this entity can be identified and located) or a well-organized community garden squat, those areas with enough rainfall for effective roof catchment are better candidates for such actions than districts in the Southwest which depend on distant water diversions for survival. The highest point of leverage in this slow-moving social disaster would be to keep foreclosed or delinquent residents in their homes. The government should have done this, but chose to side with banks against the people. Homeowners and their neighbors needn't accept this outcome as inevitable, however. As Congresswoman Marcy Kaptur of Ohio has urged, "You stay in your homes!" Many owners have successfully resisted foreclosure by demanding to "see the note." These weirdly sliced-and-diced mortgages have become abstract bond obligations, often with no clear chain of title, thus nothing on which (or nobody) to prosecute a foreclosure. And as the sheriff of Wayne County, Michigan, reminded us last year when he declared that he would no longer pursue evictions in the Detroit area, "Possession is ⁹⁄₁₀ of the law."

However you secure it, access to land is essential to garden farming. The US has been the land of opportunity for tens of millions, and while top-down opportunities such as a new urban homesteading law may emerge from the present crisis, there's no need to wait for a government handout. Land is available now almost everywhere, and for the enterprising, it can be surprisingly cheap.

## Qualities of Land

There is no perfect piece of land, but all other things being equal, you are better off with land that has some elevational change, but most of which is level to gently sloping. The house, if there is one, shouldn't be at the bottom of the landscape but is better in the middle of a sloping parcel. Compact shapes are preferred to long or irregular parcels for ease of access. Land on narrow ridges is difficult to use, difficult to access, often subject to harsh winds and storms and at greater risk of fire. Land on broad ridges, especially in non-mountainous regions, can be very advantageous as cold air will drain away to either side, leaving warmer conditions on top. Aspect is important, and if you plan to live on it, at least a significant part of the land should face

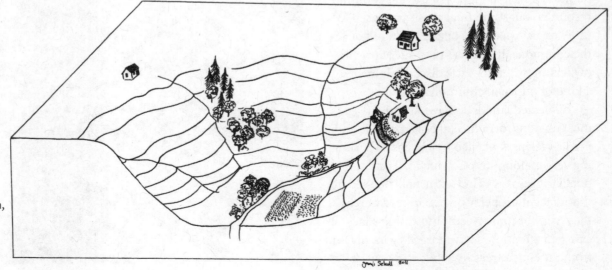

Agricultural Terrain,
Pattern #3 (See
Chapter 6)

[Credit: Jami Scholl]

to the southeast, south and southwest. Other than the extremes of boggy soil or nearly pure sand, soil is a relatively minor factor in choosing land, as soil fertility can be improved. Clay soils will be hard to work but have great potential to build and hold fertility over time, while lighter soils, though easier to work, are more fragile and require more careful management.

If the land you are considering is wooded, you will have the advantage of available fuel and materials, but the very strong disadvantage of having to remove trees in order to build or garden. Generally, it is not a good idea to settle in a forested tract. Find a piece of land that has been developed or cut over. It will be cheaper, and you can plant the trees you want. Having a few trees on part of the land may be the best circumstance, as you can cut from these for fuel while planting most of the land as you wish.

Consider access to the land. Is it located on a paved or public road — the best and most secure access — or must you enter over an easement from another landowner? Are there any easements over the land, such as utility lines, neighbor's driveways or old rights-of-way? Do you have the subsurface rights? This can be very important in areas where mining or drilling is underway or may be soon. Is there an access road at all, or will you have to create one? How secure is the legal access? Is it traditional or deeded? The more complicated the access, the greater your risk of the land becoming worthless. Long driveways are a liability: they are costly to build and to maintain. On the other side of the issue, if you are located on a busy road, how easily and safely can you enter into traffic or turn into your driveway? What is visibility like at your frontage?

In broader terms, is your parcel within easy reach of neighbors, a supportive community or markets? Will the neighborhood be suitable for a garden farm? Will your neighbors be tolerant, or is there enough autonomy that you could shape the landscape to your needs? Could you farm the front yard, for instance? How close are services, and what are transportation networks like in the area? Is there public transit, or will you be dependent on an automobile? Could you bike safely to do business, shop or reach services? Does the property have any commercial potential? Is it visible to passersby on main roads, or is it tucked into a cul-de-sac? Does it have a location in the front that could become a seasonal market (see Pattern #68 Fruit Stand in Chapter 6)?

Consider internal access and movement also. Is the land steep, or might you be able to move a vehicle to most parts of it? If you have building to do, or any earthworking or major tree removal, this can be critically important. Are there any internal divisions to the land that make it hard to integrate, for example, a ravine, rock outcrop or sinkhole? These features might be charming and add character, but they could be serious economic liabilities. Of course the presence of odd features on a larger rural parcel is of less concern than on a small suburban lot where every square foot may have to count.

If there are buildings, consider their placement. Are the buildings central or near the road? The home mediates between public space and private, so a house located near the road will preserve more outdoor privacy than one set far back behind a vast lawn. A house close to the road will, other things being equal, have a shorter driveway and thus less space wasted on pavement. Do the buildings create shadows on the land that compromise the value of part of the ground? What about wind tunnels, narrow passageways, alleys or other difficult-to-access or awkward territory?

If there are no buildings, is there a suitable area for one or more structures? Generally, you will want some level-to-gently-sloping, south-facing ground above the lowest place in the landscape, but not on a steep ridge. There should be some level ground in at least one direction from the potential building site and good access from the house site to all parts of the land. It's preferable that the house site lie to the north side of the parcel — so that most of the land lies south of it, but also not distant from the road access. The building site should be wide enough that you can align the long axis of the house east-west. If you have a hill, can you locate the house to take shelter in the lee of it against winter winds?

Think too about auxiliary buildings — a barn, a woodshed, a garage or workshop, a greenhouse. How and where might you create these spaces? You are likely to want some form of all of these for a garden farm and, though they needn't all have good solar exposure, the workshop and the greenhouse certainly should. Access to these structures will be important, both from the main house and from the road, so that materials can come and go, including construction materials, harvest and value-added products from the land. The woodshed should be near the heated buildings, and the barn should be well-located in relation to any pastures and main grow-

ing areas. The greenhouse should be within 100 feet of the house and not far from main gardens.

The assessment of land is complex and layered, but if you can give a satisfactory answer to most of the questions I pose in this chapter, you should be in a position to make a good decision about your present property or one you are hoping to buy. The energy descent world coming at us likely means that property will be less saleable in the future than it is today. Already the US housing market is depressed because of foreclosures and overbuilding during the past 30 years. It may not recover. If you are able to sell your present home and relocate in the near future, you should expect it to be a permanent move, as there can be no certainty about the nature of the real estate market in years to come. That means the fundamental economic qualities of the location and ecological qualities of the land will take on greater and greater importance with the passage of time and the contraction of the economy, while social restrictions on land use may ease.

# CHAPTER 8

# Labor — Can You Lend a Helping Hand?

*It's only work when there aren't enough people to make it fun.*
— Andrew Goodheart Brown

Garden farming is a scale of land cultivation and an economic activity matched to the capacities of the household and the use of low-energy technologies. The traditional farming household of recent memory, as in countries of the global South today, secured labor by having children and by building an extensive network of social relations, some by blood or marriage, many more by neighborly exchanges of mutual aid. Demographic transition — the tendency for the birthrate in developed societies to drop with increased wealth — and the compelling need to restrain population growth mean that large families cannot be the source of farm labor for the foreseeable future. Of course, if you have a large family already, you may be inclined in coming years to put them to work growing food (I recommend that you start sooner than later), but for the rest of us, the strategy of building social alliances makes the most sense. Call it barn raising for the 21st century. Undoubtedly, you can go it alone as a garden farming household, but you are much more

likely to succeed and prosper if you have a network of friends and family.

The average household in the United States in the year 1900 was 4.6 persons. A century later, it was 2.6 persons.[1] These small numbers mask a revolution in the social order. The decline in household size was continuous over the 20th century, sometimes faster, sometimes slower. During the early decades of the century, industrial employment drew laborers from marginal farms to cities and towns where they often lived in small quarters. Mechanization replaced farm labor first in prosperous regions. Appliances replaced the labor of servants and children in the 1920s. As Social Security (after 1937) and Medicare (after 1964) increased their effective incomes, elderly Americans, especially widows, chose to live independently of their children. College education from the late 1940s onward drew tens of millions out of the labor force and extended the childless years of adolescence and youth. The boom years of the 1960s and 1970s also distributed wealth to the young. The baby boom generation, coming of age and rebelling against their parents in part because of the Vietnam War, scattered from traditional communities toward more metropolitan regions. Education and The Pill changed the politics of sexuality and the family, empowering many women to make choices other than marriage and children. Declines in real income after the 1970 peak in US oil production put economic pressure on women to work and further changed the politics of the traditional family. Divorce rates shot up. With each of these social changes, the household shrank.

The decline in household size was slowed but not stopped by numerous economic downturns of the 1950s, 1960s and 1970s. It only reversed (briefly) during recession years of the last quarter-century (1983, 1993–94). These transient increases in household size prefigure a long-term trend of the 21st century. From the 1980s onward, hard times — and there are more to come — have meant that more people share housing; adult children linger or return to live with parents; the elderly join in with others; couples who might have divorced stay together a little longer. Everywhere in the world, the poor are more reliant on each other for survival than on money. North Americans are now in the throes of a historic shift of momentum from ever-greater autonomy toward increasing mutualism.

We are now probably at or past a historic low in household size. Forty-nine million Americans still live in multi-adult, multi-generational families, and the number rose by 2.6 million from 2007 to 2008.[2] The long-term trend is certain to be upward, and for the sake of successful garden farming as well as energy efficiency, that's a good thing.

## Household Sizes

The household of one person is a modern phenomenon. Beyond a few recognizable social roles — the witch, the woodsman, the shepherd, the hermit — solitary living made no sense in any traditional society running on the limits of solar energy. It presented no economies of scale, no benefit from division of labor, and it imposed harsh limits on the capabilities of the individual. Redistribution of wealth from the fossil fuel economy (both as passive income and as energy slaves in the form of appliances and machines) has made single living possible and indeed attractive for hundreds of millions. Modern communications and transport have enabled far-flung social networks to replace some functions of the family and close-knit communities of the past.

But households of one considering garden farming, or any aspect of subsistence living, are under many of the same limitations of solitary people in traditional cultures: they face doing a lot of work that our society has made uneconomic, and they face it without help. To attempt anything more than a small measure of self-provision would seem to entail considerable sacrifice. I know that some people living alone today still grow a big garden, keep animals, can food or practice crafts. And more power to you! Most of these individuals are living in rural areas or small towns where the culture of self-reliance has not entirely been swept away and where they have neighbors and friends in family households who are living in similar ways.

There are adaptive strategies, but generally they involve getting help. Someone with a large house might consider taking in a boarder. There's always the option of finding a partner or joining another household with an interest in garden farming. You could organize your neighbors to share work, hire a teenager to help you for summers and weekends or lease out your unproductive lawn to a local garden farmer in exchange for a share of the crop. If you are retired, you may have the time to manage up to a ¼ acre of garden with only spot help, perhaps more if you emphasize perennials or use some of the space for animal forage, but sales for income may be out of reach.

So, while the organizing principles and ideas of this book are applicable to garden farms in a range of sizes, I recognize that it will generally require households of two or more adults to implement them effectively. And more is better. Multi-adult households (three or more persons) have a significant labor advantage even though they are now a small minority. Outside of institutions—the military, prisons, retirement and nursing homes, halfway houses, group homes for the disabled, homeless shelters, religious orders—group living situations are mostly confined to student co-ops, both formal and informal, a few urban cooperatives and a residue of more traditional extended family households with adult children, a mother-in-law or dependent relatives.

For reasons that I will explain below, a household of three adults would, in my estimation, be a nearly ideal number for a small garden farm (¾–2½ acres), except that the number is often socially unstable. Jealousy and asymmetry among the various dyadic relations tend to be corrosive. Of course there are exceptions. Family ties can strengthen personal bonds to enable three adults to cooperate durably. Three adults sharing a home can be supported on a full-time income earned by one of them, leaving two people to work the farm, with the third lending a hand in his or her spare time with household chores and bigger projects. The two, working the land together daily, can accomplish more than twice what either could do alone. And after a few years, the farm can contribute a second income, enabling the employed person to reduce hours, change jobs or shift into home-based employment. It takes considerable care to manage the strains created by this kind of division of labor, but gains in security and rapid development of the farming system are well worth the effort to do so.

If you are not already a household of three, but a couple thinking about garden farming, consider your circle of friends: Is there anyone that you could live with who would like to join the adventure? Perhaps you have room for a young partner, someone without land who'd like to gain experience and can help you get going: a daughter, a nephew or someone you contact through the farming interest lists online. The extra boost of a second full-time farmer can dramatically accelerate the evolution of the system. Of course, you cannot expect unrelated persons to contribute labor to a long-term project without compensation in the form of pay or equity.

Households of four or more adults are capable of almost anything described in this book and would find it worthwhile to be working two acres or more. Indeed, the advantages of four adults cooperating would suggest that it's worthwhile for couples with similar interests to find each other and locate next-door or in the same neighborhood even if they don't share a house.

In evaluating your effective household size, you must of course consider dependents. If you have children under eight years of age, or an elderly or special needs dependent, one of the adults must be subtracted from the farming labor force. Children from 8 to 16 years of age can contribute substantially to the farm work—on average to the value of a half-time adult—depending on their schooling regime. I know of homeschooled

Communal Labor,
Pattern #20
(See Chapter 6)
[Credit: Jami Scholl]

eight and ten-year-olds who raise their own animals and crops and take them to market (with the family, of course).

But with or without children, the most common arrangement remains that of two adults living together. So just what can a household of two adults do?

### The Farming Couple

Bob and Bonnie Gregson, farming in the Seattle region, have demonstrated that two people of middle-age in good health can manage up to two acres and make a living as farmers provided they have access to urban markets.[3] The Gregsons committed themselves to hiring no help, preferring their own companionship and fewer complications. Their working season as farmers ran about eight months. In the winter they pursued other activities that also generated some income and provided a change of pace. They pioneered a new business model and took eight years to achieve what they viewed as success, meaning adequate net income (near the median in their area) and stable markets. They believe that with better guidance others could do the same in four years. They admit that their land use, even after eight years, was

Skill-sharing like this ferrocement tank construction can help muster extra hands for big projects on the farm.

[Credit: Keith Johnson]

not as efficient as it might become, and that most of their income derived from about half an acre. It should be said that they were able to devote themselves to farming because they had savings from the sale of a home, their land was paid for and their children were grown.

My own experience working a property of ⅔ acre with a partner and part-time interns is that, as the cultivation system expands and becomes more complex, the potential yields appear to increase, while the size of the land isn't yet a limiting factor. Output (and thus income) is restricted by the available labor (and to some extent by infrastructure and the maturity of the cultivated ecosystem). Like the Gregsons after eight years, but with a smaller acreage, we after five years haven't fully exploited our land capacity.

Yes, two people can run a garden farm successfully and have a good life doing it. As we will see, there are still choices to be made about how to distribute the labor and other resources — and how fast to develop the property.

### Labor on the Farm

Garden farming is work, but also it involves managing complexity. Done right that work can be virtuous and rewarding, entertaining, creative and social, but it remains an occasion of physical exertion. Aside from some heavy haulage, mowing, tillage and wood cutting, it's difficult to mechanize work on the small farm. If you do find yourself needing to dig 600 feet of trenches, or as many post holes, or to chip up the top of a fallen tree, rent a mechanical tool for a day or a week. I'll talk about technologies more extensively in Chapter 18. For now let's consider the organization of labor.

The main repetitive and labor-intensive elements of garden farming are tillage, planting, mulching and weeding, pruning and coppice, harvest and processing. The management of animals also involves moving them, providing their food and water and lesser amounts of time spent with breeding, butchering and protection, including maintenance of stalls or pens and medical care. Infrastructure also needs maintenance and repair: fencing, water and irrigation systems, buildings, tools and machinery. The permaculture approach in regard to all types of repetitive labor is to keep it to a minimum by design.

### Design to Save Labor

Most North American households are oriented to consumption, not production, but in garden farming the opposite is true. The garden farmer, like every farmer since the beginning of agriculture, is working to capture solar energy through the growth and harvest of plants and animals. Many permaculture principles and design methods are specifically intended to help the householder conserve labor, optimize system function and thus enhance yield. Zone-and-sector design is one of those methods. As we have seen, its essence is to concentrate beneficial energies on the site, scatter hostile forces and conserve your own energies by using gravity, proximity and connection to avoid unnecessary steps, transport and work. Garden farms gain significant advantages from being small: they provide an economy of scope that enables us to manage production and resources more easily and economically because they are close to home.

Any way you look at it, however, there are two sometimes conflicting demands on the garden farmer: to grow and harvest crops and to build the farm. These reflect important permaculture principles: (3) Get a Yield, and (7) Design from Pattern to Details, respectively. It's difficult to tend crops and nurture animals, to say nothing of holding down a job, preparing meals and raising the kids — demanding daily activities — at the same time as you are contouring landscape, laying irrigation pipes or building a barn. If you haven't got the swales and the pipes in place, then come July or August you are going to be watering as part of your chores. And if you haven't got the barn built, then you're going to store your tools in a closet or under an eave where they'll be lost, rusted or inconvenient. And you'll be cobbling together inadequate and temporary shelter for animals and repairing the chain saw on the living room floor or in the rain. You can't succeed without both Harvest and Design/Build, and there is no one solution to the dilemma: each imperative task makes the other harder before they become mutually supportive. The answers range from farming in the warm season and building in the cold season — as Dave O'Neill and Lee Sturgis of Radical Roots Farm did (see Case Study D) — to starting with significant savings (see Ian Graham's tale in Case Study C), to working an outside job while you build up the farm and only slowly moving into cropping.

This last is the strategy my partner and I have followed. It's possible to succeed at either end of the range or at a number of points in between. How you deploy your time will depend on your financial resources (money is stored labor), your age, vigor and your tolerance for multi-tasking. Look at it this way: the

Porch building had to wait until late autumn when the harvest was complete.

[Credit: Keith Johnson]

land use pattern and the economy we have today were built up during a period of energy surplus and low material prices. But these patterns are now dysfunctional and must change to reflect new and permanent conditions of scarcity. The change will take some years. A good approach is to get clear about the demands this transition will place on you, adjust your expectations and then get to work.

The alternating pattern of crop and build and crop and build is physically and psychologically demanding. But if your income depends on the sunshine of March to September, then you'll find all the ways you can to use the down time from farming for things that can be done in the cold and dark. Dave and Lee understood the huge demands of the farm well enough that they even planned their family so their two children were born in the winter when Mom and Dad would have more quality time with the newborns without the conflicting demands of managing a crew and getting to market.

Like raising children, creating a garden farm and a productive household is a big task but not an endless one. The most intense effort goes in early on, and over time the systems become more and more stable and self-regulating. This should be true not only for

The Long Meadow, Prospect Park, Brooklyn, NY

[Credit: Alex L. White]

capital infrastructure (buildings, storage and distribution systems) but for the cultivation system as well. Annual crops will always play an important role in meeting our food needs, but the garden farm is an ecosystem first. And that means it must have a perennial backbone.

The basic elements of a permanent agriculture are those ecosystems that build soil: forests, unplowed grasslands or meadows, ponds and lakes and mulched gardens. In each of them we use to our advantage the tendencies of nature to enrich ecosystems over time. By designing the placement and composition of these systems, we can manipulate their yields to meet our needs while gaining all the many ecological advantages they offer. We will look at each of these kinds of systems in later chapters, but for now let's consider them in terms of their impact on the labor design of the farm.

### Get Rid of Your Lawn

The most conspicuous and potentially useful feature of our suburbs is their vast expanse of lawns, a form of grassland. Lawns represent a massive expenditure of energy and money that produces no crop; more fertilizer is used on North American lawns than by the entire agriculture of India. The industry runs to tens of billions of dollars a year.[4] Why do we perpetuate these cultural palimpsests? The psychology is largely one of keeping the deep dark woods at bay, being able to see predators coming at us from a distance and emulating the rich. Some of these reasons have instinctive roots that we must respect, but this manifestation of our feelings of fear and envy needs a good shaking out.

The genuine psychological need for open vistas can be met with very little grass and strategically placed long views. Much of the genius of Frederick Law Olmsted, the 19th-century designer of New York's Central and Prospect Parks and many famous urban venues across North America, was his realization that city dwellers needed an expansive view over a meadow for relief from the confinement of the built environment, but it doesn't have to be a very wide prospect to deliver the goods.

Looking out over your garden can be just as liberating psychologically. And while we're gardening, we're going to embrace the spirit of the deep dark woods, tame it and bring it home where it can continue to protect us. We'll plant trees and shrubs and keep them to a manageable size.

Grasslands and meadows grow grass and *forbs* (edible broadleaved herbs). This is food for grazing animals and one of the best productive uses for broadacre, cool-climate and especially semi-arid landscapes. Grass hay can also be used as mulch or stored for winter fodder. If you don't pasture animals on it, however, grassland has to be mowed or burnt periodically to prevent it turning into woodland, and long before then it becomes impenetrable shrubland if not managed. Burning is anathema in urban and suburban areas for many good reasons; mowing is the acceptable but extremely wasteful alternative. All over North America from April to October, graceful (and well-to-do) experts cut swirling figure-eights with riding mowers, while the rest of us yank and rassle desperately with smoking, belching, noisy contraptions that are always threatening to cut a foot, gouge an eye or simply give up the ghost. The ecological footprint of lawn mowers isn't small either.[5]

Apart from small areas for amenity—a play space for children, a picnic table, sunbathing, a volleyball court or a spot to pitch a tent in the backyard—there's no reason to grow grass except to feed livestock. And they can harvest and carry it themselves; you shouldn't have to. Grass makes a poor edge for garden beds because it will always be trying to spread into the lovely loose soil. Eliminate this maintenance problem by mulching your garden pathways with old carpet, wood chips or perhaps even plastic covered with gravel. For main pathways near the house, you are

better off with pavement (stone, tile, even concrete) which can take foot and cart traffic. Grass can be a suitable surface for little-used farm lanes—places where you take a vehicle a few times a year and never in wet weather—and for some work areas, but otherwise, getting rid of it counts as a major labor saver.

## Stop Churning the Soil

One of the biggest and most dangerous tasks in conventional farming and gardening is tillage: plowing the soil. It's dangerous because it exposes soil life to destructive forces and increases erosion dramatically. Tillage is hard, sometimes back-breaking work. The modern machinery that "saves labor" in tillage is also dangerous to humans as well as to the soil, whether it's rototillers or tractors. We'll see, when we look at soil in Chapter 11, how this practice has developed and why healthy ecosystems almost eliminate it. But from a purely economic point of view, if something is dangerous, destructive, time-consuming and mostly unnecessary, common sense would suggest we stop doing it, or that we do absolutely as little as we can while still achieving our main aims.

The design of the garden farm should reduce tillage over time to the minimum

A sheet mulch made with cardboard and straw. Pockets of soil allow immediate planting.

[Credit: Edward Carter]

Raised beds benefit from early spring warming and better drainage in wet seasons.
[Credit: Creighton Hofeditz]

Concentrating seedlings in one area for special care and attention reduces labor, risk and expense.
[Credit: Keith Johnson]

on woody and other perennials which you prune, coppice and harvest. It may have some small areas of lawn for amenity and permanent grass pasture for animal forage. The garden farm will have dedicated raised beds (or in dryland areas, sunken beds) for growing annuals and smaller perennials.

Something that you have selected will be growing on every square foot of your ground throughout the season, and most things will need to be cut, grazed or tended only once a year. The exceptions are small fruits (from which you may harvest repeatedly), perennial mulch plants and annual vegetables. These crops may rotate or be harvested two or three times in a year.

If you don't till, you can leave a permanent (or continuously renewed) mulch cover on your garden beds, hastening the development of fertile soil and virtually eliminating erosion. However, this requires a different approach to planting. Much of tillage is meant to eliminate plant competition for the new crop and to break up the soil surface so that small-seeded crops can germinate. Instead of doing this disturbance on a broad scale across the entire surface of your garden — which merely invites new weeds to drift in and germinate — concentrate all your seeding into flats of prepared soil mix made from compost and other select materials. Prepare these in a nursery or greenhouse where the seedlings can be protected in a favored microclimate and where you can more easily manage and support their growth. They will be gathered together for easy watering, and you can tell at a glance just exactly which ones are ready to be moved into the garden. They will have well-developed root systems so that when they do go into the beds, they'll take off.

For the cultivation of annuals, we can also learn methods to overseed one crop into another, encourage self-seeding plants and varieties (which we can move around to suit our convenience) and to let livestock do much of the residual soil preparation for us. By

required for planting perennials and transplanting annuals. How? By building a fertile soil, keeping it covered almost all the time, seeding annual crops in trays and transplanting them to beds, cultivating perennials and developing mulch systems and permanent ground covers to reduce weed pressure and improve tilth. Admittedly, this takes several years, but the benefits are cumulative and start immediately. As a garden farmer, you are aiming for an ecosystem that has a stable structure and dynamic edges. It will be based moving animals onto ground to eat pests and weeds, drop manure and stir the soil surface layers prior to planting crops and then again after harvesting to clean up plant residues, we can meet many needs while limiting soil disturbance. By grazing beds, plots and fields as part of livestock rotations, we can also diminish or eliminate the need for both tillage and mowing.

### Plant Once, Harvest Many Times

Perennials are the heart of any ecosystem. Nature uses annuals (what ecologists call weeds) to cover bare soil. Period. Annuals are opportunists that lurk on the fringes of more stable systems, waiting for a disturbance. In nature they blow in, drift in or germinate in the new sunlight, exploiting the suddenly available resources of solar energy, water and nutrient. Seen in another light, weeds are nature's paramedics, first on the scene to repair damaged soil. They accumulate minerals dynamically in their tissues. As they die, these become available to other plants through the action of soil organisms. Once the soil begins to recover fertility, the annuals give way to longer-lived plants, a process called *succession*. Conventional agriculture has adapted its practices to create disturbance and prevent succession so that our field crops, which are mostly annual weeds, can continue to grow. Our ancestors domesticated many of these plants by selecting them out of midden heaps. Thus, gardening began.

Perennials plant themselves, or rather we plant them once. Then they grow and flower and fruit for years thereafter. Perennials are better able to handle fluctuations in weather; their roots are deeper, and they will grow in almost any warm season regardless of when rains or frost come. Most importantly, perennials don't have to be replanted. They get stronger and more vigorous and yield more each year over decades. They do need to be managed, but pruning is much easier than turning the soil, and far less risky. Moreover,

the surplus material from pruning, on top of leaf drop and regular root replacement, builds soil where tillage depletes it.

### Labor Through Time—the Big Picture

Permaculture systems save labor by design. You can and should make labor-saving choices in the design of your system; however, the fruit will not pick itself. A couple working full-time during the growing season can manage up to an acre, though perhaps only half of that under annual crops, a little larger acreage if rotating animals on pasture or sticking primarily to orchard crops. That farming couple can make a living, and they can feed themselves. There will be more work in the early years to establish the system, and perhaps slightly increased work in harvest and processing in later years as the perennials mature.

There are also life cycle and generational issues to think about in labor. I'm encouraged when I read of the Hunza people of Central Asia riding horses well into their tenth decade, and I like to think of myself pruning and picking fruit trees in my 80s or 90s, but I hope to be well through with construction by then or to have passed those responsibilities on to younger partners. Good planning also includes considering succession in human

Kids as young as eight can be responsible and productive gardeners, so get them started early.
[Credit: woodleywonder works via Flickr]

terms. Who will benefit from the careful husbandry of your soils and trees, and who can help you with strenuous jobs as you age? This may seem a long way off for someone in her 20s or 30s, but it is never too soon to consider getting help.

## Working with Family

If you are fortunate enough to have children who want to do what you are doing, who like you and who are enchanted enough by the dream of the garden to stick around (or return after wandering a while), then your job will be to help make a place for them at home or nearby. This may seem a vision from the remote past, contrary to the modern dream of a corporate job in the big city, but even I, growing up in the 1950s with parents who were not farmers but college-educated town dwellers, lived only a few doors or blocks away from aunts, uncles and grandparents. Had we been more determined gardeners, we could have provided plenty of hands to can grape jelly or pick green beans. I know that millions have lived some part of this vision, and millions more are hungry for reconnection with home, family and the land. But it won't happen if we don't envision it. Nor will it come about without care for the transfer of responsiblity and authority over time.

If you are working with younger children, investing them with responsibility begins as early as they can do simple chores, and their increasing competence should not only elicit your frequent praise but deserves more formal recognition around the age of eight or nine when middle-age children are perfectly capable of tending garden crops, milking goats, working a market stand or managing small animals, or for brief periods, their younger siblings. By the age of 12 or 13, they may well be in a position to manage a whole subsection of the farm from planting or breeding to harvest and marketing and should certainly share directly in the benefits of doing so.

To engage older sons or daughters, nieces or nephews in the family farm involves more complex considerations. Young adults nesting with partners and perhaps with children will need a measure of private space. To cooperate effectively in garden farming, however, they ought to be able to walk to work in just a few minutes. Most newer homes (post-World War II) are not designed as multi-family residences, though some of the larger ones might be adapted. Can your house be remodeled or expanded, a split-level made into two units? Is there a second home on the property or could you create an apartment in the barn or garage? Or perhaps there's a neighboring house for sale or rent that might be incorporated into the family holding eventually. This close-knit society may seem anachronistic in North America's cosmopolitan world, but it really isn't so distant in either time or space. Many African families still live in compounds and manage their herds and fields collectively. Sociological studies tell us that large numbers of people marry someone who grew up within 500 yards of their own home, as counter to the myth of modern mobility as that may seem.[6]

Beyond family bonds, what seems to be required to support enduring intergenerational cooperation are conditions of stable work that allow at least a modest amount of surplus time or leisure for mentoring and sociability. Garden farming could well provide these conditions. The advantages of mentoring are profound and go beyond the transmission of subtle mastery in work to the real growing of human beings, a process our society has substantially abandoned over the last century. In my own observations of community, it is just these vertical ties — elder to younger and vice versa — that are most needed to balance out the natural perturbations, conflicts and changing viewpoints that occur between the generations.

Horizontal bonds are everywhere in our society. Same-aged classmates throughout school, small families with siblings of proximate age, social ties from work between peers ranked largely by age, senior living arrangements and childless neighborhoods all conspire to keep us separated into age cohorts. Denied the sense of continuity with ancestors and descendants, of growing up with elders telling tales of the past, our lives are stripped of much of their essential meaning.

## Finding and Forging Partnerships

The mechanisms for recruitment can be as simple as what Rob Hopkins, originator of the Transition Towns movement in the UK, describes as "a dating service for gardeners." People with gardens who are too old or too busy to work them get matched with would-be gardeners lacking land. Young urban farmers have initiated such lot-share arrangements in Eugene, Portland, Oakland and other US cities. At the intimate scale of a neighborhood, a start-up urban farmer can live a few doors down from the farm or around the block in a rented apartment. With success should eventually come the opportunity to buy a property.

Landownership is one of the key issues because farming is the occupation that marries land with labor most closely. There's wisdom in a new farmer renting land while learning the ropes of the business. But in the long term, optimal productivity is likely to require a transformation of the landscape, investment in buildings, fencing and permanent plantings. And these in turn all require security of tenure that will incline most farmers to want to own their land. At the suburban scale, this is not out of reach, but an optimal property for garden farming may still cost more than a young person or couple may be able to afford. Thus, there's an incentive to working out land-share arrangements. Working directly with an owner-partner can make the necessary investments and changes for productive farming easier for everyone.

Of course, it's not only the young who are starting to farm. Many middle-aged and

once middle-income couples will turn to self-provision and specialty farming—if only for the security it can bring—as the economy continues to shed industrial and service jobs. But the process is new. The advantages of vertical partnerships will become apparent if people can be honest with themselves and not fall into the cultural habits of rugged and usually dysfunctional individualism. The initiative can come from either end of the age spectrum, but the responsibility lies more on the older partners to reach out. They have social rank over the young as well as the privileges of greater wealth and land ownership (generally) and need to make the barriers to entry lower for younger partners. That doesn't mean giving away the farm or making foolish concessions to inexperience. It does mean anticipating the real needs of younger people, reaching out and entering into genuine partnerships that move over time toward equality.

## Creating Ties that Bind

Many older people hold most of their life savings in their homes. Thus, it's important that any land-sharing deal recognize the property value as an asset that must be conserved or augmented, and which, in any long-term exchange or vesting process, must be paid for.

Our society needs many more opportunities for older and younger generations to help each other directly.
[Credit: Roger Mommaerts]

Of course, do a thorough property analysis and design before you make any changes. Good design will not compromise the integrity of sound structures on the land — you won't want to plant large trees near the house or plant evergreens on the south side of a building where solar gain may figure into the energy equation. Good building sites should be preserved, as should scenic views.

Real values and apparent values are divergent today, however. In a suburban setting, farming is unfamiliar, and many will think it a detriment or a degradation of property values. Within a decade, these market disjunctions will begin to correct themselves, though only, I think, for properties that are well-designed ecologically. Having a fertile soil, a mature edible landscape, water catchment and energy harvesting buildings will come to be understood as real value. When it comes to actual farming, one of the first issues, therefore, may be appearances. The early stages of transforming a suburban property into a garden farm can be messy, especially if the farmer uses salvage materials to save money. Suddenly, there's a pile of manure in what used to be the lawn. Presumably any landowner considering to farm or partner with a farmer will have a commitment to fertility and productivity that can bear the temporary discomfort of messy materials, but plant flowers along the road —

**Pardon our mulch.**

[Credit: Jami Scholl]

lots of them — and keep things blooming throughout the season. Most people can live with chaos for a while if they are reassured that it's being contained and managed.

The deeper layers of a working farm partnership will rest on an equitable distribution of rewards and the building of trust. Most of the cash returns from farm labor should go to the cultivators along with the costs of seed, stock, soil amendments and machinery, while the cost of permanent improvements to the property (which enrich the landowner) should be shared: materials provided by the owners and labor by the working partner(s). Utility costs should be shared equitably, and the farm should contribute something toward the property taxes. A friendly arrangement that has frequently been used is a loose share-cropping agreement where the landowner gets a full household share of produce from the farm, but in more committed relationships, this and other exchanges should be more carefully spelled out.

Trust starts with clear agreements and expectations and is maintained by communication and by the regular meeting of obligations. Begin your partnership agreement by creating a mechanism to end it fairly. Specify who owns what, how values that are sunk into the property may be extracted and what should be treated as expenses. Make sure that

each party's responsibilities are carefully articulated. Speak to any expectations you each carry. In addition to a contractual agreement about sharing costs and benefits, assets and liabilities, you will want to establish a protocol for communicating about ordinary operations (e.g., a weekly meeting) and specify a mechanism for conflict resolution, ideally one that attempts to preserve the agreements and to avoid litigation. Around that structure, create opportunities to interact socially — in the garden, over lunch — with all working partners and other members of the community.

Family Table, Pattern #17 (See Chapter 6)

[Credit: Jami Scholl]

## Working Through the Seasons

Because of the intense seasonality of the climate in much of North America, farming tasks are strongly ordered by time of year. Fortunately, a life spent out-of-doors (as in the garden) helps align internal and external rhythms.

The rhythms of the year become ingrained over time, and their familiarity is a comfort. An intimate knowledge of the ebb and flow of life energies through the world around us forms the basis of culture. The caraway blooms and we know that it's time to plant potatoes. Eventually, our allies amongst the plants and animals, by their flowerings and stirrings, will provide all the cues we need to play our own part properly, to enter and exit, to prance and bow on the garden stage, orchestrating the movements and sometimes taking the credit.

## Transient Help

Garden farming is an emerging — or re-emerging — endeavor that holds promise for millions in our rapidly devolving society.

### The Garden Farming Year

Timing is dependent on climate zone. These mid-month targets below are well-matched to zones 6–7.

| Labor demand | Month | Tasks |
|---|---|---|
| Low | January | seed and plant orders, planning, woodcutting, ice cutting and fishing |
| Low | February | start seeds indoors, crafts and repairs |
| Moderate | March | transplanting to flats, mushroom inoculation |
| High | April | bed prep, transplants, plant trees/shrubs, young animals born/hatched, vegetable marketing begins |
| High | May | construction, harvest for market, haymaking, transplant/seed summer crops |
| Moderate | June | harvest for market, summer pruning |
| Moderate | July | harvest for market, haymaking, earthworking, pond building, foundation repair, excavation |
| Moderate | August | seeding/transplants for fall, food processing, canning/drying/storage, construction, excavation |
| High | September | seed winter cover crops, harvest and storage, food processing, construction projects |
| Moderate | October | gathering leaves, hay, mulching, mushroom inoculation, plant dormant perennials, spring bulbs, pruning, fruit/nut harvest, cellaring, wine making and other ferments |
| Moderate | November | hunting, slaughter and butchering, smoking meats, coppice and woods work, crafts, fruit/nut harvest, cellaring, prepare systems (drain pipes, troughs) for winter, clean gutters |
| Low | December | holidays, cooking, visiting, merriment, forestry, crafts |

Credit: Peter Bane

The committed farmer can attract a steady stream of visitors from voyeurs to customers to eco-tourists to future farmers. Almost anyone can fall somewhere on that continuum. Your prime advantages are proximity to people—if you live in or near the city and accessibility (by which I mean the ordinary exoticness of turning your yard into a farm, and the basic hunger for knowledge and nourishment that drives a growing segment of the population). Millions will need to learn what you know and do what you are doing. Your three or six months of experience will be invaluable to someone who knows less—and there are mobs of such future farmers and customers.

From at least the early 1980s onward, the movement toward organic farming has grown exponentially, driven by concerns about food safety and the rise of corporate power. Sustainable agriculture organizations and networks have well-organized channels to move volunteers into farm service, among which the venerable WWOOF (World Wide Opportunities on Organic Farms) has perhaps the greatest name recognition.[7] WWOOF-ing has, in fact, entered the language, at least informally, as a verb meaning to travel as an itinerant farmworker with no financial aspirations beyond keeping expenses low. In North America, most of the major state and regional sustainable agriculture organizations—such as NOFA, OEFFA, PASA, Oregon Tilth, CCOF, CFSA—have programs to match visitors with farms in need of help and willing to take short-term (from three days to several months by arrangement) interns.[8] The arrangement is usually provision of board and some shelter—which may be in a barn or camping—in exchange for part of a day's labor at whatever tasks are being done on the farm.

Well-established programs vet candidates for you, and they streamline the flow of help. There's usually a small registration fee required of all parties. As a farm host, you need never take anyone you don't like the sounds of—and you have the opportunity to look them over in advance—but you can be reasonably sure that if you do say yes, you'll be getting someone who's a genuine traveler/student/seeker and not a con artist.

If you have an active community pres-

Rooms for Guests, Pattern #66 (See Chapter 6)

[Credit: Abi Mustapha]

ence — that is, you attend various local societies, schools, clubs or churches or you have contact with high-school or college students — you can simply let it be known, as we do, that Wednesday (or whatever day of your choice) is volunteer day at the farm and take whomever shows up. The young mom from around the corner with a two-year-old to entertain may not be able to accomplish a great deal, but she'll appreciate your openness and may become a customer. Surprisingly energetic and capable people will just drop in, and whatever a visitor's level of skill, having several working at the same time allows you to accomplish a lot. You must be prepared to do some teaching. People don't need a lot, but most of them want to learn something or share something. If you aren't teaching, you may need to listen to others tell their stories. It hardly matters what the subject — you can hold forth about plant taxonomy, seed collecting, the weather, pest problems and solutions or simply commiserate about the lousy state of healthcare in the US. Give a little tour of the farm, but do get down to work pretty directly. Most people will stay less than two hours, but some will hang in there for the whole day. Send regulars home with lots of thanks and a handful of salad greens, some scallions, apples or a potted plant — whatever is surplus at the season.

Successful farms develop their own word-of-mouth networks, which can be quite extensive. Your farm website (if you have one, and you should) can be a big help, but your reputation in whatever circles you operate can draw helpers to your door or your mailbox. You may find, as many small farmers do, that you can choose whom to take. If you like someone's work, you can encourage them to stay or return the following year. Make it worth their while: add a stipend or give them some responsibility to manage a market stand or run an animal operation, just as you might offer a young person in your own family. This is why people approach organic and ecological farms — to learn the art.

If you take farm apprentices, make a point of formally structuring education into their weekly schedule. Organize field trips to enrich their experience (and perhaps your own), and be sure to teach what you know in something like seminar settings, not just over-the-row-of-beans advice. Farming is a great deal about technique, but it also involves planning and strategizing and it deals with many factors besides those on the site. There are markets, neighbors, climate and weather, suppliers and the farming community. There are also your own story, the farm's history and its unseen qualities as subjects for learning.

Whatever the basis of your labor recruitment and exchange, be sure to set out clear expectations about schedules, hours, compensation, privileges, responsibilities, visiting friends, time off, flexibility, training, technique, performance and communication.

The first and often unspoken rule of permaculture is, "Get help!" Through the exchange of labor that it will elicit, garden farming offers the promise of helping to heal our broken communities, even as it helps us to cope with the predicament of an energy-constrained future. Contrary to conventional economic thinking and many modern attitudes, people as workers and people as eaters are not the problem. They're part of the solution. Creating a new way of life, or resurrecting the best parts of an older culture and offering to share that treasure with others — your family, your neighbors, young people in your community — could just be the adventure you've been looking for. If you set out on that yellow brick road though, be sure to pick up some allies along the way — you'll need them, and they'll need you.

# CHAPTER 9

# Running on Sunshine

The sun and the turning of the earth generate the basic elements of climate on which plants and animals depend and to which we as both farmers and Earth dwellers must adapt. Humans undoubtedly influence the climate in ways we are still learning about, but there's little any one of us can do to change the big picture. We can, however, learn intimately the microclimates of our landscape, and to a considerable degree we can manipulate them to our advantage. Indeed, doing so is one of the basic strategies in permaculture design. Having many microclimates within a small area provides the basis for diversity, as each distinct environment will support different plants and animals. It should come as no surprise, therefore, to learn that regions with a wide range of microclimates — particularly subtropical and subtemperate highlands such as those in central Mexico, Peru, Ethiopia, the Caucasus, southwest China and the Himalayas — are rich in biodiversity. The Russian biologist Nikolai Vavilov, studying crop origins in the 1920s, concluded that these special regions provided the conditions for domestication of most of our economic plants.[1]

*Climate* on the broad scale is an interaction of latitude, altitude, proximity to coasts and local topography — factors that together shape winds, modulate solar gain and influence air layering. *Microclimate* has been defined as the climate near the ground. Microclimate is further influenced by the aspect of the land (which direction the ground faces), its degree of slope (and thus solar intensity and air drainage), the presence and shape of vegetation, local water bodies and the nature of the soil surface or pavement, as well as by people-made structures. We can make environments warmer, cooler, wetter or drier within a range determined by larger geographic factors. Our aim as designers is to remove or mitigate the limits to growth and health of our crops and ourselves.

In most of the United States and Canada, the primary factor that limits plant and animal growth and health is cold weather, specifically freezing temperatures. In some areas and in some seasons, heat can be limiting. And at some time in almost all regions, moisture is a limiting factor — either too little or too much.

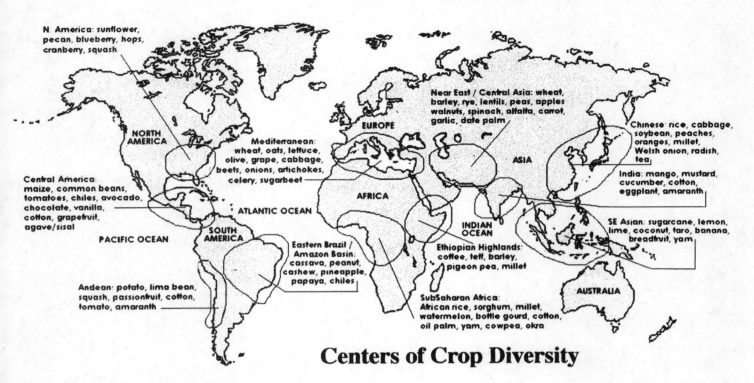

N. America: sunflower, pecan, blueberry, hops, cranberry, squash

Central America: maize, common beans, tomatoes, chiles, avocado, chocolate, vanilla, cotton, grapefruit, agave/sisal

Mediterranean: wheat, oats, lettuce, olive, grape, cabbage, beets, onions, artichokes, celery, sugarbeet

Near East / Central Asia: wheat, barley, rye, lentils, peas, apples, walnuts, spinach, alfalfa, carrot, garlic, date palm

Chinese: rice, cabbage, soybean, peaches, oranges, millet, Welsh onion, radish, tea

India: mango, mustard, cucumber, cotton, eggplant, amaranth

SE Asian: sugarcane, lemon, lime, coconut, taro, banana, breadfruit, yam

Andean: potato, lima bean, squash, passionfruit, cotton, tomato, amaranth

Eastern Brazil / Amazon Basin: cassava, peanut, cashew, pineapple, papaya, chiles

Ethiopian Highlands: coffee, teff, barley, pigeon pea, millet

SubSaharan Africa: African rice, sorghum, millet, watermelon, bottle gourd, cotton, oil palm, yam, cowpea, okra

NORTH AMERICA · ATLANTIC OCEAN · PACIFIC OCEAN · SOUTH AMERICA · EUROPE · ASIA · AFRICA · INDIAN OCEAN · AUSTRALIA

## Centers of Crop Diversity

Centers of Crop Origins and Diversity Identified by Vavilov

[Credit: *Permaculture Activist #43*, Peter Bane]

Air layers on a slope stratify by temperature.

[Credit: Peter Bane]

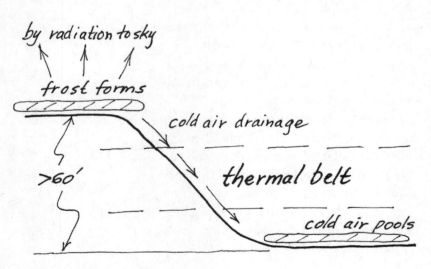

by radiation to sky

frost forms

cold air drainage

>60'

thermal belt

cold air pools

### Read the Microclimates

As with other aspects of our natural and cultural surroundings, observation is the key to understanding microclimates. Some seasons of the year and times of day provide better opportunities for detecting small differences in temperature and humidity. March is the premier month for learning about microclimates in temperate regions of North America. As the sun moves north of the Equator and days lengthen, stronger solar energies warm the ground and the air above it rapidly, but thermal inertia means that soils and nights are still cold. The alternation between night and day generates a powerful stratification of air layers, just as it begins the pumping of sap up and down the cambium of trees. This is maple sugaring season in the north, and in the south it's the month when the grass turns green and the first flowers emerge.

Walk about in the late afternoon of a clear day in March (or November). The sunshine will have created warm air, but as the sun drops toward the horizon, that heat will rapidly radiate into the clear sky, just as it does at night. Wherever there is some up and down to the landscape, the radiating effect will create pools of cold air, and that cold air, being heavier than warm air, will flow like a syrup down to the lowest parts of the landscape.

Notice where the cold air accumulates and which areas remain warmer. Your body can detect temperature differences of as little as 2°F. And you can step from a cool air layer into a warm one in a single pace if you are walking up or down a slope. You will be able to feel these microclimates. These same effects operate at other seasons, but your ability to perceive them is greatest in the shoulder seasons.

A late winter or early spring snowfall can also provide dramatic evidence of ground temperatures: the snow will melt rapidly away once the sun emerges, leaving behind an infrared snapshot of the terrain.

Similarly, a cool night in late August or September can reveal frost pockets. If the day has been humid and temperatures fall after dark, patchy fog will form in low areas that later in the autumn will be subject to early frost; these areas will also stay colder into the springtime.

Plants too can indicate warm or cold microclimates by where they grow naturally. In urban areas it's especially useful to observe annuals that are acutely sensitive to frost. Look for members of the nightshade family: wild tobacco, ground cherries and escaped domestics like tomatoes. All these plants are true tropicals. They only grow in temperate zones as annuals. Cold soils prevent them germinating, and frost kills them off reliably. So their emergence in spring will indicate warm spots in your landscape.

## Analyze the Visible Landscape

Observation is invaluable, but you don't have to see evidence of microclimates to find them. Some things are just logical. Land that faces toward the sun will be warmer and drier than ground that slopes toward the pole. Wind directions are predictable in broad terms over the course of the year. Local changes in elevation generate microclimates in ways that you can anticipate. All bodies of water have a thermal impact in both hot and cold seasons. Buildings, forests and hedges deflect winds, accelerate them or create turbulence based on their shape, orientation, length and height. Big, broadleaved trees transpire large amounts of moisture during warm weather and thus create cooler conditions around them.

To start making sense of the visible elements, take an inventory of the environmental influences on your property and note them on a simple sketch map. Particularly, take note of the following:

- aspect and slope in relation to sun
- vegetation
- buildings
- prevailing and storm winds
- elevational differences (for air and frost drainage)
- bodies of water
- pavement and soil surfaces

March snow reveals differences in surface temperatures due to aspect, shading and plant cover. Bare soil is colder. This view looking south.

[Credit: Peter Bane]

If you have access to a contour map, you'll be able to see the aspects of the land, but even without contours, you can indicate the direction of drainage with arrows to give yourself an idea of aspects. All environmental influences, or *sectors* as we call them in permaculture design, are directional. They come onto the property from elsewhere: the sun, the wind, storms, noise, fire, pollution, views, migrating animals, pollen and seed drift. For the purposes of microclimate analysis and design, we're primarily concerned with sun and wind. We also need to know about structures, water bodies and trees on the land itself. And it will help you to indicate anything nearby — such as tall trees (especially evergreens), hills or neighboring buildings — that may block sunlight, particularly in winter, or affect local winds. These objects will cast shadows that help to create cooler microclimates; some may deflect airflows. Once you've assembled this information on a sketch, you're ready to analyze it to locate microclimates. And having puzzled out the locations, you can then refocus your observations to confirm or question your analysis.

Let's review the basic science of microclimates so that we can understand and influence the environmental conditions around our homes and farms to save energy and create comfort.

## Finding the Sun

The sun drives all living systems on Earth, and knowing where it will be in the sky on any given day has been a subject of keen interest to humans for the whole of recorded history and probably much longer. In most of the US and Canada, the sun rises in the east, makes an arc through the southern sky and sets in the west. The Earth's axis is tilted to the plane of the celestial ecliptic, which is an imaginary plane roughly defined by the movement of the Earth and most of the other planets around the sun. The angle of the tilt, which varies slightly over periods of tens of thousands of years, is presently 23.5°. That's an important number, which you can remember because it's the same number as the latitude of the Tropic of Cancer (23.5° North) and the Tropic of Capricorn (23.5° South). Those invisible

circles around the globe mark the northern and southern limits of the sun's apparent movement through the seasons.[2]

Everywhere on Earth, the sun appears to oscillate 47° between its high and low extremes at the summer and winter solstices — roughly June 21st and December 21st. If you live at the Equator, that looks like the sun at noon being half the year in the northern sky and half the year in the southern, but always pretty high overhead at noon in whichever direction. This is true for at least some part of the year at all points in the tropics, but north of the Tropic of Cancer and south of the Arctic Circle (about 66°N latitude) — which includes most of the US and Canada — the sun is always in the southern sky at midday. It's just higher in the sky between March 21st and September 21st (the Equinoxes) and lower during the fall and winter months. The farther you get from the Equator — the higher your latitude — the more impact on climate that difference in the sun's elevation makes.

Knowing the upper and lower limits of the sun's daily traverse of the sky for our home place is important to the design of buildings and plantings, so we need to know how to calculate or estimate those limits and the resulting arcs that the sun makes. The whole point is to know where the energy is coming from so that we can capture it (or shelter from it, as appropriate). To find the sun's upper limit at summer solstice noon, determine your latitude (check an atlas). Here's the formula for the sun's angular elevation at midday at different times of the year:

90° – (your latitude) = Equinox elevation
Equinox elevation + 23.5° =
    Summer Solstice elevation
Summer Solstice elevation – 47° =
    Winter Solstice elevation

Simple calculators that tell you the sun's midday angle at your latitude on any given day of the year can be found online.[3]

The other important element of the sun's

Sun angles are determined by the Earth's tilt, latitude and season of the year.

[Credit: Peter Bane]

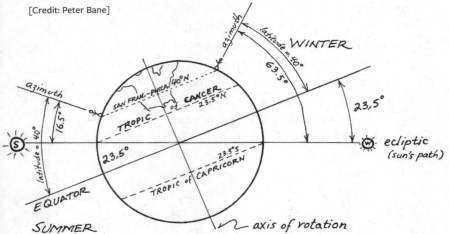

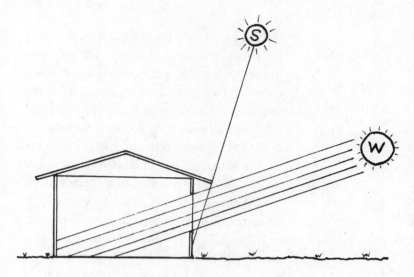

movement is where it rises and sets at different seasons: always in the east and west, of course, but sometimes northeast/northwest and sometimes southeast/southwest. At the equinoxes, the sun rises due east and sets due west, but in summer throughout temperate North America, it rises in the northeast and sets in the northwest, making a long arc lasting many hours. In winter, the converse is the case: the sun rises in the southeast, makes a short traverse of the sky, not rising very high, and sets quickly in the southwest. At the latitude of Nashville or Oklahoma City, that summer arc is about ⅔ of the sky at solstice. As you move farther north, it increases. In Vancouver, Duluth or Portland, Maine, the arc at midsummer is about ¾ of a circle. In Churchill, Manitoba, or Juneau, Alaska, it comes closer to the full 360° that you reach above the Arctic Circle.

### Bringing the Sun Indoors

The sun's angle has many applications. The most central of these is dwelling comfort. Studies in the US and Europe have shown that people in industrialized countries spend more than 90% of their lives indoors.[4] Home comfort is therefore critical to well-being, and energy gained from the sun in cool seasons means fuel saved; that translates into time and money, whatever your heat source. All buildings for winter occupancy should be designed or modified to capture heat from the sun. The very few exceptions are buildings in climates where winter skies are nearly always overcast — some parts of the Pacific Northwest and Atlantic Canada. In places like northern Michigan, Ontario, northern New York, the St. Lawrence Valley and the Cleveland-Erie region, which have cloudy skies much of the winter, the shoulder seasons from March to May and September to November are much sunnier and still cool, so solar gain is well worth collecting.

The best way to do this is to orient the long axis of a building east-west, so that the south

side is one of the long sides. In this way, more floor and wall surface can be warmed directly by sunlight. Since the common ceiling height of eight feet effectively limits the penetration of even low winter sun to about 14–18 feet into the dwelling (depending on latitude), almost all solar heating will take place in rooms to the south.

If these rooms can be fully half the house and enough mass and insulation are present, then sufficient heat can be gained during

The sun's rays enter a building more deeply in winter, yet only reach 14–18 feet through typical window heights.

[Credit: Peter Bane]

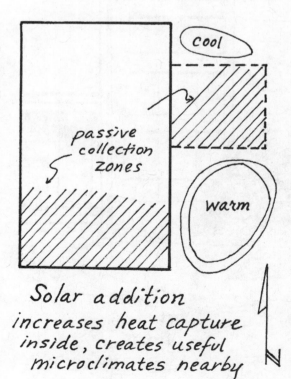

Solar addition increases heat capture inside, creates useful microclimates nearby

Adding a room can change the solar potential of a building.

[Credit: Peter Bane]

the day to hold the building at a comfortable temperature through the night. For existing houses with a different orientation, or which are square, it's often possible to add a room that can turn a rectangle into an "L" shape, creating a sunny courtyard and effectively widening the solar catchment of the structure.

Without moving or adding to your house, it's often possible to enhance solar gain along the existing south side, as we did with two small ranch-style dwellings on our property in Indiana. Twenty-four feet wide at either end and about 40 feet long, they faced the street, as most houses do. Each had dead-wrong solar orientation. Still, 24 feet is a pretty big wall. We added more and larger windows and built thermal mass into both houses in the rooms on the south sides. On a sunny day in winter we can add 4–5°F to the interior temperature just from sunshine. That's about ¼ of what we could get if the

houses were turned toward the sun, but it reduces our need for fuel.

### A Comfortable and Functional Home

The home is a midpoint between the public space of the street and the private space of the household. It needs to be as close to the street as can be tolerated, but somewhat central to the land around it, so that all parts of the farm can be reached easily. Having the house toward the north of the property presents an advantage insomuch as most of the land then would lie on the sunny side of the buildings, creating south-facing outdoor space.

Elevation and drainage are additional factors of importance. The house should be located at least a few feet above the lowest part of the landscape, and preferably at the midpoint of any slope of substantial length, to take advantage of the thermal belt created there. Of course buildings for occupancy should be placed on south-facing ground, though a little southeastly or southwesterly can be serviceable. Shelter from the wind may be created with windbreaks, but if the property offers a well-placed hill into which you can shelter from winter winds, by all means consider using it. Avoid ridges in mountainous country, unless they are closely flanked by higher ridges that shelter you from winds and wildfire. And if you have a spectacular view, try to preserve it for some place in the landscape to which you must walk — don't build the house there, but consider creating a meditation hut, a gazebo or a small bench in a grove of trees as a destination. Put the house where it will harvest and conserve energy, repair damaged ground, access the property and the road easily — and where it can be connected to other functional structures and cultivated ground. After the selection of the land, the decision about where to place buildings is the most important you will make and certainly the one with the greatest potential impact on your economy. Take your time and choose wisely.

South-Facing Outdoors, Pattern #36 (See Chapter 6)
[Credit: Jami Scholl]

On larger properties, the interplay of landform, access and available sunlight may offer more than one possible building site. If so, make your best choice, but preserve the other site or sites for future use — they're all of great value. Don't make any permanent plantings, earthworks, pavement or developments that would prevent a house or other building occupying those choice locations later.

### The Greenhouse Principle

The principles of microclimate design are similar whether you are building a house, setting up a chicken coop or locating an orchard — but the particulars vary. The model for our living structures is the Earth itself, which is a round greenhouse. The atmosphere, just like windows in a home or plastic or glass coverings over a greenhouse, admits light, which striking the surface of the earth, the floor, the walls or the beds in the greenhouse, changes frequency and is absorbed or reradiated as infrared light or heat. That longer-wavelength light can't travel through the glass, plastic or the atmosphere as easily as the visible light coming in, so it bounces around inside and warms the air. This warm air absorbs more heat radiating from the now warming surfaces within the structure, or near the surface of the earth — and voilá, we have (temporarily) captured the sun's rays.

Three elements of a structure work together to capture heat from the sun: glazing to admit the light, thermal mass with surfaces exposed to the light and a tight, insulated envelope to hold the resulting heat within the structure. The same three elements work in a greenhouse, but even two of them can make a difference out-of-doors when put together. For example, the eaves of your house on the south or west side act as hats over a narrow band along the building. As the sun strikes the south side of the house and the soil or pavement under the eave in the midday to late afternoon, it generates heat that warms the air in this narrow band. The warm air rises and the heat radiates out toward the sky, but the hat-like eave keeps some of that energy in place beneath it to warm the building and the garden beds below through the night. This can extend the growing season under that eave by weeks or even months. If the wall is dark or made of brick or stone, even better, as these surfaces will absorb and reradiate more heat. Even though there is no window holding the air within this narrow sheltered band, the wall itself acts as a windbreak to slow the approaching wind, allowing warmed air to linger in this special microclimate. You can do the same thing on the southwest or west side of a building. Create a small pond in this area, and you can reflect more energy onto the warm wall and at the same time add a thermal battery, because water is the very best storage medium for heat. Stone, brick and dark clay masonry are also very good thermal batteries, provided the sun strikes them directly.

We can apply this understanding of simple physics throughout the landscape.

Dark brick facing south creates a narrow but pronounced microclimate for this fig in a Danish courtyard.

[Credit: Peter Bane]

A tree creates a warm microclimate beneath its canopy on the sunny side.

[Credit: Jami Scholl]

The ground beneath the south side of a tree canopy will have a sheltered and warmer microclimate beneficial to tender plants.

An arc of trees facing the south will reflect light and heat toward its focal point. We call this a *sun-trap*, and if it's a true parabola, there will actually be a *hot spot* at the center.

The main thing to remember is that the sun is the source of energy, so you need to know where it will be in the sky and arrange your buildings and landscape to receive its rays at the right time and season. That radiant solar energy can be stored for short cycles in massive bodies like the earth, ponds, water tanks, stone walls, pavements and stone mulches from which it is released slowly. And the effects of solar warming can be enhanced by using trees and shrubs to shelter buildings, plants and animals from the wind. To achieve natural cooling, we use the same elements, but respond to them in the opposite way: we shelter from the sun, and we focus the wind for cooling ventilation. Since the sun and the wind come from substantially different directions in winter and summer, it's possible to adapt effectively even to a dramatically bipolar climate with strong swings from cold to hot.

### Heat-Absorbing Surfaces

By studying the sun's movement during the course of the year, we can determine how best to capture or deflect it according to our needs. Of course the earth itself warms during summer months from the radiant energy of the sun, but nowhere does that warmth reach much farther below the surface than about 20 inches, which is why permafrost forms in Arctic regions. There the earth itself freezes very deeply from the effects of long winters and persistent cold; during the brief Arctic summers, only the surface foot or two of soil thaws out.

Surfaces intercept light and convert it to heat more readily if they lie perpendicular to the direction of the rays. In summer, when the sun is high, more of its energy is converted to heat as it strikes the surface of the earth. In winter, when the sun is low, the light comes obliquely through the atmosphere. More of it is scattered (which is why October skies are so blue), and the light that does reach the surface hits at a low angle so more of it bounces off unchanged. Conversely, and perhaps counter-

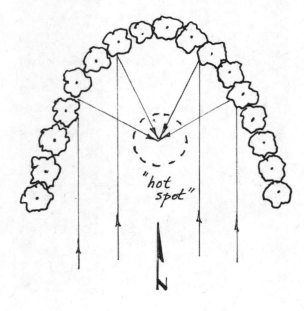

A parabolic planting creates a hot spot to favor a shelter or a tender crop.

[Credit: Peter Bane ]

"hot spot"

N

intuitively, a steep, south-facing slope can actually intercept solar energy faster for the few midday hours of a sunny winter day than it can in summer, or just as much.

Sloped ground responds to light in much the same way. Each 5° that the slope tilts toward the sun effectively moves that ground one climate zone southward (about 300 miles). And conversely, slopes that tilt toward the pole are cooler.

Of course, these effects are highly localized—they may hold over an entire mountainside or only exist in narrow bands of a few feet or in small patches, but they can be critically important nonetheless. Since frost and cold temperatures are most often the limiting factor to the productivity and health of plants, animals and ourselves, most of our interest lies in warm microclimates and learning to avoid cold pockets. But there are exceptions.

A productive use for cool slopes, besides providing a good location for root cellars, is to chill the soil around the roots of fruit trees, thus delaying their time of blooming in spring. If you live in regions where winters are mild and the onset of spring is highly variable, it can be difficult to get reliable tree fruit crops. (More northerly regions will be similarly affected over coming decades as climate zones continue to shift poleward.) After receiving a certain minimum number of hours between 28° and 40°F (called chilling hours, different for each variety), most tree fruits will respond to warming temperatures by ripening buds and opening flowers. A warm spell in February or early March can set blooming in motion, but often this is followed by a sharp freeze, wet snow or heavy frost in late March or early April that can damage the blossoms and reduce or eliminate the fruit crop. Bush fruits and nuts are less susceptible but still affected. The plants get their signals primarily from soil temperatures. We can plant our fruit trees on a north-facing slope, or on the north side of a building—where shaded soil will stay cooler longer—to give

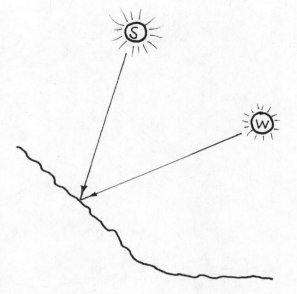

Steep south-facing slopes can gain as much heat in winter as in summer.

[Credit: Peter Bane]

South- and north-facing slopes on opposite sides of a New Mexico valley show dramatically different thermal conditions at the same moment in February.

[Credit: Peter Bane]

North of the house, two plums flanking a columnar pear with their roots in shaded soil and canopies in full sun will bloom later but still ripen well.
[Credit: Creighton Hofeditz]

them the best chance of holding off blooming until the weather is settled. Placing the trees where summer sunshine will strike their limbs but winter and spring sunlight won't heat their feet is ideal. This means planting near the top of a north-facing slope (which also allows cold air to drain away, helping to prevent frost) or a little away from the north wall of a building. The trees will grow into the ripening light, but their roots will remain safely anchored in cool territory (effectively much farther north).

### Aspects Around the Compass

We've seen how south-facing ground is warmer than north-facing, but what about the in-between slopes: east, west and all the variations? North and northeast slopes are the coolest and are suitable for the growth of trees and shrubs. North-facing slopes are often a source of local groundwater, as surplus moisture can break the surface to form springs. When the sun rises and strikes

easterly slopes, they are usually covered with dew and the air is cool from nighttime radiation. It takes a lot of energy to evaporate the dew, which keeps the temperatures lower for a while. East-facing ground continues to warm by conduction and convection from the movement of warm air as the day goes on, but not from direct radiation, which is by far the most powerful process. Southwest-facing is the driest, but west- is much drier than east-facing, and so northwestly is drier than northeasterly.

Just as early morning light through east windows is welcome and gives a helpful boost to start the day, afternoon light through west windows is almost never helpful. In winter it will make little heat difference because the sun is so low, and in summer it will heat up the house fiercely. So it is with sun on sloping land. While we can avoid west solar exposure for our buildings by design and by using adaptive structures such as trellises, awnings, shrubbery and window coverings, we can't entirely avoid sun on our southwest- and west-facing land. We have to use those hot, dry slopes to advantage. It turns out that northwesterly slopes, being drier and a little cooler, make good pastures and grain fields, as broadleaved weeds are at the disadvantage of grasses which have fine root networks to gather soil moisture effectively. Westerly slopes are favorable for some orchard crops because the drier conditions reduce fungal problems and the heat helps to ripen fruit. Another reason to put fruits to the west rather than the east is to reduce the danger of early spring frosts. Orchards on east-facing slopes are more vulnerable than those to the west where, before the sun strikes them directly, the blossoms and young fruits can thaw slowly with the rising air temperatures of the day.

Some fruits like grapes and figs, which originated in the dry summer conditions of the Mediterranean, prefer a south-to-west exposure, though you can certainly grow them elsewhere if you must. Subtropicals like tomatoes and squash also favor hotter conditions, and even like it a bit dry. You can put your laundry lines (Pattern #41) and your drying yard (Pattern #40) to the south and west, as well as your woodshed (Pattern #21), though not if that is also your fire sector. You can also design systems to balance out moisture conditions. Create a pond, install drip irrigation or use hardy tree cover to intercept the hot sun and keep soil temperatures cooler. Knowledge is power.

### Microclimate at the Edge

Localized effects is the name of the game in microclimate design. European gardeners have long recognized the value of the tiny bands of microclimate right next to south- and west-facing stone or brick walls. Trees can be trained to grow flat against such walls. This practice is called *espalier* (which means shoulder in French and connotes outstretched arms). It's not unlike growing grapes on wires

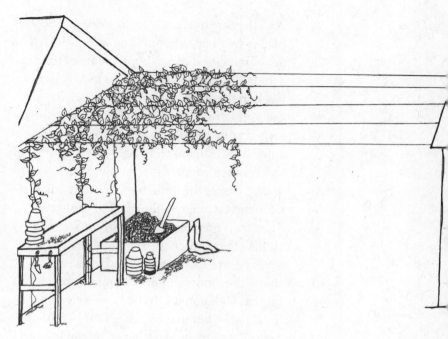

only against the edge of a building. In fact, many trees can be grown on wires too, and are then called *cordon*. This is especially valuable where space is limited.

Pergolas and trellises (Pattern #43), like porches and awnings (Pattern #42), are extremely useful for shaping outdoor spaces to create microclimates of shade or warmth. They can form windbreaks if vertical, or shade and overhead insulation if horizontal. Whole outdoor rooms (Pattern #37) can be "grown."

Trellises for Shade, Pattern #43 (See Chapter 6)
[Credit: Abi Mustapha]

Deciduous vines provide shade in summer; they can be pruned away for winter.

[Credit: Peter Bane]

And they are quick. Vines on trellis can be placed very near to a building with little danger of structural damage and plenty of time to avoid it. They can cover an area in a matter of weeks rather than decades. Moreover, deciduous vines drop their leaves and can be pruned away during the cooler months to allow sunshine back into the building.

Another important edge effect we can learn to use is the influence of water bodies. Climate is strongly influenced by the ocean. Palm trees grow in the south of Ireland, the climate there moderated by Gulf Stream waters. Cape Cod, surrounded by ocean, shows the same minimum annual temperatures as Oklahoma City, 600 miles closer to the Equator but in the middle of the continent. The same moderating effect can be seen along the shores of large lakes. The south shores of Lake Erie and Lake Ontario in Ohio, Pennsylvania and New York and the Niagara region lying between them are famous for vineyards and orchards. So too, are the eastern shores of Lake Michigan where cherries, blueberries and many other fruits are raised commercially. Any body of water will exert a microclimate influence on the air around it to about ¼ of its width from the edge, but the strongest effects are felt in the closest half of that band. Wind direction should be considered as well. In the case of Lake Michigan — which averages about 80 miles wide, there is some influence on the Michigan side (downwind) to about 30 miles inland, but on the Wisconsin side, the effect is limited to about 15 miles. A pond on your land will exhibit similar influences. In addition, a pond can be a source of extra reflected light. We'll look further at the many functions of water bodies and aquaculture in Chapter 10.

In general in the Northern hemisphere, we want to create south-facing outdoor spaces (Pattern #36), both for living and for growing, as these will have the best light and warmth for human comfort and plant health during much of the year. We can modify these sunny spaces for hot summer conditions by using awnings, trellises and pergolas, or in sunnier and more southerly latitudes, create additional outdoor rooms to the north and east of buildings, such as porches for sitting in the shade. Land in the shade of buildings tends to be of more limited value, so keep these areas to the edge of your property, avoid them as much as you can and make use of them for special cold functions such as storage and to delay the blooming of fruits.

## Sunshine Moves the Wind

Closely related to sun and shade, climate and microclimate are much affected by winds. Both global and local wind patterns are shaped by the differential heating of the Earth's surface by the sun and by the regular rotation of the planet. The dominant weather systems of the mid-latitudes (between 30° and 60° North, or roughly from Miami to Anchorage) are driven by a series of 16–18 large cyclonic air masses that turn counterclockwise — like very large and slow hurricanes — and move around the Northern hemisphere about every 90 days. These get their spin from the continuous falling of cold polar easterlies, and at their southern extremes fade out on the fringes of the tropical tradewinds. The effect of these large spiraling air masses is to move weather fronts from west to east across the continent on a 7–10 day cycle that is more pronounced during the cooler months. In summer, warm ocean air from the Gulf of California and the Gulf of Mexico pushes north against these continental systems to disrupt the regular movement of storm fronts. As this phenomenon develops with warming weather in the spring, tornadoes, hail and storm phenomena are generated from the clash and shear of warm and cold air masses sliding against and over each other. These fronts typically meet along a broad mid-continental band stretching from the Texas panhandle up to Lake Erie, with Kansas at its epicenter. Absent local topographic variations, the prevailing winds over most of the continent are westerly.

Of course the picture is much more complex than this. Santa Ana winds blow into southern California from the Mexican desert in winter. The shape of Lake Erie creates a prevailing northeasterly-southwesterly wind dynamic. Coastal influences bring hurricanes and nor'easters to the Atlantic seaboard, and the Rocky Mountains show dramatic variation over short distances. But most parts of the continental mid-latitudes see weather move from west to east. Knowing that prevailing winds are more likely to be westerly gives us a point of departure for learning local variations.

Some local air movements are especially significant. Coastlines get daily onshore and nightly offshore breezes as the land heats and cools faster than the sea. Hills and valleys generate daily updrafts and nightly downdrafts for the same reason: convection brings warm air onshore and up-valley with the heat of the day. Cooler air pushes downslope and offshore at night. Wherever mountains rise sharply from a valley floor or plateau, there are usually turbulent winds along the range because of the mixing of warm and cold air layers.

Anywhere cold air falls from a high ridge, it can create a potential frost pocket. Narrow valleys are especially vulnerable to frost for this reason, as a bend in the valley, a band of trees or even a cluster of farm buildings can create a *frost dam*. Above the valley bottom and below the ridge wherever there is 60 feet or more of elevation difference, a band of warmer air, called a *thermal belt*, will form. This is a favorable environment for tree crops and a good location for dwellings as it will be warmer in winter and also has the advantage of good drainage. Where we live in southern Indiana, the landscape is formed by a series of broad ridges and steeply dissected but shallow valleys carved out of limestone. We live on a ridge, and though our land is higher than

most of the ground around, it doesn't rise up in a visible way. We get surprising easterly breezes in the early evening almost every day from spring through fall as warm air, displaced by falling cold air, comes up the slope. As a result, the microclimate of our property and the few houses near us is 3–4°F warmer than land 300 yards down the road. This makes a great deal of difference in the length of our growing season and the crops that will flourish for us. We also enjoy cooling breezes on warm summer evenings, even when the day has been calm.

You may not be able to move your house into a more comfortable spot, but if you are looking for a property to farm, you can be aware of and look for these conditions. And you can use this model of microclimate to modify the conditions of your land to make them more comfortable. If your house is forming a frost dam, for example, you may be able to plant an arc of trees upslope to deflect the cold air coming down toward you.

Or if a band of trees forms a frost dam below your buildings that leaves them in a cold spot, you can open a drain by selective removal. Sometimes it's adequate simply to

An upslope tree planting can divert cold air drainage.

[Credit: Peter Bane]

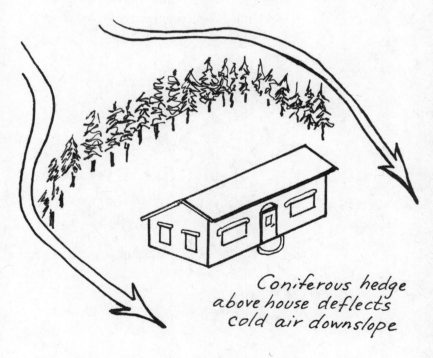

*Coniferous hedge above house deflects cold air downslope*

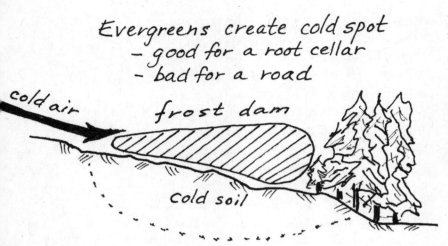

Evergreens create cold spot
– good for a root cellar
– bad for a road

frost dam

cold air

cold soil

A frost pocket
is created by
damming cold air.
[Credit: Peter Bane]

limb up the trees to allow the cold air to flow along the ground beneath them. And if you want to create a cold area — suppose your property is all south-facing, warm and cozy but without a good spot for a root cellar — plant a copse of evergreens to make a frost pocket behind them.

After sunshine, wind energies are one of the most important sectors affecting any property. Cooling breezes are one of the few low-cost ways to keep comfortable in hot weather, while cold winter winds stress livestock, increase heating bills and can harm crops and tender plants. The best gauge of wind influence is your own collected observa-

tions and those of long-term residents around you, but you can usually obtain a diagram of prevailing wind directions and intensities — called a *wind rose* — from the local airport. Remember that your winds may be significantly different from those at the airport 15 miles away, but a wind rose can give you a general idea. Once you know what to expect from prevailing winds, you can design, build and plant to deflect or funnel the wind according to your needs.

### Windbreaks

Farmers in territories as distinct as North Dakota and New Zealand have embraced windbreaks to enhance the growth of their crops and shelter their livestock, and many of us in urban and suburban landscapes could benefit from well-placed hedges and shrubberies.

Windbreaks follow some simple principles of aerodynamics. The best protection comes when the windbreak, formed by lines of trees and shrubs, lies perpendicular to the prevailing wind direction. Winds that strike against a barrier at an oblique angle, rather than straight on, tend to speed up along the edge, causing more damage. Where winds are variable or the property line doesn't permit ideal placement, the windbreak should be crenellated or wave-shaped.

Windbreaks work by creating higher air pressure on the windward side: as the wind approaches, it encounters resistance, and drawn by the lower air pressure above the tree line, lifts up and over the barrier. Surprisingly, this creates a windbreak effect even on the windward side, though less so than on the lee. You can determine the magnitude of that effect by estimating the height of the tree line. A line of trees 20 feet high will slow the wind starting 100 feet on the windward side and continuing to 600 feet on the leeward side, though most of the effect is in the first 200 feet downwind of the trees. If you are planting successive lines of windbreaks, therefore,

A crenellated
windbreak buffers
variable winds.
[Credit: Peter Bane]

Crenellated
windbreak
intercepts
Variable winds

## SUGGESTED WINDBREAK AND HEDGEROW SPECIES BY SIZE

### Large Trees (to 50' or more)

American Beech (*Fagus grandifolia*) (doesn't coppice well)
Honeylocust (*Gleditsia triacanthos*)
Pin Oak (*Quercus palustris*)
Black Walnut (*Juglans nigra*)
Black Locust (*Robinia pseudoacacia*)
Red Oak (*Quercus rubra*)
American Basswood (*Tilia americana*)
Hackberry (*Celtis occidentalis*)
Sweet Gum (*Liquidambar styraciflua*)
Slippery Elm (*Ulmus rubra*)
Chinese Elm (*Ulmus parvifolia*)
Eastern Hemlock (*Tsuga canadensis*)
Blue Spruce (*Picea pungens*)

Chinquapin (*Castanea pumila*)
Lilac (*Syringa vulgaris*)
Hazel (*Corylus americana, C. cornuta*, hybrids)
Autumn Olive (*Elaeagnus umbellata*)*
Russian Olive (*Elaeagnus angustifolia*)*
Willow (*Salix* spp)
Siberian Pea Shrub (*Caragana arborescens*)*
Chinquapin and Scrub Oaks (*Quercus prinoides, Q. ilicifolia*)
Osage Orange (*Maclura pomifera*)
Viburnums (*Viburnum prunifolium, V. trilobum, V. lentago*)
Dogwoods (*Cornus stolonifera, C. florida*)
Sumac (*Rhus typhina, R. glabra*)
European Yew (*Taxus baccata*)

### Medium Trees, Large Shrubs (8–35')

Mulberry (*Morus rubra, M. nigra, M. alba*, hybrids)
Persimmon (*Diospyros virginiana*)
Apple and Crab Apple (*Malus* spp)
Alder (*Alnus rugosa, A. rubra, A. incana, A. crispa, A. cordata*)*
Redbud (*Cercis canadensis*)* in full sun
Hornbeam (*Carpinus caroliniana*)
Witchhazel (*Hamamelis virginiana*)
Mountain Ash/Rowan (*Sorbus americana, S. aucuparia*)
Plums (*Prunus virginiana, P. pumila, P. americana, P. spinosa*)
Apricot (*Prunus armeniaca, P. mandshurica*)
Pawpaw (*Asimina triloba*)
Hawthorn (*Crataegus* spp)
Arborvitae (*Thuja occidentalis*) evergreen

### Smaller Shrubs (3–8')

Bristly Locust (*Robinia hispida*)*
Rose of Sharon (*Hibiscus syriacus*)
Quince (*Chaenomeles* spp)
Spicebush (*Lindera benzoin*)
Rose (*Rosa* spp)
Chokeberry (*Aronia melanocarpa, A. arbutifolia, A. floribunda*)
Indigobush (*Amorpha fruticosa*)*
Buffaloberry (*Shepherdia canadensis, S. argentea*)*
Elderberry (*Sambucus nigra*)
Bayberry (*Myrica gale, M. pensylvanica*)*
Barberries (*Berberis canadensis, B. vulgaris, B. thunbergii*)
New Jersey Tea (*Ceanothus* spp, including many western spp)*
Canadian Yew (*Taxus canadensis*)

* = nitrogen-fixing

place them at intervals of 15 times the height of the tallest trees to get the most benefit from the least planting. Also, because trees and shrubs lift the wind up and over, there is a potential for turbulence on the leeward side. This is especially so if the tree band is very narrow—a single line of tall trees is the worst shape. Better are several lines of trees and shrubs planted close together to form a gentle curve, in profile not unlike an airplane wing, that can lift the wind and let it drop gently on the other side. Planting three to five rows of trees and shrubs, with the tallest trees in the center and the shortest shrubs on the outside, will give the best effect. It also allows you to choose multiple species.

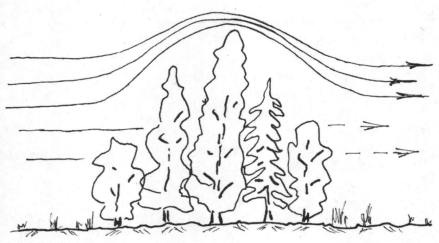

Multi-species, multi-row windbreak

In profile, a well-designed windbreak shows a gradual rise and fall.

[Credit: Peter Bane]

Windbreak and hedge species should be selected for their multiple yields, not only shelter from the wind, but timber for fuel and polewood, livestock and bee fodder, floral products and reserves of emergency food. A hedgerow of seedling apples or crab apples may yield mostly bland fruits, but these can be fine for animal fodder or cidermaking and may well throw up the occasional wild sport that has excellent qualities. These plantings should be inexpensive and easily managed. They should also be diverse. Choose many different species. In that way, if an insect pest or a disease affects one species in the

windbreak, it won't destroy the whole. The aim is not for a solid barrier, but for about 50% reduction in the volume of the wind. Choose deciduous species that have horizontal branches, hold their leaves into the winter (like oaks and beeches) or sprout multiple stems (like hazels) so they will be effective when dormant. Use some evergreens, especially where their shade won't inhibit other species or hold snow where you don't want it. Plant a variety of fruiting or mast species which will feed wildlife or your household in a pinch. And plant densely, planning to take an early harvest of poles, animal fodder or firewood from the thinnings.

## Small Structures for Microclimate

We have seen how land aspect, slope, surfaces, vegetation and bodies of water can affect microclimate, but we can also create microclimates using simple structures to protect one or a few plants. These can be as simple as a tent of sticks or grasses thrown over some tender garden plants. They include row covers, plant cages and jars over seedlings — or can be more complex, from cloches and hot boxes all the way up to greenhouses.

Tires have been used to grow potatoes and strawberries, adding extra warmth in the

Jars and greenhouses trap warmth from the sun and the earth.

[Credit: Peter Bane]

Left: A single straw bale shelters a young tree against wind and later decomposes into mulch.
[Credit: Peter Bane]

Right: Straw bales form a suntrap and windbreak in the garden, give off compost heat and provide a planting surface for lettuce at the same time. Design by Chuck Marsh.
[Credit: Peter Bane]

early spring as well as structural support to get these cool- into warm-weather crops going. Straw bales make a quick and easy windbreak for a single tree.

Woven mats, temporary fencing, haystacks and simple pavilions or sheds have all been used to protect crops and livestock from wind, rain and sun. The emphasis with such emphemeral elements must always be on low-cost, ease of construction and disposal, reuse of industrial materials and where possible, multifunctions: a straw bale windbreak creates habitat for worms and will transform into excellent mulch. It can also double as a planting surface for light-feeding crops such as lettuce.

Whatever strategies you employ, your property design begins with an analysis of the impacts of climate on your site and the possibilities to moderate and enhance the microclimates you find there. This requires familiarity with sun, wind and water—and knowledge of a variety of techniques for harnessing their energies. Permaculture teaches us to use and value the gifts of nature, but also to catch and store energy—these are the terms of a working partnership between people and the land.

# CHAPTER 10

# Water from Another Time

*It don't take much, but you gotta have some*
*The old ways help the new ways come*
*Just leave a little extra for the next in line…*
*They're gonna need a little water from another time.*[1]

The retreat of the glaciers left behind a water-rich landscape across much of North America. The great rivers upon which we depend for our commerce were once nourished by tens of millions of acres of wetlands. These expanded the capacity of watersheds to accommodate both surplus and shortage of water.[2] Yet both farmers and city officials have invested heavily in systems to remove water from the landscape, draining wetlands, accelerating runoff and paving over recharge areas even as they extract ground and surface water to meet their thirst. Both approaches are short-sighted.

Most of the steady flow of the Earth's accessible rivers has been tapped for irrigation, water supply or navigation over the past 60 years.[3] Large dams impound most major watersheds and have displaced millions of people and flooded some of the world's best farmland. Groundwater is expensive and increasingly polluted. What remains to augment this critical resource in times of growing shortage is flood flow or rain runoff, which amounts to over 70% of annual streamflow. Because climate change is making both rainfall and drought more intense, it is ever more important that we store water at the household and neighborhood level everywhere to take the strain off public water supplies and to support food production.[4]

The city dweller or suburban gardener in almost all parts of North America except the Southwest has access to relatively unlimited amounts of low-cost water. However, this should not lead us to complacency. Farming requires a lot of water. A 5,500-square-foot (⅛ of an acre) garden requires over 3,000 gallons of water a week during the growing season. To keep crops alive during a two-month dry spell could cost hundreds of dollars, and you might face usage restrictions imposed by public water authorities. Climate patterns are changing; drought is becoming more common in the once well-watered regions of the Southeast and Midwest, and water rates are

Water—Source and Force of Life, Pattern #5 (See Chapter 6)

[Credit: Jami Scholl]

of water are needed, but even redundant supplies will be insufficient in the worst of times without the ability to conserve and reuse water. These practices must be developed and maintained by habits of attention even during times when water is abundant; in many areas flood and drought can be as little as a week apart.

Water we fail to collect and water we use and release all returns to soil and streams. To drought-proof our landscapes we must develop the soil as a water storage by building up its carbon content. Healthy soil acts as a sponge to absorb and release rainfall and runoff. But water in soil is not available for some special and very important purposes: drinking, washing, stock watering, aquaculture and fire control. For these and a myriad of related purposes, we need fluid water in tanks and ponds.[5] And because of the weight of water and the energy cost of pumping it, we should design for a substantial supply of water that can collect and flow by gravity.

bound to rise along with the price of all other energy-linked commodities.

You should not attempt any commercial or large-scale garden farm enterprise without planning and creating substantial onsite water resources that will allow you to keep your crops and livestock healthy over a wide range of climate conditions. And your design should provide more than one source for this essential element. Two or more good sources

## Sources of Water

Except where artesian conditions bring water to the surface under pressure, or in hilly or mountainous country where water breaks the surface at high elevations, underground water must be pumped, involving a considerable energy cost. Traditional economies spent enormous amounts of scarce energy in the form of human and animal labor to do this; they also devised wind-driven pumps to move water up and out of the ground. Until the 1930s saw the widespread expansion of the electric grid to rural areas, many farmers and ranchers in the US supplied their households and their animals with water from windmill-driven pumps.

No agricultural civilization has arisen or persisted anywhere without access to water flowing by gravity. Even the peoples who invented agriculture on the desert coast of what is now Peru — one of the driest regions of the world — were able to draw water down from

[Credit: C.K. Hartman]

the Andes through subsurface channels.[6] Farmers in the Canary Islands receive no rain, but water their crops by collecting and channelling moisture from dew and fog collected on high volcanic slopes.[7]

Only our access to cheap fossil fuels enables us to pump huge volumes of water today (roughly half of US cities are supplied from wells), but agriculture dependent on deep wells is a doomed proposition.[8] Water tables are falling from too great a demand and inadequate recharge.[9] Groundwater too is being polluted by a wide range of careless industrial practices from sewage and effluent injection wells to new methods of hydraulic fracturing and the use of toxic solvents in search of natural gas.[10]

Because of its dependence on electricity, the supply of which will become irregular and expensive over coming decades, well water should be considered the least resilient of all water sources. If you have a well, consider pumping directly from windmills or using solar- or wind-generated electricity. Even if these methods can't yet provide for all your needs, they might provide a buffer that could make the difference between hardship and disaster. Either alternative will require storing water at the surface, or preferably at some elevation above the use points. You should also consider creating large surface storages for those times when electricity is unavailable, especially if renewable energy or direct mechanical pumping aren't feasible in your area.

Running surface water offers many advantages—it's often available in large volumes and can sometimes be directed by flood flow without pumping. But it usually comes with a raft of restrictions in the form of water rights, public regulation and upstream pollution, over which a single landowner has little control. In any case, most small properties don't have any surface water to speak of. Though many of those could and should be enhanced with surface water storages in the form of small ponds, any such storages would be pri-

marily supplied by our third potential source: rainwater and short-lived runoff.

Because it is so little understood and at the same time widely applicable and relatively cheap and easy to create, I will emphasize rainwater catchment as a primary source of water for the garden farm or suburban household. Larger properties in the arid West are the least likely to be able to depend on rainwater, though they too should make greater efforts to harvest it from roofs and pavements, and to slow and retain rainwater by check dams, gabions and other water-holding structures in ephemeral drainages.

For small holdings in arid regions, catchment of roof and other runoff may be sufficient to provide most of the water needed for garden-scale agriculture. A redundant supply will always be needed, however. In our household we both catch and store roofwater—using only gravity to distribute it, and use a pressurized public water supply for those household tasks for which it is valuable: dishwashing, showers and sometimes lifting water to clean the roof or gutters.

## Moving and Storing Water from the Sky Down

Storing water means catching, concentrating, filtering and storing it securely, then releasing it safely and economically. Water falling from

Two check dams will slow occasional stormwater in this Colorado mountain drainage, causing it to drop sediment and form terraces behind them. They were constructed of stones, brush bundles of Canada thistle (cut for weed control) and old car parts found nearby.

[Credit: Peter Bane]

Roof Catchment, Pattern #25 (See Chapter 6)
[Credit: Jami Scholl]

Students inspect a sod roof at the Boyne River School outside Toronto. Rain is held by the sod and released slowly, reducing storm surge. The runoff is clean enough to drink without treatment.

[Credit: Peter Bane]

the sky is free, clean and full of potential energy. Our aims in storing water are to capture some of that free resource, preserve some of its potential energy and restore and maintain its cleanliness. To have access to adequate water, especially in times of dearth, is the essence of security.

Any impervious surface will shed nearly 100% of the rain that falls on it (some is lost to evaporation). This includes roofs, road and parking lot pavements, patios and sidewalks. In the absence of adequate pavement or roof surface, an artificial collector can be created by laying plastic sheeting on the ground.

Much less water is shed from vegetated ground, about 10% from closed canopy forest and 20% from grassland, but this water too is suitable for collecting in ponds or swales. Cropland and bare soil surfaces shed more of the water that falls on them, but it is usually laden with sediment, making it difficult to use. The primary sources for rainwater collection in our urban and suburban areas are roofs and pavements. Of these, roofwater is cleaner, available at higher elevation and easier to purify.

To use roof runoff we need to consider the surface. Not all roofs are equally clean. The best roof surfaces for water collection are enamelled metal, clay or concrete tile, wooden shakes or sod. These give a very high-quality runoff, which may still need some filtration. Sod or green roofs give the cleanest runoff for they act as a filter too.

Asphalt shingles are regrettably common. They are typically treated against the growth of mold. However, after a few years, the biocides embedded in the shingles have been substantially diluted and degraded by sunlight, so that their primary limitation over time becomes the slow but steady release of grit from the breakdown of the surface. This can be filtered. Some older metal roofs were anchored with lead washers under the nails. If you face these or similar heavy metal risks, consider replacing the roofing. Galvanized steel makes a good roof, but the zinc plating leaches into the runoff, especially if your rain is acidic, and zinc can be toxic in high concentrations. Don't drink water collected from

Enamelled metal panels form a durable roof with very high-quality runoff. [Credit: Creighton Hofeditz]

Pulley-and-pivot systems are one way to divert the first few gallons of dirty roofwater flushed by rain before directing runoff to storage. [Credit: Peter Bane]

galvanized roofing regularly; a little won't hurt you. It can be used for irrigation, laundry, toilets and similar purposes. I would be cautious, however, about providing it to livestock as a sole source.

The presence of airborne pollutants in roofwater is a function of your proximity to sources such as busy highways, industrial smokestacks and agricultural sprays. Trees will help to scrub dust and pollutants from the atmosphere, so living within a wooded area or surrounded by a screen of trees offers some protection from unwelcome aerosols. A cordon of 20 feet or more of clearance between the edge of the roof and large tree canopies should be maintained where possible; a larger gap would be preferable on the upwind side. Keeping trees away from roofs will improve safety, and it helps to reduce the load of insects and bird manures, as all these small creatures find habitat and refuge among tree branches, leaving behind their detritus. If birds are inclined to roost on the ridgeline or gutters, consider running a wire just above the peak to discourage them.

Airborne contaminants will accumulate on the roof between rains. The first few gallons of a rain event to hit the roof will thus carry much more debris than those hundreds or thousands that follow — the rain itself cleans the roof. A green or sod roof will redeem airborne debris by capturing or digesting it in the soil medium, but from all other roof types dust and debris will be carried away in the first flush. To prevent this material from accumulating in your tank, where it can decompose and potentially trouble the waters, some way must be found to divert the first runoff selectively and filter the remainder to maintain a high quality.

First-flush diverters are not difficult to create. Their designs tend to fall into two broad categories: pulley-and-pivot or double downspout.

A variant, not strictly a first-flush mechanism, uses a cutaway downspout. The lower section has a steeply angled top which is screened. Water flushing from the gutters into the upper downspout falls a few inches in the open air, debris hits the screen and is flung to the side, while the runoff cascades through the screen into the lower pipe.

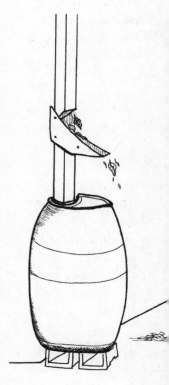

Cutaway downspouts are a commercial item or can be homemade.
[Credit: Abi Mustapha]

The simplest connection between a roof and a tank is a single pipe from a shed-style roof dropping directly into the cistern through a filter basket.

[Credit: Peter Bane]

With one moving part, this double down-spout diverter intercepts first-flush water flowing from the gutter to a standpipe that is connected underground to an above-ground cistern. The vertical downspout tube allows the first 5–6 sediment-laden gallons to be kept out of the collecting system. A removable plug at the bottom of this tube permits discharge of the collected flush water.

[Credit: Creighton Hofeditz]

This is a variation on gutter guards, intelligent insomuch as it is somewhat self-cleaning, but it doesn't actually divert the first gallons. Its effectiveness as a filter depends on the size of the screen and careful alignment between the upper and lower sections of the downspout, which can be connected on up to three sides.

At the tank, secondary filtration (after gutter screens and first-flush if used) is by some combination of simple sieves, screens, fabric or changeable filter media. This architecture, however, is not always possible, as tanks may need to be located many scores or hundreds of feet from the source roofs. It may not be practical to run overhead pipes across a driveway, road or through congested areas of the landscape. Aesthetics may militate against exposed plumbing in plain view of the dining room or other sensitive locales. In that case, the collecting system needs to run underground. In order for the water to go down, then come up into the tank, the collector plumbing must be sealed from a height near the gutter outlet to the spillover point inside the tank.

## A Water Tank for Every House

Rainwater tanks may be made of ferrocement, plastic, fiberglass, masonry such as block or tile, rot-resistant woods such as redwood or cypress or stainless steel. As I have had to inform several of my students over the years, cob (building mud) will not serve as it is porous! The simplest and cheapest form of large water storage tank (barring spectacularly good salvage) is a cylinder made of welded steel reinforcing mesh and lined with a carpet sandwich.

I have built several large and small ferrocement tanks, improving on the design each time, and prefer these to other types of tank for a variety of reasons. They can be built to a range of sizes according to need. The materials — steel rebar, welded mesh, wire ties, cement, sand, gravel, and PVC plumbing fittings — are still widely available in the industrial world.

Ferrocement tanks are durable, pleasing in appearance, need little maintenance and impart alkalizing benefit to the stored water — usually acidified rain. Technical mastery is not required for a good result, though proficiency with a trowel is desirable, and the engineer or supervisor overseeing the job must be conscientious. Most domed cylindrical ferrocement tank roofs will easily support a platform or provide a pleasant venue for stargazing, fruit picking or sleeping out of a summer night. Our 10,000-gallon tank has spawned an attached aboveground root cellar. The chief disadvantage of ferrocement tanks is that they require a full year or more of planned and steady attention to complete. The actual labor needed for a large tank (200 to 400 hours for 5–10,000 gallons capacity) can probably be accomplished in less than three weeks with enough hands at certain points, but in my experience, the building process is rarely continuous. Tank building of this type has become a specialty business in and around Austin, Texas, where lengthy dry spells punctuated by heavy rains make household water security a major economic issue.[11] If you are contemplating the building of one or more ferrocement tanks, I recommend you consult someone with experience, read several descriptions of the process and, most importantly, install the tank(s) and any plumbing as early as possible in your development, as they have a large footprint and you do not want to tear up landscape near gardens or other buildings more than once.[12]

Plastic tanks (polyethylene) are safe for drinking-quality water, and they are quickly available off the shelf from farm suppliers. The tanks come in sizes from rain barrels of 50 gallons — which are now common and provided with fittings — up to farm tanks of 2,500 gallons or more.

Tanks made of polyethylene must be coated, painted or buried to prevent them cracking. If buried, they must be filled with water before backfilling the earth around, as they can collapse if left empty. On the other hand, manufactured tanks are easy to plumb and can be combined to make larger storages. They are often available used or as salvage. A common form is the soft-drink concentrate container of 250 gallons. These are larger than

A 1,600-gallon ferrocement tank as its slab is being poured. A "spider" of rebar reinforces the slab and sticks up to form the lower reinforcement of the walls. All work on this small tank was done by hand, including concrete and mortar mixing.
[Credit: Peter Bane]

the now ubiquitous rain barrel and come with threaded fittings, thus are more practical for the garden farm. If you prefer not to store your drinking water in plastic, I share your concerns, but remember that drinking water is a very small portion of all the water you use (no more than two gallons per person per day), and this quantity can easily be filtered.

Fiberglass tanks are stronger than polyethylene and can be made quite large. They are much less common. Gasoline tanks for underground storage are now usually made of fiberglass, which suggests that these kinds of tanks, if properly manufactured, can be buried while empty.

Another commonly available and portable masonry tank is the septic vault, widely available for under $1,000 delivered in a range of sizes to about 1,500 gallons. Suppliers will customize the plumbing fittings to your specifications. The tanks are large enough to be used above grade without freezing, provided climate is not extreme and some insulation is used. They can also be buried. The result

A precast concrete septic vault makes a fast and easy water tank. [Credit: Peter Bane]

Three pipes positioned through the slab will provide inflow from gutters/standpipes, outflow to hydrants and a drain for cleaning the tank. [Credit: Keith Johnson]

can be an almost instant water system. I had a 1,200-gallon concrete vault set next to a house in middle Tennessee some years ago, surrounding it with straw bales for winter insulation.

### Design Factors for a Tank

Whatever their construction, all water storage tanks need similar design features. There must be a source of water adequate to fill the tank at least every year. This means annual supply must exceed annual use by at least the capacity of the tank. Our water tank holds 10,000 gallons, and we have an average roof runoff of over 50,000 gallons per year; hence we could increase our storage. Whatever runoff source you capture should be filtered to remove gross particulates and contaminants. Design the filters so they can be easily and regularly maintained. Every tank needs a drain with a valve or stopper, enabling it to be emptied completely. This is important should you need to make repairs or clean out sediments.

In subtropical climates, where frost is a minor issue, larger tanks can be built without a roof. Sometimes a tarp is suspended on top of the water to prevent the accumulation of leaves or falling debris and to exclude animals and their excrement. Sometimes a detached roof is built over the tank but not in contact with the walls. This can double as a collecting surface to supply the tank. Warm climate tanks may also be furnished with plants and fish (if the water is to be used for irrigation and not domestic purposes) as a means to control the growth of algae and insects. Where climate or use does not permit ecosystem controls on these potential problem vectors, they must be physically excluded, which means a completely closed top, screened inlets and outlets and a fully dark interior.

In cold climates, tanks must be protected from freezing. This is more critical with tanks of less than 3,000 gallons or climates colder than zone 5. In southern Indiana, nominally zone 6, our at-grade tank of 10,000 gallons is bermed up two feet of a total wall height of 6½ feet. Winter temperatures cause a scrim of ice from 1–3 inches thick to form on the water inside the tank. This doesn't negatively impact either its structure or function.

All tanks that collect water passively need an overflow so that surplus can be released safely. The inlet should be above this overflow level, and the supply outlet (where water is taken from the tank for use) should be slightly above the floor of the tank. This small gap between floor and outlet creates a sediment trap along the tank bottom which helps to protect the distribution plumbing.

All tanks, regardless of size, need to sit level on a well-built platform or on solid, undisturbed ground. Water weighs about eight pounds per gallon, so any sizeable water tank will, when full, weigh more than almost any human-made object in or around your house and if not well supported could endanger buildings or people. Ground-level runoff and tank overflow should be directed away from the tank foundation so as not to undermine it.

## Preparing for Overflow

Storing water is a way of buffering your farm system against drought, but at the other end of nature's pendulum, excess water can be very destructive. Design for water requires that we prepare for both too much water and too little, conditions that recur almost everywhere from time to time. Fortunately, there are some synergies to be captured from this two-pronged approach. The cheapest and largest storage for water is the soil. Reshaping garden beds and fields to hold water appropriately both ensures better use of available precipitation and helps to reduce flooding even as it protects crops. Cycling water through crops can build organic matter in soil and thus increase its water storage capacity and the resilience of landscapes.

Storing water in the earth is the last backstop to any system, natural or human. Water in tanks should overflow to ponds, ponds may overflow to small earthworks, which in turn recharge groundwater to support the growth of plants and trees, themselves a water storage. Forests help to make more rain, so we can reach a virtuous cycle where storing water brings more rain, allowing us to store more water. When the soil is fully saturated, water will flow on the surface, creating more and more resources downstream.

Whether carried out with shovel and rake or executed with a bulldozer or tractor, design for water requires us to shape the surface — and sometimes the subsurface — of the earth. By creating level areas where before there were slopes, we allow water to soak in rather than run off. By creating small barriers to overland flow, we can direct where water will move, and by creating small basins and swales (ditches on contour), we can allow runoff to accumulate briefly while it infiltrates to groundwater. The thoughtful placement of many small earthworks is a fundamental strategy in permaculture work, enabling us to rehumidify and restore soils and landscapes with very modest investments of time and energy. The degree to which we reshape landform is primarily a function of scale. On a farm of 200 acres, it may be worthwhile to create contour terraces in sloping fields or grass buffer strips in low areas or where the slope is gentle, while a garden of 20,000 square feet may warrant the careful construction of raised beds throughout so that virtually every square foot of the landscape is sculpted.

Nothing more perfectly illustrates the meaning of the familiar permaculture saying "Make the least change for the greatest effect" than muddling around with a hoe to redirect water from one channel to another during a rainstorm. Vast amounts of energy can be tapped, productivity released or

Swales between raised garden beds collect rainwater and act as pathways when dry.

[Credit: Keith Johnson]

damage averted by small, simple and timely interventions in the flow of water. Water is easily persuaded to move sideways according to our wishes, as long as it is allowed to drop even a little bit. Water remains inexorable in its response to gravity — down it goes, into the ground and out to sea. It will reliably flow over a slope that no eye can detect: one inch in 100 feet (1:1200).

## Design for Water

Design for water rests on these five pillars:

1. At least two sources.
2. Store it high.
3. Release it clean.
4. Match quality to purpose.
5. Slow the flow and follow the wave.

Water has its own intelligence, reflected in its patterns. In nature, moving water finds the channels of least resistance, escapes through the lowest point of release and always moves in wave patterns. A healthy flowing stream takes the shape of a serpent: it meanders back and forth along its path, always curving, always undulating. When we look closely at the water in a channel, it too is oscillating, spiralling up and down, back and forth. The effect of water's meandering is to reduce the force of its descent, which serves to limit its destructive power and allow the living systems along its path to derive maximum benefit from its life-giving moisture.

Most of our job in relation to water is to intercept and direct its flow into a variety of storages — tanks, ponds, swales, rain gardens, livestock, plants, the soil — and only release it once it has done a whole series of duties.

By storing water high in the landscape, we make it available to flow by gravity when and where we need it. We can either collect water from high surfaces or we can pump water to high reservoirs using low-cost energy such as solar- or wind-power systems, which in this application need no electric batteries. The high tank itself acts as a battery. Sometimes water will flow from higher elevations and so can be diverted from a stream and stored

### Water's Duties in the Landscape

Water should be used as many times as possible in the landscape before being released. Broadly, forestry occurs above the keyline and agriculture below, with trees and water storages at every level. From top to bottom:

| FUNCTION | ACTION | STRUCTURE | LOCATION |
|---|---|---|---|
| 1. hydration of soil and biomass | catchment | forests and swales | ridge and skyline |
| 2. generation of energy | storage | dams and drains, pipes | upper slopes |
| 3. storage and saturation for fire control | storage | dams, swales, and tanks | upper slopes |
| 4. microclimate buffer | storage and flow | surface bodies | mid-slopes and lower |
| 5. domestic supply | usage | tanks and plumbing | keyline |
| 6. livestock watering | usage | tanks and plumbing | keyline and lower |
| 7. recreation, amenity | usage | pools and ponds | throughout |
| 8. irrigation of crops | usage | tanks, ponds and plumbing | lower slopes |
| 9. nutrient transfer/ waste treatment | cleansing | wetlands and channels | lower slopes |
| 10. aquaculture | revitalization | ponds and raceways | keyline to lower slopes |
| 11. wildlife habitat | distribution | surface bodies | throughout |
| 12. groundwater recharge | retention | groundwater | valleys |

Credit: Peter Bane

at the upper end of the property for later use below. But the most readily available source of high water is rainfall.

To be stored in useful amounts, rainfall must first be concentrated. Over most of North America, an average rain event will drop less than one inch, and events of more than two inches are fortunately infrequent as they almost invariably hold the potential for erosion or flooding. But even five inches of rain from directly overhead won't fill a bucket, let alone a tank or a pond. We have to calculate the runoff from the watershed — be that board, roof, pathway or hillside — that will fill our bucket, tank, garden swale or pond.

### Calculating Runoff

Runoff is a function of area, surface quality and the volume of the rain event. A very important formula allows you to estimate water yield: for each 1,000 square feet of roof receiving an inch of rain, 625 gallons of water have fallen on it. Another way of saying that is:

⅝ of the roof area (in square feet) ×
the depth of rain (in inches) =
the potential volume of runoff (in gallons)

This formula is applicable to other impervious surfaces such as a roads and parking lots. Collection is always imperfect, however. Some rain evaporates from the roof, some splashes off, some overshoots the gutter; on pavement it finds cracks and drains away underground. Especially from light or sporadic rain on warm days, evaporation can almost eliminate runoff.

Pervious surfaces, usually the ground, but also thatch, sod and wooden shingle roofs, will absorb some portion of the rain and release it only slowly, most often by evapotranspiration. In the case of wooden shingles, this is a fairly trivial amount; with a thatched roof, a little more. Sod and other living roofs are specifically designed to absorb a significant amount of rain before releasing any. They also transpire and evaporate some of what they

catch. In this way, they help to regulate as well as purify runoff and reduce heat, especially in urban areas, and have been mandated for this purpose in Copenhagen, Stuttgart and other central European cities as an alternative to increasing investment in storm sewers.[13]

### Going with the Flow

If you store water, you must also provide for its release and continued flow. You can only slow water down. You have a right to use it, but you don't own it. All of it must sooner or later be released, if not to the stream, then to the soil or the air. Slowing, holding onto and letting go of water is a little-appreciated art that every gardener and farmer must practice.

On our property, almost all of the water flows from north to south, but it divides along a subtle spine as it leaves: some flows southeast and some southwest. Eight roofs intercept and redirect some of the rain. Each of these roofs offers us choices about where to collect and how to channel runoff. On the two largest buildings, we have hung the gutters to drain counter to the land flow (that is, from south to north) in order to bring the roof runoff closer to the high end of the property where we have our storage tank. This makes it easier and cheaper for us to distribute the collected

Runoff from roads and parking lots can be harvested for irrigation. This example in Catalonia drains to the open concrete tank visible behind the pavement.

[Credit: Peter Bane]

Water bars on the driveway divert surface runoff into garden swales while adding minerals to nearby soils.

[Credit: Keith Johnson]

water. The driveway, which has a large asphalt parking pad, drains mostly to the east. Shortly after we bought the property, a heavy rain event gouged a large pothole in the end of the drive near the street. We took note and created water bars, angled shallow berms that act as speed bumps both for vehicles, and most importantly, for runoff, moving water into the garden rather than out to the street directly. This has both prevented further scouring of the driveway and provides a source of irrigation water and minerals for a large part of the garden, an example of what permaculture calls turning a problem into a solution.

### Matching Quality to Need

For drinking and cooking, of course, water of highest quality is needed. Though if the cleanest water available is still subject to significant bacterial contamination, adequate amounts of water for drinking and cooking can be boiled, filtered or treated. The traditional East Asian practice of drinking hot tea reflects those cultures' adaptation to high population densities and the consequent widespread presence of bacterial pathogens. Our household takes treated public water from a pipeline, then further filters it to suit our tastes and concerns.

Of lower quality, but moderately suitable for domestic purposes, is roofwater coming off of clean surfaces. We make an effort by various means to filter our roofwater and store it cleanly. This is our primary irrigation water, but we also use it for laundry, toilets and outdoor washing, including washing vegetables. We drink it from time to time, even though it isn't treated to kill bacteria. It would certainly be suitable for livestock.

Below roofwater are various grades of runoff and reuse water. Concrete pavements can be quite clean if not subject to heavy traffic. Asphalt is slightly less desirable, but still acceptable, especially if older (the more volatile toxic compounds have leached or evaporated over time). Gravel roads and driveways are good and represent an opportunity to introduce minerals to the landscape. As a response to the acidic soils and high rainfall environment of the western North Carolina mountains where I used to live, we chose to buy road gravel from the one local quarry that was mining a deposit of dolomitic limestone, thus helping to balance and remineralize our roadsides.

Pond water makes an excellent source for irrigation and livestock watering, providing the volume is adequate. Surface water running off of grassland or woodland is highly suitable for filling ponds and is the most common source for these water bodies. Some basic measures should still be taken to reduce the amount of sediment carried by whatever runoff fills the pond; this will extend its life. If the pond has a clear inlet channel, this can be widened and deepened just upstream of the pond to create a small basin. By spreading out the flow, this miniature sediment or filtration pond slows down the influx of runoff, and as the moving water slows, it will drop much of its sediment load.

For garden or field irrigation, all forms of runoff (unless heavily contaminated by persistent pollutants) are suitable, along with pond water, greywater and, of course, all

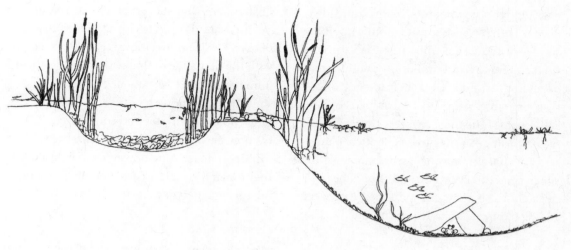

A small basin upstream of the pond inlet traps sediment from runoff; the sediment may be recycled to gardens and other crops.

[Credit: Abi Mustapha]

cleaner forms. For example, driveway runoff is adequate — even though it may carry trace amounts of petroleum and heavy metals. Traffic on driveways is slow and light, and oil leaks can be monitored and mitigated. Road runoff contains petroleum distillates leaking or spilled from vehicles, small amounts of cadmium (released by the abrasion of tires running on the pavement) and asbestos from brake pad wear. The asbestos is dangerous only if inhaled. The amount of cadmium is a function of traffic speed, weight and volume. Petroleum distillates can be broken down by fungi into harmless constituents useful to plants. In countries where leaded gas has only recently or not yet been eliminated, lead is a major pollutant, especially on roadsides. At Renaissance Farm, we make driveway runoff safe by directing it into shallow pathway ditches between the raised beds of the garden, so that sediments accumulate there and any petrol-based toxins are broken down in a matrix of wood chips and fungus or are sequestered by the high levels of organic matter present. If the flow is larger (as from a parking lot), passing it through a small woodland, grassland or wetland as a filter might be warranted. We use greywater for irrigation as well, especially that from our laundry and bath, which carries primarily soil, soap and flakes of skin. Various forms of reuse water can be safely applied to soils. Avoid persis-

tent pollutants and harsh chemicals in your drains (lye is acceptable for cleaning clogged drains, though mechanical action or enzyme solutions are preferable). Small amounts of soap, fat, food waste, animal manure, dust and grime are safe to apply to soils when carried by wastewater. If the volume of greywater is large, special structures may be appropriate to assist the breakdown process.

The most polluted forms of runoff — including sewage waste and waters containing some petroleum distillates — can be directed onto woody perennials harvested primarily for fiber and fuel. This should not be seen as

Greywater from a bathhouse (not shown) will alternately fill these trenches which distribute it to irrigate the hillside orchard. Perforated drainpipe will rest on the blocks and be covered and surrounded by wood chips which will compost while filtering the effluent.

[Credit: Peter Bane]

a writ to ignore pollutants, but in some low-energy and low-resource economies, spilling sewage or industrial effluent onto tree plantations for energy and timber production may be an improvement over their uncontrolled discharge into waterways, farm fields or dense settlements. The most severely contaminated waters, as for example from chemical plants or mine tailings, may be irredeemable without specialized intervention, and should be sequestered as toxic waste until their discharge can be treated at source or terminated.

## The Water Cascade

The principal reuse water on the garden farm will be household water in various grades. (Pond water can, of course, be used first by fish and then by other livestock or for irrigation.) By learning to move water from highest quality to next highest in small increments, we activate what I call the household water cascade. This is an essential concept to understand and a primary ethic for conserving water and other resources. The most absurd violation of this ethic occurs billions of times every day in North America: the use of the

flush toilet. Almost everywhere, drinking quality water is used to flush toilet wastes into sewers. Public authorities struggle to keep up with the demands of this cultural folly, but at home we needn't perpetuate it.

Here is a rundown of typical grades of household wastewater and some suggestions for how best to use each type, with the reminder that any type of water can be used further down the scale at any time that moving or storing it becomes seriously inconvenient:

### Discarded Tap Water

Water lost when warming a shower is suitable for all washing purposes. In theory you could even cook with it or wash vegetables, but practically, the bucket you collect it with may not always be free of soap, so it easily goes to handwashing, bathroom cleaning, laundry or toilet flushing.

### Bathwater

Warm and a little sudsy, it's ideal for hand-washing delicate items or prewashing very dirty clothes. Just throw them in the bathtub before you get out, stomp around a little bit and rinse later.

### Laundry Water

Produced in copious quantities, it carries some soap, soil and trace amounts of body wastes. It may also contain a useful amount of heat energy and is quite safe for use on trees, fruit crops, lawns and ornamentals. It's helpful to pass it through a simple wood-chip filter.

The chief hazard is putting too much in one area that drains poorly, and so creating waterlogged soils. Households in dry areas should minimize the use of detergents and monitor for salt buildup in the discharge area.

### Dishwater

Suitable for flushing toilets or spot watering of garden beds. Use two dishpans if you want to make the best use of your water and conserve energy.

Water Cascade,
Pattern #47
(See Chapter 6)
[Credit: Abi Mustapha]

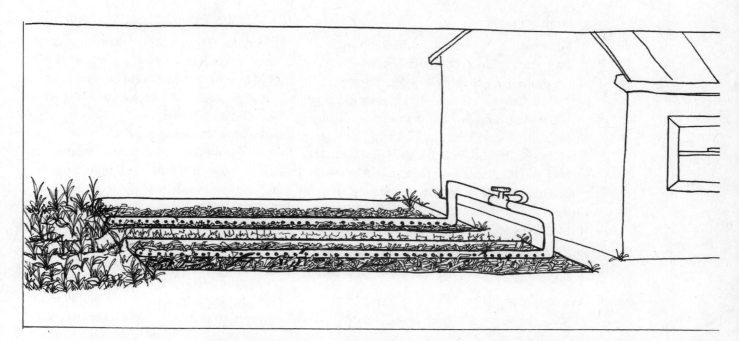

Greywater Trenches,
Pattern #48
(See Chapter 6)

[Credit: Abi Mustapha]

## Desert Dishwashing

If you have greasy plates or pots covered with food residues, pre-wash these with a little water and a scrub brush (or your hand). Discard your pre-wash water into a five-gallon bucket next to the sink. Clean the basin of grease with a little soap. Soak utensils in a small bowl or basin if the flatware is covered with dried or sticky food. Sort the pre-washed items into glassware, ceramics, flatware, plastics and cookware. Wash them in that order (from cleanest to dirtiest). Unlike common practice, start with a small amount of hot water and a little soap in a basin for washing. Clean the glassware, rinsing into the empty basin with a thin stream of hot tap water until you have about an inch in the rinse basin. (If you don't have piped water, you can simply start with a small amount of hot water in the rinse basin.) Run a thin stream of water only when you are actually rinsing an item. After rinsing several items, you will have enough water in the rinse basin to rinse other items without running more water. After a dozen items or after you get through the plates and bowls, the wash water will have cooled and may be some-what soiled and the soap ineffective. Discard this small amount of washwater into the bucket. Pour the still-warm, relatively clear rinse water into the wash basin, add a few drops of soap and continue cycling through items in the same way until finished. We wash plastic bags in our household to reuse them, and I usually let these accumulate for a few days until there are enough to run a separate wash-rinse cycle for bags only. They need the cleanest water, but I like to do them last, so I can lay them on top of the already clean dishes and later take them outside (or in winter by the stove) to air dry. We find it helpful to use a small spray bottle filled with dilute dish detergent for spot application and portion control. The bucket of accumulated wash water can be used for flushing toilets or can be poured onto garden beds near the back door. The last basin of rinse water can be left for pre-washing the next set of dishes or pans, or for scrubbing vegetables. The last load of washwater, if not too soiled, can be used for cleaning the kitchen floor or the stovetop.

### Shower Water

Put a bucket under the tap when you are showering to intercept water headed down the drain while the shower is warming up. The bucket won't interfere with your washing and rinsing, and some of the water cascading from the showerhead and bouncing off your body and the shower walls will wind up in the bucket if it's placed over the drain. Also, the dribbling of the showerhead when it is shut down by the water-saver valve will be caught too. You can use this shower water for flushing the toilet. It's also helpful if you need to scrub the shower stall or tub after bathing.

### Greywater

So-called greywater from sinks, showers, dishwashers and laundry carries a variety of wastes plus soaps down the drain. Most of this can go into the landscape with little or no treatment. Its use for irrigation is sanctioned in California and other western states where water is scarce.[14] Laundry water is the easiest to divert. Washing machines have flexible drain hoses that fit loosely into open vertical collecting pipes. The washing machine comes

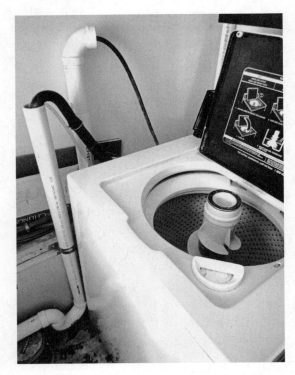

Two drain pipes and a flexible drain hose offer most householders the option to move laundry greywater to a biofilter trench or into septic lines for conventional treatment.

[Credit: Creighton Hofeditz]

equipped with a pump designed to lift the effluent eight feet or more (in order to get it out of basements where many washers live). It's easy to move these drain hoses into new plumbing lines that take the greywater where you want it (to a tree or the garden) instead of to the sewer or septic tank.

Accommodation for greywater reuse can be easily designed into a new plumbing system and can even satisfy the requirements of the code if necessary. A friend building a new home in Asheville, North Carolina, a few years ago worked with the city inspector to design a greywater holding tank with sump pump and float valve. Under normal operations this small tank (about 70 gallons) would empty itself regularly into a series of upslope garden terraces through subsurface perforated drain pipes. However, the tank was also plumbed to carry any overflow to the city sanitary sewer, so that if the pump failed, power went out or if there was any other mechanical malfunction, the fail-safe condition would send the greywater to the city's treatment plant. There was no surface discharge. While this system involved regular electric use (by the pump), it was a good compromise under the circumstances. In less tightly regulated areas, simpler solutions are available.

These are examples of reuse that can move resources down the cascade to make water go further. You can also avoid using water in toilets by collecting urine in sealable containers for direct application to the garden. Urine — what author Carol Steinfeld calls liquid gold[15] — is basically free of bacteria when it leaves your body (unless you have a urinary tract infection). Any bacteria that may get carried with it won't easily live outside the human body and are suppressed by natural soil organisms. And it's a superb source of plant food. Of course you can leave urine to your toilet waste stream and conserve water by flushing with greywater, but the release of nitrogen that accompanies urine is a serious contaminant for waterways and groundwater when concentrated through sewer systems and leach lines.

By using conservative household practices (desert dishwashing, greywater for flushing, low-flow showerheads), it's easy to bring water consumption levels down to about 10 gallons per person per day (about 10% of the US average) while still washing clothes by machine, enjoying hot showers (though perhaps not every day) and having a high quality of life. At that modest level of use, you could easily store enough water to supply your entire household for six months to a year, even on a small property. This is the level of water security for which we should be aiming. Since water pumping takes 25% or more of the electricity used in our cities (44% in Tucson), a reduction in the demand for water would lead directly to power savings, lightening the load on municipal budgets while reducing carbon emissions from power plants.[16] No one needs to be uncomfortable, unclean, unsafe or unhealthy from any of these practices; we just need to lose a lot of our wasteful, indulgent habits. I can say from personal experience that a hot shower is no less pleasurable for coming through a low-flow valve. And if that used shower water then heads out to the garden to help grow food, a whole realm of other satisfactions comes into play.

## Water Reclamation

Water can be purified of organic wastes by settling and filtration, by the bacterial breakdown of pathogens and by allowing the roots of plants to take up the excess nutrients that it carries. These methods, which can be incorporated into simple structures on a small scale, can handle all the substances that need to go into household wastewater. Water that is also contaminated by toxic organic chemicals such as biocides, oil or solvents needs a further step: it must be passed through a filter medium, such as wood chips, supporting the growth of fungi. These organisms are able to digest the long-chain carbon molecules

found in organic toxins.[17] Heavy metals are best excluded from water, but where they are a contaminant, filtration or sequestration in non-agricultural soils are the best answers.

Wastewater treatment principles are based on the biology of the organisms that break down the waste and capture the nutrients from it.

### 1. Garbage In, Garbage Out

Put nothing in the system that can't be eaten by organic life. Avoid harsh chemicals even if they can ultimately be digested. Food and body wastes, soap, vinegar, salt, household ammonia, hydrogen peroxide, baking soda, dirt and grime are fine. Oil, biocides, large amounts of chlorine bleach, strong acids, paint, solvents and industrial chemicals are not OK.

### 2. Something Must Eat the Waste

You must create habitat for worms, millipedes, pillbugs, plants, bacteria and fungi. Build it and these organisms will come.

### 3. These Are Aerobic Ecosystems

They use oxygen-loving bacteria to digest microbial pathogens and use small organisms — from worms to pillbugs and fungi — to eat larger pieces of organic matter. In wetlands, the oxygen is provided by the macrophytic and emergent plants (swamp dwellers) which release it through their roots. In composting greywater systems, oxygen diffuses from the atmosphere, so the absorbing medium must be uncompacted (with a lot of pore space) and be allowed to drain periodically. Influxes of liquid should not be continuous. Holding tanks are employed to collect intermittent flows, and pumps or dosing siphons release them all at once, resulting in a pulsed flow.

### 4. Time Is Required for Breakdown and Digestion

The system must have enough capacity to absorb and hold the waste stream for several

A wetland water treatment cell showing cattails emerging from a gravel matrix. The cell is lined and in this case raised above grade level to facilitate outflow toward ponds.

[Credit: Creighton Hofeditz]

### 6. Use All of the Capacity as Evenly as Possible

The system should distribute the effluent efficiently throughout the entire medium, avoiding soggy spots and dry patches.

### 7. Adjust Flows and Levels

In systems that retain water (artificial wetlands), it's helpful to be able to regulate the (always subsurface) water level by means of a valve or flexible drain. Lowering the water level, for instance, can prevent freezing in winter.

### 8. Keep the Pipes Clear

If the waste stream is chunky, some simple, renewable system of filtration and cleanout is required at the upstream end to keep pipes flowing freely. Your bathtub drain screen or sink strainer is the first line of defense. In a septic tank, two chambers are used, and the first allows suspended elements to sink or float, while the remainder of the liquid waste flows out from the middle of the chamber. The same principle of design can be applied to a greywater system where a lot of food waste, for example, may be discharged. Alternately, a sand filter or wood chip or straw filter can be designed into the discharge line in a way that allows the filtered material to be collected periodically and composted.

days at a time. This is usually accomplished by creating two or more parallel cells that can be switched every few days by use of valves or movable drain tubes.

A cell of about 4 × 4 × 16 feet should be able to hold the effluent from a water-conserving household of four people for up to four days. Composting systems can be alternated or refreshed at longer intervals.

### 5. Keep It Out of Reach til It's Clean

Greywater should not be exposed to human contact more than 12 hours after initial use. Obviously, we've all had our hands in used dishwater at some time, and no harm has come from it. But greywater that is not being actively purified by plants, animals and soil microbes will culture pathogenic microbes quickly, becoming black water or sewage. Retention should therefore be done below the soil surface until such time as the pathogens have been substantially digested. Pooled, nutrient-rich water is an invitation for pest organisms such as mosquitos, too. The direct discharge of fresh greywater to soil is usually harmless provided it soaks in within a short time.

### 9. Renew Biological Components

In a composting system, the medium needs to be changed at intervals between six months and two years, depending on flow rates. In artifical wetlands, the plants need to be harvested (for mulch or animal fodder) several times during the growing season.

### 10. Pay Attention and Learn as You Go

These systems are well understood in principle. In practice, since they are living ecosystems, they will be dynamic and will respond to inputs. How yours behaves should inform your management, so remain observant and flexible.

## Using Water Economically

A water system consists of sources, storages, distribution lines and control points. Our permaculture principle of integrated design reminds us to let every element serve multiple functions, so we should recognize the possibilities that water bodies provide as firebreaks, air conditioners, barriers and heat sinks. Open water reflects light, and ponds can be placed to enhance light to greenhouses or homes. Ponds are also habitat for fish and amphibians which suppress mosquitoes, flies, slugs, snails and other garden pests.

Water sources, which include all rooftops, wells and ponds as well as the mains supply, should be interconnected by the farm distribution system so that water can be sourced from where it is available most economically and moved to where it is needed by pipes, hoses, troughs and trenches. If your rainfall is less than 40 inches a year, or if you already have a significant dry season, you should install irrigation lines that will enable you to sustain gardens and intensive plantings throughout the growing season. In 2010, Indiana and much of the lower Ohio Valley and mid-continent endured three months of hot weather with very little rain, the most severe drought locally in a generation. We can expect both rising average temperatures, greater extremes of heat and longer dry spells, as well as more severe rain events and flooding as climate change continues to bite. This means both storing more water and being prepared to use it to support yields.

A plumbing system is essentially a tree or distributary pattern with stem, main branches and smaller lines fanning out from there. A well-designed system should be protected from frost over its entire length, should have valves at every major branch and at many smaller ones to facilitate maintenance and repair, and it should use the minimum length of pipe consistent with necessary traverses. Underground pipes should follow easily accessible routes, such as pathways, so that they can be dug up, repaired or altered if necessary. It's not a good idea to plant large woody perennials atop plumbing, nor to pave or build over it except for the limited extent necessary to get water lines into buildings. If you are burying water supply lines, take the opportunity to run other utilities such as power or waste distribution lines in the same trenches as appropriate. To the extent possible, keep valve boxes and hydrants near the edge of buildings or pavements in locations where they are easily accessible but not vulnerable to traffic. If you must have a valve box or hydrant in the middle of a garden bed or small field, create an island around it with a shrub or a small structure to afford protection. Be especially careful about water infrastructure around the stalls or paddocks of big animals, who can easily damage it.

Whatever your source, some water should be stored at elevation on property so that it can be released to flow by gravity. Tanks will have drains and outlets, but ponds should also be designed with drains so that water in them can be drawn out easily. In the event of emergency, this can be critically important, but even in ordinary times it is always preferable to use water that flows by its own weight. If you must pump, do so from a source into a high storage, then draw down the stored water by gravity. This arrangement gives you many more options for pumping economically: solar, wind or even animal power.

Fire control may be an infrequent use of water, but after domestic supply it is arguably the most important. Once you have identified your fire sector, if you don't place a pond there as a firebreak, you should at least make it possible to flood-flow water from a pond or tank into swales in the fire sector. Of course, water can also be pumped under pressure to fight fire, but one of the best uses of stored water in the event of wildfire is to keep vegetation

Entrance and interior of a root cellar built between two concrete water tanks made from pre-cast septic vaults. (Design by Darren and Espri Bender-Beauregard) [Credit: Keith Johnson]

A peninsula in a pond presents a choice location for a greenhouse. Not only will there be more reflected light, but frost will be mitigated by the thermal mass of the water. An island in a pond or a floating raft on a tether can be used as nesting ground for waterfowl. And the same barrier function can be deployed by arranging a pond, even a long and narrow one, around one or more sides of the garden, like a moat.

All livestock need water, even bees — which, like children, require a shallow access so as not to drown. (Put a few sticks, rocks or reeds into the shallow end of the pond, and the bees will take care of themselves.) But larger animals should be excluded from open water because their hooves can do so much damage to sensitive habitat along the banks. Manure too, should be kept out of ponds unless designed as a specific feed input (as, for example, from confining poultry or pigs above a pond as a source of nutrient). Rainwater, whether stored in tanks or in ponds, is the best source for livestock, but may need to be supplemented by water from mains or wells.

## Food from Water

Aquaculture is of growing importance worldwide to supplement static or declining supplies of wild-caught fish. This trend will only accelerate. The small farm would do well to develop some aquatic systems for yield. Some easy crops to grow on a small scale include water chestnut (*Eleocharis dulcis*) and watercress (*Nasturtium officinale*). A small kiddy pool of four-foot diameter is sufficient for a small crop, which can be fertilized periodically with a little urine. Yields of water chestnut can reach 0.4 lb per square foot. It can also be grown in combination with channel catfish, the health and yield of which it seems to improve. Wild rice (*Zizania aquatica*), while not actually a rice, is a commercially viable specialty crop of northern regions and a staple of the traditional Ojibway diet. Cattails

around the home and other structures green, succulent and incombustible.

Open water can be a source of cooling in the landscape. We all appreciate a good dunk in hot weather — and better yet a swim. Ponds placed to intercept summer breezes can cool the airstream moving toward the dwelling. And if large enough, ponds can provide a sink and source for heat pumps conditioning the air in the house. Tanks can also play a thermal role. Friends nearby have placed a root cellar between two partially buried water tanks.

A floating duck house provides the fowl protection from predators.
[Credit: Peter Bane]

(*Typha* spp) and arrowheads (*Sagittaria* spp) also produce starchy food parts and might be developed into suitable crops over time.[18]

The most important freshwater food crop is, of course, rice (*Oryza sativa*), and in Asia it is classically managed in polycultures involving fish, ducks and floating plants that provide pest control and extra yields. Takao Furuno describes an intensive, integrated system for smallholders based on many years' practice in Japan. He and his family derive over $100,000 income from five acres of land. The high income is based on two strategies: direct marketing to consumers and processing all products on farm. Japanese agro-technology has been miniaturized to enable all manner of farm and food processing operations to be mechanized economically at the scale of 5–7 acre farms: rice is husked and polished, miso is fermented and much more. Yields of rice in Furuno's system exceed 4,400 pounds per acre.[19] The vast bulk of US rice growing is done by large farms in southern regions, but with changing climate patterns, Mississippi-like conditions can be expected to migrate to southern Illinois by the latter part of this century, opening up new regions to the crop. Cold-hardy varieties of rice are even now being grown in trials in Vermont.[20] Based on the principles Furuno sets out, and with the import and subsequent domestic development of appropriately scaled machinery, small-scale, integrated rice-duck-fish polycultures could become an economic crop for the North American garden farm.

The multiple yields of integrated Asian aquatic systems (up to 6,000 pounds per acre per year of fish plus many other benefits) are intriguing and season the literature of both permaculture and forward-thinking aquaculture texts. These models, which depend on a suite of carp species that are prohibited in many jurisdictions, should continue to inspire us even if the productivity of our native and adapted species polycultures cannot quite attain the same levels. It is clear, however, that flesh foods offer far greater economic potential from water than plant crops for most North Americans. In Louisiana, crayfish are intercropped with cypress trees,[21] and in South Carolina channel catfish have been grown successfully with water chestnut,[22] but in most situations, the extensive plant cover of water bodies required for economic yields suppresses the carrying capacity for fish and makes harvest of animals impractical.

Present opportunities for small-scale food yields lie in two directions: intensive and extensive aquaculture. The latter is best represented by the classic warm-water farm pond of from one-half to five acres. Typical stocking would be with a combination of largemouth bass (*Micropterus salmoides*) and bluegill (*Lepomis macrochirus*). Frogs, turtles and perhaps minnows would stock themselves; minnow eggs often arrive on the legs of wild waterfowl or in water from other ponds introduced to inoculate a new pond. A few garden farms may have existing ponds of this type, and others may have the opportunity to create them. The bass-bluegill pond ecosystem is easy to stock, but it is somewhat tricky to manage — especially in ponds of under ½ acre — and has low productivity (up to 100 pounds per acre per year). Colder regions

Linear pond for trout fingerlings in New Mexico permits ease of harvest and restricts access by larger fish.

[Credit: Peter Bane]

may have the right conditions for trout, and in warmer areas catfish are sometimes stocked either with bass-bluegill or as an alternate crop fish. Most farm ponds are harvested by hook-and-line recreational fishing, a pleasurable but inefficient method.

### Improving Pond Yields

A typical farm pond could benefit from improved design and management. Crop fishes, while chosen to complement each other, do not occupy all the available food niches, and so use the pond's basic ecological productivity poorly. There is little use of pond plankton by any crop species, nor of edge or floating vegetation, nor of benthic (bottom detritus) resources. Catfish, some species of which will consume benthic material, can partially rectify this defect, but bring their own problems, as their roiling behavior stirs up the bottom mud and can inhibit the carnivorous and visual hunting fishes from feeding well. Secondly, the population balance between bass, which prey on small bluegill, and the bluegill, which reproduce rapidly, is difficult to maintain. What often happens is that the larger bass are overfished, the bluegill population spikes and bluegill development is thus stunted. The remedies are obvious: eat smaller bluegill (they are no less tasty!) or limit the harvest of bass so they can control the bluegill population.

Some possible improvements to the conventional farm pond would include linear shapes, the inclusion of interior dikes or peninsulas making harvest easier, the use of flowing water (from irrigation ditches or springs), including non-finfish species such as clams, mussels, frogs or turtles and, of course, creating a more complex aquatic ecosystem.

Simple additions such as electric bug lights or pheromone attractants to draw more insects to the pond would improve feeding conditions for bluegill. Mollison, McLarney and others have suggested floating rafts with rotting logs (termites) or roadkill carcasses

(maggots) to supply feed organisms.[23] Bluegill, being surface feeders, would readily consume these as they fell into the water.

   McClarney makes some intelligent suggestions for an improved North American pond
polyculture to include freshwater clams and
bigmouth buffalo (which are efficient native
filter feeders), young common or mirror carp
(plankton-eating Eurasian species that are not
everywhere permitted),[24] bullheads, channel
catfish, redear sunfish, chain pickerel (native
benthic feeders) and grass or mirror carp to
eat edge vegetation (sterile hybrid grass carp
are widely available). He also proposes that
cage culture could greatly improve the yield
of the typical pond without altering the basic
bass-bluegill system significantly.

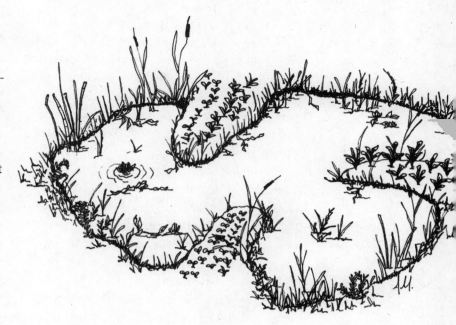

Interior dikes or
peninsulas can be
added to a con-
ventionally shaped
pond to support
more efficient
management and
increase habitat.

[Credit: Abi Mustapha]

## Aquaculture Species for Ponds

| COMMON NAME | LATIN NAME | ECOLOGICAL NICHE |
|---|---|---|
| Largemouth Bass | *Micropterus salmoides* | carnivore |
| Smallmouth Bass | *Micropterus dolomieu* | carnivore |
| Bluegill | *Lepomis macrochirus* | insectivore |
| Green Sunfish* | *Lepomis cyanellus* | eats insects, crustaceans, molluscs, small fish |
| Redear Sunfish | *Lepomis microlophus* | eats insects/molluscs |
| Pumpkinseed Sunfish | *Lepomis gibbosus* | eats insects/snails |
| Chain Pickerel | *Esox niger* | carnivore |
| Yellow Bullhead | *Ictalurus natalis* | detritivore |
| Brown Bullhead | *Ictalurus nebulosus* | detritivore |
| Black Bullhead | *Ictalurus melas* | detritivore |
| Channel Catfish | *Ictalurus punctatus* | omnivore |
| Blue Catfish | *Ictalurus furcatus* | omnivore |
| Common Carp | *Cyprinus carpio* | filter feeder |
| Mirror Carp | *Cyprinus carpio carpio* | filter feeder |
| Grass Carp** | *Ctenopharyngodon idella* | eats vegetation |
| Mozambique Tilapia | *Oreochromis mossambicus* | filter feeder |
| Blue Tilapia | *Oreochromis aureus* | filter feeder |
| Rainbow Trout | *Oncorhynchus mykiss* | insectivore |
| Brown Trout | *Salmo trutta fario* | insectivore |
| Brook Trout | *Salvelinus fontinalis* | insectivore |
| Fresh Water Prawns | *Macrobrachium rosenbergii* | filter feeder |
| Red Crayfish | *Procambarus clarki* | omnivore |
| White Crayfish | *Procambarus blandingi* or *P. acutus* | omnivore |
| Mussels | | filter feeder |

*green perch or crapet verte
**white amur

Yields from feeding of cage-cultured fish could reach to 1,000 pounds per year or more in cages of 64 cubic feet. This is, of course, dependent both on feeding rates and on oxygen levels in the water. Generally, oxygen levels are not a problem except in very small ponds or tanks, but maintenance and a regular feeding regime are required.[25]

Fish and other aquatic animals offer us valuable protein in a form that is relatively easy to harvest and prepare. Many crop species are available that can take advantage of system wastes: plankton, insects, small fry, weeds. Aquatic animals, which are cold-blooded and float, are more efficient in their conversion of feed to flesh than warm-blooded and terrestrial creatures who must build bones, fur and feathers to move about and regulate their bodies. Nutrient transfer in water is more efficient than in soil. The per-area yields of aquatic systems can be several times those of land-based forage systems. Chinese pond polycultures, for example, can yield 6,000 pounds of fish per acre per year.[26]

Water Gardens and
Fish Crop, Pattern #34
(See Chapter 6)
[Credit: Abi Mustapha]

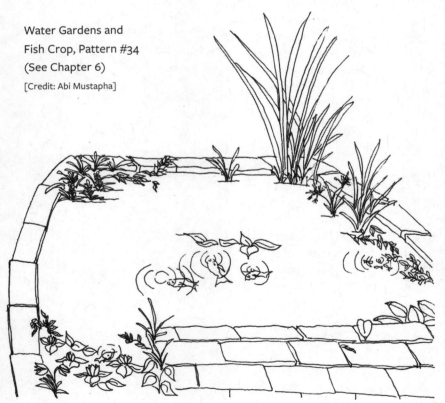

### Raising Fish in Tanks

For most small properties, tank culture offers the greatest opportunity for an economic yield. It also enables us to culture non-native species (e.g., tilapia) that occupy more appropriate feeding niches and grow faster. Using warm-water species, Steven Van Gorder describes in great detail a tank system for raising up to 100 pounds of fish in a five-month season.[27] Using passive solar collection principles, domed covers on the tanks allow the system to function well outdoors in climates to at least zone 5 (Iowa and southern Wisconsin across to southern Ontario, Pennsylvania and Massachusetts plus the southern Plains, many areas of the Southwest, the Pacific Coast north to British Columbia and selected Mountain West microclimates). Obviously, in colder regions, a second layer of thermal insulation could be provided by a plastic hoop house, and the thermal mass of the tank water could be used to help regulate greenhouse temperatures throughout the year for enhanced vegetable production.

Tank culture need not be monoculture. Various species of tilapia combine well, and may in turn be grown in the same tank with channel catfish. Carp mix well in these and a wide variety of polycultures. Remember that these fish species are primarily plankton and algae eaters (channel catfish are omnivorous and predatory on small fish and crustaceans), so predation is a minor issue, provided the species are of comparable size when introduced. Crustaceans such as crayfish and freshwater prawns, which are productive aquaculture species in their own right, can be grown in separate tanks using the same recirculating water. The bottom-dwelling crustaceans take advantage of the detritus from the larger tank, eating dead algae, fish wastes and uneaten fish food.

The labor burden of tank culture is not great in comparison to the yield. After setup, the principal duties include feeding the fish two to three times daily (more frequent feedings result in better consumption of the available feed) and weekly cleaning of drum clarifiers to remove accumulated solid wastes. These chores provide ample occasion for regular monitoring of the levels of dissolved oxygen, ammonia and nitrites in the water. There are, of course, additional harvest and seasonal duties, such as regulating temperatures in the tanks during extremely hot weather (by opening and closing the domed covers, for example).

### Limiting Factors

One vulnerability of tank culture is the need for a constant supply of electric power to run the pumps, which maintain oxygen levels and move wastes through the drum clarifiers to remove nitrates. The fish can survive for a number of hours without circulating water, but not for days. The economics of running small electric pumps to produce high-value crops are likely to remain favorable for the next decade or two, even if carbon taxes are finally implemented. Recirculating fish culture systems may be too costly of power for off-grid situations, however, at least those without a constant supply of wind or hydro-power. Also, some species are not easily spawned by the small operator, so that an external source of fry may be required. But the difficulties of using commercial fish foods are more serious. While readily available, these materials are almost all compounded with either wild-caught fish protein, soybean meal or both. Wild-caught fish protein is an extravagant feedstock from a diminishing resource base and is therefore ethically questionable. Soybean meal in the US and Canada is substantially produced from genetically modified soybeans, and many readers would rightly question the wisdom of using such material at all, let alone in one's primary food supply. Organic fish feed is some years away from commercial deployment. Since the appeal of homegrown fish from a small area is considerable, and the technical capacity, financial

investment and time and energy required are not onerous, we are left wondering how to make our own feeds.

Laurence Hutchinson recommends a form of trout culture he calls ecological aquaculture.[28] His system of raising trout or other salmonids is based on creating a naturalistic food chain within a structured pond architecture. The limitation of this system is the requirement for a good source of flowing water. Oxygen requirements of trout are such that ½ gallon per minute of flowing water from a natural source is required to sustain roughly one pound of fish. He estimates that ¼ acre would be required for the series of connected aquaculture ponds in his design. The trophic levels include algae, *Daphnia* (a kind of zoo-plankton or water flea), *Gammarus* (a small freshwater shrimp) and minnows (in North America these would be small members of the carp family). The smaller feed organisms (especially *Daphnia* and *Gammarus*) can be raised and sold in the aquarium trade or to scientific labs, as they are both in demand for feeding captive fish. Also, aquatic plants such as water chestnut and watercress as well as waterfowl are easily integrated with the design.

Readers in northern regions with good surface water resources would do well to consider raising trout. Native species are available, and the systems for culturing feed

Malaysian prawns are grown as an annual crop in open ponds in Kentucky. This filtration box between the ponds facilitates harvest in September.
[Credit: Keith Johnson]

Joint cropping of lettuce in floating trays with trout makes better use of nutrient.

[Credit: Peter Bane]

organisms are well articulated. I have seen simple but effective systems of trout culture in the mountains of North Carolina.

But for much of the middle and southern US, for all areas with limited water supplies and increasingly for northern regions facing climate change, warm-water species are more likely to offer success.

### Warm-Water Intensive Aquaculture

The most ecologically productive feeding systems are those in which filter-feeding crop species are grown on blooms of phytoplankton. It is thus possible to go from a waste such as animal manure through algae to crop fish in two trophic levels. The energy saved is transformed into fish flesh. This efficiency is reflected in the famously productive traditional Chinese carp polyculture, and has been more recently validated by experiments at the New Alchemy Institute with blue tilapia. They demonstrated that a once-a-year infusion of five pounds of cow manure in a 9,600-gallon tank would generate a large algal bloom that would provide the primary food source for the tilapia. The algae also aid in denitrifying and oxygenating the water and, by darken-

ing it, help with solar heating. Supplementary feeding with garden plants and an external filtration system enabled the New Alchemists to harvest 55 pounds of tilapia in a single 6-month season. The tank was located within a greenhouse. Yield from this low-intensity system was about ⅙ of that reported for Van Gorder's Recirculating System on a water volume basis, but is twice the maximum yield per square foot achieved in experiments with extensive outdoor polyculture conducted by the Agricultural Extension Service in southern Illinois. There, more than 1,300 pounds of fish were harvested from a 13,000-square-foot pond in a two-month season from a seven-species polyculture fertilized by the manure of eight hogs. The bulk of the yield in that Illinois experiment was from three Chinese carp species, all filter feeders.[29]

Clearly, filter-feeding fish are the way to go if you have limited space and want to reduce dependence on external inputs. McLarney suggests the native buffalo fish as a suitable filter feeder, though it, like the lowly carp, is considered a trash fish by many and has received little technical attention. Native crustaceans and molluscs also occupy

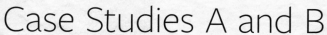

# Case Studies A and B

### Case Study A: Renaissance Farm

*See text description at the beginning of Part II.*

1. The farm as it appeared in March 2006 when we acquired the property.
[Credit: Keith Johnson]

2. The center of the system (between the two main buildings) as we begin to sketch out walkways, garden beds and a future patio.
[Credit: Keith Johnson]

3. Early work required extensive pruning and removal of older trees—an opportunity to lay in firewood.
[Credit: Keith Johnson]

4. The addition of a 48 × 20-foot high tunnel in the second year enabled us to begin year-round growing.
[Credit: Keith Johnson]

5. Water bars harvest runoff for gardens in the former front lawn. A young forest garden emerges upper center.
[Credit: Keith Johnson]

6. Young plums and a pear anchor fruit tree guilds north of the main house. Comfrey in the herb layer provides mulch and fertilizer.
[Credit: Keith Johnson]

1. After three years the oldest garden beds begin to quicken.
[Credit: Peter Bane]

2. Early to flower in the spring and fast to fruit, quince enlivens the understory.
[Credit: Keith Johnson]

3. Small heads on heavy-feeding cabbage reveal that the soils are still underdeveloped.
[Credit: Peter Bane]

4. Taking advantage of mild autumn weather, we erected a new porch to carry solar panels, and to provide a dignified entry and outdoor room.
[Credit: Peter Bane]

5. The summer following, photovoltaics were providing 100% of annual power use as the porch enhanced the utility of both house and garden.
[Credit: Edward Carter]

1. Unheralded work goes on throughout the growing season.
[Credit: Peter Bane]

2. In 2010, new ponds converted a former Zone of Accumulation for incoming materials into a productive area. The ZA—marked by strawbales—now lies farther south.
[Credit: Creighton Hofeditz]

3. New solar glazing on the south end of the house enhanced energy performance, while a new door brought easier access to laundry lines and a woodshed. A sour cherry blooms in this warm microclimate.
[Credit: Keith Johnson]

4. Sheltered from NW winds and benefitting from solar gain and heat from the adjacent building, this fig flourishes after five years in the ground.
[Credit: Keith Johnson]

5. Pink and orange mangels, an older type of fodder beet, yield heavily even in clay soils.
[Credit: Peter Bane]

6. Zinnia flower heads gathered for seed processing.
[Credit: Peter Bane]

1. Stacking functions: the Drying Yard accommodates wood cutting, a small pond, two beehives, and the laundry lines. All benefit from open ground, good solar exposure and strong SW breezes.

[Credit: Peter Bane]

2. Pears, tomatoes, figs and hazelnuts from the September 2011 harvest—all planted after we purchased the property.

[Credit: Peter Bane]

3. Pruning back, caging and insulating fig trees in the late autumn ensures their hardiness in a climate zone slightly too cold for them.

[Credit: Creighton Hofeditz]

4. A star of the 2011 garden was purple cauliflower which bore heavily for two months in fall and produced side shoots after primary heads had been harvested.

[Credit: Peter Bane]

5. By the sixth year, the system center had acquired pavement, a summer kitchen, generous garden beds all around and a pergola that will carry grape and kiwi vines. Contrast photo #2 on the first page.

[Credit: Peter Bane]

## Case Study B: Jerome's Organics

*See text description following Chapter 10.*

1. A narrow winding road connects Jerome's mountainside homestead with the town of Basalt, Colorado two miles away.
[Credit: Keith Johnson]

2. Construction of the new Phoenix greenhouse shows a main adaptive strategy for the challenging climate on Basalt Mountain.
[Credit: Jerome Osentowski]

3. In a terraced corner within the protective perimeter fence, students create new garden beds using straw, worm castings and other animal manures. The cut bank above the road shows the undeveloped nature of the original red clay soils.
[Credit: Jerome Osentowski]

4. The mature forest garden undergoes regular controlled disturbance as some plants are removed or die and new types and cultivars are planted into openings.
[Credit: Joshua Choate]

5. Hollyhocks have naturalized in the understory, offering near continuous blooms throughout the short growing season.
[Credit: Joshua Choate]

6. Herbs dry in this homemade device under bright Colorado sun.
[Credit: Warren Kirilenko]

1. One of many plums blooming in the forest garden.
[Credit: Jerome Osentowski]

2. Jerome's garden always has a profusion of beneficial insect life, be it domesticated or wild.
[Credit: Warren Kirilenko]

3. Several varieties of apricot fruit well in this high, cold desert climate.
[Credit: Jerome Osentowski]

4. A mid-season fruit in Colorado, gooseberry responds well to the dappled light through multiple stories.
[Credit: Jerome Osentowski]

5. Small and compact because of the dry climate, a typical fruit tree guild is seen in terraced beds: comfrey and horehound occupy the ground level, goose-berry fills the shrub niche (center), a plum (with white tree tube) anchors the assembly.
[Credit: Jerome Osentowski]

6. This small pond is a hub of life, resplendent with dragonflies. Cattails at the inlet (right) filter runoff from a road above the garden.
[Credit: Warren Kirilenko]

7. One summer day, a pair of ducks took up residence on Jerome's pond, the only body of still water for a mile in any direction.
[Credit: Warren Kirilenko]

1. Jerome demonstrates grafting, one of his key skills.
[Credit: CRMPI]

2. In 2009, this small flock of heritage turkeys lived happily on the surpluses of the forest garden.
[Credit: Peter Bane]

3. This thermostatically-controlled intake plenum pulls air from above the greenhouse beds and pumps it into perforated pipes laid beneath the soil where its heat and moisture are released.
[Credit: Warren Kirilenko]

4. Double-thickness polycarbonate panels now glaze all the greenhouses.
[Credit: Creighton Hofeditz]

5. After the fire, Jerome increased water storage by adding tanks to his downspouts. This one provides water to the small nursery around it.
[Credit: Peter Bane]

6 .The back of Phoenix greenhouse shows the protruding sauna (center) and a small workshop (right). The workshop roof will become a deck in this tiny enclosure where every square foot counts.
[Credit: Warren Kirilenko]

1. Greenhouse Mana provides a Mediterranean climate with well-established fig, pomegranate, rosemary and artichoke (center) among annual plantings of tomatoes and peppers. [Credit: Warren Kirilenko]

2. Rabbits are integrated with the garden in a cut-and-carry system. They provide a steady supply of clean meat and excellent fertilizer. [Credit: Warren Kirilenko]

3. The south aisle of Phoenix reveals how activities are stacked; even in the walkways, pallets protect habitat for worm beds below. [Credit: Warren Kirilenko]

4. "Pebble Beach" is the heart of Phoenix greenhouse where the designer becomes the recliner. This space doubles as Jerome's conference room and access to the sauna (at rear). [Credit: Warren Kirilenko]

5. Better climate battery technology and insulative design enable Phoenix to harbor true tropicals. Papaya fruits in the high Rockies in an atmosphere that never freezes. [Credit: Warren Kirilenko]

6. In the second year after planting, Phoenix had become a jungle where subtropical, Mediterranean and desert ecosystems blend. [Credit: Warren Kirilenko]

the filter-feeding niche and can make pond polycultures more productive. Conventional aquaculture follows the energy-intensive, centralized, chemical and corporate pathways, piling contamination into the food chain and perpetuating monocultural practices. So clever researchers, mostly backyard tinkerers with little funding, have turned to exotic species (such as the various subtropical tilapia) that are productive but are not banned because they cannot survive in native bodies of water.

### Alternate Feeding Strategies

Most fish are opportunistic feeders, and even herbivorous fishes such as tilapia are known to eat small animals when these are available in good supply. Insects, insect larvae, worms, grubs, mealworms, zooplankton and phytoplankton may be cultivated either in the fish tanks or nearby. Soldier fly larvae, crickets, small grasshoppers, manure worms and a number of other invertebrates are routinely cultured as bait organisms or for gardeners and composters and could become a staple source of protein for fish in tanks. *Daphnia* and *Gammarus* grown in small ancillary ponds with water plants in the manner described by Hutchinson, and harvested regularly, could supply a significant portion of feed for warm-water fishes, especially catfish. There are many advantages to this approach. Fish fed on natural foods almost never develop diseases as their nutritional needs are met. However, the challenges of feeding bulk vegetative material to fish in tanks are, besides labor in handling, that the density of nutrition is lower than in prepared commercial feeds, so the fish eat and excrete more. This can easily foul the filters in a recirculating system. And feeding is less efficient because of the bulk nature of the feeds: more will be offered than the fish will eat, and the residues must either be quickly removed or will contribute to fouling the tank water. Culturing small animals would seem to be a

better strategy to achieve adequate protein levels in tank-raised fish, though supplementing with vegetable feeds is certainly possible and recommended.

Tilapia and carp have been shown to accept (and even relish) high-protein leafy vegetables such as duckweed, azolla, amaranth leaves, carrot tops, comfrey, alfalfa, hairy vetch, purslane, sheep sorrel and squash leaves. As a way of addressing the problem of bulk, these materials, especially the water plants azolla and duckweed, might be dried and crushed before feeding in a tank culture situation. Good design would place these feeds and their drying equipment close to the fish-growing tank for ease of handling.

Tilapia have been shown to accept pelletized rabbit feed (basically alfalfa and some vitamins), though the pellets fall apart quickly in water. The ingredients of catfish feed are well researched and tested. They or similar feedstocks are relatively easily grown at home, many substitutes for corn and soybeans are available, and the technology for pelletizing may be more accessible than assumed. Pelleting machines are becoming more widely distributed as interest in pellet stoves increases. (In our southern Indiana community, for example, a pellet producer has recently set up shop and is looking not only for fuel customers, but for alternative economic activities in the off-season.) It would not take many customers to justify a run of custom pelletized fish feed. Organic fish food is still not available in the US, and early entrants into the market could realize a significant premium.

Though the answers are not yet certain, the need is clear, and the path toward homegrown fish food is well marked out. Once you have stored water, it only makes sense to enrich both it and yourself with fish and all their by-products. Indeed, dedicated fish tanks create an opportunity for even more water storage to the ultimate benefit of the land. The smart garden farmer will begin experimenting soon.

# Jerome's Organics, Basalt, Colorado, USA

[*Color Insert #1 includes photos of this property.*]

**LOCATION:** Mid-elevation, south-facing cove above the Roaring Fork Valley on the western slope of the Rockies, elevation 7,200 feet. Immediate neighbors barely visible: about five houses within ½ mile. Historic climate zone 4a, rapidly trending toward zone 5b with 17 inches precipitation per year, much of it as snowfall.

**DISTANCE TO URBAN CENTER:** Two miles over a winding mountain road.

**ECOSYSTEM:** Pinyon-juniper forest with Douglas fir in moister locations and copses of scrub oak. Alkaline red clay soils formed from igneous rock. A full suite of wildlife: mountain lion, bighorn sheep, elk, mule deer, turkey, black bear and smaller creatures range through an extremely vertical landscape dissected by steep ravines. Small permanent streams swell to torrents briefly in a handful of rain events during the year.

**SIZE OF PROPERTY:** Eight acres, of which about ½ acre is fenced and gardened under irrigation.

**OPERATOR:** Jerome Osentowski, assisted by regular interns throughout the year.

**ESTABLISHED:** 1982.

**PRODUCTS SOLD:** Salad greens, fresh herbs, nursery stock, figs from greenhouse cultivation.

**SERVICES OFFERED:** Education and design.

**MARKETS:** Onsite visitors to educational programs and demonstration garden, local food shops.

Jerome grew up in a family of smallholding gardeners, foragers and hunters in Nebraska but moved to the central Colorado Rockies after service in the Air Force. He saw many gardens while in Europe in the 1960s and came home with a taste for fresher food, envisioning a more sophisticated form of growing than he had known as a youth. The rugged mountain landscape where he settled sits astride the old coach road to Denver, but now only local traffic negotiates the deep clay ruts and blind hairpin curves of a narrow track.

Jerome enjoys hiking the meadows above his land and the proximity of wildlife, yet town is literally within walking distance. The area enjoys a favorable microclimate with some 300 days of sunshine a year. Natural hot springs at the mouth of the Roaring Fork Valley made Glenwood Springs an early tourist attraction, but the area's export economy was based on coal mining at Carbondale. The past half century has seen explosive growth driven by skiing and tourism anchored on the very wealthy town of Aspen at the upper end of the valley. Fruit growing is still economical in the sheltered valleys of the Colorado River mainstem and over the mountains at Paonia.

Sunshine, water and soil are the minimum requirements for growing plants. This site has an abundance of the first, a slender but steady supply of the second and began its life as a garden farm with very poorly developed soil. Jerome's operations have always depended on pumped water from a small spring below his house. Ultimately the source of the spring is snowmelt from a lens of water under scree-covered slopes at the top of Basalt Mountain. This oasis or riparian pattern of cultivation matches the profile of drylands around the world.

Virtually everything productive about the site has come from Jerome's persistence and ingenuity. Frequent frost during the growing season and long winters with lots of sunshine cried out for cloches and greenhouses to enhance the microclimate of this wind-sheltered cove. Spoiled hay from the ranches in the valley below and leaves and food wastes from the town of Basalt became the basis for compost that built garden soils where none had existed. A sturdy fence, put together from scrap materials, keeps the abundant wildlife at bay and the garden plants and small animals Jerome raises for meat and eggs safe from predation—most of the time. From the earliest days when he grew salad and greens under cloches for winter sale and developed mixes of baby greens before they were commercially common, every delivery to town gave him an opportunity to backhaul scrap lumber, roofing or soil-building materials.

After working for a few years as a carpenter in the summer to support his winter ski habit, Jerome developed skills to build his own structures. His present dwelling is located along the main road, above the property's largest area of moderate slope and near its water source. A narrow, vertical house stair-steps up the south-facing hump of a central ridge. Draped in gardens on three sides, the shed-roofed wooden house cradles many tons of native rock wall behind its big glass windows. Over the years, greenhouses, porches, storage rooms and slender additions have added more living and growing space, and the wooden siding has lately been clad in steel roofing panels as protection against wildfire and the harsh effects of sun and snow. What ennobles this house is the uplifting view from the main living area where an open plan connecting kitchen, eating and sitting areas overlooks wooded slopes across the valley and a massive fig tree in the oldest greenhouse.

Emblematic of Jerome's approach, the fig (a Mediterranean native) has been happily bearing fruit high in the snowy Rocky Mountains for two decades, nurtured in the shelter of a long, unconventional wooden and plastic greenhouse that generates heat for the dwelling and a variety of herbs, salads and fruits for its owner. In a quarter century the fig has grown to all but fill the 20-by-25-foot space, though by persistent pruning, Jerome manages to make room for tomatoes, sprawling rosemary bushes, a pomegranate, artichokes, a water tank and a changing array of salad and vegetable crops. One of Jerome's ancient cats, nearly as old as the tree, lives out her dotage napping under the spreading boughs. The figs begin to ripen in late August and continue bearing through October as frost settles on the golden leaves of the aspen-covered slopes outside. The harvest has reached such proportions that not only can Jerome and an annual influx of permaculture students eat their fill of figs during the peak of the season, but there are usually a number of boxes to sell to special customers: Colorado mountain-grown figs are an exotic delicacy that commands a good price.

Like some alpine flowers, Jerome has developed and refined a tiny but persistent niche in this mountain community. The wealth upvalley at Aspen has provided not only trickle-down materials, but a market for high-quality produce. Winter salads, baby greens, edible flowers, fresh herbs, wheatgrass and its juice and botanical preparations, each in their time, have been some of his crop successes. The rich crowd are fickle, and competition for their dollars has been fierce. Winter salad mix and baby greens built Jerome's first greenhouse and launched his business, but eventually California growers moved into that market and undercut the premium he had been getting for his novel and well-grown product.

Greenhouses gave Jerome an early economic advantage, enabling him to raise crops that couldn't otherwise be grown for a thousand miles around. The logic was compelling: a south-facing site; cold, dry, clear air; frosts at any month of the year and sunny days at all seasons just begged for solar design. Early collaboration with then-Colorado-based designer John Cruickshank led them to develop innovative ideas for in-ground greenhouse heating.

The systems employed at Jerome's site draw warm moist air from the upper reaches of the greenhouse down through large tubes into a network of perforated pipes running under the growing beds. A fan located in a central plenum and controlled by thermostats moves the air. An important principle of physics that makes these active solar systems so effective is the phase change of water: vapor in the air high in the greenhouse condenses when that air is pumped into the cooler ground. The energy thus released is enormous, many times more than that required to cool the air by a few degrees. This effects a much greater heat transfer. Both moisture and heat are absorbed by the soil, so that cooler, drier air is returned to the greenhouse atmosphere. This prevents daytime overheating and keeps the plants healthier. At night, thermostats can reverse the flow of air so that warmed air from below the soil is circulated into the cooler greenhouse to buffer its fall in temperature. By planting crops able to bear cold temperatures, Jerome can grow throughout the year. With their roots in soil at 65°F, the plants can easily take air temperatures in the greenhouse below freezing, though with the latest designs, even these light frosts have been virtually eliminated.

Each iteration of greenhouse design improved on the previous one, and each had a different mix of plants, but over the years all acquired perennials. Each time Jerome would make a foray into the tropics (where he was fond of spending part of the winter) to do a permaculture project, he would learn more about warm-climate crops. Sometimes he would return with seeds or cuttings that would take root in one of the houses. Little by little, the Roaring Fork Valley became home to bananas, chayote vines from Central America, pomegranates, jujubes, aloes, papaya, citrus, agaves and a myriad of smaller plants.

Long before tropical ecosystems began to flourish on Basalt Mountain, however, permaculture influence helped propel Jerome into planting more and more perennials, and as the forest garden meme gripped him, he saw the possibilities of a permanent outdoor garden. The search for hardy varieties of fruits for a climate that naturally stunted most deciduous trees presented a comprehensive design challenge and led Jerome on an extended tour through nursery catalogs and germplasm repositories. He learned that he needed not only tough rootstocks, but standard-sized trees for their vigor. Even so, many 10-year-old bearing trees stand little taller than the diminuitive gardener.

A pond helped soften the microclimate of the small bowl west of the house, but in order to exploit the valuable environment thus created, Jerome needed to terrace the steep slopes around the pond to create planting areas. He brought in many tons of stone to face the terraces. These in turn garnered more heat from the sun and warmed the area even more. The stones also provided abundant habitat for garden snakes, lizards and other pest predators. Gradually more and more of the garden was given over to perennials. At the same time that nursery plants became more of a cash crop—in conjunction with landscape design services— Jerome moved his annual vegetable and salad production into the marginal spaces between groups of trees and shrubs. The young perennials grew somewhat slowly, and the intense Colorado sunshine meant that these semi-shady areas around and underneath the young trees afforded excellent conditions for brassicas, spinach, lettuce, orach and other vegetables, many of which became self-seeding endemics in the evolving ecosystem.

One of the lessons taught by 15 years of forest gardening is the need for periodic disturbance from the gardener. As the older areas closed canopy, the diversity of understory plants dropped, making it apparent that pruning and removal of some plants would make the garden as a whole more productive. The need for periodic pruning of the overstory found a regular expression as Jerome developed a passion for nitrogen-fixing trees and shrubs. The extra fertility they harvested helped the interplanted fruits and nuts to grow with more vigor, and this was accelerated when they were coppiced or pruned back.

Jerome's strategy was to plant these nitrogen-fixing plants throughout the garden so that their roots interpenetrated with those of the crop trees, shrubs and vines, then to coppice the nitrogen-fixers regularly for mulch. Each cutting of the tops results in a release of nitrogen underground as some of the microbial associates which actually harvest nitrogen from the atmosphere are starved of sugars that the plant had previously supplied. This dieback of bacteria releases stored nitrogen which can then be nabbed by the plants' hungry neighbors.

Experience with various guilds in the forest garden led naturally to creating a perennial backbone for the greenhouse ecosystems with guilds of subtropical, desert and Mediterranean plants in different zones of the four structures. Jerome spent so much time in his two large greenhouses during the winter that he began to find them a welcome subtropical relief from Colorado's long, cold winters. He set up a hammock and an archery range (dubbed Pebble Beach after the famed California golf resort) in the free-standing greenhouse which he named Pele, after the Hawaiian goddess of the volcano. The greenhouses had become more than simply adjuncts to the farm; they were living rooms full of plants, food and sunshine—a kind of tropical paradise burgeoning on the harsh slopes of the Rocky Mountains.

 ## NITROGEN-FIXERS USED OR TESTED AT JEROME'S ORGANICS

1. Siberian Pea Shrub (*Caragana arborescens*). Best coppice.
   The following varieties and related species can be used in a design as ornamentals, but are not necessarily good for coppicing:
   a) Thin leaf (good ornamental)    b) Pink flowering    c) Dwarf and Semi-Dwarf
   d) Weeping    e) Johnson Calimer
2. Mountain Mahogany. (*Cercocarpus montanus*). Native to six feet. Jerome coppices these.
3. Bitterbrush (*Purshia tridentata*). None in garden but plenty on the mountain. Deer love it. Low profile.
4. New Mexico Locust (*Robinia neomexicana*). Native to 25 feet.
5. Goumi (*Elaeagnus multiflora*). Slow grower; edible berries.
6. Russian Olive (*Elaeaganus angustifolia*). Vigorous. Good bee plant. Four trees onsite. Colorado has now banned planting these.
7. Sea Buckthorn (*Hippophae rhamnoides*). Undergoing trials.
8. Wild Indigo (*Baptisia tinctoria*). Grows well. Coppicing questionable. Nice ornamental. Also used as dye.
9. Thinleaf Alder (*Alnus incana ssp. tenuifolia*). Grows down at the river. Not yet established at the site but well-adapted.
10. Leadplant (*Amorpha canescens*). Small shrub to three feet, low profile. Not good coppice.
11. Amur (*Maackia amurensis*). From Russia. Medium tree to 40 feet. Needs acidic soil. No success at this location.
12. Broom (*Cytisus praecox*) Spreading shrub to five feet. A little tender and potentially dispersive.
13. Chinese wisteria (*Wisteria sinensis*). Expansive vine.
14. Black medic (*Medicago lupulina*). Inconspicuous native groundcover. Very helpful in revegetation efforts.

Like the winding and hazardous access road, water has been a sore point over the years. The spring catchment below the house never failed to run water, but maintaining a pump in conditions where any rainstorm might bring a wall of mud down the ravine was challenging. Over time, fluctuations were ameliorated. Mulch on all the garden beds kept evaporation to a minimum. The pond helped stabilize immediate supplies, but pumping was always required. After a greenhouse fire, Jerome installed plastic cisterns to collect roofwater (never enough in this dry climate, but helpful) and added another large tank up the hill into which he could pump extra spring water for subsequent gravity flow irrigation. In recent years, he has acquired the use of ditch water running around the mountain onto his land from above. And though one can never take water for granted in an arid region, Jerome seems to have as secure a supply now as anyone could hope for, backed up as it is by at least three sources.

If you tried to measure the outputs of Jerome's garden farming efforts, you might at first glance wonder where the evidence was for success. The productive area is about ½ acre. He no longer sells large amounts of salad greens. Herbs are a minor crop. No large trucks full of produce are pulling out of the parking lot. But understood as a complex system with a succession of many crops and yields, and as a 30-year demonstration of resilient living amidst harsh conditions, Jerome's Organics and his Central Rocky Mountain Permaculture Institute (CRMPI) stand as exemplary features of the local and national landscapes. Organic soils have been built and maintained from local resources. Mature apples, grapes, plums, pears, cherries, apricots and walnuts are bearing in the garden, as well as berries of several kinds. He raises small animal protein and secures local supplies of cultivated and wild meat for the freezer. The home food system is robust. Subtropical and desert fruits

pour out of the greenhouses every year, enough for sale of surplus. Groundcovers like the modest little spilanthes cover many greenhouse beds and yield botanicals for sale. Heritage varieties of local fruits have been collected, catalogued, propagated, sold and planted into public and private landscapes up and down the Roaring Fork. CRMPI has started and continues to run a CSA-farm school that trains young people as future farmers. There are forest gardens at the Basalt and Carbondale high schools and middle schools, and since 2010, a climate-battery grow-dome at the Basalt High School. Hundreds of design students—many of them now teachers in their own right—and thousands of other visitors have seen paradise in this cold desert, and been inspired to do amazing things across the valley, the state, the mountain region and the continent. CRMPI has produced a DVD, *Natural Controls for Noxious Weeds*, addressing the growing problem of widespread herbicide use,[1] and Jerome is at work on a book about indoor growing spaces. He continues to host a regular feature on local radio station KDNK introducing his listeners to permaculture and food system issues. Young people still beat a path up the hill to visit and learn from the old man on the mountain who seems to be having more fun with every passing year.

For a household without children, however, the question of succession is ultimately paramount. How does the value created over a lifetime transfer into new hands and leave the original owner with sustenance for retirement? The next turn of the wheel here is to spin off a number of interdependent livelihoods from the farm's rich resource base that can be taken over by younger partners, leaving Jerome to step back and pursue creative endeavors such as design, writing and making films. From a steady crop of young annual recruits, the old-growth designer has been quietly scanning and selecting for a few hardy new pioneers.

# CHAPTER 11

# Soil—the Real Dirt

*The soil is an animal which is everywhere a mouth.*
— Emilia Hazelip

*The truth is that nobody has ever exposed a scientific reason to till.*
— Edward H. Faulkner

Soil is the edge between the living and the dead. The top few inches or feet of the earth beneath us holds the key to all terrestrial life. Recycling waste matter and the dead bodies of plants, animals and countless microbes, soil organisms are the proximate source of the building blocks for all organic form.

Paradoxically, soil both retains impact, remembering events from the deep past, and it can change rapidly. Nearly five centuries after European explorers first saw expansive agriculture and dense settlements along the banks of the Amazon River, the deep, black, carbon-rich and human-created *terra preta* (dark earth in Portuguese) soils of native farmers continue to grow abundant crops.[1] Yet of all the elements that contribute to and limit agriculture — climate, landform, water, vegetation, cultural factors and others — soil is the most changeable and the most rapidly modified. This is a condition fraught with hope and peril.

## Soil Is Alive

Soil is visibly composed of mineral particles, but these are only the outward habitation of vast numbers of bacteria, fungi, protozoa, nematodes, earthworms, ants, termites, pillbugs, springtails, millipedes, rodents and hundreds of other kinds of creatures, the bulk of them invisible to the naked eye. Living organisms in the top foot of soil may amount to 11 tons per acre.[2] While the physical condition of mineral soil (its surface contour, porosity, texture, compaction and parent rock) can dramatically affect the quality and abundance of soil life, without a vigorous community of soil organisms, no plant growth can be sustained.

Soils are created from an interaction of *energy* (chiefly from the climate and living organisms which transform it into action) with *resistance*, expressed as a function of geology — or parent rocks — and the resulting landforms that they reveal over time. The weathering of rock creates habitat for

187

soil organisms at the same time that it grinds minerals into finer and finer particles which can be eaten by plants and microbes, and even by small soil animals. Nutrients cycle up and down, and in healthy ecosystems they are held closely by the embrace of living organisms and soil particles. Though some nutrients are always being lost through erosion and leaching, in every terrain material is also accumulating from wind-borne and alluvial deposition. Animal migration works in both directions carrying nutrient across porous boundaries. Generally, these processes remain in balance over very long periods of time.

Soil in a natural state is by no means quiet. Nor is it solid. Rather it is filled with the action of plant roots, fungal hyphae and burrowing animals which distribute living and dead organic matter from the surface to deeper layers, pull minerals from parent rock and open channels for air and water to circulate. Small soil animals, searching for food, consume dead plant and animal tissue and excrete pellets, casts, frass and other manures which bind bits of inorganic material with organic waste and digestive secretions to form *colloids*. Earthworms are particularly effective and powerful in this way, leaving behind them not only burrows that become channels for air and water, but nearly perfect packets of plant food in the form of their castings. Their digestive secretions, called *mucopolysaccharides*, act as a glue to bind soil particles into clods or

lumps, giving the soil as it develops over time a structure that is hospitable to its plant and animal inhabitants, resistant to erosion and filled with niches for microbes to occupy.

The action of soil organisms is manifold, subtle and immense in its impact if allowed to proceed undisturbed. Hordes of bacteria proliferate by gobbling up soluble carbohydrates, released into soil by the profligate photosynthesis of green leaves, and also available from the dead tissues of plants, animals and microbes strewn over and entering into the soil surface. These microbes become in turn prey for larger but still invisible protozoans and nematodes. Fungi, also benefitting from the banquet of sugars and starches offered by plant root exudates, extend their digestive apparatus — which consists of long, single cell-thin filaments called *hyphae* — through the soil in search of minerals. They mobilize these nutrients and can transfer them directly into plant root tissues across wide areas that may extend over thousands of acres. Fungi, of which there are many types, can also pass chemical signals from one plant in the network to many others. These mycelial webs have been compared to the Internet in their reach and coordinating function.[3]

Both plant roots and the soil microbes that flourish around them produce weak acids which help to break down the parent minerals in the soil, making them soluble and available for uptake. At the same time, rainfall and heat act to dissolve, leach and oxidize these mineral ions. All forms of soil life depend on relatively stable conditions that allow the percolation and retention of water and the steady infiltration of air into upper layers of the soil.

## Soil Layers

All of this busy activity on and beneath the soil surface is stratified by layers.

The soil, which is a product of both weathering and digestion, is built from the top down. The *Organic Layer* on the soil surface consists of undigested and unconsolidated

Soil is alive with countless microbes, fungi, plant roots and small animals whose actions build healthy structure and create fertility.

[Credit: Jami Scholl]

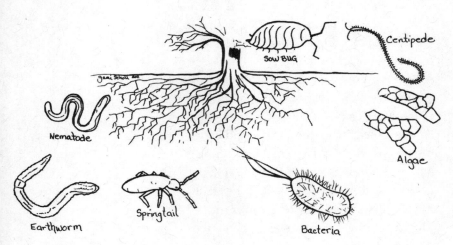

detritus: the dead parts of plants and animals that have fallen to the earth. It functions as a cover for the soil, breaking the fall of raindrops and preventing erosion. It is also a layer of food for decomposing organisms, and many creatures — from squirrels and pigs to earthworms and pillbugs — forage here. Any soil without an organic layer is vulnerable or already damaged. Even in deserts, where there appears to be little material on the soil surface, algae, mosses and fungi form delicate cryptogamic crusts of organic matter that retain the soil particles below them.

Lying just below the Organic Layer is the *Accumulating Layer*. It is the "A" or topsoil horizon of conventional soil science. Here the organic residues of decomposition accumulate. Many organisms have their homes in this layer, from earthworm burrows and anthills to mole tunnels and groundhog dens. Aside from the organic detritis directly above it, this layer has the highest fraction of carbon of any of the soil layers. When organic matter content is measured in a soil test, the sample is taken from this layer. In temperate regions healthy soils may have from 10 to 20% organic matter in the "A" horizon, while in tropical regions, for reasons we will see below, 5% organic matter is quite good.

Below this lies an *Elluviation Layer* from which minerals are being leached by the percolation of water. Tillage effectively mixes the Accumulating and Elluviating Layers, leading to a leaching or washing out of fine soil particles and the creation of a plow pan below. Next is the "B" or *Banking Layer* in which some of those leached minerals are trapped and other mineral nutrients are held by clay particles. The "B" layer is usually called *subsoil*; it has low levels of organic matter and many fewer roots. Below the upper subsoil lies a "C" or *Chemical Layer* of broken and weathered rock and coarse, unconsolidated soil particles. Below this is the bedrock or *Durable Layer* of the soil. Bedrock is not always solid but may be fractured and permeable,

either due to its chemical composition (limestone is easily dissolved by water) or due to seismic activity, faulting or geologic uplift.

Most plant roots spread vigorously through the upper layers of the soil, and this is indeed where most of the nutrient is made available. It is also where the gardener and farmer have the greatest influence both directly, through mulching, tillage and planting — and indirectly, through our influence on plants and animals that work the soil. Since soil fertility is directly related to mineral balance and organic matter content, it behooves us to understand how organic matter gets into soil. We can alter mineral balance too — and we will examine how this is done, but it is harder and also, as we will see, less often necessary. Particles of organic matter (a measure of carbon content) along with clay particles hold tightly to mineral ions, so by building up the carbon in our soils, we create a bank of available minerals from which our crops can make the withdrawals they require.

As well it should be remembered that 97% of the bodies of plants, animals and microbes consists of four elements: hydrogen, oxygen, carbon and nitrogen. All of these occur as gases in the atmosphere and so are cycled and recycled out of that immense reservoir — everywhere available with few limits. This is the good news and the main reason why soils can remain self-fertilizing over very long periods of time. The remaining 3% of organic form consists of phosphorus, potassium, calcium, sulfur, magnesium and trace minerals, and it is these that typically constitute limiting factors to plant and animal growth and health. So our focus in soil science, as in soil repair, is to follow the minerals. Healthy ecosystems retain them; most landscapes and especially

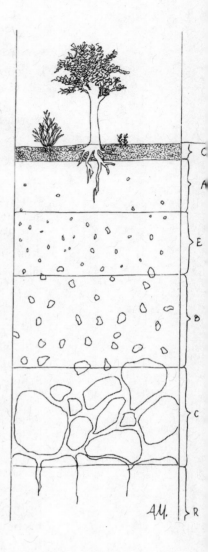

Proper soil management leaves soil layers intact and uncompacted.

[Credit: Abi Mustapha]

agricultural fields have suffered moderate to severe mineral loss, which is reflected in unhealthy plants and an epidemic of degenerative disease in modern human beings.[4]

## Nutrient Digestion and Assimilation

Under forests and grasslands not broken by the plow — in other words, in healthy soils — organic matter accumulates from the growth and decay of plants. Plants not only live and die; they shed roots, leaves, twigs and limbs with regularity. All of this material is food for microbes. The rate of growth and decay of plants, and the resultant cycling of nutrients, is a function of moisture and temperature: the warmer and wetter it is, the more growth and rot will occur, and the faster nutrients will move.

Though it is everywhere threatened or damaged by agriculture and urbanization, soil is not uniform over the globe, neither in its physical makeup nor in the dynamic of its life. Temperate soils act as a reservoir of nutrients for the plant life that colonizes them because the green cover is dormant for at least half the year, most of its life force withdrawn below ground. In tropical regions, plants only become dormant where there is a pronounced dry season. Photosynthesis, growth and decay are continuous in the presence of warmth and moisture, and so is the uptake of nutrients. Thus, tropical soils hold very little nutrient — little more than 20% of the mineral fertility in the ecosystem, contrasting with 95% or more in temperate regions. In temperate climates, soil nutrients pulse upward into plants during the growing season, and subside into soil during winter, whereas in tropical regions, they move rapidly both up and down, spending little time idle in the soil. The most important consequence of this insight is that very different forms of agriculture are needed in tropical regions than in temperate countries.

In temperate climate regions, soil is built by the development of durable carbon in the form of *humus*, long-chain molecules created out of biomass by living organisms in the soil. Soil nutrient attaches to humus and also to clay particles as we will see below. So building soil fertility in cool regions means adding organic matter to soil from the top down as mulch. In tropical countries, nutrient is accumulated and held primarily by living plants. Mulch is also important here to keep soil temperatures cool, but it is not enough. It becomes imperative, in restoring degraded tropical ecosystems, to plant vigorous pioneer trees — many of them nitrogen-fixing — plus large herbs and shrubs such as papaya, banana, agaves and castor bean to create soil cover and a plant community that begins the cycling of nutrients. That forest cover must be expanded, enriched and maintained. The plant biomass is the primary reservoir of nutrients; it is vain to imagine that nutrients remain in devegetated soils. They are baked and leached away very rapidly.

Bacteria are the most common organism in soil: a teaspoon contains millions of them. They fall broadly into two categories: aerobes and anaerobes, which flourish in the presence or absence or oxygen respectively. *Anaerobic bacteria* are of an older evolutionary lineage. They evolved during a period in the earth's development prior to photosynthesis and the creation of an oxygen-rich atmosphere. Although there is a common belief that anaerobic soil conditions are harmful and to be avoided, this is only partly true, and the misunderstanding is the cause of much mischief. All healthy soils have many microsites where at any given time anaerobic conditions prevail. This is a normal part of the cycle of digestion and assimilation of nutrients.

All the bacteria that consume dead organic matter are *aerobes*: they need oxygen to function, just as we do. Being rugged little beasts, however, they do not expire when deprived of oxygen temporarily, but merely slow down their metabolism and become inactive. That is, they cease to eat. All plant pathogens are also aerobes, but of all the aerobic bacteria, they are the most sensitive to oxygen deprivation, falling asleep more quickly than run-of-the-mill mulch munchers.

## The Oxygen-Ethylene Cycle

Oxygen is always infiltrating soils from above—more readily when they are not compacted. But as aerobic bacteria digest organic matter, they use up the oxygen in the soil around them. It is as if everyone in the gym is huffing and puffing away on exercise bikes or pressing weights or running laps, but the ventilation is shut down. Oxygen levels drop, and as they do, first plant pathogens and then other digesting bacteria abandon their work and go to sleep. As oxygen levels continue to fall, anaerobic bacteria become active, multiplying in number. Their metabolism produces a gas called ethylene (the same gas given off by brown paper bags and that ripens apples and bananas). Ethylene further retards the aerobes. It is present in all healthy soils at trace concentrations of one to two parts per million, and its production depends on the presence in soil of a precursor, found primarily in old and dead plant matter, reacting with ferrous iron, about which we will learn more below.

One of the most common soil nutrients is nitrogen, a component of all proteins and amino acids, essential for green plants and often a limiting factor in their growth. Nitrogen occurs in two common forms in soil, *ammonium* nitrogen ($NH_4^+$) and *nitrate* nitrogen ($NO_3^-$). Aerobic bacteria digesting dead organic matter liberate ammonium nitrogen. Ammonium is the dominant form of nitrogen in undisturbed soils, but is almost never present in tilled agricultural soils, where nitrate prevails. Ammonium is positively charged and bonds tightly to particles of clay and organic matter, so it is held in soils until needed, but when more is accumulating than is being used (because, for example, a tree in the area has died, ceased to take up nitrogen and also is being decomposed, releasing extra

nitrogen), then the surplus ammonium is converted by other soil bacteria (called nitrobacter) into nitrate, which may be lost back to the atmosphere but is also readily soluble in water. In this way, unused nitrogen can dissolve and flow away to where it may be used by living plants.

In the absence of nitrate nitrogen—that is, when soil digestion is normal rather than accelerated, and when soil conditions locally become anaerobic (because digesting bacteria have temporarily exhausted the oxygen in the soil)—iron, which is common and abundant (between 2–12%) in all soils, changes its valence. It transforms itself from ferric (red or rusty) iron to ferrous (black or reduced) iron by releasing an atom of oxygen into the now reduced atmosphere of the soil. This has tremendous implications for soil fertility and agriculture because it initiates the assimilating phase of soil nutrient transfer to plants.

Ferrous iron reacts with a precursor found in leaf litter to convert it into ethylene gas, helping to further inactivate aerobic bacteria. Iron in its ferrous form also sheds its bonds to phosphate, sulfate and ions of trace minerals, dumping them unceremoniously into solution. And this new and rather promiscuous ferrous iron then forms bonds with particles of clay and organic matter. As it does so, ions of magnesium, calcium, potassium and ammonium which had been occupying those sites are displaced, also into solution. The highest concentration of this activity is in the rhizosphere or tiny envelope of water surrounding the root hairs of plants.

Thus, as soil microsite conditions change from aerobic to anaerobic, all the plant nutrients—nitrogen, phosphorus, potassium, sulfur, calcium, magnesium and trace minerals, which had until then been held tightly to soil particles—enter into solution and can in that form be taken up by plant roots. This does not occur when the soil is constantly aerobic, nor does it occur when nitrate nitrogen—the form typically applied as chemical fertilizer—

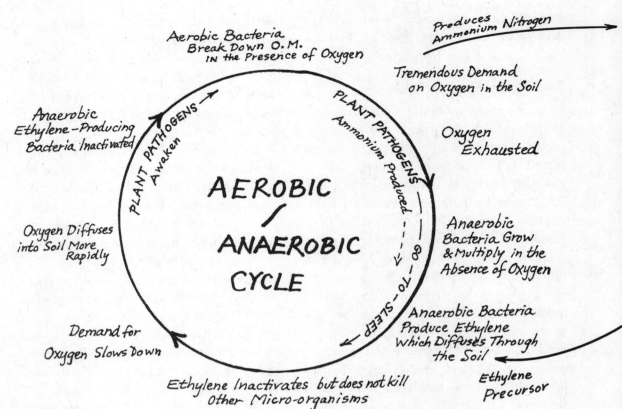

The Oxygen-Ethylene Cycle, an essential function of healthy soil in the rhizosphere.

[Credit: Peter Bane]

is present in large quantities. So industrial farming, by tilling soil and applying nitrate fertilizer, is preventing normal processes of soil nutrient assimilation by plants.

In turn, plants feed soil organisms at their roots in order to enhance water and nutrient uptake and also to stimulate respiration by aerobic bacteria which cyclically exhaust oxygen in the rhizosphere, initiating anaerobic conditions, the production of ethylene gas and nutrient assimilation.

This entire process is cyclic, a rocking back and forth at millions of microsites throughout a healthy soil between aerobic and anaerobic conditions, between digestion and assimilation. Both phases are necessary for dead matter to be consumed by microbes and resurrected by living plants. Conventional agronomy fails in every way to respect this normal and natural process. Indeed, it deliberately disrupts soil feeding work, substituting industrial inputs for the normal processes of soil self-fertilization.[5]

## Agriculture Consumes Its Own Resource Base

Since its inception at the close of the last Ice Age, agriculture has advanced, with few exceptions, by the destruction of forests in temperate climates. Under these woodlands, the farmers of Sumer and 50 centuries later their cultural descendants crossing the Alleghenies into Ohio both found deep reserves of nutrient-rich soil, charged with thousands of years of accumulated organic matter which enabled them to prosper without regenerating or properly feeding their soils — for a time. In many cases — across the black earth belt of Ukraine into Poland, Germany and France, and in the Great Lakes and northern prairie regions of North America — glaciers deposited mineral-rich topsoils from farther north many feet deep, and when the ice withdrew, these were colonized by prairie grasses or forests which continued building organic matter for thousands of years until the first farmers broke the sod or burnt the woodlands. In

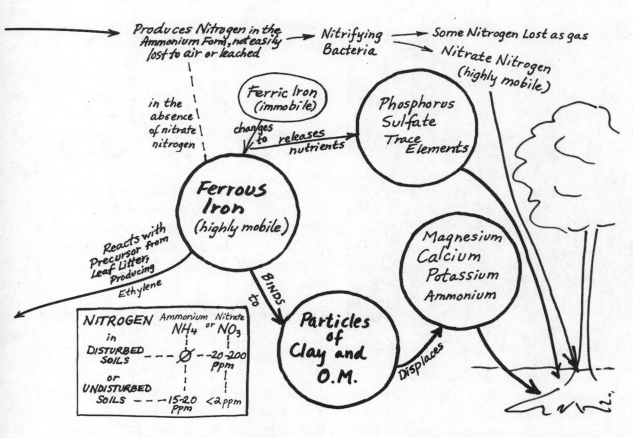

other locations, such as Mesopotamia, Egypt and the Yellow River basin, agriculture began in well-watered valleys where great rivers had deposited layer upon layer of alluvial silt. In all cases that we know, agriculture began with a rich resource of soil created by nature, but, failing to understand how soil fertility was created and maintained, early farmers degraded the land, often leading to the collapse of civilizations. This ignorance was perpetuated and its consequences deferred by westward expansion from the Mediterranean basin into Europe and again from Europe to the Americas and the Antipodes.

Shifting cultivators in the tropics practice *swidden* where population densities are low enough. Swidden farmers clear small patches of forest — usually by burning — and plant crops into the openings thus created. The ash released by the fire provides a quick burst of mineral fertility, but this is soon exhausted, and within a couple of years, the farmer must open another patch, repeating the cycle in a new location. Thus the need for large territories, as forest regeneration takes 15–20 years to accumulate minerals in the bodies of trees to a level that can sustain another burn-and-crop cycle. This was possible in Africa, Central and South America and Southeast Asia for many centuries where population densities remained low, but has ceased to be a viable option nearly anywhere since the early 20th century.

But swidden can also be incorporated into a more sophisticated form of perennial agriculture by observant growers. The Lacandon Maya of Guatemala and southern Mexico practice slash-and-burn agriculture, but they incorporate two variations that help to sustain their land use. Firstly, they char much of the felled wood from tree clearing using *pyrolysis* (heating in the absence of oxygen) to produce charcoal which can permanently enrich the soils with stable carbon. The charcoal serves much the same function as humus, which is itself a complex of long-chain carbon

Traditional charcoal-making in Central America.

[Credit: Robert Frigault]

A soil sample agitated in water will gradually settle out to reveal soil particle composition. Sand drops first to the bottom, silt next and clay last, forming the top layer. Organic matter floats.

[Credit: Creighton Hofeditz]

Secondly, the Lacandon follow (or rather lead) the natural succession of the ecosystem from clearance back toward forest by cultivating perennial and woody species of fruits, nuts and other economic plants, so that they enjoy a prolonged harvest over a generation from each disturbance, gradually converting native woodland into a semi-cultivated food forest.[6]

The conclusion that is forced upon us is that agriculture must now become responsible not only for growing food but for growing soil. The remainder of this chapter will focus on the methods for building living soil, beginning with assessment.

## Taking the Measure of Soil

Soils are classified according to texture, structure, series and slope. Other measures we apply to soils include portion of organic matter, color, acidity or pH and mineral balance.

*Soil texture* is a product of geology and weathering and can be assessed by the hand and other simple tests. Soils may be gritty, spongey, buttery, smooth or sticky, reflecting the relative proportions of sand, silt, clay and organic matter present. These can also be measured directly by mixing a small amount of the soil in a quart jar of water, shaking and allowing the resultant slurry to settle out. Particle composition affects pore space — and thus nutrient and water-holding capacities and drainage.

*Soil series* are names (e.g., Haworth, Bartholomew or Niagara) associated with soils of a distinctive texture in a given locale. Across the US and Canada almost all soils have been mapped, classified and evaluated as to their suitability for agriculture and other purposes, and this data is a matter of public record available through government agencies, the Natural Resources Conservation Service of USDA and Agriculture Canada.

*Soil structure* is a function of organic action and traction. It also affects pore space. Soil may be platey, prismatic, blocky or

molecules. Both charcoal (also called *biochar*) and humus molecules have large surface areas, stable chemical bonds and a negative electrical charge which gives them an affinity for the mineral ions used by plants. Their large surface area also provides excellent habitat for soil microbes.

granular in a continuum from compaction to good tilth, which despite its name is not intrinsically connected to tillage. Healthy soils are more granular, that is, they form larger clods. This reflects the long, uninterrupted work of soil organisms digesting and aggregating soil particles with organic gels—and the presence of large amounts of roots and root hairs.

*Organic matter* is all the living and dead organisms and their residues in the soil. In temperate zones, 1% organic matter represents about 20,000 pounds per acre in the top foot of soil; this percentage reflects a desperate level of soil degradation.[7] Three percent organic matter is poor but workable, and most agricultural soils in North America have less than 3% today. Five percent is a fair level, 10% is good, and as much as 20% can be achieved in cooler regions.[8] Organic matter levels are lower in areas of high rainfall and higher average temperatures. Thus organic matter in Georgia Piedmont soils might be good at 6%, while in the St. Lawrence Valley, 10% would be moderate, reflecting lower rates of oxidation.

*Soil pH* (percent Hydrogen) is a measure of acidity or alkalinity on a logarithmic scale of 1–14, where 3 is about as acid as lemon juice, 7 is neutral and soils above about pH 9 are very difficult on plants. Good garden soils are found in the range of 6.0–7.0, but in dryland areas good results may be had in soils up to about pH 8.5.

Levels of pH have many implications for soil fertility and health. Acid soils (below 5.5) are potentially dangerous in older urban areas, old orchards and around industrial sites where heavy metal contamination is more common. Though all cultivated crops will do well in the range of good garden soil, some such as blueberries and potatoes have greater tolerance for acid soils. All soils will tend to come toward neutral with the addition of organic matter. Soil pH is also amended by the addition of minerals.

## Mineral Composition and Balance

Most readers will know that nitrogen, phosphorus and potassium—primary nutrients (NPK by their chemical symbols)—are the most heavily used by plants. Three secondary nutrients—calcium, magnesium and sulfur—are also important, and their shortage may limit the growth of plants. The role of trace minerals is less well appreciated, but they too can be limiting factors for the health or productivity of crops or livestock, and obviously for human beings as well. Trace elements may be as many as 40 in number, but at least zinc, manganese, copper, boron, iron, molybdenum, cobalt, chlorine, iodine, sodium, chromium, fluorine, selenium, vanadian, germanium and silver are known to occur in the human body and must be presumed to have a metabolic function, however poorly understood.

University of Missouri soil scientist Dr. William Albrecht studied mineral nutrition in soils, plants and animals. By the middle of the last century, he had established what he regarded as an optimal mineral balance based on the percentage of cation exchange capacity (CEC)—the ability of the soil to hold and exchange positively charged ions—that each occupied.[9] Calcium takes up the greatest part of CEC in a healthy soil (68%), followed by magnesium (15%), potassium (8%) and ammonium (3–4%). Excesses of any nutrient above the ideal profile would be unusable, Albrecht found, while shortfalls would result in deficiency and ultimately disease. Because of the primacy of calcium, it assumes a key role in unlocking mineral fertility. Farmers of many traditions, both chemical and organic, have observed the value of regularly applying lime, or in the case of alkaline soils, gypsum (calcium sulfate), to balance both pH and, perhaps unbeknownst to most of them, mineral availability.

The art and science of balancing minerals in soil is complex, and practices around it range from regular liming to the foliar

# SOIL
## MINERAL BALANCE

Ca    Mg    K    NH₄   |    P    S    TRACE

68%    15%    8%    3-4%   |    3%    2%

BOUND to CLAY and O.M.      BOUND to Ferric Iron

CATION (+) EXCHANGE CAPACITY    ANIONS (-)

STRONGLY ---- BOUND -- WEAKLY ----→

Nutrient minerals form bonds of varying strength reflected in the optimum ratio of their saturation in the clay-humus-iron complex of the soil. More tightly bonded ions should take up more of the sites.

[Credit: Peter Bane]

application of ancient sea minerals, fish emulsion or seaweeds to the use of specific herbs in biodynamic preparations by what are essentially homeopathic procedures. Assessing mineral balance begins with an environmental analysis and proceeds to soil sampling, tissue assays and the long-term observation of plant regimes and animal health.

## Environmental Analysis

Our assessment begins with a series of questions.

### 1. Is this soil likely to be acid or alkaline?

The answer will depend on the geology and geography of the site. Soils east of the 98th meridian (a line passing roughly between Winnipeg and Dallas) are generally inclined to be acidic, while those west of that line are more likely to be alkaline. Soils derived from limestone will be less acidic — other conditions being equal — than those derived from non-calcium-based rocks such as granite, gneiss, shale or basalt. High-rainfall areas on the West Coast will tend toward acidic soils like much of the East, while some parts of Missouri and Iowa may be slightly alkaline despite lying east of the rainfed line because their soils derive from limestone parent rock. Desert soils and irrigated soils in sub-humid and semi-arid regions may be not only alkaline but highly salted.

### 2. Is the soil new or old?

Regions of glacial deposition will have newer soils. For example, in southern areas of Canada and northern states of the US, soils have been enriched by glacial transport, but in northern Ontario and Quebec, the glaciated areas of the Laurentian Shield were impoverished by the same scraping process. The Rocky Mountains are relatively newly uplifted and show evidence of vulcanism. Both processes add minerals to soil. Volcanic deposition has enriched soils in eastern Washington and continues to add minerals to other parts of the Pacific Northwest and Hawaii. The Appalachians, on the other hand, are about 200 million years old and much eroded. Some parts of coastal California rest on very old rocks that lack fertility, and this can be compounded by a long history of leaching from high rainfall. In areas with very cold climates, soils may be poorly developed simply because soil life processes are retarded in the absence of heat.

### 3. What is the nature of soil particles, its texture?

In general, soils rich in clay particles with their huge surface area and strong electrical charge will hold minerals better than silty or sandy soils, which can be more easily depleted or leached.

### 4. Do atmospheric conditions impact soil pH?

While government-mandated removal of sulfur from fuel, the use of low-sulfur coal and the addition of stack scrubbers have reduced sulfur dioxide pollution, coal plant emissions remain a major source of acidity which washes out of the atmosphere with rain and snow and impacts soils. The Tennessee and Ohio valleys and New England are major downwind areas for coal emissions from the Midwest. Locally, acid rain may also be a factor in some western regions such as Colorado, Texas and New Mexico, where large amounts of coal are burnt to generate power.

Soils subject to acid rain must be continually buffered by the addition of organic matter and calcium.

### 5. Has the land been farmed chemically?

Most conventional agricultural practices deplete organic matter and the soil minerals that it holds, not just NPK, but also secondary and trace minerals. These are rarely replaced. The longer land has been farmed with machine tillage and chemical fertilizer, the poorer its fertility is likely to be.

### 6. Is there a history or present source of heavy metal contamination?

Older urban areas are suspect; be wary of salvage yards, metal plating facilities, tanneries, chemical plants, former gasoline stations and wood treatment facilities. All are sources of heavy metals and other contaminants. Lead was only removed from gasoline and paint in the US after 1976, and many older buildings are still carrying a legacy of lead. All soils near roads and highways more than 35 years old will have a burden of lead near the pavement. Particles of lead, being heavy, do not move far from where they are generated, but neither do they go away unless specifically removed or buried. Cadmium particles from the abrasion of tires are being generated near all roadways even today. Mercury is present in the atmosphere wherever coal is burnt, which is to say in most of the US and some parts of Canada. Arsenic too is emitted by coal plants, but high levels are also associated with former orchard sites where arsenic-based pesticides were heavily used during the first half of the 20th century. Radioactive isotopes are thus far not deposited in high concentrations in the US except around some of the 108 commercial nuclear reactors and a dozen or more sites involved in the production of fuel or nuclear weapons. Of nuclear power plants, about a dozen sites are the most seriously contaminated; three dozen more to a lesser degree. Five nuclear generating complexes in On-

tario, Quebec and New Brunswick, as well as a number of Canadian research and industrial sites routinely release substantial quantities of tritium, which is taken up by living organisms nearby and poses risks that are not adequately acknowledged by official standards. These releases also impact US residents of the Great Lakes region.[10]

### 7. What plants are growing in the soil?

Some plants indicate mineral-rich soils. Nettles prefer rich loamy soils, and beech are said to grow on richer soils. Other plants have a competitive advantage on poor soils. Rhododendron and mountain laurel, for example, are typically found on highly leached, acidic soils. Early successional species (weeds and pioneers) work to accumulate minerals, and the buildup and maturing of the soil biotic community aids mineral uptake. So other things being equal, ecosystems in advanced stages of succession will have accumulated more minerals.

### 8. How well are the plants growing?

Deformities, chlorosis (yellowing or streaking of leaves) or pest problems indicate mineral deficiencies, as do poor flowering or fruiting. At a more subtle level, Brix readings (the measure of sugar in plant tissues) can indicate indirectly the presence or absence of adequate mineral nutrition in soil.[11] When fruits are deformed, prone to spoilage or insipid in flavor, suspect that mineral levels are too low.

These last two steps (questions 7 and 8) are also part of our long-term observation process to monitor soil health.

## Repairing and Building Soil

The essential formula for the repair and maintenance of healthy soil is a mantra, easily remembered by its holy sound:

> Add
> Organic   or   "AOM"
> Matter

Organic matter increases food for soil microbes that create humus. Increasing the carbon content of soils increases their water- and mineral-holding capacities. Organic matter increases healthy soil structure and porosity; it also helps to aggregate soil particles, reducing erosion.

It is important to add this organic matter in two ways. The first is to grow plants, which put their roots down into the soil layers where they introduce photosynthates or sugary and starchy root exudates to feed soil microbes, and where these roots regularly die and are sloughed off to decompose. The second way is to lay green and brown vegetative material on the soil surface for the soil animals and fungi to consume and incorporate into the soil layers they inhabit. The soil's mouth is at the surface, not 3, 6 or 10 inches down. Plowing to incorporate plant residues is not unlike forcing unchewed food down someone's throat with the notion that this will help them to eat. The digestion of plant residues requires oxygen, which is most readily available to aerobic bacteria, fungi and soil animals at and near the surface. By continuously adding organic matter on top of the soil, we can ensure a steady supply of food to the organisms that are really building the soil. When we grow plants in soil and lay green and brown plant residues (and dead animal parts, compost and other soil amendments) on the surface, we create the Organic Layer, just like the leaf mold of the forest floor. No actions of the plow or of the spade can simulate all these results.

The most direct way that we can introduce organic matter directly into deeper layers of soil is to grow roots in it by cultivating plants. These can be short-lived crops, green manures, shrubs, trees or even annual or perennial weeds. Plants sustain their own root systems (as well as sharing food and energy with nearby plants and soil organisms) by producing sugars and starches through photosynthesis. They pump these carbohydrates down through their phloem cells from the inside of their leaves to the outside of their roots. If soil disturbance damages or destroys some of the roots, a portion of the plant's leaves and stems must die because there will not be enough moisture and minerals returned to nurture the aboveground parts. The converse is also true. If the plant's leaf canopy is eaten by a herbivore, pruned by the gardener or damaged by a blast of wind, portions of its root mass must be sacrificed to keep the whole system in balance.

Energy supplies will be diminished by aboveground disturbance, so belowground structure must contract. This material, of course, does not simply retract into the plant's other roots. It is abandoned and begins to decompose, just as our own bodies slough off skin and hair cells all the time.

This process of plants growing roots and abandoning them is one of the primary ways in which organic matter builds up in soil. It is also one of the reasons soil organisms cluster around plant roots. The other is to feed on the sugars the plant is deliberately releasing there. Some forms of fungi called *endorhizal* (within the root) actually penetrate into the plant roots to facilitate this nutrient exchange. One quarter or more of the energy a plant

Plants abandon roots routinely but also when their leaves are grazed or cut.

[Credit: Jami Scholl]

captures from sunshine is delivered to soil organisms directly for their use. By feeding soil microbes, plants gain superior access to soil moisture and minerals and to a kind of external immune system that they do not have to create.

All the organisms that live in soil need moisture, and for the most part, they need protection from sunlight and drying air currents. Exposing the soil surface by removing leaf litter and other organic matter simply forces both soil organisms and roots to dive deeper and work harder to gain the protection and habitat they prefer. This reduces their ability to enrich the soil, both because soil life is disturbed and thus less productive, and because the deeper layers of soil are less supportive of soil life due to lack of oxygen and organic matter. It is astonishing how poorly understood this most fundamental truth of soil biology is amongst agronomists and industrial farmers.

When I first toured farmlands in Hawaii in the early 1990s, I saw macadamia nut orchards managed in the typical way. Macadamias grow with an expansive network of feeder roots in the top few inches of the soil. The roots have to be near the surface to get any nutrition at all. The orchard had mechanized the harvest of the nuts, which are round and have a thick shell. Powerful vacuum machines sucked all the nuts off the ground beneath the trees, shooting the material into screened hoppers that sorted the nuts from leaves and sticks. As I saw with my own eyes, these machines were strong enough to suck up stones the size of golf balls, and so they took almost everything off the ground except large rocks and the trees' anchoring roots, which were raised in profile like the veins on an old person's hand. On the very day I visited one orchard, a scientist from the extension service was there to investigate a widespread new problem: sudden tree death. As he explained carefully, mac nut trees across the island were developing evidence of a fungus,

and within 90 days of its appearance in the bark, the tree would be dead. It was obviously a source of great consternation. He and his colleagues were trying to figure out what could be done about the fungus. As he talked, the mac nut harvesting vacuums droned on in the background.

What astonished me was the inability of the scientists or the growers to put together the puzzle pieces that were so obviously visible: the cultivation practice of vacuuming the mulch layer off the ground was starving the mac nut trees. That fungus was just the plant pathogen that was getting rid of the weakened trees. The growers were fixated on the scientist and the "research-based" extension advice and, of course, on their own investments in mechanization. At almost 30 years of age, the trees were past their period of youthful vigor, had big canopies to support and were simply being worn out by the chronic abuse of their root systems.

## Soil Needs Mulch

The short lesson in all of this is that we need to keep a mulch cover on soil at all times. It protects the soil from erosion by wind and rain, protects soil organisms from drying and light exposure and conserves soil moisture. Mulch also feeds the soil, which feeds plants.

Seven tons of topsoil are lost from an acre in Iowa for each ton of grain grown, and that is considered a "moderate" level of soil loss.[12] The US Soil Conservation Service, established in 1935, built terraces, planted windbreaks and promoted contour plowing in response to the Dust Bowl brought on by the plow. Since World War II, the use of chemical herbicides has reduced plowing under, a system broadly called "no-till." It sounds good until you realize that it means more poison on the land and in our food. Pest problems are spreading, and environmental toxicity and cancer have become epidemic. Herbicide use on crops in the US just keeps going up, which should not be surprising considering how powerful the

Keyline furrows leave only minimal disturbance at the soil surface—typically they run near to contour, but allow water and air to reach deep into the soil to aid in building humus. (inset) The shanks of a Keyline plow with slipper implement points can be seen here as retracted from the soil with the pulling bar raised.

[Credit: Brian and Ann Marie Bankston, (inset) Peter Bane]

chemical industry remains and how much profit is to be made from selling agrichemicals.[13] What is disappointing and a little surprising in view of the terrible costs is that productivity gains have reached a plateau.[14]

A more thoughtful and targeted approach to fertility management has been taken by the Rodale Institute in Pennsylvania, which has developed machinery to enable organic farmers to grow and mulch green manure crops which support soil fertility. The implement can be pulled behind a tractor and is designed to roll, crimp and flatten plants grown for mulch. This technique leaves roots in place to decompose with no soil disturbance; it knocks them back with mechanical action rather than poisons. Being annuals, these plants — from cereal rye to vetch to buckwheat among a broad palette of possibilities — do not grow back from their roots, and if rolled before they go to seed, they give all their growth to enrich the soil without becoming weeds themselves. The cash crop to follow can then be drilled directly through this thick mulch of green manure (which effectively suppresses other weeds at the same time that it protects the soil surface).

## Learning from Large-Scale Rehabilitation

Though this book focuses on smaller-scale garden farms, it is instructive in exploring soil repair processes to examine the work of P. A. Yeomans, an Australian agricultural engineer. Yeomans designed a number of large farm properties in the semi-arid agricultural districts of New South Wales in the 1940s and 1950s based on a system he dubbed Keyline. The *Keyline* marks the highest elevation in a valley where water can be economically stored behind earthen dams. By finding the Keyline, Yeomans could locate farm ponds (or dams as they are called in Australia) where farmers could both hold significant amounts of water, and whence water could be released to reach the broadest areas.

A Keyline is formed by connecting a series of key points; each key point is the point of inflection in a slope where the convex upper slopes invert and become concave. All humid landscapes take the characteristic shape of an S-curve or sine wave from the erosive action of water over long periods.

After analyzing a farm property to identify the Keyline, Yeomans would build a series of earthen dams along it in the side drainages of a main valley — where moisture and runoff would naturally concentrate, starting at the highest drainage and working down. Each dam would be built with an overflow and a drain that connected it in series with all the lower dams. Also, from the drains of each dam, many parallel and subsurface furrows would be plowed extending out from the small side valleys onto the adjacent ridges. These furrows were designed to carry water from the dam drains into the surrounding pastures and fields. The proper functioning of the furrows was so critical to the success of the whole system that Yeomans designed a special plow, or *slipper implement* as he called it, to create them.

With a water infrastructure established, the farmer was poised to orchestrate a series of soil climaxes that would simultaneously build topsoil very rapidly, grow nourishing forages to fatten livestock and prepare the farm ecosystem to succeed rapidly from grassland to savanna. Here's how it works.

Even as early as the first passes of the sub-soiler, a seed box attached to the back of the plow scatters seeds of deep-rooted legumes (e.g., alfafa and sub-clovers) into the newly cut furrows. There the hard-coated legume seeds sit and wait for moisture. Once the furrows have been seeded and water has accumulated behind the dams, the furrows are flooded at the beginning of the next growing season to stimulate germination of the legumes. These then root at the bottom of the furrows, grow up through the slit trenches to the surface and above and flourish from the reserve of water soaked into the surrounding soil. The floodwaters flow completely below the surface, and so are not subject to loss by spillage or evaporation. The legume crop has its growth point well below the surface too, helping to protect it from drying out. And the selected legumes have the characteristic of putting out very extensive root networks. Alfalfa, for example, is a native of Egypt and adapted to dryland conditions; a single plant can spread its roots over 30 feet.

As the plants mature, the farmer introduces cattle or sheep to graze them down. By timing the grazing (or mowing if animals are not available) to coincide with early flowering of the legumes (about 10% flowering), the maximum volume of forage can be harvested above ground, and at the same time the maximum growth of roots is achieved. And, as we have already learned, when the tops of plants are eaten, a corresponding volume of roots is abandoned to die and decompose in the soil…in this case, in the subsoil. And these plants' roots carry their own nitrogen. The nitrogen-fixing bacteria also die back, causing nitrogen to be released and facilitating the decomposition process. Drawing air, moisture and soil organisms down to their roots, the legumes create ideal conditions

for a rapid buildup of organic matter in the subsoil, enabling it to take on the character of topsoil—filled with pore space and organic matter—quickly. Since the forage plants are perennial, they can be managed over several seasons in this way without replanting. Each climax of growth and grazing results in a new flush of organic matter created in the subsoil with a corresponding increase in soil life activity, soil carbon and water-holding capacity. While growing beef and mutton for market, the farmer simultaneously grows an organic sponge of fertility beneath the pastures.

After several years of soil development, it has often proved worthwhile to draw the Keyline plow through the furrows a second time but to take it several inches deeper, below the newly formed topsoil, then repeat the seeding-growth-grazing-soil climax cycle, using flood irrigation as needed to maintain growth. The ponds would provide a buffer against drought as well as a source of water for the livestock. Eventually, the deep soil built—often as much as several inches of organic soil in a few years—would be holding such a lens of water beneath the pastures and fields that they would be effectively drought-proof. At that point, rows of hardy trees could be introduced on contour lines downslope from each set of furrows for forage, nut crop or timber production. As the trees matured, their roots protected and watered by the irrigating furrows and anchoring the hill slopes even further against erosion, the whole farm

Every hill and valley typically has convex upper slopes, concave lower slopes and, between them, a keypoint.

[Credit: Peter Bane]

ecosystem would be transformed into a more humid and stable landscape, protected against drought, wind and erosion alike and yielding ever more forage, meat and wood product by the year.

Yeomans applied modern technology and fossil fuels to systematic land repair and regeneration for the first time. He crossed the threshold at which agriculture had been stalled for 10,000 years; he devised a way to build soil fertility and depth on a broad scale while harvesting economic yields from the same land, and at the same time, largely without erosion and buffered against drought. And he did it in the same kind of semi-arid climate in which agriculture first emerged. In so doing he prefigured our 21st-century needs to moderate climate through soil building and tree planting, while raising more food.

## Structures to Harvest Water and Build Soil

The practical import of Yeomans' work to garden farmers lies in the way it structures the process of improving soil within a nested hierarchy of design decisions about the farm. To apply the lessons of the Keyline process to small properties, we need to prepare landforms so that they exploit and augment existing water resources and act to prevent or limit soil erosion. Just as Yeomans first stored water in his landscapes and then distributed it strategically from those storages to stimulate plant growth and soil building, so should we. On a small suburban property with a high proportion of paved or roofed surfaces, the points of collection and concentration may be obvious. But even here, the landform too will have natural or human-made drainages that must be observed and assessed. Only when tanks and ponds are full and the soil is completely saturated should there be any runoff from the farm. In order to achieve this, we will need to construct many small earthworks.

It is not likely for many small properties that Keyline plowing would be the most appropriate technical solution to water storage and distribution, but there are numerous ways to retain and infiltrate water that the garden farmer should learn to apply. Foremost among these is the *swale*, a ditch dug on contour. A variation of this is the *water bar*, which is a berm raised in a pathway or road to divert water, often into a swale or drainage ditch. *Pits* both large and small are frequently used to store and infiltrate water, leading up to ponds of every size. In sloping land, *terraces* are often used to retain soil, infiltrate or hold water and improve the practicality of cultivation.

### A Swale Way to Hold Water

Swales are small earthworks that capture the pulses of rainwater running from a broad area, hold the water for a time and allow it to infiltrate the soil. They consist of a berm on the downhill side of a ditch, and can be dug by large or small machines, built by people with mattocks, spades and rakes or by animals pulling a drag bucket. However you dig them, they must be built on contour. The ditch should have a broad, U-shaped bottom — in very wide swales, nearly flat — with gentle gradients on the back slope. The berm may be wide or narrow according to the needs of the circumstance.

Small Earthworks,
Pattern #32
(See Chapter 6)
[Credit: Jami Scholl]

Every swale should be built in a way that allows it to overflow when full to another catchment in a controlled manner, minimizing erosion. Swales are often constructed in a series over a whole hillside or terrain, and if so, the first built should be located near the top of the local watershed, following the principle that it's easier to control runoff when the quantity is smaller. The fall from one swale to the next is best determined by looking at the site; however, a rule of thumb I have often used is to allow five to six feet in elevation between swales.

Swales are often constructed to spread out the concentrated flow of a drain pipe or culvert; these are sometimes then called *spreader drains*. They can similarly take the overflow from a pond or a tank, or any other collecting surface such as a road or parking lot. A swale should hold the localized runoff from a one-inch rain event. Another way to increase the volume of a swale is to furnish it with pits, excavating deep spots where water may stand longer. Whatever the volume of the swale, it should infiltrate the water it holds within 48 hours of the end of a rain event. It is important, therefore, not to compact the bottom of a swale. Swales, which conveniently run on or near contour, may be used for walking paths, but should only infrequently get any vehicle traffic.

Swales will collect not only water but sediment, seeds, manures and other detritus washed along with runoff. As such, they become a resource that may be harvested periodically for mulch and nutrient. A swale and berm (Brad Lancaster in his books calls them berms-and-basins) offers several distinct niches for planting.[15] The basin itself is subject to periodic inundation and will have wetter soils. The top of the berm has drier soils but also allows plants growing on it to tap into the reservoir of water that accumulates below the swale. Downslope from the berm, water resources are increased, and over time this effect may reach for hundreds of feet. One

A broad ditch on contour, the swale harvests everything in motion—runoff, snowmelt, seed, manure and mulch—to enrich the landscape. In steep country, swales will be closer together. Their berms should always be vegetated and their ditches uncompacted.
[Credit: Peter Bane]

way to increase the yield of a swale system is to plant it with perennial trees, shrubs, herbs and grasses that can be cut for mulch or forage. In a dry region, a swale concentrates runoff into a zone that has higher productivity than surrounding land.

Swales can double as firebreaks or composting zones. They can become garden paths on the uphill side of raised beds, providing passive irrigation water, access and a place for coarse mulch material to break down. I have designed garden terraces where greywater was discharged into swales at the back and allowed to infiltrate downslope beneath the crops.

Over time, swales will tend to fill in, so they must either be maintained by periodic harvesting of sediments — or a plan must be developed to use the resulting terrace as a planting area. In flatter areas where machines must sometimes pass, the berms should be made very low and gradual to facilitate large wheels rolling over them.

Swales and Terraces,
Pattern #35 (See
Chapter 6)

[Credit: Jami Scholl]

### Terraces for Sloping Land

The incidence of soil erosion in steeply sloping land is so great, and the importance of preventing it so urgent, that the need to create terraces has been self-evident to almost all traditional mountain agricultures. In small areas, however, the North American garden farmer may be faced with backyard hillsides every bit as steep, and may find herself confined to precious scraps of urban real estate that warrant similar intense efforts. If so, it is best to establish terraces at the outset.

In rough country where trees are already established, the beginning of a terrace system can be constructed by felling or clearing trees, then laying the logs, twigs and branches on contour into rough windrows behind which lighter material is piled and soil allowed to accumulate. Pigs have also been penned and moved along contour to help create terraces by their rooting behavior. Where stone is not available to face the terraces, trees or shrubs for coppice may be planted close together along the banks. Dense-rooting perennial grasses such as vetiver (which does not seed) have been recommended for tropical climate

terrace edges, but most cold-hardy grasses are more problematic to introduce to cropping zones — they either run by stolons or seed vigorously, sometimes both. Comfrey and other deep-rooted herbaceous perennials offer a good way to retain soil while limiting competition with other crops. I grow a row of comfrey at the bottom of my garden for just this reason: it captures nutrients carried in runoff before they leave the system. The soft leaves can be harvested several times a year and make an excellent, high-protein fodder or mulch. If the terrace face plants are nitrogen-fixing, then they provide a built-in fertilizer for the growing zone, releasing nitrogen for the benefit of the crop each time they are cut.

Ideally, terrace construction would begin with the ultimate retaining walls, either as dry-stack stone or densely planted shrubs, herbs or trees, but interim or sacrifice material may be the only thing available with which to start. Then you have to grow the rest of what you need. The levelling of the terraces can follow once soil retention structures are in place.

With terraces and even with swales, it is important to consider at the beginning the foot and vehicle circulation that will be needed through the area. How will animals be brought in, if at all? If tree crop is to be planted, how will the harvest of eventually large volumes be accomplished? Can you get a cart or a wheelbarrow through? What about a truck or a wagon? Think about the flow of materials, people and of water, and provide helpful interfaces.

### Pits and Water Bars

Water bars primarily shed water from roads and pathways on sloping ground. The harvested water needs to be directed into a garden bed, swale, pond or other catchment structure to disperse its erosive force. Water bars are best constructed of limestone gravel, which will pack more and more tightly over time, somewhat offsetting the natural erosion of traffic. Where limestone is not available, a base of heavy clay can be partly mixed with coarse ballast and then topped with smaller stone. An alternative sometimes used is to lay a log or two boards or straight saplings across the travelway in a shallow trench. The trench acts to divert the runoff while the wooden members hold up the wheels of passing vehicles. Most water bars are more easily negotiated and maintained where they form a diagonal to the direction of travel.

Pits have broader application. They can be used to collect a source of runoff to irrigate a single tree or provide moisture for composting. If infiltration is the primary purpose, a pit can be filled with rubble or coarse gravel, but pits can also be filled with sticks and rough woody biomass or with such waste material as junk mail, broken chipboard and cardboard scrap, in which circumstance fungus can be cultivated. Spread out into a broader basin form, pits become rain gardens, designed like swales to hold and infiltrate runoff. They may then be planted with their own mix of wetland and emergent plants.

Pits in a garden pathway/swale hold extra water longer to favor nearby plantings.
[Credit: Creighton Hofeditz]

### Check Dams and Gabions

In more steeply sloped or dissected landscapes, gullies may form from deforestation or poor farming practices. Anywhere runoff concentrates to erode soil may benefit from these small and porous structures, which act to slow runoff and capture sediment. *Check dams* are leaky dams. They can be made from stones, logs, saplings, fencing, brush bundles or any non-toxic materials that can be stabilized in a gully, wash or dry drainage. They are meant to check the flow of water. Check dams need not completely block the channel. In larger streambeds or where flow is light but more constant, they are often built as a series of offset wing dams, extending from alternate banks where they are anchored, into the middle of the channel. This encourages the water to take a meandering course, slowing it without preventing drainage. In ephemeral watercourses, check dams are often built across the drainage and tied into both banks. They then serve to create a series of terraces. As runoff slows, sediment is captured behind the check dams, gradually building up a sponge of silt and organic debris that holds water during dry times and can eventually stabilize stream flow. Of course, in times of high water, check dams can be breached or blown out, so it is important to reduce this risk by making sure that they be leaky and well tied into the banks, allowing most of the water to pass through and around, albeit more slowly than before.

*Gabions* are steel or sometimes woven bamboo baskets into which large gravel or cobbles are piled after the baskets have been placed and wired together. Gabions are frequently used as porous retaining walls along highway or river embankments, but they can also be used as more durable check dams where the seasonal fluctuation in stream flow is large.

Two gabions above slow occasional heavy flows in this dry drainage while trapping sediments to repair a damaged watershed. The check dam downstream benefits from protection by the gabions; a partial terrace has already formed behind it.
[Credit: treesftf via Flickr]

Eventually, after some silt and moisture have accumulated, check dams or gabions may be planted with willows, alders or other water-loving species of trees and shrubs to further stabilize the stream channel. Over time, gullies can be substantially filled in, flooding reduced and stream flow restored by this process.

## Feeding Soil

Not all soil repair requires such dramatic intervention as keyline plowing, terraces or check dams, but water management is the beginning of any soil building process. Feeding soil — and most soils are starved — consists of a systematic program of planting and harvesting fertility crops and of supplying water sufficient to grow them.

We take crop rotations for granted today because many and complex types — from four-year to seven- or even nine-year patterns — were developed during the 19th and early 20th centuries. However, European agriculture, from which our North American systems evolved, only adopted legume *leys* (perennial rotations with arable crops) in the 17th century. This mixing of annual crops with pasture on the same ground provides some of the most resilient forms of crop rotation today. Not only through nitrogen-fixation, but also by the use of short-lived perennials, the soil's mineral balance and self-fertilizing processes can be partially restored. The prostrate, white-flowered perennial legume *Trifolium repens* is known as Dutch white clover because it was first used in the Low Countries for this purpose.

Chemical farmers have drastically simplified their rotations in recent years, often merely alternating corn with soybeans. Organic farmers cleave to the traditional pattern that predated fossil fuels, interleaving fertility with cash crops, usually one-to-one. Cold-hardy fertility crops such as vetch, turnips or mustard are favored because they can be grown during the winter to restore fertility for spring or summer grain or vegetables. Biodynamic gardeners practice a form of rotation within beds by growing in succession crops that bear from different parts of the plant: root, leaf, fruit and seed. Certainly rotations are a valuable technique and were one of the first steps Western agriculture made toward enhancing yields. But they require long-term commitments that can limit cropping responsiveness to weather and markets. More integrated approaches have now been developed that increase flexibility.

Fertility can also be provided by *spatial intercropping*, the pattern of which can vary based on scale. Large farmers grow strips of legumes between strips of grain, allowing the extra fertility from nitrogen-fixation by the legumes to feed the nitrogen-hungry grain crops.

At the garden scale, intercropping may look like companion planting or alternating rows of different vegetables. A yet more sophisticated approach is to integrate perennial fertility crops as islands among or along the edges of fields or garden beds and to harvest them periodically to support the yield of main cropping areas. Perennials stabilize the soil's self-fertilizing processes, prevent disturbance and mine Banking and other deep soil layers for extra nutrients. One variation of this is *alley cropping* where rows of productive trees or shrubs form alleys in which arable crops (grain or vegetables) are grown.

The perennials not only draw water from different soil layers than the annual crop — effectively making better use of the space and water resources — but are often nitrogen-fixing and may be coppiced for chop-and-drop mulch, an action that simultaneously releases nitrogen into the soil. Fertility crops should be selected to give as many yields as possible, not just of soil nutrients and mulch but also habitat for beneficial insects and birds, appropriate windbreaks and shading, aromatic compounds, livestock, wildlife or bee forage and even emergency food. When perennials bound the cultivated plot, they also help to contain and remediate soil erosion.

Restoring the self-fertilizing abilities of a soil may take three to five years of deliberate work, longer if more serious degradation has occurred. If land has been cropped by conventional methods of tillage and especially if chemicals have been used, deliberate remediation should be undertaken by the application of mulch and the growing of fertility crops prior to any attempt to grow cash or food crops. If you have the circumstances to do it, planting degraded land to a mix of pasture grasses and legumes and then rotating animals over it intensively is an excellent way to kick-start soil regeneration. We will examine animal feeding and movement in Chapters 13 and 14.

Alternating strips of grain and legume follow contour to reduce both erosion and the need for fertilizer.
[Credit: Photo by Tim McCabe, USDA—Natural Resource Conservation Service]

## Applying Mulch

Large amounts of biomass for mulching are still available in North America's resource-wasting societies. Mulch is abundant in and around cities and in humid regions, but is often difficult to come by in drylands. All organic materials are potentially useful, but not all are equally valuable. Cardboard and paper are widely available for the hauling. Junk mail is delivered to your box, and while I might not seek it out, I never throw this away, but turn it into worm food by laying it down beneath sheets of cardboard or other top mulch in pathways and on new garden areas. Sheets of newspaper and cardboard are valuable for suppressing unwanted grasses or weeds in establishing new beds. If shredded, of course, paper and cardboard make a fine animal bedding or even a top mulch. Wood chips are regularly available if you will take a bit of time to contact utility crews working in your area. They are usually happy to have a convenient spot to drop their loads, and you can easily wind up with 10–20 yards of shredded wood waste. Almost all of this is tree tops and limbs, and during the warm season it is often rich with green leaves. Horse manure is commonly free for the hauling from stables, including often police stables. In many areas where countryside is nearby, rabbit, llama, goat, cattle and even chicken manure may be had for the asking. Straw bales have risen in price in our area, but may still be cheap in wheat-, barley- or rice-growing regions. You will get some volunteer grain in your beds, but this is easily controlled and may even be welcome. Spoiled hay is ubiquitous, but usually brings with it a raft of unwanted weeds the control of which may be more trouble than the hay is worth. You could always compost the hay. In many metro regions there are cocoa processors who may have hulls available, while every town has multiple coffee dispensaries where grounds may be collected. These make excellent worm food. In the Southeast, baled pine "straw" or needle is not uncommon. Look also for rice or peanut hulls if you live in a region where these are grown or processed.

Working with areas of more than an acre requires growing your own mulch in place — in other words, planting fertility crops. Mulch crops may be perennial or annual, woody or herbaceous. The long-term success of any garden farm will depend on the development of woody perennials and animal cycling of vegetation to maintain fertility, but during the early years it may be necessary to draw much more on herbaceous annuals grown for mulch. These must be matched to the season and moisture available.

## What About Compost?

*Compost* is a dark, loamy organic mould produced from the accelerated alchemy of soil organisms. It resembles a very light and porous silty topsoil with very little mineral soil content, and contains large amounts of humus and a rich diversity of healthy soil life. Most regions now have some enterprising farmer or business person who sells finished compost to gardeners for upwards of $80 per cubic yard. You can, of course, make it yourself.

Compost is specifically made by the gardener or by large-scale operations combining animal and vegetable remains in a ratio of between 20 and 30 parts of brown or dead material of low fertility to 1 part of green or highly nitrogenous matter such as manure, food waste or freshly cut plants. These components are moistened and layered with small amounts of native soil to introduce microbes, sometimes salted with additional mineral amendments such as rock phosphate, greensand, lime or wood ash and piled in heaps or in windrows with a minimum diameter of three feet. This thickness of the pile ensures adequate insulation to hold heat given off by rapid bacterial digestion of the biomass. A well-made compost pile may reach temperatures over 160°F — hot enough to kill most seeds — at which point only certain thermophilic or heat-loving bacteria are still active. After the pile cools, other soil organisms proliferate. Composting reduces the volume of the original material about ⅔.

To make compost requires a good volume of material at one time. Piles that are too small will fail to achieve the high temperatures that kill pathogens and hasten breakdown, though all heaped organic wastes will tend to decompose over time, whether or not they achieve critical heat. Cool piles attract many soil animals, and if your local supply of biomass is limited, you can enhance your composting process by introducing manure worms, sometimes called red wigglers (*Eisenia fetida*). Compost may be made at any season — piles in winter should be made a little larger to provide more insulation — but the greatest volume of material is typically available in late summer and autumn as crops and weeds

## Cover Crops

Any crop planted for the specific purpose of soil enrichment should be slashed, rolled or grazed just after it begins to flower. Crops are listed in groups by type of plant, and with some variability, from cold-hardiest to most tender. Members of the same genus are grouped together, e.g. *Trifolium* spp. Some suggestions are made as to regional suitability, uses, characteristics, tolerances and rotations or companions. A = annual; B = biennial; P = perennial.

| Name | Type | Latin | Comments |
| --- | --- | --- | --- |
| **Legumes** | | | |
| Kidney vetch | BP | *Anthyllis vulneraria* | extremely cold- and drought-hardy; taprooted; w/barley, oats |
| Milkvetch, Cicer | P | *Astragalus cicer* | cold- and drought-hardy; rhizomes; Great Plains, West |
| Vetch, hairy | AB | *Vicia villosa* | winter cover to z.4; shallow roots; top N-fixer |
| Vetch, common | A | *V. sativa* | winter hardy to z.7 |
| Bell/Tick beans (favas) | A | *V. faba* | cold-hardy to 15°F, cool temp.; top N-fixer, w/corn |
| Peas (field) | A | *Pisum sativum* | USA/Canada winter forage, seed; hates heat; shallow roots, w/oats |
| Clover, white | P | *Trifolium repens* | cold-hardy, moist; durable, spreading; forage |
| Clover, red | B | *T. pratense* | temperate, N/N Central; used in rotations; w/timothy |
| Clover, alsike | BP | *T. hybridum* | cold-hardy, all moist soils; shallow roots |
| Clover, Ladino | P | *T. repens* | temperate to z.6+, prostrate stems; hay, forage |
| Clover, rose | A | *T. hirtum* | arid, winter crop in the South and California |
| Clover, crimson | A | *T. incarnatum* | winter sown to z.7, summer annual in the North |
| Clover, subterranean | A | *T. subterraneum* | semi-arid, Medit. climates to z.7, self-seeding |
| Sweetclover | B | *Melilotus officinalis* | semi-arid to humid; all soils USA to S. Canada; top bee forage, P and K accumulator; w/cereals or flax |
| Lupin, large white | A | *Lupinus albus* | cool, moist, N Central and NE to Gulf Coast (winter); fast growth, deep roots; cooked seed edible; poor soils |
| Lupin, yellow or fragrant | A | *L. luteus* | cool moist, spring/fall sown; vigorous; sandy soil; hay w/potatoes, winter cereals |
| Sainfoin | P | *Onobrychis viciifolia* | dry, temperate, alkaline soils, deep roots; good w/barley; hay/forage |
| Kidney beans | A | *Phaseolus vulgaris* | warm season, moist; all soils (MI-NY and southward); w/corn, potatoes, wheat, clover, cotton |
| Soybean | A | *Glycine max* | tolerates frost, WI-AL; late types yield most biomass |
| Alfalfa | P | *Medicago sativa* | semi-arid, temperate to hot; deep roots, hay/forage |
| Alfalfa, black medic | A | *M. lupulina* | aka yellow trefoil; cool, alkaline soils, w/clovers |
| Alfalfa, bur clover | A | *M. polymorpha* | winter grown, South and West; self-sowing |
| Cowpeas | A | *Vigna sinensis* | summer edible, OH-MO and South; in rotations, shade |
| Hyacinth bean | A/P | *Lablab purpureus* | warm, arid to humid, all soils; vigorous, edible cooked |
| Rough pea, Caley pea | A | *Lathyrus hirsutus* | SE, winter cover, all soils; seeds toxic to livestock |
| Lespedeza (Sericea) | P | *Lespedeza cuneata* | summer growth, USA except NW; biomass, poor soil |
| Lespedeza Korean | A | *Kummerowia stipulacea* | shorter-season, north of *K. striata*, more alkaline soils |
| Lespedeza Japanese/ Common | A | *Kummerowia striata* | SE/Gulf Coast, frost-killed; yields in late summer/fall |
| Hairy Indigo | A | *Indigofera hirsuta* | TX-GA-FL; suppresses nematodes; sandy loams |
| Sesbania, hemp/common | A | *Sesbania herbacea* | SW, subtropical, arid irrigated; deep roots; all soils |
| Pigeon pea | P | *Cajanus cajan* | subtropical, semi-arid; deep roots; edible, 150 days |
| Sunn hemp | A | *Crotalaria juncea* | subtropical; good N-source; kills nematodes |
| **Grasses** | | | |
| Rye | A | *Secale cereale* | very cold-hardy, dense roots; winter cover; grain |
| Wheat | A | *Triticum aestivum* | cold, semi-arid; winter grain in mild regions |
| Barley | A | *Hordeum vulgare* | cold, semi-arid; winter grain to z.8; salt-tolerant |
| Oats | A | *Avena sativa* | cool, moist; good hay; poor soil; winter cover in South w/vetch |

| Name | Type | Latin | Comments |
|------|------|-------|----------|
| Millet, Proso | A | *Panicum miliaceum* | warm moist, catch crop (60 days); grain, N. Prairies |
| Timothy | P | *Phleum pratense* | cool climate bunchgrass, all soils; best hay |
| Bromegrass (Field) | A | *Bromus arvensis* | N/N Central winter cover; bunchgrass, extensive roots, fast growth; forage, hay or green manure; rots well |
| Bromegrass (Smooth) | P | *B. inermis* | cool, semi-arid, W/NW; deep roots, fine top growth |
| Orchardgrass/Cocksfoot | P | *Dactylis glomerata* | wide range, extensive roots; tolerates shade; w/Ladino |
| Ryegrass, annual (Italian) | AP | *Lolium multiflorum* | summer annual, winter-hardy z.7+; fibrous roots, quick |
| Pearl millet | A | *Pennisetum glaucum* | warm, sub-humid grain (MD south); forage, smothers weeds, tolerant of poor soil |
| Foxtail millet | A | *Setaria italica* | warm moist; quick growth, catch crop, fodder |
| Sorghum | A | *Sorghum bicolor* | sub-humid; summer grain, forage; w/cowpeas |
| Sorghum (Sudan grass) | A | *S. bicolor* var. *sudanense* | summer forage; catch crop, large biomass yields |

## Forbs

| Name | Type | Latin | Comments |
|------|------|-------|----------|
| Kale | A | *Brassica oleracea* | cold-hardy; edible, winter forage S of NJ |
| Rape | AB | *Brassica napus* | cool summer or southern winter; smothers weeds |
| Mustard, field or turnip | A | *Brassica rapa* | very fast, winter-killed |
| Chicory | P | *Cichorium intybus* | cold-hardy; deep roots, top forage, fast growth |
| Spurry | A | *Spergula arvensis* | cool moist; sandy infertile soil, fast-growing (60 days) |
| Sunflower | A | *Helianthus annuus* | USA/Canada; large biomass producer, oilseed |
| Yarrow | P | *Achillea millefolium* | wide range; drier soils; high-protein, healing herb |
| Mangel | A | *Beta vulgaris* | fodder beet producing large amounts of biomass |
| Burnet | P | *Sanguisorba officinalis* | cool, moist; edible, tall (5 feet) |
| Daikon radish | A | *Raphanus sativus* | edible; "vegetable crowbar" opens hardpan |
| Charlock/field mustard | A | *Sinapis arvensis* | useful weed; opportunity forage, silage |
| Buckwheat | A | *Fagopyrum esculentum* | frost-killed; 60 days to seed, smothers weeds; edible |
| Borage | A | *Borago officinalis* | low-growing, fast spreading |
| Lambsquarters/Fat hen | A | *Chenopodium album* | useful weed; edible, magenta variety is larger |
| Creeping thistle | P | *Cirsium arvense* | healing, nutritious weed of poor soil, mow just before it seeds to control; silage w/alfalfa |
| Persicaria/Red shank | A | *Polygonum persicaria* | useful weed; silage |
| Fennel | AB | *Foeniculum vulgare* | culinary, medicinal; include in forage |
| Caraway | A | *Carum carvi* | culinary, medicinal; include in forage |
| Dill | A | *Anethum graveolens* | culinary, medicinal; include in forage |
| Chickweed | A | *Stellaria media* | edible, forage; early in spring |
| Groundsel | A | *Senecio vulgaris* | useful weed; fodder for small livestock |
| Sheep's parsley | B | *Petroselinum crispum* | edible, medicinal; forage relished by livestock |
| Cleavers/Goosegrass | A | *Galium aparine* | useful weed; medicinal, poultry fodder |

## Fertility Plants

| Name | Type | Latin | Comments |
|------|------|-------|----------|
| Comfrey | P | *Symphytum officinale* | superb fodder, mulch and fertilizer N, K, Ca |
| Burdock | P | *Arctium lappa* | edible root, large leaves, shade-tolerant |
| Horseradish | P | *Armoracia rusticana* | culinary root, vigorous growth |
| Cup plant | P | *Silphium perfoliatum* | tall sunflower relative; fast, biomass producer |
| Jerusalem artichoke | P | *Helianthus tuberosus* | edible tubers, windscreen; large biomass yields |
| Nettles | P | *Urtica dioica* | mineral rich medicinal |
| Plantain, common | P | *Plantago major* | forage of high nutritive/medicinal value |
| Plantain, ribbed | P | *P. lanceolata* | forage on good soils, superior to clover or grasses chop-and-drop mulch for garden |

Credit: Peter Bane[16]

ripen and leaves fall from the trees. Compost piles are often placed in shaded parts of the garden in order to prevent drying from direct sunlight (and to take advantage of ground not supportive of many plants). By making piles in different parts of your farm or garden, you create multiple reservoirs of soil organisms, especially earthworms, from which you can inoculate new garden beds, fields and paddocks.

Compost piles should be kept constantly moist but not wet—like a wrung-out sponge—and in rainy climates may need to be protected by a roof or a coating of mud. The pile must be loose, not compacted, for which purpose sticks are sometimes driven or built into it. Breakdown can be hastened by turning the compost, folding the outer material into the center of the newly turned pile.

Any organic waste can be composted, including cheese, meat, bones and viscera of animals; human, pet and other animal manures; cotton and wool fibers, leather, wood chips, paper, oils and fats, provided the energy- and nitrogen-rich materials are kept in balance with the carbonaceous materials. Urban composters are often advised not to add meat or bones to their piles lest rats infest them or other animals dig into them. This is less a problem when the composting is done in a closed chamber. The composting of humanure—which is a safe and sanitary process when done correctly—should only be done within piles that are protected from animal intrusion and screened against flies.

Composting preserves most of the mineral nutrients available in the original biomass, but much of the nitrogen may be lost to the atmosphere. At the same time, composting creates large amounts of humus, which can endure in soil for long periods. Compost also contains a myriad of beneficial soil microbes and thereby gives a boost to any soil to which it is introduced. Many gardeners make a virtual religion out of composting, and recipes for special composts have proliferated. While

Mulch and Compost, Pattern #62 (See Chapter 6)

[Credit: Abi Mustapha]

there can be no doubt that compost is wonderful stuff, its availability is by nature limited. Though vast volumes of biomass are presently wasted in North America, there is ultimately a limited amount from which to make compost. And in the nature of that material, it is highly dispersed, bulky and expensive to concentrate. Compost, therefore, should be seen as a dear and special resource for selective application and not as a panacea for soil repair or fertility maintenance. I find the best use of compost to be the enrichment of soil for seedling flats and as a light top dressing on newly planted garden beds or perennials.

Compost, the chief value of which is its microbial life, may also be made into a "tea" by dispersing finished compost into water and stirring or aerating it with a bubbler. The resultant product can then be sprayed onto soils to inoculate them with its teeming millions.[17]

## Mineral and Microbial Amendments

Mineral amendments are helpful to achieve and sustain high levels of productivity and should be on the checklist of any new gardener. Remembering the optimal balance of primary and secondary nutrients, we can look at some common types of mineral soil amendments. A good way to disperse rock minerals is to mix them into potting soils used for seedlings.

*Lime*, or calcium carbonate, is widely useful in all areas with acid soils or acid rain. It comes as ground limestone and is widely available from bedrock underneath many parts of the Ohio and Tennessee Valleys, where it is quarried. Near seacoasts, ground oyster shell may be the best local source. We save eggshells from our kitchen, dry and crush them and distribute them to the garden. Calcium carbonate is a necessary additive in poultry feeds for laying hens and ducks. Lime is also found in the shells of snails, molluscs and crustaceans and of course in the bones of fish and other animals. Some deposits of calcium, called dolomitic limestone after its discovery in the Dolomitic Alps of southeastern Europe, contain high levels of magnesium. This mineral may or may not be needed in your soils. Calcium hydroxide, or quicklime, is a builder's product. It is highly caustic and should never be applied directly to soils.

*Wood ash* is a valuable source of potassium (potash). It also contains useful amounts of phosphorus and calcium, plus other minerals that may be present in local woods. Wood ash is highly alkaline and should never be piled on soil. Treat it as a salt and sprinkle lightly. Small amounts of wood ash may be added to compost. The common country practice of pouring wood ash into the outhouse after each deposit is a waste of both ash and manure. The ash suppresses the decomposition of the manure, and both are buried too deeply to be of use to plants or soil organisms. Earthworms don't like wood ash, so be careful not to heap it on anywhere. We apply ash from our woodstoves on snow surfaces in the winter to hasten melting on sunny days as well as to dilute the ash.

*Urine* make an excellent soil amendment. It contains virtually all of the nitrogen, potassium, secondary and trace minerals that leave the body. The urine from an adult human provides enough fertility to grow the food for one person. Urine is practically free of pathogens unless the body is suffering kidney or bladder infection and is only a concern if taken from an individual undergoing chemotherapy, whereupon it becomes toxic waste. Urine should be spread onto coarse mulch material, not poured directly on plants. It should also be diluted or dispersed widely, not concentrated. It can be poured onto cellulosic materials such as wood chip, sawdust, spoiled hay, straw or leaves to hasten decomposition. We collect urine in used plastic milk jugs, then take it out to the garden just before or after a rain, spreading it late in the afternoon or evening when the smell is less likely to offend neighbors. Urine, especially that from carnivores (including humans), is helpful in repelling deer and other mammalian pests. Arguably male urine contains more aromatic compounds that are intended to aid the animal in marking his territory.[18]

*Rock phosphate* remains available for the time being as a mineral mined from old seabed deposits, but prices will become more volatile as major industrial interests begin to acquire the world's limited resources. Phosphate is essential for good fruiting and flowering in plants. The rock form is like a soft, tan-colored soil. If your soil is deficient, take advantage of current availability to apply some as soon as you can. Small amounts added to soils regularly are better than large amounts in one shot. Phosphate tends not to move rapidly once present in soils. Applying rock phosphate as a dusting allows the normal action of soil animals to draw it down toward the roots of plants.

*Phosphate* can also be attracted in the form of wild bird manures and insect frass. One of the benefits of trees and shrubs near garden beds is the habitat they provide for small creatures whose manures are deposited near crops. Dovecotes and other bird houses were traditionally placed in gardens to help control insects and concentrate nutrients. All small animals, which must eat frequently and have high metabolic rates, pass relatively phosphate-rich manures.

Seaweeds, fish emulsion, crushed crab shells and other ocean-based products are the most reliable source of *trace minerals*. Seaweeds and fish emulsions, being readily dispersed in liquid, can be sprayed as foliar feeds directly on plants as well as applied to soils. Routinely applying a source of trace minerals is an excellent prophylactic for the health of your soils and crops. Iodine especially is scarce in the continental interiors, far from seacoasts. Trace minerals are almost impossible to overdo, but tiny amounts can make a big difference. Expense is the chief limit to applying them.

## Mycorrhizal Fungi

Fungi and bacteria are the primary decomposers in soil but they operate in very different ways. Bacteria dominate in alkaline conditions and tend to produce more alkaline soils. Disturbed soil environments such as farm fields and tilled gardens are more often dominated by bacteria, which eat many of the sugary and soft cellulosic wastes of dead plants and animals. Fungi, on the other hand, tend to acidify soils and to associate with humid woodlands, wherein soils are rich in plant lignins (the tough binding tissue of wood). Fungi attack these brown and less accessible forms of food with relish. Bacteria are quick to multiply and to exploit available food resources, particularly "green" or fresh protoplasm. Fungi take longer to reproduce, but have more enduring effects. Fungi are more mobile and their hyphae (feeding strands) have a wider reach, being able to capture small and large living soil organisms and to penetrate their bodies as well as the tissues of woody detritus, dead insects and animals. As a consequence, fungi can mobilize nutrients over a broad area. Fungi have also developed powerful enzymes and acids for breaking down resistant substances: chitin (the cellular coating of insect bodies), bones and sinews of animals and long-chain carbon-based molecules such as petroleum, pesticides and even nerve gas.[19]

Fungi also develop relationships with plant roots and appear to be critical for plant nutrition. Plants only moved out of water and onto land some 400–460 million years ago, after aquatic plants developed symbiotic relationships with fungi. As many as 95% of all plants, possibly more, have root associations with fungi, called *mycorrhizae*. As they break down dead tissues, fungi take minerals, including nitrogen in the ammonium form, into their bodies and hold them, swapping them in trade with plant roots for sugars, or releasing the whole lot upon death. Fungi appear to be particularly important in making phosphorus, which is tightly bound to iron crystals in soil, available to plants.

Fungi and bacteria occur in almost all soils, but when soils are less often disturbed and where woody material begins to accumulate (as in forests or woodland gardens where succession has advanced), fungi will come to dominate over time. They can be understood as the "old-growth" giants of the soil. Their long thin hyphae are particularly vulnerable to tillage and also suffer from too much exposure to agrichemicals.

We can assist our crop plants by helping them to form mycorrhizal associations, which have been much disrupted by conventional agriculture and gardening practices. Minimizing tillage and eliminating chemical usage are the two most important steps to take, but where drought stress or other conditions of poor nutrient availability obtain, we may want to inoculate our crop plants with mycorrhizae that we culture. Mycorrhizae occur with greatest frequency around the roots of perennial plants in undisturbed native soils whence they can be harvested.[20] Mycorrhizal inoculants may also be purchased commercially and applied to newly planted perennials or crop seedlings. The fungi and the plants know what to do when they come into contact with each other. They need only be brought together physically. In healthy soils, these associations will form naturally.

## Building Living Soil

The living community in the top few feet of the Earth's surface is complex beyond our imagination or knowing, and we are utterly dependent on its continued healthy activity. What we have learned, much of it in the last 50 years, is enough to say that we must completely rethink our agriculture. Apart from the violence that it does to our bodies and the living fabric of ecosystem relationships, chemical use destroys the soil's ability to maintain fertility; the application of fertilizer is in every way an addictive and deadly game. In a future sane world, chemical fertilizer will be reserved — like drugs for an emergency room patient — for use on severely damaged soils at the first phase of reconstruction and not after that. Tillage too must become a tool of very occasional and special use, primarily as in Keyline, to enable permanent soil building processes. Soil fertility requires a supply of water, earthen structures for erosion control and the cycling of plant material through growth and decay. We must develop cultural systems that enable us to lay down significant quantities of old and dead leaf litter and other organic detritus on the surface of all soils; most of this must be grown in place or on the farm. And our technology and our design thinking must be focused on methods for managing complexity: intercropping, polycultures, perennials mixed with annuals — and for slashing, rolling, crimping or otherwise sacrificing fertility crops without uprooting them. Trees and shrubs must play a much larger role as harvesters and distributors of mineral fertility. Controlled pyrolysis of biomass to produce biochar for soil enrichment and gas for fuel offers great promise for the widespread and small-scale capture of carbon from the atmosphere. And as we will learn in greater detail in subsequent chapters, animals of all sizes and types can be powerfully employed to build soil, not only through the application of their manures, but through their controlled and intensive grazing and foraging on both permanent swards and among annual and perennial crops. Our goal for the next century must be to raise carbon levels in the soil as we lower them in the atmosphere.

# CHAPTER 12

# Plants, Crops and Seeds

It is not known how many plant species may be found on Earth, but the rate of discovery in remote places is probably not keeping pace with the rate of extinction in areas impacted by humans. It is estimated that between 500,000 and 1 million species of plants are alive on Earth, a tiny fraction of all those that have ever lived. This is significantly fewer than the number of animal species, mostly insects — of those mostly beetles — which number nearer to 10 million.

It has been claimed that humans have used some 100,000 species of plants over our 1,500 generations of gardening.[1] Perhaps 10% of this number have been eaten. Of these, about 3,000 species used for food and drink are available commercially today or have been collected in arboreta or botanical gardens. Only about 150 crops have been developed extensively for agriculture.[2] A mere 20 species provide 80% by weight of the food consumed by humans. And among those 20, a relatively slim number of cultivars provides the bulk of production in each crop.

## The Family Tree

Plants, belonging to *Plantae*, one of the six great kindoms[3] of evolution, begin to appear in the fossil record about 800 million years ago. Some of the earliest evolved plants, mosses and liverworts, which have no roots, are still with us. Indeed, the direction of evolution is strongly conservative, and as biologists are coming to understand, the higher or more recently evolved kindoms, orders and families of organisms incorporate complex cooperations of older, earlier beings. Our human bodies, for example, contain ancient associations of once free-living species that have become our mitochondria and other cell organelles and structures. So it is with Plants, Fungi and all other animals.

Among the plants, humans — ourselves just two to three million years old and one of the newest mammalian species — have intimate relations primarily with the Angiosperms, or flowering plants, which are the most recently evolved phyllum or division of the plant kindom. Flowering plants emerged over the past 80–140 million years during a period when the great supercontinent of Pangaea had begun to rupture into smaller continental and island land masses, which we now know have drifted over the surface of the Earth, forming various enduring or transitory combinations. This movement of land masses, together with the emergence and differentiation of flowering plants, itself largely a response to new niches of cold and arid terrain resulting from the divergence of landforms, has created the amazing green world we know today.

All organisms, plants included, are categorized by their kinships or evolutionary affiliations. The six Kindoms are subdivided further into Phylla (plural for phyllum) or Divisions, Classes, Superorders, Orders, Families, plus Tribes, Genera (plural of genus) and species. Species may further be differentiated into subspecies — and among domesticated plants, cultivars and varieties, or among animals, breeds. Taxonomists recognize many intermediate ranks as well.

Biologists everywhere now use Latin

## 31 Crops That Feed Humanity

| Crop | Family/SuperOrder | Region of Origin/Greatest Diversity |
| --- | --- | --- |
| Rices | Grass | China, West Africa |
| Wheat | Grass | Middle East/Mediterranean/Central Asia |
| Maize | Grass | Mesoamerica |
| Potatoes | Nightshade/Borage | Andes |
| Barley | Grass | Middle East/Ethiopia/China |
| Cassava | Malva/Cotton Alliance | Amazonia/Caribbean |
| Oats | Grass | Middle East |
| Sorghum | Grass | West Africa/Ethiopia |
| Soya | Rue-Pea Alliance | China |
| Sugar Cane | Grass | SE Asia/India |
| Citrus | Rue-Pea Alliance | China/SE Asia/India/Mesoamerica |
| Sugar Beet | Goosefoot/Caryophylls | Mediterranean |
| Beans and Peas | Rue-Pea Alliance | Mesomerica/Mediterranean/India/Central Asia |
| Rye | Grass | Middle East |
| Banana | Ginger | SE Asia |
| Tomato | Nightshade/Borage | Mesoamerica |
| Millets | Grass | Africa/China/India |
| Cottonseed | Malva/Cotton Alliance | Andes/Central America/India |
| Sesame | Mint | Sub-Saharan Africa |
| Apple | Rose | Middle East/Central Asia |
| Onion | Lily Alliance | Mediterranean/Central Asia/China |
| Mango | Rue-Pea Alliance | India |
| Palm Oil | Palm | West Africa |
| Peanut | Rue-Pea Alliance | Amazonia-Chaco |
| Coconut | Palm | SE Asia |
| Olive | Gentian | Mediterranean |
| Sw. Potato | Nightshade/Borage | Mesoamerica |
| Yam | Lily Alliance | West Africa, China, SE Asia |
| Grape | Grape | Mediterranean |
| Cabbages | Violas/Mustards | Mediterranean/China |
| Date | Palm | Middle East |

Credit: Peter Bane

binomials consisting of Genus (always capitalized) and species (lower case) so that they and we as scientists, nursery growers and gardeners can make sense of the complexities of plant diversity and communicate with each other. Maize, or corn as we call it in English-speaking North America, belongs to the kindom *Plantae*, subkindom *Tracheobionta* (vascular plants), superdivision *Spermatophyta* (seed plants), division *Magnoliophyta* (flowering plants, also commonly called angiosperms), class *Liliopsida* (the monocots or single-stemmed plants), subclass *Commelinidae*, order *Cyperales*, family *Poaceae* (grass), genus *Zea*, species *mays*. Maize may be further differentiated into flint, dent, flour and popcorns based on the relative size and hardness of the seed coat and germ, and thence into varieties distinguished by color, size of plant, cob and other intrinsic traits. As a eukaryote, corn is related to boletes and bobcats, as a plant, to mulberries and mosses, as a magnoliophyte to rubber trees and radishes, as a monocot to onions and coconuts, as a commelin to Bermuda grass and nut sedge, as a cyperale to bamboo and sugarcane, and as a member of the family Poaceae to barley, millet and teff.

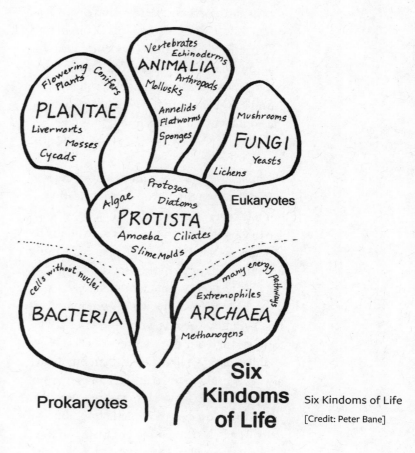

**Six Kindoms of Life**

Six Kindoms of Life
[Credit: Peter Bane]

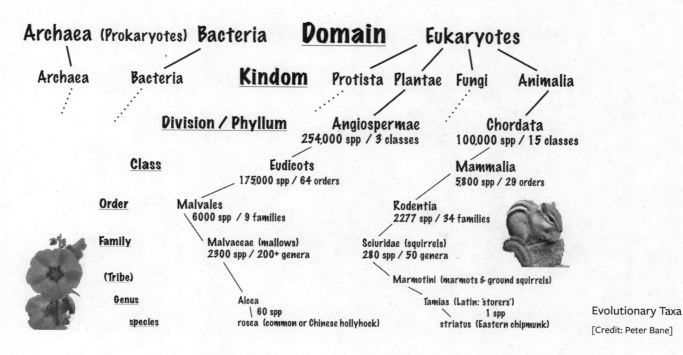

Evolutionary Taxa
[Credit: Peter Bane]

All gardeners and farmers have a professional as well as a personal interest in plant and animal genetics, diversity and reproduction, so by explaining its structures and importance, I hope this chapter will lay a foundation for the practice of diversity in the context of garden farming.

## What's in a Name?

Nomenclature, or the naming of organisms, is important to their proper identification, but it is especially so with plants. As we will see in Chapters 14 and 15, there are relatively few species of animals important in farming, but hundreds or even thousands of plants may be part of a permaculture system. They come from all over the world, and many were first named by people with different native tongues than our own. How do we know what to call them? We like the familiar, but it isn't always a guide to the truth. If a plant has not long been part of our culture, then we are unlikely to have a common name for it, but if we do, that can be even more confusing. In the central Mississippi Valley where I live, gum trees grow, but what does that name mean? Here it is likely to mean *Nyssa sylvatica*, also called black gum, a large tap-rooted deciduous hardwood tree. As you move south from Indiana toward the birthplace of Elvis Pres-

ley in northern Mississippi, its cousin, *Nyssa aquatica* (water gum, swamp gum or tupelo), becomes more common. Both grow together over much of their range and have overlapping habitats. Also found in this region is the sweet gum (*Liquidambar styraciflua*), a completely unrelated tree, which nonetheless may occur in the same patch of woods with either of the others. Sweet gum belongs to a family most of whose members come from China, though it is native here. And if you go to English-speaking Australia and talk about gum trees, everyone there will think you are referring to one of the several hundred eucalyptus species.

It becomes important, therefore, that we learn to use scientific nomenclature in relation to plants. The practice of permaculture depends on developing and sharing useful knowledge; Latin binomials, or the two-word scientific names of species, are an important part of that body of knowledge.

*Taxonomy* (the classification of organisms — which includes their names) is not a fixed science but undergoes frequent revision as knowledge expands. This is largely because discovery, and the naming that went with it, preceded classification. Science found many pieces of the puzzle of life before it began to make the whole picture. With more power-

Eucalyptus (gum), sweet gum and black gum are widely divergent and botanically unrelated species all bearing the same common name.

[Credit: Jami Scholl]

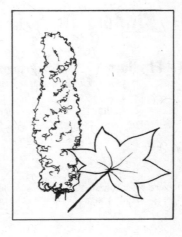
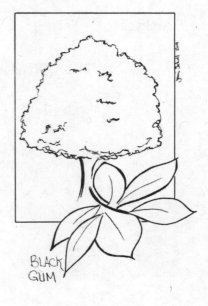

EUCALYPTUS          SWEET GUM          BLACK GUM

ful tools to analyze genetic affinities, we are now catching up. It can be vexing to grow up thinking that the carrots and parsnips belong to the family *Umbelliferae*, the umbels (wonderfully easy to remember because they have umbrella-like flowers, thus the name), only to have taxonomists change the name to *Apiaceae*. Still, most plants retain the same Latin binomial for generations or even centuries. And as scientific understanding advances, well-known plants are less and less likely to change categories. So, though some scientific names change over time, they are a more reliable way of identifying and communicating about plants than are common names. The same logic applies to a myriad of other species of fungi, insects and even tinier organisms.

## Economic Botany

The angiosperms (*Magnoliophyta*) provide most of our food and medicine. This phyllum represents about 550 plant families organized into 36 superorders. You can see from the tables on pages 216 and 220 which of those superorders contribute the most important crops, but all 36 are important to the ongoing survival of life on earth, and our garden farms should aim to include plants from all these major branches of the tree of life — an approach Alan Kapuler calls *deep diversity*. In simple terms of plant families represented in North America, Thomas Elpel singles out

7 of the 115 families found on the continent as the most common and important.[4] These include most of those with economic significance: legumes, mints, sunflowers, mallows, mustards, carrots and lilies. Each of these represents a different superorder, and though among this group of plants there are few trees and shrubs, if we add the rose family (*Rosaceae*) which includes most of our tree, shrub and cane fruits and the beech family (*Fagaceae* — part of the rose superorder) which includes most of the cultivated nuts, we would have a good set on which to focus our first efforts at learning plants.

Flowers are the main way that genetic connections between related plants are observed by scientists. For example, black locust (*Robinia pseudoacacia*), a large native tree and a member of the legume family, produces flowers that share the familiar butterfly- or orchid-like form with plants as diverse as lupin (*Lupinus* spp), a low-growing herb, and redbud (*Cercis canadensis*), a small tree.

The season and color of bloom varies among them, the ecological niche of each species varies, the leaf forms vary, but the flower parts bear a striking similarity. All are members of the legume or pea family, *Fabaceae*. Of course, plants are not always in bloom, so we also have to learn about their secondary characteristics — leaf shapes, growth habit, branching form, bark and even their insect

Flowers of lupin, black locust and redbud—all legumes—show family resemblance.
[Credit: lupin—Salicia, black locust—red.raleigh, redbud—Chris Breeze, all via Flickr]

## The 31 Superorders and Some Crops that Come from Them

Though none of these is genetically insignificant, groups marked by a single asterisk (*) produce the bulk of human food while those marked by two asterisks (**) have few if any edible members.

### Monocots

| | |
|---|---|
| *Alismatidae* | arrowhead, water plantain |
| *Triurididae*** | |
| *Aridae* | taro, calamus, calla lily, monstera |
| *Liliidae** (Lilies) | onions, garlic, leeks, camas, iris, crocus, asparagus, yams, vanilla, agave, yucca |
| *Bromelianae* | pineapple, cattails, pickerel weed |
| *Arecidae** (Palms) | dates, coconut, Nicobar breadfruit, hat palm |
| *Zingiberidae** (Gingers) | ginger, canna, arrowroot, banana |
| *Commelinidae** (Grasses and Sedges) | rice, wheat, corn, sorghum, bamboo, sugar cane, barley, rye, oats, nut sedge, water chestnut, bulrush |

### Dicots

| | |
|---|---|
| *Malvanae** (Cotton Alliance) | manioc, cotton, fig, mulberry, breadfruit, hemp, okra, hops, jujube, kenaf, hibiscus, castor bean, nettles, linden, elms, euphorbias, Thai yellow velvetleaf |
| *Violanae** (Mustards and Cucurbits) | cabbages, squash, mustard, cucumbers, melons, papaya, passionfruit, radish, violets, capers |
| *Primulanae* | Primrose, sapotes, persimmon, starfruit, abiu, eggfruit, silver bell tree |
| *Asteranae** (Composites) | sunflower, lettuce, artichoke, marigolds, Jerusalem artichokes, burdock, chamomile, echinacea, campanulas |
| *Solananae** (Nightshades and Borages) | potato, tomato, capsicum peppers, tobacco, eggplant, sweet potato, comfrey |
| *Lamianae** (Mints) | sesame, sage, basil, oregano, lavender, foxglove, mullein |
| *Gentiananae** (Gentians) | coffee, olives, quinine, cleavers, milkweed, noni |
| *Myrtanae** (Myrtles) | pomegranate, evening primrose, fuschias, rose apple, jaboticaba, clove, allspice, eugenias, Surinam cherry, feijoa, eucalyptus |
| *Proteanae** (Proteas) | macadamia |
| *Polygonanae* | buckwheat |
| *Caryophyllanae** (Chenopods and Cacti) | amaranth, quinoa, beets, spinach, chard, cacti |
| *Nymphaeanae* (Lotuses and Pepper) | water lilies, lotuses, pepper spice |
| *Magnolianae** (Magnolias) | avocado, annona, cinnamon, spicebush, camphor, bay laurel, sassafras, American wild ginger |
| *Theanae* | tea |
| *Aralianae** (Carrots-Ginseng Alliance) | carrot, celery, dill, parsnip, parsley, cilantro, fennel, cumin, anise, caraway, lovage, ginseng |
| *Balanophoranae*** | |
| *Cornanae* (Heaths) | blueberries, kiwi, cranberries, madrone; elderberry, honeysuckles, valerian, corn salad, hydrangeas, columbines |
| *Loasanae*** | |
| *Podostemonanae*** | |
| *Rosanae** (Roses and Beeches) | apples, almonds, peaches, pears, plums, cherries, chestnuts, walnuts, apricots, currants, gooseberries, pecans, hazels, hickories, beechnuts |
| *Vitanae* | grapes |
| *Rutanae* | citrus, mango, cashew, flax, coca, litchi, longan, rambutan, akee, acerola, khat, sumac, geraniums; (Legumes) soya, beans, peas, alfalfa, clover, favas, carob, tamarind, lupins, jicama, acacia; (Sandalwoods) quandong |
| *Ranunculanae* | buttercup, poppies, opium |

Credit: Peter Bane

and animal associates and their preferred habitats — in order to identify them when they are dormant or merely not flowering.

Conventional farmers and gardeners may grow only a few dozen familiar plants of traditional sorts, but permaculture growers often deal with unusual plants and may have between 300–1,000 species on an acre or two. As we seek to meet more of our own needs locally and from things we grow, we will turn to a wide variety of compatible plants from all over the globe, just as agronomists and the nursery trade have done for generations. Many of these may be exotics or analog plants, which the science of plant origins can help us to find. An example of an analog plant is Chinese chestnut (*Castanea mollissima*), often grown because of its resistance to the blight that wiped out most of the once widespread American chestnuts (*Castanea dentata*). It fills the same economic niche — though a slightly different ecological one — as its American cousin. Thus, it is an analog to the native chestnut.

## Where Plants Come From

Plant origins are not important, in my view, for the purpose of keeping ecosystems in some form of "native" purity. We are not only past the time when that is possible outside of very small areas, but climate change has handed us a mandate to accept and to create new combinations of plants, animals, insects and fungi in the interest not only of meeting our own needs but to preserve the diversity that is the basis of life itself.

Plant origins matter because they have shaped to a considerable degree the expectations of the organism. A plant that originated near the equator will not likely respond well to very short or very long day lengths in the temperate regions, though there are exceptions. Plants will have coevolved with insects and animals; they will have pollinators; they will have ecological niches that tell us much about how they should be cultivated. Studying plant origins can tell us where to look for analog plants, those that can function in the place of a native species, but that may have preferred characteristics, e.g., disease resistance, better flavor or higher yields.

In most parts of the US and Canada, we grow cool-weather crops in spring and fall, and hot-weather crops for two to four months in the summer. While farmers and gardeners may give this little thought because they have internalized the planting of peas, onions and spinach in spring, beans, corn, tomatoes and squash in summer and cabbage, chard and broad beans in autumn, we should recognize that in doing so they are marshalling a sequence of genetic resources, first from the north temperate regions of the planet, then from the tropics, to achieve optimal growth

---

### Annual and Biennnial Vegetable and Oil Crops by Season

Cool-season crops are in general frost-tolerant, and originate in temperate regions of Europe, temperate Asia or North America. Those indicated with an asterisk (*) grow well in warm seasons as well as cold. Most warm-season crops originate in regions of Mesoamerica, the Andean highlands, Africa, India or tropical Asia and are frost-intolerant.

---

### Cool-Season Crops

onions, carrot, peas, lettuce, cabbage, beets*, spinach, kale, chard*, garlic, broccoli, celery, broad beans, cauliflower, parsley, collards*, brussel sprouts, radishes, mache, kohlrabi, turnip, arugula, endive/chicories, napa, pak choi, leeks, fennel, tatsoi, mizuna, rutabaga, dill, flax, parsnip, mustard, coriander

### Warm-Season Crops

tomato, tepary beans, corn, summer and winter squash, pumpkins, potato, capsicum peppers, peanut, amaranth, sweet potato, cucumber, cowpeas, watermelon, eggplant, lima beans, okra, melons, soybean, cotton, sunflower, sesame, basil, ground cherry, pigeon pea, tomatillo, yams, malabar spinach, roselle

---

Credit: Peter Bane

and productivity based on a wildly variable climate.

If you try growing lettuce in most parts of the southern or midwestern US or in central California in summer, it will likely bolt before you get much of a crop. There's too much heat. In the same conditions, while lettuce is wilting or going to seed prematurely, squash and tomatoes feel right at home. Annual crops pose one set of challenges, while perennials offer another. Not only do apples, plums or figs need the right temperature range, they have additional temperature constraints during periods of blooming and ripening, and they need certain periods of dormancy or chilling hours in order to start their annual cycles properly. Knowing the origins of crops provides insight to their needs and thus helps us to grow them more easily and with less stress.

## The Importance of Breeding

Plants are capable, as are humans, of modifying their environment. But unlike humans, who reproduce in small numbers with long cycles of maturation, plants — even large trees — can change their biology to adapt to the environment in significant ways. As gardeners and farmers, we want to exploit this capacity of plant genomes to express different characteristics. By selection and even by deliberate breeding, we can develop varieties of many useful species better adapted to our local conditions.

What introduces variability into a genome (the genetic reservoir of a species) is diversity of environmental conditions. With variability comes resilience (the capacity to meet a wide range of challenges). That variability builds up over time in the genetic memory of plants, as generations accrue that have survived drought, flood, insect plagues, viral diseases and fire. And variability, or diversity, builds up faster where plants can reproduce more frequently — in the tropics and subtropics. There, plant growth is seldom slowed by limiting factors such a drought or cold. As a measure of this, consider California. At 150,000 square miles, it represents about 2% of the land area of the US and Canada combined, yet is home to more than 6,700 plant species of the 20,100 that occur across the two continental nations.[5] The absolute measure of diversity — the number of species per acre — is greatest in the tropics, where evolutionary change has continued unabated for tens of millions of years, but the rate of variation may in fact be greater in subtropical regions, and especially in mountainous ones where relatively benign growing conditions are punctuated by disturbance: winter frost, landslides, earthquakes, floods, fires and windstorms. This might seem counterintuitive, but think for a moment about the opportunities to reproduce. It's not enough to have lots of sunlight, mild temperatures and water. A plant also needs access to soil, and in the tropical rainforest most of the real estate is already spoken for. Finding a niche for one's seedlings to take root and grow without being eaten is a perilous endeavor. Where vegetation and soil are more often disturbed, the chances of a new variety succeeding are greater. In any case, humans have exploited those subtropical highland experiment stations to create the crops that feed, clothe and heal us, and we did most of that work more than 3,000 years ago without benefit of written language.

Before Gregor Mendel inferred the laws of genetics by applying statistical analysis to observations of peas in his garden (work published in 1866), farmers had long and successfully selected vegetables and fruits for qualities they desired, without understanding how those variations occurred. But Mendel's discoveries opened the door to much of modern plant breeding. What Mendel saw in the variation of his crops were the results of a random recombination of traits from the parent plants by the mixing of chromosomes. Every living organism that reproduces sexually has an even number of chromosomes in every cell of its body. Humans have 23 pairs, or a total of 46, half of which come from the mother, the other half from the father. Each pair may control one or several obvious physical traits, and Mendel was able to determine that some traits were *dominant*, others *recessive*. Mendel's laws of genetic variation allowed breeders to reach into plant and animal genomes in search of more desirable traits, accelerating the selection of varieties to suit new tastes and environmental conditions, rather than waiting for nature to throw up whatever variations might occur.

For example, farmers selected grains the heads of which did not shatter but held tightly to their seed — a trait of little value to the wild grass that wants to scatter its seed far from the mother plant — but that greatly helps increase yield under cultivation and simplifies harvest. Under natural selection pressures, these non-shattering phenotypes would have been uncommon and would have left fewer descendants, while their shattering cousins would have disseminated widely and had a better chance of surviving. But grains were brought into a coevolutionary relationship with humans, wherein they agreed not to fling their seed about loosely while we provided favored conditions for the plants to grow, allowing us to harvest, select and replant their seeds for them. This is a very deep intimacy indeed. Corn has reached such a state of interdependence that its seed will scarcely germinate at all unless removed from the cob by humans.

What scientists learned was that new traits were more likely to be expressed in plants that had widely divergent types of parents, a phenomenon called *hybrid vigor*. The late 19th and early 20th centuries saw the epoch of cultivated diversity as plant breeders swarmed all over the possibilities of genetic recombination and selection facilitated by Mendel's discoveries and applied them to all the cultivated crops of humanity, which in the age of the railroad, steamship and telegraph were becoming almost universally available.

Markets for novelty also boomed with the increasing number of farmers and gardeners in North America and the other New Europes of South America, Africa and Oceania, as well as through the rapid expansion of urban wealth and the growing taste for exotic foods that came with the Industrial Revolution. Unfortunately, this explosion of diversity was undercut by the mechanization and concentration of agriculture that began in the first decades of the 20th century and continues today.[6] Where consumers of 1900 saw new colors and qualities of foods they had never seen before and were as amazed by them as consumers today are about 3-D movies and mobile phone apps, commercial farmers soon began demanding crop varieties that were better suited to mechanical harvest and later to chemical management, while industrial processing and marketing through advertising and retail outlets began to separate desirability from the time-honored standards of flavor and nutrition. Fewer people gardened or farmed as the 20th century ripened, so the preference for flavorful, nutritious or practical varieties that suited the small grower or the home table went by the way. Selection focused instead on durability, keeping qualities in storage, uniformity and ease of mechanical manipulation — all traits more suited to commerce.

## The March Toward Seed Monopoly

As large economic interests penetrated further into agriculture, which had traditionally been a domain of common life, the monopolistic tendency of the market increased pressures for control of market share. This led first to strategies of vertical integration, followed increasingly by mergers, and ultimately toward political control of markets through lobbying and regulatory capture. With the rise of the chemical industry following World War II and the near universal mechanization of farming in the advanced industrial countries which it abetted, the breeding of seed for commodity crops took on new importance. Farm labor diminished, scale increased and farmers assumed more specialized roles, manipulating chemicals and machinery with less biological and cultural diversity on the farm. Where their great-grandparents would have kept seed from year to year, post-war farmers were convinced by Agricultural Extension to modernize, which meant purchasing hybrid seed that promised higher yields, applying bigger machinery and more chemicals to exploit those yields — and in the words of US Secretary of Agriculture Earl Butz, to "get big or get out."

There were, in the years after World War II, justifiable anxieties among world leaders about the rapid growth of global population and the ability of food producers to keep pace. The war itself and the violent decades of revolution and upheaval that preceded and followed it as old empires collapsed and former colonies threw off their masters created a lot of real hunger. Hunger has always been understood by governments as dangerous to the established order. So research was launched into high-yielding varieties of wheat, rice and corn, and programs were set in motion by the Rockefeller Foundation and other international bodies to disseminate their results to areas of the world where population growth threatened to outstrip agricultural productivity — among them Mexico, India, Indonesia and the Philippines. Thus the Green Revolution was born.

## The Green Revolution

Books have been written about this initiative, from inception to outcome, and it remains a subject of considerable controversy, which I am unable to settle here. At a simple level, seed dependency increased because the new Green Revolution seeds were all hybrids whose ownership was not in the hands of peasant farmers but of large corporations, interlocked in their management and later by mergers with other large corporations

that sold fuel, tractors or chemicals or that marketed grain internationally. These corporations were, of course, the familiar names, and their ownership was tightly integrated with the elites of the former Western imperial powers. The patenting of life forms was, through intense political pressure applied behind closed doors, legally enabled in the 1970s so that profits from hybrid seeds could be maximized. This was promoted — when it was discussed publically at all — as providing incentives for innovation.[7] What it actually provided was increasing private control — now approaching monopoly — over common property resources of humanity.

The mechanism for securing hybrid seeds from simple reproduction or "piracy" was not patent rights, though that helped the corporations to keep tabs on each other and on potential seed-breeding competitors, but rested in the laws of genetics itself. Coming from dissimilar parents, hybrids have a highly diverse genetic makeup. Put simply, they are unpredictable. If two similar hybrid plants are crossed, their F1 or "first filial" offspring will include a wide mix of types of which only a few will resemble the parents. It is possible to stabilize hybrid seeds so that they will reproduce true to type, but this requires about six or seven generations of backcrosses to achieve. From each generation, the breeder must select those offspring most like the desired type, cross them with each other, reselect for the desired type and so on; in each generation, most of the resultant crosses are rejected in favor of the few that exhibit the preferred traits. Seed companies have the resources to do this on a mass scale, but choose for commercial reasons to maintain pure lines of the parent seed from which to generate the desired F1 hybrids. Most small farmers cannot do either.

Hybrid seeds, in and of themselves, are not evil. But the expansion of their use to support increasing intensification of agriculture worldwide, combined with plant patenting

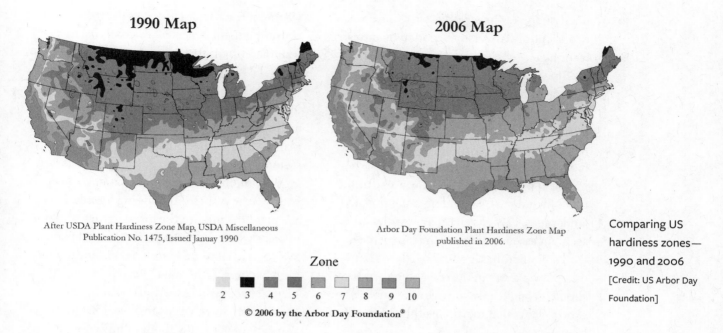

**1990 Map**

After USDA Plant Hardiness Zone Map, USDA Miscellaneous Publication No. 1475, Issued Januay 1990

**2006 Map**

Arbor Day Foundation Plant Hardiness Zone Map published in 2006.

Zone

2 3 4 5 6 7 8 9 10

© 2006 by the Arbor Day Foundation®

Comparing US hardiness zones—1990 and 2006

[Credit: US Arbor Day Foundation]

laws and the destruction, intended or otherwise, of landrace diversity in staple crops throughout many of the largest agricultural centers of production in the global South, saw a massive increase in monoculture and a dramatic narrowing of the genetic diversity underpinning agriculture itself. This has placed the very future of humanity in jeopardy. The earliest evidence of widespread risk from the hybrid breeding strategy came with the outbreak of Southern Corn Leaf Blight in the 1970 growing season in the US. Much of the US seed corn crop was genetically identical (which was not a result of hybrid breeding alone, but of its use as a commercial lever and of mechanization leading to vast monocultures), and it proved fatally susceptible to a disease organism that first appeared in Florida and then spread throughout the Corn Belt. Only emergency efforts to breed up and replace seed with new resistant varieties prevented absolute calamity. The recovery was dependent on having access to wild ancestral strains of corn from the center of corn diversity in Mesoamerica. Yields were down about 15% at the end of the year, nonetheless — a harsh blow to farmers. At the same time this was going on, leafhoppers

were spreading Tungro disease among Green Revolution hybrid rice in the Philippines and leaf rust was devastating genetically identical coffee plantations in Brazil.[8] These kinds of risk from monoculture persist today with the FAO warning that wheat rusts are spreading from the Middle East to other wheat-growing regions as I write, and that Sigatoka (black rot fungus) is threatening many of the world's genetically similar banana plantations.

## The Ethics of Diversity

The cultivation of diversity, which leads to resilient ecosystems, is a principle in permaculture but is also expressed in our ethical injunction to "Care for the Earth," which means to respect and conserve all species. Every garden farm should strive to introduce and maintain more plants than it can crop. In part, this is a strategy for selection, but at a deeper level it is a strategy for survival, for diversity is the life raft of life itself.

Climate change has shifted North American climate zones northward by an average of 15 miles a year over the past 20 years. If humans do not now disseminate existing species to new and appropriate habitats, we will see massive losses among our native flora piled on top of the already severe devastation of agricultural diversity begotten by industrial farming.

This is not, of course, an indulgence to destroy intact native plant communities, but an exhortation to use our already disturbed settlements, roadsides, farms and managed forests to extend the ranges and expand the habitats of native and useful species. As a practical matter, we should always be mindful to include natives among our cultivated landscapes, gardens and fields for the unseen but essential relationships they make with soil microbes, pollinators, birds and other key players in the ecosystem. And we should thoughtfully consider the qualities of exotic plants that may make them difficult to manage.

People choose crops to cultivate in order to get yields, either of materials or services, and for that purpose we have always selected vigorous plants. Plants that have developed the ability to grow rapidly or to produce large quantities of fruit, seed or biomass make good candidates for domestication. Given a choice, we tend to favor plants that grow tall, set large fruits, ripen quickly or offer the farmer similar advantages. Consequently, it becomes our job to reign in that success, limit plant growth or design systems in which rampancy is either contained or useful. It is a complete failure of imagination to blame plants that we neglect to harvest for their exuberance or to attribute to them human motives of aggression, mischief or cunning. Plants don't create narratives about right and wrong, observe invisible political, cultural or conceptual lines, develop strategies or issue declarations of war. They can never accurately be said to be invasive. It is we, not they, who have invaded territory and unhinged ecosystems.

In our Indiana garden we deliberately favor opportunistic plants, especially those that will take root from broadcast seed. Then we take advantage of their willingness to germinate without attention to fill up spaces we might not use or to save us the trouble of sprouting them in special microclimates. We want perennials that grow rapidly, and we expect to trim, prune and coppice them regularly. These cuttings become nursery stock, fodder, mulch and craft materials. But there are practical limits: we do not put expansive plants — those that spread by rhizomes or stolons — into areas where their often unseen growth might become problematic. We keep young trees away from building foundations. We plant the mint in the forest garden and not in areas where we cultivate small salad plants. We confine running bamboo on islands or behind barriers in the soil to keep it from colonizing spaces where we don't want it. And we remove self-sown weeds that we don't want. You cannot garden without killing a lot of plants deliberately. Just as we should take no offense that plants move onto "our" turf, we should have no guilt about chopping, mowing or grazing them down as it suits us. Out of common sense and self-care, we should, however, refrain from poisoning plants because poison knows no boundaries; anything we put into the garden, the forest or the field, we are agreeing to put into our bodies.

## The Politics of Diversity

Rearing its ugly head 35 years ago with the introduction of plant patenting laws, the corporate drive for control and uniformity has accelerated dramatically in the last decade. The same few global corporations, known colloquially as Big Pharma, have gathered into their control an integrated product line that includes agrichemicals, seed, genetically modified organisms and pharmaceuticals. And while these market sectors are often identified as "the life sciences," the primary focus of all their products is death in one form or another.

The genetic engineering of crops by these very large companies and the political and economic bullying that has gone into marketing them have dispersed exotic and dangerous toxins into billions of widespread crop plants in North America and beyond. Bacteria have already been shown capable of moving these novel genes — for resistance to pesticides, for example — into field weeds, making them in various ways superweeds. Bacteria in our guts may well be able to "learn" from GM foods we consume how to manufacture pesticides in our bodies. Species barriers appear to be no obstacle to the movement of genes. A farmer in Australia growing organic wheat and oats sued Monsanto for contamination of his crops with Roundup-resistant genes that came from canola in adjacent fields.[9] Opposition to GM foods and crops has been widespread around the world. Reports from Haiti indicate that hybrid seed corn "donated" by the St. Louis-based Monsanto Corporation has been burnt by Haitian farmers who saw it as a clear ploy to displace native landraces of corn and colonize even further Haitian agriculture.[10] For many years in North America, farmers whose crops, including heritage seed varieties, were inadvertently contaminated by pollen from GM crops have been sued and had their crops seized for patent infringement on novel and completely preposterous legal grounds nonetheless upheld by appellate courts.[11] And though the anodyne studies released by industry claim there are no substantial harms, independent research has shown strong links between the consumption of genetically modified foods and allergies, degenerative illnesses and reproductive failure.[12]

At this point there can be no doubt that a full-court press is underway by agribusiness to institute GM crops across all major agricultural sectors and in all world markets, ensuring that these genetic forms overwhelm or supplant all other varieties, not because they are superior (yields of GM crops are in fact lower than those of conventional varieties of the same crop while nutrition is inferior if not negative), but because they are proprietary and therefore profitable. This is one of the gravest threats to humanity's well-being, food security and ultimately our freedom. Genetic resources are nothing if not for the long term. And extinction is forever.

Erosion in the number of farmers brought about by the intensification of agriculture has meant a loss of plant and animal breeders and seed savers, which must now be made up by an expansion of garden farmers cultivating crops and nurturing diverse plants. While it is daunting to undertake the stewardship of humanity's heritage of food, fiber and medicine crops, no one has to do it all. You can begin by focusing on the crops of greatest importance to you and your household, choose a few varieties that hold special meaning and gradually add others as you become more comfortable with seed saving.

## The Challenge of Polyculture

All of our farms, however vast their extent — and all of our gardens, whether in cities or countryside — exist in a matrix of wild plants and animals, forests, grasslands and waterways. Nature is already a polyculture, so we can have some confidence that this strategy will work when applied to cultivated systems. Our job is to fit into the larger logic of nature and to shape our work in her image. We have come to think of our garden and farm crops so much as ultimate arbiters between sun and soil that we have forgotten that most plants have some other kind of relation to the land. They are members of far-flung genomes occupying habitats throughout the region and around the world. Those habitats can tell us what our cultivated landscapes need to look like: most of our crops are naturally subordinate to larger plants and in turn nurture smaller ones; they provide habitat for a host of creatures large and small, some of which may provide us with excellent yields of food, fertility or cash. And they each have one or several organisms that keep them in check. Bereft of these myriad connections, our crops are literally out of place. The challenge of polyculture and the key to healthier landscapes is bringing elements of the larger wild matrix into every part and level of the farm and garden: from hedgerows, windbreaks and strip forests, to browsing animals, bird fertilizers and orgies of soil microbes.

## Planting in Guilds

The *guild* (an assembly of cooperating plants) is an important permaculture concept that embodies the practice of polyculture. A polyculture is not merely a collection of different species brought together in a small space like some elevator car filled with strangers, all attempting not to intrude on each other. It is much more like an office or a shop floor of coworkers or a neighborhood of familiar households whose lives are intertwined. The designer finds or creates guilds that are functionally integrated. This means that the elements within them are mutually supportive and that as a whole these small ecological groupings stand a better chance of flourishing

The Three Sisters:
corn, beans and squash
[Credit: Jami Scholl]

and producing together than their individual members do by themselves. Guilds enrich the evolutionary story of *symbiosis* whereby organisms, formerly separated, become deeply implicated in sustaining each other's lives.

We can draw insight to this creative process from the familiar Native American guild, a combination of corn, beans and squash planted together, called the Three Sisters.

Besides yielding abundant energy food, corn provides a central stalk onto which the beans vine. These in turn supply extra nitrogen for the hungry corn and also complement its amino acid profile to provide a complete dietary protein, implicitly involving the farmers who must carefully calibrate the planting times in order to ensure that the sisters remain in balance. Squash, with its prickly leaves and sprawling habit of growth, covers the ground, providing insect deterrence and a living mulch as well as a third valuable food rich in calories, beta carotene and important minerals such as zinc. Recent research suggests that cleome, a showy garden annual, may have been a fourth sister grown as a trap crop for squash beetles.[13] By planting these crops as a guild, the yields of each per acre are diminished compared to monoculture, but the sum of yields per acre is increased over what could be achieved from any one of them alone. Moreover, the work of weeding, staking and controlling pests is reduced. When supplemented by small animal protein in the form of eggs or poultry, the Three Sisters represent a complete staple diet, needing for variety only such greens, berries, fish and game as could easily be cultivated or gathered from the wild.[14] This guild has been passed down through the ages because it successfully sustained many widespread cultures on this continent. The opportunity exists to identify more such guilds as a basis for cultures of the energy descent future, and in particular to design guilds of perennials that can address the need to regenerate ecosystems as well as to sustain cultures.

Designing guilds is also a favorite permaculture parlor game, and the pursuit of ideal guilds is not unlike the quest for another five-letter G-word, the Holy Grail, often sought and seldom found. The important point is to combine complementary species for synergistic effects so the whole is greater than the sum of the parts. But there is a larger purpose to this game with many answers. Guilds represent a designer's shorthand, one that promotes success with plants by providing for their needs from the outset.

Typically, a guild will be organized around a single tree or small group of trees of a single species, for example, a pear or a walnut. Either of these tall woody perennials would need shorter shrubs and herbs as companions to occupy the space around it, provide shelter and ground cover, build soil and repel pests; at a minimum, to nurse the young standard during its sapling years: the list of potential candidates is long. One might choose currants, gooseberries, hazelnuts or autumn olives for the shrubs. All provide food and have a thickety habit, but gooseberries are thorny; autumn olive fixes nitrogen. Comfrey, rhubarb or even burdock with its big leaves, edible root and tolerance for shade might provide a source of mulch and other yields.

The observant designer would not only organize the space around the central element, but also the time, staging the blooms, maturation or fruiting of subsidiary elements to abet the growth, pollination or harvest of main players. Several permaculture designers, Holmgren and Rick Valley among them, have remarked on the synergies that emerge between ruderal (escaped or self-planted) apples and brambles as the latter form both a protective barrier around the young tree and later yield space to it as it matures, moving outward by tip-rooting to provide an expanding thicket of living fence that helps the apple survive predation by herbivores. They are describing a restoration guild for a food forest.

In our pear or walnut guild, some vining plants such as nasturtium, groundnut (*Apios americana*) or even hardy kiwi might be added once vertical growth is well advanced. Annuals such as scarlet runner beans or cucumbers might be used for a few seasons while perennials are getting established. Some bulbs would be appropriate, including daffodils to repel voles and alliums for general pest control. A ground cover of hardy mints and perhaps some clovers would be preferable to grasses of any sort. And opportunities would exist for a few catch crops — turnips, daikon radish, ground cherry — to grow in temporarily open patches of understory. Obviously, there are no definitive answers to the question of what should accompany a pear, an apple or a walnut planting. But with the concept in mind of using all the space above and below ground while providing for as many of the needs of the various elements as possible from within the guild, a mutually supportive community can be designed that will be more readily established and maintained than a monoculture.

Less often recognized in the quest to identify guilds is the role that insects, animals and fungi may play. Of course pollinators will be required with many flowering

Fruit Tree Guild, Pattern #53 (See Chapter 6)

[Credit: Jack Heimsoth]

plants, especially fruit trees of the rose family, and the design of a guild to include mostly flowering plants can address this. Something should be in bloom through every part of the growing season, providing bee forage and nectar for beneficial insects at all times. (See Bee Forage lists in Appendix 2.) Species with many small flowers (such as carrot, cabbage and sunflower family plants) will meet the needs of beneficial wasps, syrphid flies and other pest predators. Birds, small rodents, amphibians and reptiles are particularly important and can play helpful roles if regulated. Insect pests of all stripes may laugh at pesticides, but have little resistance to chickens, ducks or frogs. Habitat for some snakes and toads can keep slugs under control. But should you have leopard frogs or tree frogs? Garter snakes or black snakes? There is no right answer at this stage, but after the guild partly assembles itself, the designer may be able to describe it more completely and thus help himself and others to use it again. The right pest predators will come if you provide some logs, stone piles or terraces and a little water. Fungi can be encouraged by providing a lot of woody mulch, including coarse branches and logs; some mushrooms can be cultivated. Most of these non-plant elements will come and go as they please. Your role as the gardener is to encourage the ones you prefer and scatter or repel those that may be harmful. As a designer, you must take note of what works.

## Finding and Filling a Niche

Central to the concept of guilds is the ecological *niche*, of which guilds are collections or assemblies. A niche, as we learned in Chapter 3, is a job description. Every plant, indeed every species, has its niche, and to use the plant well we must investigate what role or roles it plays in its surroundings, look into its size, shape and relationships and ask how it grows and bears. Will it vine and climb, or sprawl over the ground? Does it expect to be browsed or grazed? Will it dive deep and store sugar in a root or tubers, or will it put its energy into a compact form around a central stalk? Must it have pollinators? Does it grow to seed in one season, or require several? The answers to these questions help us describe the plant's niche, the distinctive combination of food resources, disturbance, soil moisture, light levels, pollination, air movement and symbiotic links that help it to thrive. Our job as cultivators then is to find or create that niche for each species in a guild suited to our particular conditions of landform and climate. Just as often, we must survey the niches created by our landscape designs (growing out of our own needs) and find the right productive crops to occupy them.

The central role of plants in polyculture stems from their rootedness. Once planted, they can only move slowly by incremental growth or by distributing their seeds, tips or stolons into new territory. So spatial relationships to other species become critical to

Roots take many forms underground, but most are found in the top few feet of soil where air and nutrient are more plentiful.

[Credit: Abi Mustapha]

the design of cultivated environments. This architectural aspect of gardening has visible and invisible components. Up to half of the biomass of most vascular plants, the vast majority of species, lies underground, out of sight. For trees and all large plants, the roots not only harvest soil moisture and nutrients, but provide stability against wind, animals and gravity.

Roots can crack open bedrock, growing even where there appears to be no soil. Dryland trees can send roots very deep in search of water: mesquite (*Prosopis juliflora*) has been measured at more than 100 feet below the soil surface.[15] But most plant roots will be found in the top three to four feet of soil, with larger specimens of plant species reflecting more developed soil horizons and greater fertility.

As rooted beings, most plants must hold their ground, maintaining a connection to both water and nutrient. They do this by two basic approaches, reflecting different niches: some live long and hang on, modifying the environment around them to their advantage; others make lots of seed and spread to new ground. Long-lived plants tend to grow *tall* or sprawl widely to claim sunlight and soil moisture. If a plant is *short* — and virtually all of our vegetable crops and many of our fruits are short — it has one of two ecological niches. It may be a fellow traveller, as many of our garden vegetables have been, a short-lived and opportunistic weed that hopes to find a patch of sunlight long enough to set seed. The seeds of these easy riders travel on the wind or on the fur or feathers or in the gut of some animal or bird to another patch of bare soil. In doing this, a fast-reproducing plant accepts a role in *succession* (the development of more complex and durable plant communities) as an outsider, a first responder, a hit-and-run artist. Growing tall is not in its interest, for this wastes metabolic energy that needs to go into seed production. Alternately, a plant can adapt to growing beneath or on the edge of trees and shrubs, taking such bits of nutrient, water and light as may be filtered through the canopy, along with the protection and support the taller plants around it provide. Bulbs are so adapted, as they reproduce by division, spreading to form clumps. Many herbaceous perennials adapt to this niche by sprouting early in the spring and doing much of their annual photosynthesis before trees leaf out. Brambles move toward the sun from year to year by tip-rooting, keeping themselves in their favorite habitat on the sunny, protected edge of the woods.

## Aiding Succession

*Weeds* are able to establish in very harsh conditions while also accumulating minerals in their tissues that become available to surrounding and following plants as the weeds decay. Their quick life cycle acts as a kind of band-aid for bare soil, providing minimal, low-cost protection so that erosion can be staunched and repair set in motion. Most weeds are annuals, some are short-lived perennials.

Ecosystems change over time to favor larger and longer-lived perennials. Short-lived plants play specific roles in repairing soil disturbance and building fertility. Mid-sized pioneers are the hinge for cultivators seeking to accelerate regeneration.

[Credit: Peter Bane]

## Canopy Bearers
large, woody, long-lived

## Pioneers
woody or herbaceous, fruit- and seed-producing, perennial, often N-fixing, dense- or large-leaved

## Weeds
annuals or short-lived perennials, seedy and sun-loving herbs and grasses

*Pioneers* are mid-sized plants that live longer than annual weeds, from two or three years up to thirty or forty for some small trees. They may be soft or woody. Some large herbs like cup plant (*Silphium perfoliatum*) and pokeweed (*Phytolacca americana*), or in the tropics banana and papaya, are pioneers. The spreading canopies and greater biomass production of pioneers modify ground temperature and moisture levels, helping to accumulate more carbon and accelerating soil formation. Many of these plants also produce small fruits and nuts which attract birds and rodents to deposit manures, bringing phosphate and nitrate to enrich the soil. Some pioneers are thorny, some are toxic or irritating and many are dense and multi-stemmed. All of these characteristics help to exclude large animals from the area, reducing soil compaction. Pioneer plants frequently fix nitrogen or have mycorrhizal associates that aid in mobilizing phosphorus. And under their dense cover, seedlings of taller and longer-lived trees may sprout and grow. These will in time form the canopy of a new ecological community. While we do not want every part of our cultivated ecosystem to be forested, it is important to know which role in this inexorable process any given plant is meant to play.

## Strategies for Growth

*Annual* plants grow from germination to seed production within one year. *Biennials* (two-years) germinate and grow in the first year but pass through a dormant season before setting seed in the second. Many common garden biennials belong to the cabbage and carrot families. These originated in climate regions like the Mediterranean, where winter rains aid seed germination and mild cool-season temperatures slow growth but do not kill the plants. Full maturation and flowering take place in the following year with the return of seed-ripening solar energy. The cycles of these plants may sometimes be better suited to fall sowing in summer-wet climates. *Peren-*

*nial* plants regrow from their roots each year and may live for centuries. Some trees don't fruit or set seed for decades after they sprout, and may seed at irregular intervals thereafter: many nut trees, for instance, set nuts sporadically every 2–5 years. Called mast years, these are part of the tree's strategy to evade seed predation by rodents: a *mast year* leads to a boom in squirrel and chipmunk populations, followed by a bust. After seed predator numbers reach a new low, the trees set more seed so that some will survive.

Perennials may be woody or non-woody, herbaceous, shrubby or tree-like. Walnut trees (*Juglans* spp), Jerusalem artichokes (*Helianthus tuberosus*), orchardgrass (*Dactylis glomerata*) and daffodils (*Narcissus* spp) are all perennial but have very different growth forms. Soft-tissued herbaceous perennials typically store energy in woody roots or bulbs during dormancy. Taller plants grow new stems each year while conserving most of the prior years' growth. These new stems and branches, which lignify (turn woody) late in the growing season, endure over winter. On them, new buds form that in the spring will sprout leaves and more stems. On all woody perennials, the living tissue of the plant consists of a relatively thin layer of cambium just beneath the outer bark, plus the leaves. As woody perennials grow, their stems and branches thicken; each year they grow a new layer of cambium over all previously living tissue, which then ceases to metabolize. In healthy trees and shrubs, most of this incorporated tissue remains moist and durable, but wherever the bark is damaged or a branch or stem is broken, fungi and insects may enter and begin decomposing the interior wood. Trees and shrubs cannot move but must endure whatever weather and animals inflict upon them, so they have evolved mechanisms to repair themselves, not as animals do by replacing tissue, but by entombing diseased parts of themselves, growing around wounds and rot. Very old trees are often completely hollow.

Because of the need to cover all of the previous year's growth with new cambium, all trees and shrubs left to grow reach a maximum size and age within the limits of their soil and environment. When they reach this limit, which is defined by their ability to photosynthesize enough carbon to expand their cambium layer fully, they begin to abandon limbs to balance growth of cambium with leaf surface area. This tends to be a slowly degenerative process, as each limb abandoned further reduces leaf surface area for energy capture, and less energy means less cambium growth the following year. The forester or gardener who regularly prunes back a shrub or a tree can, by keeping it smaller, extend its life almost indefinitely.

Perennials may be short- or long-lived; some have lifespans that are indeterminate and may be extended by management. Short-lived perennials may diminish slowly from local exhaustion of limiting minerals, or they may spread in rings, seeking new supplies of soil nutrient. Periodically cutting trees back — a process called *coppice* — keeps them in a permanently juvenile condition where almost all of their tissue is new, potentially prolonging their lives for centuries beyond the normal span. Shrubs have coevolved similar relationships with browsing animals: most are of a size that places them right in the line of eating by herbivores, yet the shrubs are able to regrow vigorously after their tips are eaten. A certain measure of browsing (or pruning) pressure keeps them vigorous. Most smaller perennials benefit from periodic disturbance, which may distribute their rhizomes, divide their bulbs or split their roots to encourage new growth.

A main aim in garden design is to reduce unnecessary labor and soil disturbance, so that perennials become a kind of backbone if not covering a majority of space in the garden over time. Annuals are more flexible, bearing their edible parts within a few months of sowing. This is important to people who

Very old trees are often completely hollow.
[Credit: Howard Dickins]

must move about in a territory or when food reserves are poor, but annuals are very demanding of particular conditions of moisture, heat and light, so that during periods of erratic climate they may perform poorly. Ask any farmer with flooded fields in May or no spring moisture about the vagaries of annual cropping. If you are in the business of growing annuals, then seed supply becomes an urgent matter. And you will also need to rotate your crops so that different annuals grow in each row, bed or field from season to season. This is because annual crops have particular nutrient profiles that they require and so pull different minerals from the soil. They are not meant by nature to endure in one place but to give way (to succeed) to other species.

The use of annuals should always be understood as a kind of pulse or *catch crop* in the landscape: planting a short-lived or pulsing crop to catch temporary resources of sunlight, water or fertility. We can use catch crops beneath our fruit and nut trees while these are small, growing sun-loving vegetables in the space that in a few years will be partly shaded. Or a neglected bed of weeds can become a

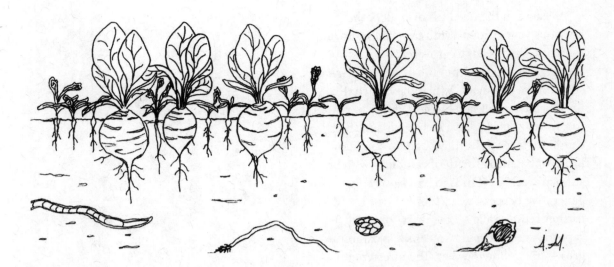

Catch Crop, Pattern #55 (See Chapter 6)

[Credit: Abi Mustapha]

catch crop when we turn it into mulch, compost or forage by moving animals through the area. Radishes interplanted with carrots are a pulse crop — we use the ground and sunlight that will be occupied by the carrots later in the season to ripen the quick-growing radishes. If we take this attitude, we will see bare soil not only as hazardous but as an unrealized opportunity, and we will plan to occupy all the niches in our garden farming landscapes at all times.

## Propagating Plants

Every farmer grows crops from seed, and every gardener has made root divisions and cuttings, but the creation of a rich and diverse agroecosystem rests on the assembly of many hundreds of species and many thousands of plants, so the diverse skills and broad knowledge of plant propagation become a gateway to permaculture success. Seeds are nature's original form of wealth — concentrated, portable and relatively long-lasting packets of useful information capable of self-replicating in the right conditions; they are completely magical. They also leverage our intelligence, skills and labor enormously in return on time invested. Seed of specialty crops may be worth several hundred dollars a pound. The nursery trade is the most profitable form of horticulture per square foot. A hundred cuttings of

grape or fig that one would have to make at the end of the growing season for reasons of good management can turn a half an hour of work and a few buckets of waste biomass into a $1,000-dollar crop two years out. Of course the potting and watering take a little time and expense, but not much. This should tell us by inference that the costs of establishing a farm ecosystem, and thus the potential savings from propagating plants for our own use, are enormous.

Plants reproduce by seed and by division of bulbs, corms or tubers. They also send out runners, root from their tips and sucker from their roots. In all these ways they create new, potentially independent plants, sometimes offspring genetically similar or distinct from the parent(s), sometimes clones. We can take advantage of the way plants grow to propagate them in all these ways and a few more.

### Vegetative Methods

Many of our common fruit trees are reproduced by combining genetic and vegetative methods of propagation. A rootstock may be grown from seed (for example a crabapple) or may be a root sucker of a *stool* (stump) of a known type, clipped and potted.

These *whips* (young seedlings or shoots) are then grafted with a piece of a woody branch end from the *scion* (selected fruit-

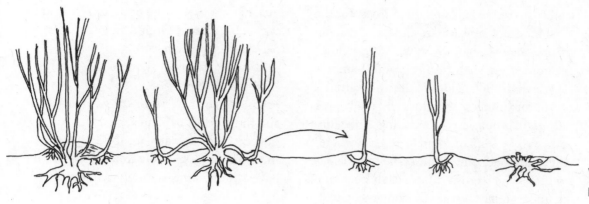

Whips and Stools

[Credit: Abi Mustapha]

ing variety) which is of the same or a closely related species as the rootstock. Sometimes just a single bud is grafted into the cambium layer of the rootstock. Several different styles of cutting and matching the cambium layers have been developed, but in every case the physical connection must be joined closely and must be held stable for several weeks while the adjoining tissues fuse.

New trees or shrubs can also be propagated by *layering*, which is a process of bending a branch down on the ground and mounding soil over it. Usually the bark is scored to encourage rooting where the branch is to be buried. Sometimes a hormone solution is applied to stimulate root formation. An overnight decoction of cut willow pieces in water provides a usable rooting solution. A stone or a log may be necessary to hold the branch in place, and it must be covered by organic soil. *Air layering* is a variation of this done above the ground using a small water-resistant package of soil surrounding the twig or branch. A bit of the bark is scratched, then surrounded by soil or peat moss kept moist by the plastic or foil barrier and taped to the branch. Once roots have formed in 10–30 days, the rooted branch can be cut off the parent and planted on its own. Many trees, shrubs and some herbs will send up suckers from their roots, which is simply a natural form of layering.

Some woody and even non-woody plants can be propagated from simple *cuttings*.

Along with grapes and figs, currants and willows are easily and successfully grown from cuttings of 10 to 14 inches taken just above a bud at the tip end and just below a bud at the base end.

You can put willow cuttings (10–20" long, ¼–1" thick) in a bucket of water for a few days or up to a month until roots form, provided you change the water periodically, not allowing it to become stagnant. Other cuttings, such as grapes or currants, can also be held in water for a short time, but will hold over winter without roots if potted by burying about ⅔ their length in loose soil or even coarse mulch. You can also plant any of these directly in the ground — again, bury ⅔ — and they will root. Be sure to water them in well

Chopped pieces of willow stem soaked overnight in water will yield an excellent rooting hormone solution. Over 70 species of willow are native to North America.

[Credit: Peter Bane]

and keep the soil around them moist for the first month. Large grasses and herbs can be propagated vegetatively as well. Sugar cane is grown from stalk cuttings, while bamboo is grown from divisions of the underground corms or runners. Tomatoes and many brassicas will propagate from cuttings. Like most other vines, tomatoes can be layered where the vines touch the ground. They can also be grafted, which growers have lately been doing to impart blight resistance from a rootstock to a more sensitive scion. Yields can as much as quadruple.

Vegetative propagation is faster and simpler than growing from seed, but introduces no new genetics. It also may be difficult or impossible in cold seasons. And replanting cannot always be delayed. Bulbs and many other monocots (such as bananas, camas, iris and cannas) are propagated by division, as are garlic and onions, though such plants may also propagate by seed. Comfrey, horseradish, rhubarb, burdock and older asparagus plants can all be divided, as can most herbaceous perennials. Not all perennials form thick roots. Mints, for example, spread in patches by extending root shoots, and can be propagated by digging a small clump of rooted stems and transplanting to a new area. Tuberous plants often grow rooted shoots which are then separated to form new plants. Sweet potatoes are grown in this manner after harvest — from so-called slips. Half submerge a tuber in a glass tumbler: rooted shoots will emerge and form leaves; detach and plant these. The vines can also be clipped from the garden and planted with or without roots. Potatoes are grown from whole or cut pieces of the spuds that have been allowed to scab over. Smaller tubers may be planted whole; the larger the tuber used, the more likely it will be to flourish, but also the fewer propagating units a plant can yield.

Root vegetables can be propagated by cutting off the crown with a section of the root and replanting this. For example, take the top inch of a carrot, turnip, beet or parsnip and stick it back in the ground after trimming back the top growth.

### Working with Seeds

Seeds, which are fertilized plant embryos and the product of plant sexual reproduction, can remain viable for thousands of years, though some (especially tropical seeds) may have a lifespan measured in weeks or even days. Among temperate vegetable crops, parsnip seed is the most ephemeral, lasting seldom more than a year, while lotus seeds commonly remain viable for up to 30 years, lying in the mud of a shallow pond. Corn discovered buried in the adobe walls of New Mexican houses has germinated after 80 years and more, while wheat found in Egyptian pyramids has sprouted after several thousand years.

Every act of saving seed is also an act of genetic selection, so one must pay attention during the growing season to the characteristics exhibited by plants. Good, diverse yet stable crop genomes are not always maintained, however, by selecting only the best and brightest. While this is a good beginning assumption, with some crops (corn among them) a wide range of plants, including those of unimpressive character, must be included among the seed crop to ensure reliable production. The nature of genetic selection is a complex subject beyond the scope of this book, but fortunately, fine and reader-friendly references are available to guide you.[16]

You can develop the habit of seed saving by noticing when plants have formed seeds; this takes place throughout the growing season, though the late summer and autumn months are the richest time for seed harvest. If you see a plant that interests you with ripe seeds (often indicated by the splitting of the seedpod or ripening of the fruit), collect them. Try to gather the ripest seeds from several of the best-looking or most interesting and desirable plants, but don't take all the seed from any one small population outside your own garden.

Seeds usually ripen from the top of the plant down because the first flowers form there and the seeds from them have received the most sunshine. Sometimes plants will have ripe seed and flowers at the same time, which allows you to select for flower color and genetic type to some degree. Ripe seed will usually shake out of a crushed or split pod; sometimes pods will shatter and spew seed all around. Ripe seeds are typically well filled out, and may have begun to darken relative to unripe seed of the same type. Reject any seed that shows evidence of insect damage or mold. If you notice or suspect weevils, you can place the seed in a freezer for 48 hours before packaging it. For convenience, we buy and use coin envelopes for seed packaging, though any new or used envelopes would do. Always record the provenance (where the seed grew), the date and any relevant characteristics as well as whatever name you have for the plant.

Seeds are alive and will breathe very slowly until they germinate or expire. Since sprouting is stimulated by warmth and moisture, and sometimes by additional factors, seed viability is inversely related to temperature and humidity. You must store seeds cool and dry. Once seed is ripe, a further period is needed for it to dry for long-term storage. Once dry, it can be kept in that condition by storing in airtight jars or tins. Of course, proper precautions should always be taken against rodents and other vermin. We use an old metal file cabinet with a variety of bins, jars, bottles and plastic tubs nested in the drawers. This is kept in a cool, dark pantry room on the first floor. Basements are to be avoided due to excessive humidity. Do not store seeds (for planting) in the kitchen for the same reason.

Most vegetable seed can be stored for several years if kept in appropriate conditions. Though germination rates decline, they do so somewhat slowly until a threshold is reached,

after which most of the seed will be inert. Old seed of dubious viability can be broadcast into an area as a cover or catch crop — if you get anything, it's a bonus.

*Sprouting Seeds*

Some seeds require pre-treatment before they will germinate. Chilling seeds before germinating them is called *stratification*, a process required by many temperate tree fruits. The extent of chilling varies a little, but typically 30–60 days at refrigerator temperature (34–38°F) is sufficient. In other cases, seeds are conditioned to being passed through the gut of an animal or bird. The scratching and immersion in acids that occurs in such a situation is called *scarification*. Gardeners can scarify seeds by abrading them with a file, nicking the seed coat with a fine blade (a fingernail clipper works well) or immersing briefly in boiling water. A very short spin in a blender with plenty of water can also do the job. The object is to scratch, swell, thin or otherwise render the seed coat more permeable to moisture and less tightly bound to the

The ideal conditions for seed storage are cool, dry, dark and not subject to significant temperature fluctuations.

[Credit: Creighton Hofeditz]

embryo. The seeds of many legume trees and shrubs are noted for needing scarification, without which they can sit dormant for years.

The growth of legume seedlings, including garden peas and beans, is aided if the seeds are dipped in a rhizobial inoculant. Common types of this powdered mix of bacteria can be purchased at most farmers' coops or ordered from some seed suppliers. A dozen or more strains of rhizobial inoculant specific to certain groups of legumes are known, the variants mainly important for perennials. If you have grown peas, beans, alfalfa or clover in your garden previously, it's likely that the nitrogen-fixing bacteria with which they like to associate are already present, but in a new bed or a new plot of ground or for more robust results, it can be useful to dust the seed before planting.

To germinate seedlings requires favorable conditions, including steady moisture and some level of soil warmth. Many gardeners provide a heated bench for seedling trays as the critical threshold is measured by soil rather than air temperatures.

Once you have planted seeds, do not let the soil dry out, and once the seedlings have put up shoots, these will probably need a light misting at least every day and sometimes more often until transplanted in the ground. We much prefer to manage seedlings all in one place, and use our solar greenhouse for this purpose. There we can easily monitor emergence and soil moisture, and can rapidly apply water, thermal protection and fertilizer as needed. There are fine reasons for planting garden seeds directly into prepared beds, but we find it easier to manage the flow of plants into the garden by starting them all in flats, transplanting as they and we and the weather are ready.

We take measures to ensure the development of good roots on our seedlings. We use a potting mix containing about ¼ native or garden topsoil, ¼ sand, ¼ vermiculite and ¼ well-finished compost. If you haven't good native soil to begin with, increase the proportions of the other ingredients. To a wheelbarrow load of this mix, we add about a cup each of rock phosphate, greensand and limestone plus some source of trace minerals. You can use alfalfa meal, powdered humates, kelp meal, crab shell, fish emulsion or whatever may be at hand. The sand and vermiculite ensure good drainage, the native soil brings in microorganisms and the compost ensures that enough nutrients are available. This mix should then be sifted through a ¼-inch screen. The presence of minerals in the potting mix enables the seedlings to develop expansive and powerful root systems even while very young. This helps ease transplant shock and ensures rapid growth after transplanting.

### Why We Transplant

Having flats of well-developed seedlings available helps reduce start-up costs in the spring and, even more importantly, allows you to cover whole beds or sections of the garden quickly without tilling, thus keeping down weed competition and supporting a healthy soil fauna. Minerals in the potting mix also supply nutrient in the right (small) quantities just where it is most needed and at the right time, something you could never do as well

Bottom-Heated Seedlings

[Credit: Peter Bane]

in the field. With transplants you can drop a few annuals into a perennial guild or surround a young fruit tree with a catch crop. Growing in flats allows you to remove and discard weak seedlings before they go into the garden, ensuring better coverage with less backtracking. When direct seeding, it's necessary to plant densely and then to thin, removing the weaker plants, meaning more trips to each plant. But this wastes seed and gives you nothing for it. If you simply put the same amount of seed into flats, later transplanting the choice half of it into the garden, you still have many viable seedlings that can be sold or traded. Transplanted seedlings are not immune to attack by slugs, voles, birds and other varmints, but if the seed is fully sprouted and turned into roots and shoots, it's much less appealing to seed predators.

### Alternate Planting Strategies

While we much prefer managing all our seedlings for later transplant (especially perennials), there are times and places for broadcast planting, and also for direct seeding by drilling or other focused methods. We manage a half acre primarily with transplants; perhaps we could double that, but anything beyond an acre would require many more hands for the same kind of management, so the labor cost makes it worthwhile on larger plots to spend more on seed, accepting some losses and the inefficiencies that come with planting directly.

A dibble stick is the most primitive tool for planting (other than one's fingers). It consists of a pointed stick of a comfortable length poked into the soil at the right intervals. Drop one or a few seeds at a time into the resulting holes. A step up from this is to scratch a furrow and drop seeds into it, then cover them up with a foot or a hoe. The Planet Junior seeding machine is a non-motorized device with two in-line wheels, a small plow, seedbox and drag chain which one pushes along a row to accomplish the same end. Seed wheels with different-sized holes regulate the planting of

various common garden seeds at the right intervals. A flying wing marks the next row as you go along.

In general, seeds want to be planted in soil to a depth about four times their diameter. A ¼-inch diameter pea should be put down one inch into the topsoil. Squash, corn or beans can be planted a little deeper, and onion sets, garlic cloves, large nuts and fruit pits deeper yet. Tiny seeds should only be sown on the surface, so one needs a well-prepared seedbed for such crops as carrots. (Lettuce is surprisingly adaptable and will sprout and root even on coarse mulch.) With very small seeds, we may sow on the surface, then cover very lightly with a sprinkling of fine sand. This reflects light and helps hold moisture near the seed. To get this kind of condition in the garden is more complex and requires baring and raking the soil, removing rough mulch and exposing the seeds and tiny seedlings to the force of pelting rain, which can easily wash them away, another reason we prefer transplants.

### Broadcast Methods

When biomass is the primary aim, as with a green manure or cover crop, broadcasting works well. Small seeds can be mixed with

The Planet Junior allows rapid and regular planting of a variety of different vegetable seeds.

[Credit: Creighton Hofeditz]

sand or a little finished compost to make them easier to distribute. The trick with large seeds is to get them high in the air over the ground to be sown. They then fall in a scatter pattern. Small seeds are more likely to be carried by the wind in streams, resulting in irregular planting patterns, so keep their distribution near the ground. For a dense and more uniform cover with small seeds (up to the size of cereal grain), a rotary seeder can be employed. This consists of a seed bag connected to a spinner carried by a shoulder strap. One walks along cranking the spinner which flings seed out and toward the sides.

If overseeding into existing vegetation or into harsh conditions, seedballs may be the best approach. A mix of clay soil and (usually) several types of seed is combined with a small amount of manure, selected minerals and any inoculants that might be needed. The mix is moistened and rolled between the palms into balls the size of hard candy. These should hold together well. The seedballs should be allowed to dry a little so that they will not disintegrate until rains wet them thoroughly in the field. They can then be tossed into weedy or brushy areas, over fences into vacant lots or wherever hardy plantings of pioneer species would be helpful.[17] The seedballs are designed to wait for soaking rains before the seeds germinate and to provide a burst of initial fertility to allow rugged plants to establish in harsh terrain.

With these methods in mind, we can turn in the next chapter to the selection of crops with an eye toward efficiencies of land and labor, and high nutrient rewards.

# CHAPTER 13

# Setting Plant Priorities

## Key Foods

No purpose for cultivating plants is of greater importance than feeding ourselves, though the making of medicine and fiber lie close behind. In choosing plants to grow, a prime selection factor is the amount of space available and the relative value of its use. Some foods or plants are of greater importance than others because of their nutritional or economic value. Of course, there are limitations of climate, but most regions of North America can provide a sustaining diet and pharmacopaeia from plants and animals. In the built environment in and surrounding our cities, land is a limiting factor, and in the economy of such places, labor may also be short. "If," as Mrs. Rai urged us, "…you need it, then you should grow it." But if we cannot grow everything, how should we make our choices?

## *Herbs First*

Where space and time are most limited, and for many other reasons, culinary and medicinal herbs should be the first crop. Most of these are weeds, many are annuals and they are as a group easy to grow, suitable for planting in pots, on a windowsill or balcony, in small spaces near the back door or amidst other plants. Culinary herbs enhance the flavor and palatability of staple foods, and some of them, like oregano, are powerful antimicrobial medicines as well. Herbs, being mostly weeds, are soil-repairing plants. Their ability to accumulate minerals makes effective medicine both for the soil and for our bodies.

Herbs also have considerable market value. When our household cultivated a garden plot of about 1,500 square feet, our cash crops were shiitake mushrooms (grown in the deep shade of hemlock trees) and basil, sold fresh by the grocery bag full (sometimes 20 pounds at a time) to a local Italian restaurant that made its own pesto. Without using much of our space, we made over $3,500 from those two crops in one season, something over $2 per square foot, while still eating berries, salad and other crops from most of the space. We also made our own medicinal tinctures, some of which we still use years later: elderberry for colds, spilanthes for toothache and sore throat, dandelion for liver cleansing, echinacea for immune support.

Many common flowers are edible.

[Credit: pink_fish 13 via Flickr]

### Fill Your Salad Bowl

The category of herbs spills easily over into salad and leafy greens. Chickweed is both a spring tonic and a fine salad green. In winter we enjoy mache, or corn salad, a member of the valerian family, harvested for fresh eating from the cracks in our tiled patio. It resows itself every year and requires no attention. Lettuce has come to be a similar, self-sowing crop, though we also plant and transplant it deliberately. We eat many leafy plants in our salads, and also like edible flowers.

An important species in this category is chicory (*Cichorium intybus*), slightly bitter greens that include the endives, radicchio, escarole and others. Young leaves of beets, purple kale and chard make colorful additions to the salad bowl.

### Flowers Brighten Everyone

Do not neglect to grow some flowers — not only the edible ones such as pansies, violets, begonias, sunflowers, day lilies, roses, calendula and nasturium (there are more) — but also flowers of every type, color and season of bloom. They buoy the spirit, represent a good potential cash crop and are important for sowing goodwill among your neighbors and passersby. They also attract butterflies, hummingbirds, bees and other pollinators, making your garden more productive. Flowers are available for every niche in the garden from full sun to constant shade, and can — if they are not for eating — be grown in contaminated soils, such as roadside strips and areas near old buildings where lead paint may be present.

### Greens Are Nutritional Powerhouses

Leafy greens transport poorly, so having a backyard supply ensures greater freshness. They are also subject to more contamination than other crops if not grown by organic methods. Leafy greens contribute significantly to nutrition by providing vitamins C and A, folic acid and many important minerals. They are, pound for pound, a good source of protein, though we tend to eat small quantities. Most leafy greens are not heavy feeders, so can easily be sustained in a small garden plot. Leafy greens such as spinach, kale, turnips, chard, collards, arugula, tatsoi, mizuna and bok choy are cold-hardy and may be grown over winter with some covering in most parts of the US and southern Canada. The range of candidates in this category is larger than this short list, and something is suitable for every climate and season. The hardest time of year to grow leafy greens is mid-summer when temperatures are highest, but New Zealand spinach, Malabar spinach, collards and sweet potato greens (yes, the leaves are edible) offer some good possibilities. You can also extend the season for leafy greens by planting in some cooler microclimates such as the shade of small trees or on the north side of tall vegetables or shrubs.

### Small Fruits, Big Dollars

After you've provided herbs, salad, flowers and leafy greens for health and well-being, introduce some small fruits to your repertoire. They deliver good amounts of vitamin C and other antioxidants and cost quite a bit in the store. Some fruits like currants and gooseberries are seldom available in commerce. And who can ever get enough strawberries? Berries are perfect kid food, providing not only nourishment but entertainment, which may be more important. With a well-selected assortment, you can eat fresh fruit of the most delicate types from late May through first frost with scarcely a break. We start with mulberries (which admittedly grow on a tree) and strawberries in May, then get juneberries, red raspberries, black, white and red currants, gooseberries, black raspberries, blackberries, blueberries, cherry tomatoes, ground cherries, tomatillos and in the fall red raspberries again. Everbearing strawberries yield throughout the season. A blueberry patch sufficient to yield a quart or two a day for several months will fit in the space of a small bedroom (about 10 × 15 feet); they fill the late summer slot right into early fall. Strawberries and cherry tomatoes can be grown in pots on walls or in vertical planters. Ground cherries (*Physalis* spp), members of the nightshade family like tomatoes, are delicious and grow wild with little attention.

If you have them, spread them around. If you don't, it's easy to get some fruits and plant the seeds in your garden. The fruits are ripe when the lanterns fall off the vine and the fruits begin to turn from green to gold. Hardy kiwis (fuzzy ones too, if you have the heat for them) and grapes round out this wonderful category, filling in the summer and fall periods when spring berry production has subsided. And they are vining plants that can fit in almost anywhere.

The five groups of plants listed above probably provide the most nutritional and dollar return per square foot of anything you can grow legally. However, they don't provide a staple diet (though I nearly live on berries in the springtime), so where you have enough space for broadscale food production, you need to think about other crops as well.

## Staple Crops

Staples are the foods that provide the most calories for the largest part of the year. In cold and dry climates they must keep well in storage through the long months of the dormant season. Among the easiest calories to grow and store are those found in potatoes. This undemanding crop is relatively pest resistant and, if stored in complete darkness at cool temperatures, will keep for up to ten months. Potatoes make a good first crop in a new bed where soil fertility levels are questionable, as they are not heavy feeders. They prefer a light soil but will grow almost anywhere. They can be planted without digging by placing cut pieces of the tuber on the soil surface and covering with a thick straw or leaf mulch. As the tops grow out, add more mulch to keep them upright. And to harvest, simply pull the mulch back and hunt for the spuds. In cool regions, this technique can be modified by growing in tires, stacking more on as the vines grow, unstacking to harvest.

Despite their long history as a staple in cold climates (where they surpass most grains in the production of calories per acre), potatoes are frost tender and must be protected during the early weeks of spring. Dark tires

Ground cherries sprawl over the ground, bearing their hazelnut-sized fruits in little paper lanterns, and freely reseed themselves.

[Credit: Creighton Hofeditz]

Tires filled with straw mulch can be stacked to force potato vines upward while providing extra warmth in spring.

[Credit: Peter Bane]

Clip several 6-inch pieces of sweet potato vine, root them in a glass of water and plant in a pot over winter to have propagating material for the following warm season.

[Credit: Creighton Hofeditz]

capture heat from the sun to give your crop an early start and to sustain it through cool nights. Potatoes are typically grown by cutting whole spuds into 2–5 pieces, each with several "eyes." We dredge these in a rock mineral mix and allow their cut surfaces to dry before planting. Potatoes, of course, come in hundreds of varieties and many skin and flesh colors including, blue, red, yellow, purple and

even black, as well as the familiar russet with white flesh.

Sweet potatoes require a longer growing season than so-called Irish potatoes but can yield well wherever five frost-free months are available. Besides having edible leaves, they are extremely rich in beta carotene, the precursor of vitamin A. Light soils are best, but if you have clay, just keep adding organic matter; the tubers will get bigger over time. An important tip is to mulch them heavily (10–18"); they flourish in shade, so try them under tree crops as a ground cover. Sweet potato is believed to be a South American crop, but stands out in West African and Hawaiian cuisines. We grow, among others, a purple Hawaiian variety with a white skin and a very rich, somewhat dry flesh that tastes almost like cake when baked. You can buy sweet potato slips commercially, or keep some vines in a pot in the house over winter to have plants to start in the spring.

We find that some varieties will keep quite happily as tubers for months in a cool room at 50–55°F, but the tubers should not be allowed to freeze.

Though not a main source of calories, onions and garlic are indispensable for cooking. Their sulfur compounds make them quintessentially savory and support good health as well. Onions, especially the walking variety sometimes called Egyptian onions, are virtually trouble-free, can fit into small spaces here and there in garden beds, keep well without special effort and make excellent companion plants. The walking variety are easy to propagate as they form bulbels at the top of their stalk. You can pluck the bulbels, pop them into a pocket and plant them wherever in the garden you want some onions. Onion and garlic flowers provide excellent forage for beneficial wasps. We find yellow onions keep the best among the common types, whites the worst. Reds are intermediate. Even with undeveloped soils and limited garden space, we have been able to provide ourselves all the onions we need from our part-time gardening efforts, using up the last of the fall bulbs as the first spring scallions are harvested. If planting onion sets gives you indifferent results, try growing from seed. Onions, though cold-hardy, must be set out in the spring in order to make bulbs, as they are day-length sensitive, drawing their cues to store energy in their bulbs from the rapid lengthening of the day from spring to summer. Fall planting will give you only scallions. Garlic is a good crop for cash or trade and, of course, a mainstay of health and flavor. It is typically set out in autumn, well before the soil freezes, to make bulbs at the following midsummer. With either of these bulbing alliums, eat the smallest bulbs and plant the largest ones to increase the size of your plants from year to year. (Oh, and keep feeding your soil.)

Beets are relatively easy to grow and now come in yellow and striped varieties as well as the classic deep red. The greens are edible, either in salads or as a potherb. We have had good success growing mangels, which are an older type of beet, now mostly used as fodder for animals. They grow large, up to a pound or more each, and are slightly more fibrous than modern beets. They still have good flavor and can be quite sweet.

Less durable than beets, but perhaps more versatile in cooking, are carrots, which also come in a range of colors beyond the familiar orange. Carrots prefer sandy soils, but if you have clay soils as we do, try the shorter varieties and mulch over them after they have well sprouted. Carrot seeds are tiny and slow to germinate. Plant the carrot seed on the surface, sprinkle a little sand over it, cover with a board and check them for germination after about 10 days. As the first seeds begin to germinate, prop the board up on one edge with a stick or stone to allow the sprouts room to grow. Once you have good germination, you can remove the board. Another trick for keeping track of your carrot seedlings is to interplant radish seeds among them. The rad-

ishes are up and gone in three weeks, which is about the time the carrot seed will take to sprout. The radishes provide a little cover and are easily noticed so you won't be tempted to step on the carrots or overlook them.

We have had good success with kohlrabi which, though not a root crop, is nonetheless the storage organ of a member of the cabbage tribe. Kale, cabbage, cauliflower, collards, kohlrabi, brussels sprouts and broccoli are all variants of the *Brassica oleracea* species. With each, a different part of the plant's superstructure has been bred to develop prominently. The edible part of the kohlrabi is a bulbous stem, which we find to be sweet. Kohlrabis produce very small leaves and put most of their energy into the bulb. We eat them raw, peeled and sliced thin, with a little mayonnaise, or just plain, though they make a fine addition to soups or can be steamed. They'll keep well in a root cellar and deliver a lot of calories. Brussels sprouts grow as miniature cabbage inflorescences along a stalk and are a classic fall crop. Many people reportedly disparage the flavor, but I enjoy it. The leaves after frost are choice, and the vegetable is also high in calories. Cabbage, of course, is Old King Cole, and for all of its plainness has great versatility. It will keep well in a cellar for many months, can be eaten raw or cooked and offers a good ration of vitamin C. This benefit can be greatly enhanced by fermenting. Captain Cook is reported to have set out on his round-the-world voyage in 1779 with 20,000 pounds of sauerkraut on board; his ship returned without losing a man to scurvy, though Cook himself succumbed to local politics in Hawaii. Perhaps he should have shared the recipe. Cabbage is the main ingredient in sauerkraut and kimchi, both lactobacillus ferments that can be made with a wide variety of vegetables and even fruits and meats.

Pumpkins and squash should be considered part of every household's strategy for staple foods, particularly the winter-keeping types for obvious reasons. The four species of *Cucurbita* that yield edible fruits give us summer and winter squash in great variety. Because these plants — which humans have selected for a vast range of flavor, color and shape — can easily cross back to make a mixed bag of hybrids, it is necessary to isolate squash or pumpkins of the same species from each other, unless they are of the same variety, in order to get seed that is true to type. You can do this by choosing no more than one variety from each species (*pepo*, *maxima*, *mixta* and *moschata*) to grow at any one time. You can also usually get 2 or 3 crops per year by starting seed at one-month intervals from last spring frost through about mid-July, and avoid crossing by leading the plants to flower at different times. We have found the solid-stemmed *C. moschata* types, which include butternut and a wonderful oddball called *tromboncino* (guess what it looks like) among other names, to be more resistant to the squash borer. They give better results in our hot, humid summers than other types of squash. For eating, I like hubbards, Long Island cheese pumpkins, kabochas and butternuts, all of which have deep orange flesh, small seed cavities and are fairly sweet. I like "pumpkin" pie and make it out of all these types. Since they are sweet to begin with, I don't need to add much sugar to get a very satisfying result. But don't stop there in your exploration of the *Cucurbita* genus — it holds amazing treasures. Pumpkins and squash are a good accompaniment to poultry and hogs, as these animals will happily get rid of your excess and oversized zucchini, cross-bred squash of indifferent flavor and simply the surplus fruits of which there are usually many.[1]

Dried beans, or shell beans, are a good source of protein that can be stored at room temperature for months or even years. Allow them to dry fully on the vine during the last weeks of summer into autumn when rainfall is low; bring them in before fall rains. Shell them out soon after harvest and allow to dry completely before storing. Beans, each seed carrying such a large packet of energy, make good candidates for direct sowing or even for broadcast into deep mulch or the standing stubble of a winter or spring crop. Some familiar legumes originate in the Old World, including chickpeas or garbanzos, favas or broad beans, lentils of many types and all peas, to which broad beans and vetches are closely related. Mung and adzuki beans are of Asian origin, while the New World beans are of many colors and types from navy, pinto, black turtle and lima beans to the many-colored cranberry, Anasazi and sundry beans grown for fresh vegetable eating (so-called snap beans or string beans). All these New World beans originate in the tropics of Central or South America and are frost sensitive. So are some of the Old World legumes, though by no means all. Peas are cold-hardy, and among the beans, favas are quite cold-hardy as well, enduring temperatures to 15°F with little trouble. If started by mid-August

 **RECIPE FOR SAUERKRAUT**

Shred, slice or grate according to your tastes 4 medium heads of cabbage and 4–6 medium apples, cored. If you like pink kraut, make one or more of the heads red cabbage. Add ½ cup of yogurt or whey, ½ cup of coarse salt, 2 tablespoons caraway and 1 tablespoon dill seed. Mix by hand in a large bowl or tub and transfer to a crock of 2 gallons or greater size. Using the fist, pack the kraut tightly into the bottom of the crock and cover with a heavy plate and water-filled jar, ceramic stones or other appropriate, non-reactive weight until the brine rises above the level of the cabbage. Cover the crock with a lid and place in a cool room at 60–65°F. Leave for two weeks, then pack into glass jars and store chilled. Makes about 9 quarts. Keeps for several months or longer, depending on temperature. If your crock has a wooden lid, you may get some scum around the top of the weight. This is harmless. Remove it before packing.

or early September in parts of North America with milder climates, broad or fava beans can grow throughout the winter and will ripen their seed in the following spring. They don't do well in hot summers. Even in our mid-continental zone 6 or 7 climate, we sow field peas in the fall to harvest as salad during the winter, giving them just a little cover to ensure good growth. The shoots of peas and favas (the outer five axils) are quite delectable and even sweet in cold weather.

### King of Grains

The Native American guild of Three Sisters would not be complete without corn, which is a grain but dramatically different from its distant cousins rice, wheat, rye and the rest. Corn uses a slightly different chemical pathway, called C-4, as distinct from the much more common C-3, that enables it to harvest more sugar from a given amount of sunlight. That means corn is much more productive of calories than most other grains. C-4 photosynthesis doesn't help corn make more protein, however, and its amino acid profile is relatively poor, though complementary to that of many beans. Corn eaten as a staple, as it is throughout Mesoamerica and was for a thousand years across much of this continent before Europeans arrived, requires supplementation with the protein of beans, eggs, milk or meat to provide a balanced ration of the amino acids needed by the human body. Most traditional diets relying on corn as the grain staple used one or more of these concentrated proteins with a complementary amino acid profile. Corn is particularly short of lysine, an amino acid well supplied by black and other tepary beans.[2]

Corn has the advantage for its cultivators of growing very large ears that are easy to harvest, store and from which the seeds are easily enough dislodged. A traditional corn farmer can raise enough food using hand methods to feed 12 people — a fantastic leverage for pre-industrial work. The grain has been bred into varieties suitable for flour, grits or polenta and for popping as well as for sweet corn, a fresh vegetable. Corn is also a very useful animal fodder, both the grain for monogastrics like pigs and poultry, and the leaves and stalks of the plant, whether green, dried or fermented (silage), as a feed for ruminants like cattle.

Corn has been the subject of intense industrial development, and in the United States and Canada, the vast bulk of the crop has been genetically engineered, most of it for herbicide tolerance. However corn is too valuable to abandon to the mad scientists of Big Pharma. The difficulty in growing corn for home use or organic markets lies in the ease with which corn cross-pollinates. The corn plant is wind pollinated, so for certain separation from other corn pollen, plants should be isolated by a mile or more. If you live in a rural part of the Corn Belt, you live in a sea of GM corn pollen for part of the growing season. But if you live where corn is a marginal crop grown primarily by hardy gardeners for home use (for example, the maritime Northwest, New England, the mountain regions and the desert Southwest), you have many more options. Interestingly enough, many urban and even suburban areas are relatively free of industrial corn production. By choosing early or late windows for pollination, by roguing or removing plants of dramatically different type before they can release pollen, by selecting ears for seed from the center of your patch and by carefully checking your seed crop ears for variant kernels and removing these before shucking the seed from the cob, you can virtually eliminate contamination of your varieties by GM and other unwanted pollen.

### Must-Have Fruits

Two other crops must be included in this section about staples, and those are apples and tomatoes. I would include grapes as well, but for the fact that the cultures of North America are not accustomed to making wine at home, and this is far and away the best use of grapes.

The traditional American home economy would have been virtually unthinkable without the apple, which provides us juice, sauce, pie, alcohol, vinegar, fodder and a vast number of calories.

[Credit: Peter Bane]

Apples are a staple food and have been for thousands of years. They ripen over more than four months in most climates, come in hundreds of varieties and can be grown on large or small trees or even on wires (cordon) like grapes. Though they do better in areas with a strong winter, low-chill varieties have been bred for climates as diverse as Arkansas, Florida and even Hawaii. As a tree fruit, multiple varieties can be grafted onto the branches of one tree, allowing even the smallest of gardens to enjoy a range of apples and to achieve good cross-pollination. Many varieties will store for nearly a year if kept cold. They are suitable for raw and cooked dishes from salad and soup to stews, dessert and garnishes. And then there is cider. If North Americans are reluctant to tackle wine, we have a long and beloved history with cider, most of it hard.

As a source of sugar for alcohol, apples are one of the best choices for the household and small grower. Apples are easier to harvest and store than almost any other fruit (pears are a close second but somewhat less productive and usually taller). Apples make a suit-

able and affordable fodder for animals of all types from rabbits and horses to hogs, sheep, chickens, goats and cattle. Being able to feed surplus and semi-spoilt produce to animals is a great advantage on the farm.

As pome fruits (which include pears, quinces, crab apples and medlars) in the rose family, apples are a genetic crapshoot. Most require pollen from a different variety of apple to set fruit and seed. As a result, the seedling is rarely similar to the parent tree. For this reason, almost all commercially sold apples are grown on grafted trees. The result is a tree that performs well in its surroundings but yields fruit of a known character.

The fruit once called love apple comes to us from Central America, but has been adopted by cultures and cuisines all over the world. The tomato (*Solanum lycopersicum*), that most famous member of the nightshade family or *Solanaceae*, is a true tropical plant that will not tolerate frost. It even needs warm nights to flourish and bear heavily, so Europeans often grow their tomatoes in glasshouses in the backyard. As versatile as the apple, tomato is the indispensable companion of pasta, soups, stews, salads and brightens nearly any course of the meal.

Tomatoes dry durably and can well in a hot water bath, being acidic. Tomato's color buoys us during long grey winters with the promise of sunshine past and future, and its redness (from lycopene and other compounds) is a valuable health support, especially for men (love apples indeed!). We eat them, drink them and even throw them at politicians.

A vining plant that needs support to get up off the ground and away from moisture and fungus, many varieties of tomato are indeterminate, which means they continue growing and fruiting until frost or some other disturbance kills them.

Breeding has helped improve resistance to verticilium wilt and other fungal diseases, but good garden hygiene and the right climate are just as important. In very wet summers, a small roof over the tomato patch can help avert blight. Consider growing a few vines under a west or south eave to protect against moisture and early frost. You can extend the growing season by up to two months that way in some circumstances. Pick green tomatoes as frost approaches and set them inside on a sunny windowsill to ripen. We have enjoyed tomatoes harvested and ripened this way straight through the holiday season. If you don't want to can them, unblemished tomatoes can be stored buried in ash in a cool cellar for many months. The main remedy for various tomato maladies such as blossom end rot and splitting is better soil with steady moisture, so feed your soil with minerals and mulch.

Tomatoes are vigorous self-seeders, and their seeds resist destruction even in a hot compost pile, so once you grow tomatoes, you'll get them coming back. The plants are fairly self-fertile, but do cross-pollinate in organic gardens where many insect pollinators are active. Left alone, tomatoes will tend to revert to their ancestral type, which is a very small-fruited indeterminate vine. While these tiny pearl tomatoes can be tasty and are a handy snack for the gardener, they are not economical to pick, so you will want to save seed and renew your preferred varieties.

Plant foods are profoundly important as a source of vitamins, minerals, medicine, color and variety in the diet, and we need to know how to grow and store important fruits and vegetables. However, in this text, I write from the assumption that humans (especially those humans who live, as most North Americans do, in regions with cold winters) will not live by plants alone, but will obtain essential fats, proteins and other nutritive values from consuming animal foods as well. The crops cited above are not the only ones that a garden farmer or home gardener might choose from, and on our farm we do not limit ourselves to these one or two dozen. You shouldn't either.

Red, pink, orange, yellow, green, purple or rainbow-striped, tomatoes come in all sizes and shapes.
[Credit: Creighton Hofeditz]

The approach to diversity is continuous and iterative: expand and filter. Add species to your diet, your garden and your seed collection as your knowledge grows. Try hundreds or thousands and winnow them down to the double handful that will really feed you. Grow the rest for interest, preservation, medicine, flavor or the many, many non-food purposes of the farm: fodder, insect habitat, soil remediation, seed production, windbreak and beauty among them.

### Why Not Cereals?

The world's top three crops and eight of its top 20 are grains. I have only addressed corn as a staple because it is easy to process at home. Obviously, many of us eat wheat bread, rice, oats and more, but almost no one in the modern world processes grain from seed into food. It is done industrially because it is drudgery without the right equipment. Gene Logsdon has written wonderfully about the virtues and pleasures of growing your own grain (though even he places corn *primus inter pares*), and for those who would reclaim this body of knowledge from agribusiness, I urge you to follow his lead.[3] He describes hulling small amounts of grain for his kitchen with a blender and threshing wheat by beating it with sticks or running over it with vehicles. These are not appropriate technologies, though they clearly work to some degree.

Grain processing was one of the first industries to be effectively centralized, and as that happened, intermediate, village-scale technologies for threshing, hulling, milling and even baking were rapidly lost, surviving in bits here and there, in museums, among the Amish and overseas among traditional farming cultures. The reinvention and manufacture, even importation of this technology, would be a noble and liberatory endeavor for some clever permaculture engineers, however pedestrian and retrograde such work might seem today. I cannot honorably recommend grain growing to households already lacking labor for cultivation of the land. (Grain's use for brewing is another economic story; perhaps we should drink our cereals?) If you want to raise Old World grains, band together with other households and figure it out, or adopt a mechanized farmer and his processors.

## Plants For Every Purpose

Plants for small farms and gardens should be selected for multiple yields and functions, just as we would design buildings to be used in many ways or choose tools or breeds of animals that are multipurpose. But they must also serve economic needs. While it is usually possible to find more than one function for any plant species, the range, depth, importance and uniqueness of those functions matter. Every tree will give shade, but only a few will give syrup or provide edible leaves or even flowers. All will help retain soil, but only a few (in temperate climates) will fix nitrogen at their roots. We must choose our plants so they can play as many functional roles on the farm as possible.

It may be helpful to list some of those functions. Besides food, plants are needed for fuel, fiber, fodder, fertilizer, fencing, ferments, flowers, forage, fragrance, fungus, filtration, "farmaceuticals" and fun. Silliness aside, we use plants for mulch (a fibrous sort of fertilizer) and insulation (fibers), dyeing (of fibers) and construction (using the fibrous tissues of trees) and any number of other jobs. The economic uses of plants are legion.

### Oil and Alcohol for Fuel

Fuel from plants consists mostly of wood for heat and some oils for lighting. In northern countries, animal fats provided the bulk of pre-industrial lighting—tallow for candles and whale oil for lamps. Hawaiians fired candlenuts (kukui), and some Hoosiers know that pecans will burn when lit. Oils can also be used for motor fuel; the chief sources of biodiesel are tropical coconut and oil palms plus temperate rape (or canola) and soya, as these have the highest yields per acre under industrial cultivation. Ethanol, of potential use for combustion, can be made by fermenting any sugary or starchy fruit or seed or from the sap of some plants. Large portions of the world's cane and corn crops are converted into ethanol fuel.

### A Symphony of Fibers

Fiber plants of interest include cotton, hemp, flax, kenaf, agave/sisal, bamboo, kudzu, grain straw, nettles, yucca, hibiscus and cattail. Aside from sisal, which is one of many useful, fibrous and edible agaves, all of these are grown commercially or occur in the wild or in home gardens in the US and Canada. The *Agavaceae* family includes dozens of economically valuable fiber and food plants— maguey, the source of tequila, among them. Most are native to subtropical and dryland Mexico but some range into the American Southwest, tolerating temperatures as low as −10°F. Flax is a northern artisanal crop that has been cultivated on the home scale and still plays a role in traditional European culture, where it is spun into linen. Flax yields too little fiber at too high a cost for mass marketing, but linen is a superior fabric from the view of durability and qualities of tailoring. Where industrial methods of spinning and weaving have been applied (now almost universally), cotton is the cheap natural fiber of greatest convenience for clothing, and it is widely cultivated. Naturally colored varieties are grown in small quantities. Hand spinning of cotton is common in India and some other cultures of South Asia, but is rare in the developed world outside the realm of fine arts. Hemp, bamboo and kenaf all figure in some fiber industries from clothing to furniture, paper and construction materials. Hemp was the traditional tough fiber from which sails were made (the word canvas is derived from cannabis). Bamboo too has recently been macerated and spun into fine fiber for weaving; it has also been used structurally to reinforce cement and to peg straw bales into walls. We keep some plants of yucca in the garden for tying vines to trellises. New Zealand flax (*Phormium tenax*) is an exotic and tender temperate plant of similar utility. Kudzu (*Pueraria lobata*) is a valuable material for basketry as well as a nitrogen-fixing forage crop and a high-value food starch (from the root). Of course the oils of cotton, hemp and flax seed are edible and the seed cake suitable for human or animal consumption, which makes for two or three main uses from each of these crops. Bamboo is perhaps the most versatile coarse-fiber crop in the world, being converted into innumerable household and craft objects from fishing poles and fencing to drinking cups, spoons, baskets, hats, matting, scaffolding, flooring and even whole houses. Where it grows it is indispensable and stronger in tension than steel pound for pound. Where the climate is too cold, willow — with more than 300 species in the temperate zone — meets many of these demands for furniture, fencing, basketry and poles. Other woods such as oak have been split for basket weaving too. The Salish peoples of the Pacific Northwest coast used spruce bast fibers to make exquisite baskets and hats. Fig roots, honeysuckle, hibiscus, kudzu and grapevines are all good candidates for basketry, being strong, abundant and relatively easy to harvest and process. Nettles, a highly nourishing plant used in culinary and fertilizer teas and biodynamic preparations, is also made into string, cloth and paper, while reedgrass, cattail and grain straw are used for thatching roofs. Straw is also used in wall construction, either loose, with clay coating (in clay-straw) or baled and stacked like bricks. Cattail has numerous other fiber uses (as well as edible parts), among them pillow stuffing, chairs, hats and matting (from the seed head and stems respectively), and its leaves are made into paper. Living cattails filter stagnant water, while cut stems are used as an absorbent medium to remove pollutants from water bodies. With edible roots and pollen that is both edible and combustible (it makes fireworks!), cattail may be one of the top ten most versatile temperate plants.

## Fertility Plants

Many plants are grown as green manures to build soil fertility, among them buckwheat, clovers, Sudan grass (*Sorghum bicolor* var. *sudenanse*), vetches and turnips. Their chief attribute is the ability to fix nitrogen or mobilize phosphorus or to produce a copious

Fertility Crops, Pattern #61 (See Chapter 6)

[Credit: Abi Mustapha]

Comfrey is a deep-rooted perennial that remains tenaciously where it is planted, but most varieties set no seed. Only careless tilling or deliberate divison will spread it.

[Credit: Edward Carter]

potassium and calcium when the leaves are applied as a mulch. It can also be made into a tea by soaking the leaves in water and applying the resulting, odorous decoction as a liquid fertilizer. Drying the leaves before making the tea cuts down the smell. Stinging nettles (*Urtica dioica*), which favor rich soil, also accumulate a wide range of minerals in their tissues, and though difficult to handle (use gloves) are useful in compost and compost teas. All weeds tend to accumulate some minerals and may be composted, mulched (if cut before going to seed) or made into teas; chamomile, horsetail and yarrow are among those favored by biodynamic gardeners for their value to soil fertility.

amount of biomass in one season. Plants can also aid fertility when used directly as a mulch or when composted or made into a liquid extract such as compost tea. The chief qualities needed for these purposes are ease of harvesting and a high mineral content. Supreme among permaculture and fertility plants for its multifunctionality is comfrey (*Symphytum officinale*), a perennial herb traditionally used as a poultice. Comfrey also makes a fine fodder and a good bee plant.

Comfrey may be cut three to five times during the growing season, and it supplies large amounts of nitrogen (2% by dry weight),

### Fencing and Barriers

Plants for fencing too are numerous, though the traditional choice of English coppice workers was hazel, made into wattles, or movable fence sections.

Willow is similarly suitable and can also be planted as a living fence; the stems can be woven together for greater strength, or even *pleached*, which involves cutting or scraping sections of the cambium layers of two or more stems, then binding these together temporarily until they grow into each other.

Many woody plants can be pleached; this increases survival rates by providing each

Wattles, panels for building or movable fences, are traditionally woven from hazel rods, a coppice product.

[Credit: Abi Mustapha]

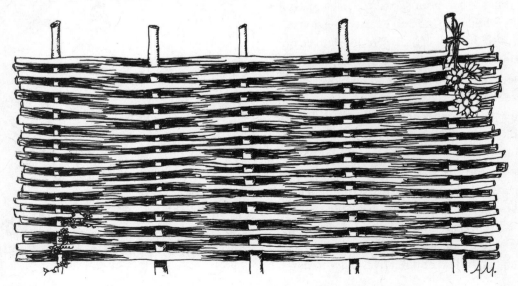

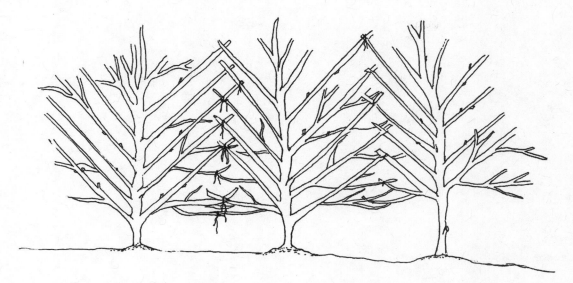

Pleaching makes a durable fence or resilient cordon fruit planting by causing the bark of separate stems to fuse.

[Credit: Abi Mustapha]

stem access to a wider root network. Hawthorns and some plums, with their prominent thorns, make good hedge plants, and can be partially cut, laid over and woven into living fences while still rooted. The most wicked fencing species is honey locust. Its 5- to 12-inch thorns can be deadly; they evolved to repel mastodons and mammoths which would otherwise have pushed the trees over to get at their large, sweet, protein-rich seedpods, still appreciated by sheep and cattle. Planted densely, comfrey — though no obstacle to animals — makes a good barrier against grasses entering the garden. For fenceposts, black locust is among the very best — reputedly used 50 years and then turned upside down for another half century. And osage orange (*Maclura pomifera*) is also extremely durable as a post, detectable up to 100 years after being placed in the ground, or as a living hedge whose dense, fibrous and spikey stems will keep out animals large and small. Jerusalem artichoke sends up a profusion of dense five-to-six-foot stems that make a fair visual barrier and windbreak for small gardens. Among traditional ornamentals, I would choose lilac (*Syringa vulgaris*, of which there are numerous sizes and several color variants) and mock orange (*Philadelphus* spp), both of which have dense shrubby growth and fantastic fragrances. Forsythia is always

inspiring for a week in the spring, but of little use at any other time. It makes a poor mulch as the cut stems and branches can root if covered while still green. For our own living fence in southern Indiana we selected hawthorns, roses, Japanese and other quinces (all thorny species), as well as hazels, plums, cotoneasters, aronia, serviceberries, sorbus, crab apples, willows, Italian alder, witchhazel and deciduous holly — about 40 nitrogen-fixing, fruiting or wildlife forage species in all. Most of these have a shrubby habit of growth with many dense stems and can be kept to 10 feet in height.

Plants are central to all cultivated and natural systems today because they have unlocked the secret of photosynthesis: they translate the sun's energy into food for all the rest of us. They build up carbon in soils, biomass above ground and food on the farm — not alone, but as the primary players. In this book, they precede animals because most farm animals depend on plants for food. In developing a cultivated ecosystem, we need lots of plants to be converted into manure and soil by many fewer animals. To bring animals into a system with poor plant resources is a formula for dependency and degradation. But of course, coming into action at the right time, animals also play a key role, as we shall see.

## Old 99 Farm, Dundas, Ontario, Canada

[*Color Insert #2 includes photos of this property.*]

**LOCATION:** A neighborhood of small farms, specialty horticulture and rural dwellers in the Lake Ontario basin, adjacent conservation land on and near the Niagara Escarpment, with intrusions of exurban sprawl at about 650-foot elevation. The glaciated landscape has a confounded geography with low hills and mounds, small wooded stream valleys, interrupted drainages and meandering local roads. Highway 403 carries high-volume traffic a few miles to the south. The dark glacial till is productive and well-drained, but agricultural erosion has degraded it in places. Climate zone 6a with 33 inches of precipitation per year. Winter snows can be heavy.

**DISTANCE TO URBAN CENTER:** 4 miles to the commercial center in Dundas, 8 miles to downtown Hamilton, a city of 520,000.

**ECOSYSTEM:** At an edge between agricultural savanna and deciduous Carolinian forest, the farm's sloping and tumbling fields display every solar aspect but have little more than 50 feet of local relief. The farm was planted to soybeans in 2007. A narrow treeline runs the length of east and west boundaries. Deer, birds and amphibians are abundant.

**SIZE OF PROPERTY:** 25 acres in a long, narrow parcel stretching from the paved local road at the north end to a stream at the back. Neighboring houses on either side of the residence and across the road are at suburban distances.

**OPERATOR:** Ian Graham has two college-age daughters, one of whom has some interest in natural living. He hopes that her part-time help on the farm may expand after she finishes her studies. Ian is separated from his wife, who had no desire for a second career in farming. He has had help from occasional resident interns.

**ESTABLISHED:** 2007.

**PRODUCTS SOLD:** Winter vegetables, eggs, dairy.

**MARKETS:** Neighbors, local shops, cow-share, farm stand sales.

While in his early 50s, Ian sold a successful small business and resolved to prepare for energy descent. He sought farm property with the potential for self-reliance and demonstration of permaculture in the region around Hamilton, an industrial city at the west end of Lake Ontario. Old 99 Farm had been cropped conventionally in recent years; soybeans remained in the field the year he bought the land. The farm had a 1903 brick home in good condition with a cellar and conventional amenities. There was also a newer two-story garage building, an old shed and a small barn accessible on two levels. A seasonal pond of between one and two acres in the center of the fields dried up most years by July. There was a small stream at the south end of the farm with a patch of woods around it.

After training in permaculture and educating himself about energy and resource issues, Ian began to create a lifeboat for himself and his family, applying his love of the outdoors to restoring and enriching a worn-down landscape and at the same time offering leadership in sustainability to his community. He knew that this would require him to gain new skills, and he relished the challenge of demonstrating competence in caring for the land, plants and animals. His vision includes modest sales of winter vegetables—a niche market—for which he has installed two large greenhouses that use climate battery technology. He also expects to operate as a mixed livestock hobby farm, plant perennials heavily, offer workshops and residential retreats and to run trials in renewable energy systems: biofuels, wind, solar and geothermal. He's deliberately chosen a middle path, seeking out old ways of living that are economical and practical, while embracing appropriate and transitional technologies. His strategy calls for reducing dependence on fossil fuels while retaining the convenience of grid systems (rejecting off-grid, primitive and unplugged approaches)—and also eschewing intensive horticultural work (such as that involved in CSA vegetable production) in favor of an experimental mode of research, demonstration and ad hoc marketing.

His first moves included establishing contour garden beds in the field along the road, cover cropping the main fields to winter rye after the soybean crop was harvested and installing a grid-tied solar electric system with a battery backup capable of running the pump, the fridge and a few lights. He insulated and improved the space above the garage as a dormitory for helpers and guests, added a straw bale greenhouse extension to the south face of the barn, created a composting toilet and made immediate energy improvements to the house. When he invited me and my partner Keith Johnson to advise him about broadscale development of the farm system, we recommended that he modify the landform around the seasonal pond to hold more water and terrace some of the north-facing slopes west of the house for polyculture orchards. A year later, both of these systems were showing good promise: the pond was full to brimming and richly fringed with wetland plants. The orchard terraces had been created and hundreds of perennials planted.

By the summer of 2009, Ian had acquired a small flock of chickens and a bred Canadian Lineback cow that had calved and was giving milk. He'd also gotten a Marama guard dog and a pair of large terriers that he was training. He was then on the verge of fencing the main farm fields into seven paddocks for rotational grazing, with the idea of expanding the livestock herd with a few more cattle, some sheep and more poultry. The two main greenhouses had been erected, and grading work around them had been substantially completed. Elements of the climate battery were in place, but greenhouse crops hadn't yet been planted. A root cellar was taking shape in the basement, and Ian was trying to determine how best to increase roofwater storage beyond a few small tanks, barrels and the existing cistern. He had also sponsored several day workshops at the farm as well as longer seminars on permaculture design, peak oil and related subjects, reaching into the Hamilton community to promote possibilities for renewable energy and sustainable urban design.

When Ian first viewed it, the farm had a number of features that appealed to him, including its proximity to city markets, substantial infrastructure and a relatively large portion of usable land. Nearby growers offered a community of interest, and neighbors with horses had extra manure. The house and barn each had a good well, and the house also had a 3,000-gallon underground cistern that collected roofwater. The hedgerows and three acres of riparian woodland offered a sustainable yield of heating fuel, and the cellar, woodshed and garage had good potential for new uses. A nearby bike trail provided 20-minute access to the city. Seven older apples were still fairly healthy; we gave him some tips on restorative pruning that should bring them back into productivity. The long, thin shape of the parcel (about 400 feet by ½ mile) was a liability that made management of the back acreage either energy- or time-intensive, so Ian's choice to put much of the land into permanent pasture and to enhance the tree cover was an intelligent adaptive design that should minimize the need for machinery. His development choices emphasize energy inputs at the beginning with increasing yields and decreasing costs over time, a classic permaculture approach. Animals cultivate the ground, converting pasture grasses into traction, meat, milk and manure while trees deepen fertility and increase biodiversity. The two systems are potentially complementary, and both restore ecological health if properly managed.

Near-term plans call for solar pre-heated water and hot water on demand systems as well as a masonry heater for the center of the house. The temporary hoop house against the barn will be replaced by a permanent earth-bermed greenhouse for starting seedlings. This will also be the hub of a nursery for perennials. In the mid-term Ian plans to obtain organic certification, and further out as fruit crops begin to bear, he plans to develop a small cold-storage facility so that he can sell fresh fruit into local markets through the winter. The pond is also planned to accommodate production aquaculture. Long-term goals for the soil include keyline subsoiling and remineralization for enhanced fertility and water capture. His vision calls for food self-reliance for the equivalent of three households, about 12 people, with surplus to sell. Home improvements will, besides the new energy systems, include developing an outdoor kitchen on the existing patio to the west of the house's central room. This will get a pergola covered with grapes and hardy kiwis. He also plans to build a direct entrance from the east side driveway into the basement for ease of loading and accessing the root cellar.

In the two years since our last visit, Ian has fenced the paddocks and now runs 16 head of Canadian Linebacks, a very rare heritage breed. He's added a dozen sheep—mostly Dorsets which will breed at anytime of the year, and a few Katahdins (a hair sheep)—and he continues to keep chickens. Taking a lead from Gene Logsdon's small-scale methods, Ian's begun raising his own grain to feed the cattle and poultry, planting and harvesting an acre of open-pollinated corn.[1] Despite dry conditions, the orchard began bearing a few apples, pears and peaches in 2010, and the 2011 season held the promise of a few bushels for home use. A handful of escaped koi have gotten into the pond and are reproducing. While this could represent a problem, it might also be an opportunity to develop a business in live fish and fish emulsion. Geese and ducks are on the horizon to exploit the water niche, and Ian has begun pulling back a bit from vegetables in favor of products that are more easily marketed and less labor-intensive. Squashes keep well and need little harvesting labor. The winter market for greenhouse vegetables remains favorable. But Ian would like to develop an ethanol project, a tree nursery and some form of aquaculture. He is still in need of farm help.

Ian's vision is clear and seems to me well-suited to make Old 99 Farm a valuable contributor to the local economy in the decades to come. His passion and energy for the work are high, and he's making good choices about infrastructure and systems. His challenge will be to develop a supportive community around the farm, including other resident farm help, and beyond that, to develop plans for a transition to his retirement. If the plan unfolds as he hopes, there will be plenty of physical activity suited to an older man to keep him engaged: moving animals, minding trees, teaching and supporting younger family members and future farmers. But the farm is far too large to be maintained or even operated by one person, with or without fossil fuels. In the present situation, with the farm undergoing dramatic changes, his executive abilities and decision-making skills, honed over a decades-long career in business, are being well used. But in the long haul, despite generally good levels of upkeep, the older buildings will need repair and modification, trees will need to be felled and pruned, and livestock care will involve some hard work. Vegetables are labor-intensive, even if you can choose your markets and arrange contract sales. And so, for these and a hundred other reasons, Ian will need to seek and develop partnerships with younger people. Ideally this will be foremost with his younger daughter, but others will need to be involved as well.

# CHAPTER 14

# Animals for the Garden Farm

Plants translate energy from the sun into food. Animals move it around. They also concentrate and deposit its residues as manure. Long before anyone thought of putting money into Irish banks — lately fallen on hard times — the cottager in that country looked upon the dung heap by the back door as real treasure. Manure makes plants grow strong. Strong plants grow strong animals. Healthy animals return that wealth to the soil under the farmer's feet.

To feed human beings, with our large brains, warm blood and considerable body mass, on the limited amount of land now available requires fertile soils. Animals are the best solar-powered pathway to achieve that fertility; they are the main way that advanced societies throughout human history have accomplished heavy tasks and built soil. In 1900, before the internal combustion engine, there were 25 million horses in the United States and an untold number of oxen, mules and donkeys.[1] Both animal traction and manure as a source of soil nutrient were largely abandoned after World War II when the use of petroleum

and chemical fertilizers became widespread in the industrial countries. But as the Cubans did in 1991, we will turn again to using animals for every motive purpose when oil scarcity forces the issue.[2]

Animals have evolved bone structures and teeth, digestive systems with acids and fermenting chambers and muscular action to peck, smash, tear, grind, mix, masticate, liquify and compact low-density vegetation and other living tissues in order to extract the energy from them. Moreover, animals can move about to find, harvest and return that energy to the place where they rest.

Of course, all this bone and muscle (and feathers or fur) diverts some of the energy that animals consume. And the digestion process is never perfect. Large amounts of mineral, fiber and even carbohydrates and amino acids are excreted. And there's the key: animal digestion is markedly incomplete. Each species derives as much nourishment as it can from its food, but never takes it all. So the remainder — waste, manure, frass, shit — is rich food for microbes and plants. It also sustains higher creatures with the right digestive

Ponies such as this one still pull carts in England. They range freely on Dartmoor in the southwest of the country during the growing season and are rounded up for sale in the autumn.

[Credit: Peter Bane]

systems — from worms and insect larvae to hogs and canines.

Animals can harvest plant matter from above the ground and even from off the property. Bees will forage pollen and nectar up to three miles from the hive; pigeons range for miles about in search of seed. The traditional practice of decamping with herds to high mountain meadows for a few months, called *transhumance*, increased the yield and nutrition of summer milk and also preserved pastures near the village for haymaking and autumn grazing. Closer at hand, animals will pull fruits and seeds off of vines, and leaves and twigs from shrubs and trees within reach. They will knock over standing crop residues, eating their fill and trampling the rest into contact with the soil while adding manure, thus hastening decomposition. All these freely given services can be made useful by design, which chiefly means organizing their placement and timing.

People keep some animals for food, fiber or work, use the wild services of many to support farming and attempt to diminish or exclude others to prevent damage to crops or predation on livestock. All insects are, of course, animals — as are birds, fish and crustaceans, not to mention earthworms — but the vast numbers of insects in their myriad forms have largely defied our control outside of very limited areas and for brief times. We have only managed to domesticate or develop complex relations with a handful, among which the honeybee and the silkworm are preeminent. Fortunately for us, most of the rest play largely benign roles in sustaining the ecosystem. At a minimum, insects are food for birds, reptiles, amphibians, fish, crustaceans and small mammals, and these are our food or feed for our livestock. Insects, in their vast numbers, also cycle large amounts of biomass into frass (their manure), laying down a fine rain of nutrient over the entire landscape, but especially where natural resources are rich.

## Livestock of Importance

Of animals we know more intimately, a half-dozen dominate the agricultural economy: cattle, swine, chickens, sheep, horses and goats. Another dozen occupy the second rank, though they may play roles on the farm that make them more important than mere numbers would suggest: working dogs and cats, ducks, rabbits, geese, turkeys, donkeys, llamas, guinea pigs, guinea fowl and pigeons. I describe the care and use of these animals with three exceptions. Because of their size and cost of upkeep, I view horses as an indulgence unlikely to produce economic returns to the garden farmer. Sheep are prone to parasites and are inclined to die with little provocation. Both of these animals require expansive areas of grass pasture or were traditionally turned out to rough browse; this book is urging that the lawns of our suburban wastelands be planted in more productive crops. Neither sheep nor horses are well-suited to polyculture. Globally, donkeys are more important to subsistence farmers than horses, but this is not the case in North America unless their role in begetting mules is taken into consideration. They are chiefly creatures of rugged terrain and will not appreciate close confinement. Honorable mention might go to water buffalo and yaks (worldwide a significant subset of bovines), camels, reindeer, elephants, ferrets, packrats, falcons and cormorants for the work they do on our behalf, and though all these are present in North America, their impact here is negligible. Research has suggested that a variety of small deer and antelope, two or more species of iguana, a handful of large tropical birds and about a dozen or more rodents not now widely used have good potential for domestication and meat production.[3] Humans keep some species of deer, elk, bison, ostrich, emu, mink, chinchilla and nutria for meat or fur, but these are specialty crops about which I know little except that for one reason or another most are not well-suited to suburban and peri-urban regions or the needs of the small producer or household.

Almost all of the animal species available to the farmer came into human company within the past 3,000–8,000 years (dogs a little earlier). The bulk of them, certainly the most economically important species, originate in the Old World. Expanding our access to and use of animal services and products is more likely to come from thoughtful and observant breeding and selection from among existing stocks than from wholly new species. As energy descent steepens, we would be better advised to preserve those resources that abundant energy has made known and available than to invest significantly in attempting to create whole classes of new ones.

I have personal experience raising or living with bees, dogs, cats, chickens, ducks, guinea fowl, horses, goats, rabbits, and have had neighbors and close friends who have kept sheep, cattle and hogs. I have observed most of the animals I write about in domesticated situations over many years, but I do not claim to be an expert about any one of them. Many growers have more experience than I do raising and caring for animals, and a few have written well about their subjects. While I heartily recommend readers to expand their investigation into animal characteristics and

behaviors beyond this survey, I find that most books on animal management are regrettably two-dimensional. They may be thorough in providing details about a certain economic model, often quasi-industrial, but there is too little attention put on more holistic approaches to livestock. My hope is to provide a useful basis for design and decision making about animals for ecological land management and for the household and small farm economy.

## The Importance of Good Breeding

It seems virtually certain that the animals humans raise for food, fiber and work will never go extinct, though from among our most common livestock, some older breeds have disappeared and many are threatened. This is largely due to selection pressures by industrial agriculture for breeds that put on weight rapidly in response to the provision of optimum feeds. Older strains of animals bred for multifunctionality, thrift and utility on the small farm have languished.

Efforts have been underway for about the last 30 years in the UK and the United States to conserve heritage or traditional breeds of farm animals that are now at risk. The Rare Breeds Survival Trust UK and the American Livestock Breeds Conservancy have had considerable success in calling attention to the plight of endangered and threatened breeds, and have pulled a number of important animals back from the brink.[4]

Heritage breeds — those older and more rugged types developed by our ancestors for their multifunctionality, hardiness, ease of birthing young and ability to forage — are readily able to gather their own feed from the wild edges of the farm or the village. On farms of the energy descent future, including garden farms of limited acreage, these time-tested qualities will again prove themselves valuable. Many of these characteristics, which involve the intelligence and temperament to avoid or defend themselves against predators, good

Confined Animal Feeding Operation (CAFO)
[Credit: Twyla Francois]

mothering capacities and herd and flock behaviors, the instincts to find wild food sources and the ability to select plants not only for nutrition but for medicinal effects have been purposefully bred out of most animal breeds now used for industrial food production.

If you intend to confine a chicken to a cage for its entire life, the abilities to hunt and catch bugs, search for greens, select good nesting sites, follow the signals of a protective rooster and remain wary of hawks and foxes are quite unnecessary and even counterproductive. But if you want to take advantage of the animal's ability to care for itself, thus reducing your work to care for it, then these qualities emerge once again as central to any stewardship of livestock.

## Ethical Dominion

I do not believe it is necessary to deny our kinship with the animals or to pretend that they have no feeling to use them for purposes of supporting our own lives. If we take animals into our care, we are accepting responsiblity for their deaths, whether by disease, accident, predation or at our own hand. I believe that we can justifiably take the lives of animals, either ourselves directly or in human community, in order to bolster our own vigor and prosperity against the

vicissitudes of a difficult fate while remaining compassionate toward our animal companions. A proper ethical stance toward animal life requires that we use what we kill and that we provide a good life to the animals in our care. This means at minimum that they enjoy adequate healthy food and water, protection from predators and the hazards of confinement and a life of companionship with others of their own and compatible kinds, with as much of their natural sexual cycle, including reproduction, as can be accommodated. It is, to my mind, unarguable that we have entered into a deep symbiosis with our domesticated animals that has prospered them as much as it has us.

Given the opportunity to eat meat, the vast majority of humans readily choose to do so. And while some people may genuinely choose to forego eating meat — and that choice should be respected — many others are obligatory carnivores, being unable to transform vegetable fats into the fatty acids required for human health, or having digestive and immune systems unable to process large amounts of grains and pulses. Our brains and nerves require the saturated fats, our hormonal systems and organs the mineral, vitamin and lipid enrichment of animal organs and muscles, if not every day, then certainly at regular intervals, and especially at key seasons of life.[5] This is neither a tragedy nor an occasion for callousness, but an opportunity for mindfulness and gratitude.

Engaging with animals requires us to be present to our carnality in ways that simple gardening does not. The keeping of animals involves either denying them as a group any sexual expression, or accepting the inevitability of surplus offspring, especially males, and of the sacrifices this requires. For animals to lead a very good life in our care and then, as my colleague Joel Salatin says, to "have one bad day" is no worse—and most likely much preferable—than to struggle for existence in the wild. Some species observe our peculiar monogamous customs, but most, especially among our domesticated animals, do not. If you are squeamish about sexual politics, confine yourself to a few geese: they mate in pairs and for life. Otherwise, enjoy the vicarious amusement of overseeing various harems, or of regulating the access of one or a few pampered males to several or many females.

## Selecting for Cooperation

All animal domestication has been a protracted process of selecting for useful qualities. Early farmers and herders chose *docility* above almost all else. Although docility and the acceptance of human presence can be stabilized in as little as three generations, rogue temperment can reassert itself with any mixing of the genetic dice. There's little use for a rooster that attacks you every time you set foot in the barnyard. He's still unlikely to be able to drive off the fox, intercept the hawk or outsmart the weasel when push comes to shove, but he'll make your life miserable. Aggression and territoriality remain deeply rooted in the genomes of all animals, even those long under domestication, and for good reason. The breeder's and the herder's aim must be to balance instincts in the animal for self-preservation with good temperment.

Other important qualities humans have

Most ewes can raise twin lambs; triplets often pose problems.
[Credit: dichohecho (Sarah) via Flickr]

selected for, and that remain important to the farmer today, are *thriftiness* (the ability to grow and prosper on modest rations and ordinary forage) and *vigor*, expressed as rugged good health, hygienic behaviors and the ability to procreate easily. The ewe that can reliably birth healthy twin lambs is more valuable than the one that has but a single offspring most years.

She may also be more valuable than the ewe who births three but can only raise two, or has poor mothering instincts that throw the burden of nurture onto the farmer. Genetic resistance to disease is invaluable, and this should be kept in mind when illness too often calls for veterinary intervention. With livestock—especially for the farm if not the racetrack—it is better to eat the weak and the wicked and to breed the strong and the calm.

Most of the animals humans successfully brought into management have social instincts that we were able to reorient toward ourselves. We see this most obviously with canines. Dogs are all descended from wolf puppies that early humans kidnapped or found and raised among themselves. Since canines have instincts to align with a hierarchy of dominance that is very insistent, it was and remains not difficult to train young pups to obey the signals of the alpha wolf. We have simply learned to speak and act enough wolf to convince the young of that species to obey commands. Once patterned to obey, dogs are relatively easily kept in line by regular light touches of reinforcement. Few other animals develop such closeness, but most of our domestic animals live in flocks, herds or small family groups, and so tolerate constraint and direction from humans once dominance and familiarity are established. This interdependence, with humans asserting a crucial role in the social life of livestock, forms the basis of our bond with domestic animals and of our success with them as well.

Animals species and breeds must be matched to the scale and conditions of the farm and to the farmer as well. Even though heritage breeds offer many advantages, including a premium on the sale of breeding stock, the serious grazier or herder of livestock will want to develop over time a breed or hybrid of breeds to suit the specific conditions of the farm. Inevitably, this takes some time and a familarity with the animals in question. You can't really know if you prefer Dominiques or Black Java chickens un-

less you've kept them, though you can take testimony from other growers and read the literature. For southern cattle raisers, an influx of Zebu genetics from *Bos indicus* cattle may be needed to withstand higher temperatures. What's most important is a flexible and empirical attitude, both with individuals, with breeds and with species.

## Animals Support the Ecosystem

The permaculture approach to livestock is what has in other quarters been called "default livestock." These are the animals who, fed on wastes of the system, take no food away from humans but add important services and harvests. They will be able to forage on rough ground that cannot easily be cultivated for other crops. They graze in pastures that are restoring fertility to arable fields or clean up pests of cultivation or of other species. They will eat spoiled or surplus vegetables, fruits, kitchen wastes, weeds or woody browse, and they will be able to harvest much of this for themselves if you want them to. A small pig that can convert kitchen slops, grubs, half-rotten tomatoes and oversized zucchinis into lovely lard and bacon, invaluable fats for the diet, would fall handily into this category.

For the smallest garden farms, some poultry and a small rabbitry may be quite enough to handle wastes and surpluses and to round out the household diet with eggs and meat. For a more expansive holding of several acres, a small dairy cow or a pair of dairy goats might be ideal for converting grass into milk, butter and cheese.

Or in a more watery landscape, geese might provide eggs and a good source of high-quality fat out of surplus vegetation, while weeding the potatoes or the strawberries.

## Integrated Design

To design the animal component of a permaculture system, we must examine the needs, yields and inherent characteristics of each animal and breed. I will give an overview

Goats prefer to eat with their heads up, browsing a wide variety of twigs and leaves for about ¾ of their feed.

[Credit: Peter Bane]

of 14 animals with more emphasis on the medium-sized and smaller species as these will be of greater importance to more readers. Before adopting any animal, you should read more thoroughly to make informed decisions. You may elect not to keep livestock. This is a completely legitimate approach; however, it will slow down soil building, require you to bring in manure from somewhere else or increase demand for fossil-fueled machinery. If livestock are not in the picture for you, skip to Chapter 15, where I take up strategies for insect and wildlife management.

## Honeybees

The smallest livestock you are likely to keep would be honeybees, *Apis mellifera*. They are social insects with highly differentiated work roles that exhibit extraordinary cooperation and group intelligence. Each hive consists of a single queen and her predominantly female offspring. Early in life, the queen takes a mating flight, receives sperm from a number of male drones (who expire after mating) and returns to lay eggs in the hive. She stores the sperm and uses it as she needs to lay fertile eggs — as many as a 1,000 a day — for up to three years. These eggs are cared for by her sexually infertile daughters, called workers, and are raised from egg to larva to new worker bee in 21 days. Fertilized egg cells turn into workers; unfertilized egg cells turn into drones.

The queen emits pheromones that regulate all activity in the hive. As long as she is alive and fertile, the workers around her will do everything to protect her and the hive and to facilitate her laying more eggs. They feed her, groom her and keep her warm during cold weather. If she dies or if they sense that she is losing her ability to lay fertile eggs, the workers will prepare and raise a new queen from ordinary fertilized eggs.

Honeybees need nest sites, and this forms the basis of their symbiosis with humans. We provide their homes, and in return we can rob

A. Bees appreciate eastern exposure to get going in the morning.

B. Smoking disrupts the bees' olfactory communication and signals possible fire, so they focus on eating and are less likely to sting.

C. Cutting the wax caps from comb honey allows it to be extracted.

D. A small centrifugal spinner can extract honey from two frames at a time.

[Credit: Peter Bane]

them of surplus honey stores, beeswax and related products.

Wild bees are constantly gathering nectar and pollen, raising their brood, filling up whatever space they occupy and moving on. The colony reproduces by a process called *swarming*. During the time they are swarming, honeybees will not sting — they are too full of honey, eaten for the swarming trip. Beekeepers who know this and are prepared can catch the swarm in a box and introduce it to an empty hive, whereupon the swarmed colony will organize itself and set about gathering nectar and pollen, laying eggs and raising larvae into workers.

Bees need access to water. They must approach water by crawling along a rock or a reed that lets them drink without drowning. They also need sunlight to warm up and get going in the morning: an east-facing hive makes the most sense. The flight path out of the hive should not cross human or animal paths for at least 20 feet in order to prevent conflict. Bees will generally exit the hive and rise into the air before heading off — in a bee line — for wherever they have been told by

their sisters that food is to be found. If you can, encourage them to gain elevation quickly, for example by placing tallish plants or a low fence a few feet in front of the hive.

The bee yard needs to have room for easy working and access to the hive or hives from the back side. This will reduce stress on the bees and the beekeeper.

Bees forage for two to three miles from the hive in all directions, so they are getting much of their food from elsewhere than your property. There are typically a small handful of main nectar crops in any given area. But you can help by providing forage plants that bloom throughout the growing season. Your yard, garden and farm should have ample areas of flowering herbs and shrubs with enough variety to ensure that something is available every day of the season. We see bees on our basil plants in October and on late-flowering radishes in November. I know a beekeeper who keeps a bag of sweetclover seed (*Melilotus* spp) in the car to broadcast along country roads near his home. If you are a beekeeper, you become keenly attuned to the sexual activity of the flowering landscape! For lists of bee forage species by region, see Appendix 2.

When in the midst of a strong nectar flow, bees are also collecting *pollen*. You can see this as bundles of bright yellow, gold or orange dust on their hind legs. Pollen is their protein, and they use it, along with their own enzymes, to make *propolis* (their all-purpose gum for sealing cracks and embalming invaders) and *royal jelly*, which is given to all larvae for three days and to queens throughout their gestation. The nectar is mixed with enzymes and layed up in open cells to lose moisture and ripen as honey. When the honey is ripe and the cells filled, the bees will cap them with a thin coating of wax. They will typically fill comb from the lower parts of the hive up and outward toward the top. They store honey, which is their main food, against the winter. They also use the cells of their comb to allow

Tall plants directly in front of the entrance force the bees to fly up when leaving the hive, lessening the risk of conflict with humans.
[Credit: Edward Carter]

the queen to lay eggs. The bees will fill up whatever space is available to them if there is nectar to be found. If the space for laying eggs gets filled with honey and it is still warm weather, the colony will prepare to swarm. Thus, the beekeeper should remain attentive and provide additional "supers" or upper hive boxes, as needed. If it is late in the season, the queen may simply slow down laying eggs, which happens in any case when cool weather sets in. As days shorten and temperatures fall, the hive boots out its drones, the queen ceases laying eggs, the workers stop flying and the colony gathers around the queen in the center of the hive, beating their wings to generate heat to maintain 92°F, which they do throughout the winter.

Bees will sting only under provocation. Each bee can only sting once, as its barbed stinger will stay in the victim and pull out of the abdomen of the bee, leaving her mortally wounded. Bees sting to defend the hive or if they are unduly harassed while flying or foraging. For most people, bee stings, while unpleasant, are very mild compared to the stings of wasps, hornets or yellow jackets, which many people incorrectly refer to as "bees." A very small number of people are highly allergic to bee stings (and to wasp and hornet stings as well). If someone in your family is susceptible in this way, you can still keep bees, but you should also keep on hand an epi-pen, or epinephrine injector. This nerve stimulant is a specific antidote to anaphylactic shock, which can occur in somewhat less than 1% of people following bee or wasp stings. If not treated promptly, anaphylactic shock can be life-threatening, leading to swelling and difficulty in breathing.

Bees need good timing and stewardship from their keepers, but a beekeeper need not pay the bees much attention for days or even weeks at a time. The beginning and end of the season are important, as is the time of greatest nectar flow and honey accumulation. Winter is a quiet season when good preparations

should allow the beekeeper to ignore the bees for several months.

Bees today suffer from a variety of mites, both tracheal mites and varroa mites, as well as less common but potentially deadly bacterial diseases.

Good hygiene can prevent problems, and while the bees are typically very orderly in their housekeeping, they need support and sometimes assistance from their keepers. Beekeepers should never feed honey from other hives to their bees — it can be contaminated with foulbrood and other diseases. Instead, when supplemental feed is needed (because of cold, rain or a dearth of nectar) sugar syrup should be offered. To this, aromatic feed supplements can be added. I blend a tincture of chamomile into the syrup I feed, and in cold weather I also use a probiotic blend of aromatic oils. But the beekeeper may be compelled to treat the hive if infestation by mites becomes severe. Many beekeepers have found that the use of miticides is preferrable to losing bee colonies.

Readers in the southern states may justifiably be concerned about the northward migration of Africanized bees. African bees were brought to Brazil some 60 years ago and have moved northward through the tropical regions of the Americas. They have been found in Texas and other southern states. The

Varroa mites (one visible on the back of the bee shown) have emerged in the last 20 years as a major parasite of honeybees. They are just barely visible to the naked eye. [Credit: Scott Bauer, USDA Agricultural Research Service, bugwood.org]

Africanized bees are notably more aggressive, reportedly following intruders (including beekeepers) for some distance from the hive and even congregating around doors when the target individual takes refuge in a building. They also seem much more willing to sting, perhaps because the colony as a whole is able to generate many more bees due to its vigor. Africanized bees are not cold-hardy at present. These same bees are widely and successfully kept in Africa and in Brazil, so people can learn to manage them. They are unlikely to be fully adapted to cold regions for some years to come, but evolution is relatively fast among insects, so it seems inevitable that, combined with the effects of warming climate, the adaptability of bees to their environment will bring more and more northerly beekeepers in contact with their traits and eventually their colonies.

The reasons to keep bees are manifold: pollination and yield are improved in many crops by the presence of honeybees. Honey, which brings a good price and keeps indefinitely, is a nearly perfect food and medicine — it is perhaps the very best treatment for burns. Beeswax makes the finest candles, burning without scent or smoke. There are many details to beekeeping, but most areas will have a beekeeping club or you may be able to locate an experienced beekeeper to mentor you as you start. Books are useful, but first-hand advice is invaluable, as is your own experience. Bees are a living system, and the results you get by working with them will vary from season to season and place to place.

## Poultry

Humans have lived with chickens for about 8,000 years in Southeast Asia, whence they have spread around the globe in the last five centuries. We also have venerable relationships with ducks, turkeys, geese, pigeons, guinea fowl, doves and quail.

The advantages of poultry are easily understood. Birds are small and readily tamed, managed and dressed for meat. They reproduce rapidly, and most of them provide eggs, which are an excellent and versatile high-protein food. They forage on scraps, wastes, insects, grubs, vermin and seeds that humans cannot or do not use, making them extremely economical and thus important to the poor. In doing their characteristic feeding behaviors, many types of poultry help to control insect populations to good advantage in the garden and fields. And with the exception of modern turkeys and some of the larger breeds of geese, they provide a meal-sized serving of meat for one or a few people — just right for the family stewpot — an important virtue for a world largely living without refrigeration.

In the permaculture landscape, poultry offer the possibility for a multifunctional fertility, food and pest control system on virtually any scale, but especially for the small property. As with all animals, their manure rapidly accelerates soil building. Not all breeds, let alone the major domestic species, are alike however, so we need to understand the habitats, yields and characteristics of the various birds.

### Chickens

It is believed that there is about one chicken for each human on Earth — some seven billion, making it one of the most successful of

Poultry: (top l–r) Pigeon, Duck, Goose; (bottom l–r) Chicken, Quail, Turkey, Guinea Fowl (not to scale)

[Credit: Jami Scholl]

warm-blooded species. While a full-grown bantam may weigh less than two pounds, most breeds range from 4 to 11 pounds when mature. The chicken is an omnivorous and hot-blooded animal, with a body temperature of about 112°F, which enables it to eat a wide range of small live protein, plant matter and seeds while foraging among dung and wastes without succumbing to the bacterial zoo that it thus encounters. Feathered descendents of *T. rex*, chickens readily consume baby and sometimes adult mice, voles and shrews, frogs, toads, lizards, snakes, earthworms, snails and whatever else moving that is small enough to slide down their gullets or be torn apart by their beaks. They will also consume meat scraps from larger animals, but they relish the pursuit of insects.

Like other birds, chickens have no teeth, but grind their food in a specialized organ called a *crop*, that is filled with small stones or grit. Consequently, they need access to a supply of this kind of mineral regularly. Chickens also need water, though they are noted for fouling standing water if confined near it, so most growers provide water from a suspended dispenser. In order to control mites that can infest their feathers, chickens need access to a dry dust bath where they can fluff themselves. A small, covered, non-muddy depression is sufficient, but hygiene can be improved by supplying diatomaceous earth.

Because the chickens' diet consists of a large amount of live protein, their manure is high in nitrogen and phosphorus. It is too "hot" for direct application to most plants, but composted or spread widely, as when chickens are allowed to range and forage, it can be very beneficial. To reduce odors, limit manure handling and improve flock hygiene, apply a deep straw or wood chip mulch in the chicken yard and coop. In searching for their food, chickens scratch with their clawed feet, which can make a considerable surface disturbance. They also peck to find grubs, seeds and bits of plant food. When the chicken yard is

Chickens kept on deep straw litter turn and manure it while searching for bugs; the result of this design by Jerome Osentowski is compost at the bottom of the slope.
[Credit: Peter Bane]

located on a slope, the birds' scratching action will result in the material thrown into their yard (food scraps, grain, garden weeds) cascading slowly downslope along with their added manure — the yard becoming, in effect, a slow-motion compost turning system. With a fence or other barrier at the bottom to retain this rich harvest, and placement near the garden, fertility management can be as simple as pitching the chicken-turned compost onto the plants.

Chickens forage most eagerly in the morning; their eating behavior near dusk is desultory. Thus, controlling their access to sensitive garden areas is imperative early in the day. For light weeding and insect control, they should be limited to short evening forays into the beds. The natural habitat of chickens is woodland where the trees and shrubs provide them protection from raptors as well as elevated roosting sites. Chickens, like other animals, naturally seek out shade and shelter during the hot time of the day and when not actively feeding. By selecting a spreading tree or bushes to surround or overtop the chicken yard, you can afford your birds both protection and comfort. And if that vegetation also provides a food for the birds, you've begun to employ good design. The mulberry is a classic choice for its abundant small fruits with many seeds (high-protein) borne over a long season, but legumes such as redbud, black locust, mimosa and acacias or small berry bushes such as autumn olive also work well.

Chickens seem to become virtually insensate when asleep and can easily be picked off by a predator if not confined within a tight dwelling. For roosting, they need bars or rods, replicating tree limbs, and the alpha rooster will typically adopt the highest perch. If that perch is just up under the ridge of the roof, and situated so the cock can't stretch out his neck fully, it can discourage early and random crowing.

The cock is not required for egg production but will, of course, be needed if you wish your birds to breed and replace themselves. If healthy and well-selected, he does play an important role in regulating flock behavior and in protecting the hens from danger. You can order unsexed chicks and, as they begin to differentiate sexually in 12–20 weeks, select the young cockerels for the stewpot, leaving one or two whom you hope will provide good service. Choose based on calm temperment and demonstrated vigor. One cock may service up to 15 hens. In a larger flock two can coexist, though one will likely remain dominant. More than this ratio is usually problematic and may suppress egg production or lead to excessive pecking, so cull the aggressive or inferior males.

Hens begin to lay at about 22–24 weeks of age if raised under naturalistic conditions and, depending on breed, may lay an egg nearly every day during warm seasons. Eggs from poults (young hens) are smaller than normal, but larger eggs are laid as the birds reach full size. If shells are thin, the birds aren't getting enough calcium. Egg quality is unrelated to color — Dr. Seuss was right, you can have green eggs with ham — but is linked to diet. If the color of the yolks is not bright yellow-orange, the birds are not getting enough green matter in their diet. As day length dwindles in the fall, the rate of laying will diminish, as it does with advancing age. After two years, most hens will be laying at about half the rate of the first year. They can continue laying sporadically for more than 20 years though most growers will replace them long before then.

Many breeds of chickens are disinclined to become broody or to sit on their eggs. Many growers discourage broodiness, as they prefer the hens to continue foraging and laying, while a broody hen will do the opposite. To raise a clutch of eggs, however, a broody hen is needed. Most flocks will have one or more hens who are so inclined if encouraged or allowed.

Chickens are delicious and vulnerable to many kinds of predators, some smaller than they are. Your fences should have firm bottoms and floppy tops to discourage burrowers and climbers. To exclude rodents and snakes, the coop itself must have no gaps larger than ¼-inch. As the birds are typically confined at night, predation by owls is unlikely, but hawks, and in some areas eagles or osprey, hunt during the day and can easily take a bird from the yard. Tree cover over the poultry yard is the best protection as raptors must glide in fast and at a low angle in order to grab a bird and escape. In all likelihood, the greatest hazard your birds may face is the overenthusiastic neighbor dog, which even if it can't reach them through cage or coop walls may harass and frighten them to death. In running around in a panic, chickens can pile on each other and suffocate. Keep dogs at a distance with electric or other fencing, and be mindful of guests bringing dogs to the farm.

An important element of protecting chickens from burrowing animals, whether snakes, rats, weasels, skunks or possums, is to skirt their yard with wire mesh laid flat on the ground. Vertical wire netting should be brought down to the ground then flared in or out horizontally about a foot away from the fence and pegged down with landscape staples. In this way, a varmint attempting to burrow under the fence will run into a metal barrier and be discouraged.

Chickens can supply much of their own feed if given access to pastures and woodland, but in the typical suburban situation or even on a small farm, they are likely to need supplemental feeds, especially in winter when insects are not available. Commercial feeds are compounded of corn, soybean meal and mineral supplements, but you can grow much of your own chicken feed. Corn, sorghum, sunflowers, amaranth, squash and pumpkins (for both seed and fruit), plus small seeded legumes such as peas, crotolaria, pigeon pea (in warmer regions), caragana or Siberian pea shrub (in cooler regions), Illinois bundle-flower (*Desmanthus illinoensis*), redbud and cassia could provide most of the supplemental energy and protein chickens need to grow and lay well. For young birds, large seeds should be cracked before feeding, though mature hens of larger breeds should be able to consume whole corn. Layers need extra calcium, of course, which is easily enough supplied in mineral form (calcium carbonate or ground oyster shell), and can be derived by recycling the shells of eggs eaten in the household. Dry the shells in the oven and crush them to forestall any egg-eating behaviors. Greens such as fat hen or lambsquarters (*Chenopodium album*) produce copious quantities of small seed and leaf that is highly nutritious. A feeding strategy that combines forage resource with soil building is to grow a grain or legume cover crop, allow the birds to gather the seed and then roll, chop or turn pigs into the standing straw to knock it down and incorporate it into the soil surface. Shrubs and herbs with small fruits such as autumn olive (*Elaeagnus umbellata*), ground cherry, honeysuckle, elder and currants are valuable as

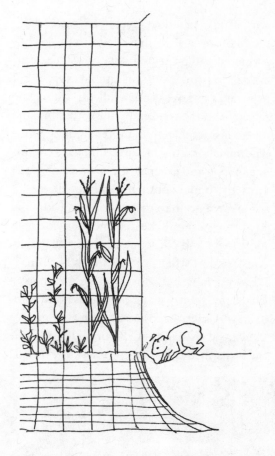

To protect against burrowing predators, bend and bury tight fencing about a foot out from the fence line. An angled and floppy top will help deter climbers.

[Credit: Abi Mustapha]

chicken forage as well. Inevitably, these plants produce more fruit than can be profitably picked by people. Let the birds do the work.

### Managing the Flock

The main management approaches to poultry are:

1. confinement
2. free range in paddocks
3. the use of mobile cages called tractors

Any one of these three strategies might be appropriate depending on the amount of land available and local conditions, but be clear that confinement represents the most work and the least use of the animals' services. Night and winter confinement is still required in most situations for the birds' protection, though bantams and other lightweight breeds may successfully roost in trees in some areas. Tractors can provide this protection without the requirement for a separate coop, though wintering over a flock in cold climates is much more easily achieved with a properly insulated and protected structure. If the birds are raised primarily as an annual crop, either as meat or as layers with a fall harvest for the market and the freezer, then the tractor can save considerably on manure handling and feeding labor. The tractored chickens can forage insects and greens from a sward or garden beds for about ⅔ of their diet but will need some extra grain to maintain best production.

Tractors should be designed for ease of movement and consequently must be of light construction. They can have open bottoms if moved daily, but consider giving them a skirt against burrowing predators if they are to stay in one area for more than 24 hours. They need a small roofed section to provide the birds shelter from rain, a nest box — and

**Poultry tractors should fit whatever garden beds or pasture terrain is to be grazed.**

[Credit: Peter Bane]

any tractor should be furnished with water and extra feed. Chickens are naturally social: a minimum grouping would be three hens, though up to 50 can be kept in a 10 × 10-foot structure if it is moved at least twice a day. On pasture, birds need access to about four square feet each per day. If forage is less dense, lower stocking rates or more frequent movement are required. To eliminate vegetation, keep the tractor in place longer. If birds are to stay overnight in tractors (which is most convenient), they will need roosts. I also suggest a snug tarp covering to preclude a raccoon grabbing one of the sleeping birds through the wire.

A modified confinement/paddock situation can be created by placing the coop at the center of a matrix of four paddocks each affording the birds about a month's worth of grazing.

These could be garden areas rotated through annual crops where the birds are allowed to glean after harvest, manure and prepare the ground for replanting as part of the rotation. This system has the added advantage of breaking parasite cycles.

Despite their hot body temperature, a small flock of hens is difficult to carry through a cold northern winter without good shelter. In those circumstances, the coop should have insulated walls and roof and a deep straw litter within. You may wish to employ heavier breeds. In far northern climates, or in regions with very wet winters, ducks or geese will be better adapted to outdoor conditions.

### Ducks

Waterfowl are among the hardiest of livestock, being resistant to disease and tolerant of very cold temperatures. They have long been valued for their fats and feathers, choice gifts by which northern peoples have kept themselves warm. Needing minimal housing, both ducks and geese tend to be calm and, with the exception of a couple of duck breeds, quiet. The differing diets and feeding habits of waterfowl also make them valuable allies for pest control in the garden or on the farm.

Ducks range in size from bantam breeds weighing about two pounds on up to Muscovies, which can exceed 11 pounds, with most breeds averaging 7 to 9 pounds at maturity. Ducks can be very good egg producers, with Campbells and Runners equalling or exceeding the annual output of the heaviest laying chickens. Duck eggs tend to be higher in protein and a little larger; their stronger albumin makes them superb in all baked goods and somewhat easier to overcook. Ducks lay their eggs by 8 AM, so it is easier to collect them if the birds are confined until after that hour. As with chickens, a male is not needed for duck hens to lay, but if you wish to breed and raise your own ducklings, a ratio of one drake to six ducks is recommended.

Ducks greatly enjoy swimming water and prefer to breed in it, but can live, breed

Rotational Grazing, Pattern #51 (see Chapter 6) [Credit: Abi Mustapha]

## Duck Breeds

| Type | Breed | M | F | Egg/year | Color | Egg size, color | Other Characteristics |
|------|-------|---|---|----------|-------|-----------------|-----------------------|
| Egg | Bali | 5.0 | 4.5 | 150–250 | white/khaki | L, white/tinted | poor mothers, good foragers |
| | Indian Runner | 4.5 | 4.0 | 225–325+ | many | L, white/tinted | poor mothers, top foragers |
| | Khaki Campbell | 4.5 | 4.5 | 250–325+ | multicolor | L, white | poor mothers, top foragers |
| | Welsh Harlequin | 6.0 | 5.0 | 240–330 | whitish | L, white | excellent foragers, lean, clean carcass |
| Meat | Aylesbury | 9.0 | 8.0 | 35–125 | white | XL | fair mothers, placid, fair foragers |
| | Muscovy | 12.0 | 7.0 | 50–125 | black/white | Jumbo | good mothers, foragers |
| | Pekin | 9.0 | 8.0 | 125–175 | white | XL | poor mothers, fair foragers |
| | Rouen | 9.0 | 8.0 | 35–150 | multicolored | XL | fair mothers, good foragers |
| General Purpose | Ancona | 6.5 | 6.0 | 210–280 | pied | white, cream, blue | active foragers, calm, rare |
| | Cayuga | 8.0 | 7.0 | 100–175 | black/green | XL, black→wh. | hardy, quiet, good mothering, foraging |
| | Crested | 7.0 | 6.0 | 100–175 | white | XL | good mothers, foragers |
| | Magpie | 6.0 | 5.5 | 220–290 | pied | L, blue/green | poor mothers, active foragers, rare |
| | Orpington | 8.0 | 7.0 | 150–250 | buff, blue, | L | poor mothers, good foragers |
| | Swedish | 8.0 | 7.0 | 100–150 | blue, black | XL, white, Bl, Gn | good mothers, hardy foragers, calm, rare |
| | Silver Appleyard | 8.5 | 7.5 | 200–275 | chestnut/white/gray | L, white | broody, lean, flavorful, rare |
| Bantam | Australian Spotted | 2.5 | 2.2 | 35–100 | mallard type | M | top mothering, foraging; scarce |
| | Call | 1.8 | 1.6 | 25–75 | many | S | top mothering, foraging |
| | East Indie | 2.0 | 1.8 | 25–125 | iridescent black-grn | S | top mothering, foraging |
| | Mallard | 2.0 | 1.8 | 25–125 | multicolored | M | top mothering, foraging |
| | Silver Appleyard | 2.2 | 2.0 | 50–125 | chestnut/wh. | M | good mothers, foragers; miniature of S.A. |

Credit: Peter Bane

and lay well without. They must, however, have good access to plenty of drinking water, which should be available in containers deep enough that they can submerge their bills and clean their nostrils. Indeed, water management at every stage of the duck's life is perhaps the most critical factor to success. You can lead waterfowl around the farm by where you place containers of water; if they are working over a large area, more than one water fountain may be necessary. On open water, from 15–25 birds per acre on ponds should not cause damage to banks or overload the aquatic system with too much manure; 60 ducks to an acre of water would be a maximum for free access.

Ducks are omnivorous, eating a wide range of terrestrial and aquatic plants, insects, slugs, snails, grain, seeds, small fish, tadpoles and frogs. They will happily consume scraps from the table, the kitchen or the garden, including baked goods, root vegetables, bits of meat and fruits. You can add eggs or milk to their diet as a protein supplement, but any eggs fed should be cooked. In providing food to ducks, it's important that larger pieces be broken up, crushed or chopped, as the birds haven't the ability to peck foods apart as chickens do. Ducks eat by dabbling at things with their broad bills, which have slight serrations enabling them to pull and tear at foods. They use their bills to noodle around in mulch in search of slugs and insects; however, they should be kept away from young plants for which they have a great fondness and to which they can be very damaging.

Ducks are most at home on the water, and the lighter and medium breeds can fly fairly well, but their short legs are easily damaged and make them slow and awkward if not comical on land. Never handle them in this way, as you might a chicken, but by their necks. They can be confined by a two- to three-foot fence, but should be closed up in a protective shelter at night as they are then quite vulnerable and easily worried. They tend not to sleep at night, but to nap during the day. Do not provide them feed at night unless they also have access to drinking water, as they can easily choke. If they are confined to a yard after dark, their diet can be supplemented by suspending a light in the area to attract insects, and they will happily snap up free protein for hours on end. Muscovy ducks have been shown extremely effective in controlling flies, catching 87% of those in a 400-square-foot cage within one hour. The egg-laying breeds have a similar appetite and ability.

Ducks need about three to five square feet of housing each at night, in a well-ventilated space, and the floor should be generously supplied with dry and absorbent bedding material. They will not flourish in dirty housing or a muddy yard. Ducks are very cold-hardy and don't need shelter from the elements so much as for protection from predators and ease of management. If kept indoors around the clock during cold weather give them twice as much space. For a yard, about 25 square feet per bird is recommended. In hot weather, ducks need ready access to shade to prevent overheating.

Ducks are efficient converters of feed (between 2.5 and 3 pounds of feed per pound of weight gain) if provided with 16% protein in their diets, reaching a good size for butchering in 7–12 weeks from hatching. Between 12 and 20 weeks, birds will undergo changes in their feathering that make them hard to pluck well. If you do not butcher by 12 weeks, it's best to wait until the birds reach a fully mature size, though feed efficiency drops during this interval. Optimum egg production continues in ducks for two to three years, somewhat longer than in chickens, and the birds can continue laying for six to eight years at a diminished pace. As with chickens, ducks will begin to lay at about 22–24 weeks of age.

Under normal breeding cycles, ducks lay and incubate eggs in the spring, become broody and stop laying, hatch out and raise their ducklings and begin to molt during the summer, being again in good condition by early fall to go through the cold weather months. When stimulated to continue laying, they need assistance to complete their molting in the second year before cold weather sets in, so many growers force molting, which consists of subjecting the birds to a significant change of habit, diet and light levels.

The challenge to starting with either ducks or geese is choosing to hatch, brood or buy. You can buy fertile eggs for hatching, a clutch of ducklings or goslings for brooding or even mature birds all for delivery through the mail. You can also seek out adult birds locally, which has the advantage that you can see them before buying and also may be able to establish a relationship with the supplier. Most people buy hatchlings, and there are many fine details to successfully brooding them. Dave Holderread's books on waterfowl are superb.[6]

### Laying the Golden Egg

Geese are distinctive in being true grazers. They prefer succulent grass and forbs and eat them close to the ground. Though they will eat a few slugs and insects, geese derive most of their diet from plant material, including bulbs, seeds and fruits. In fact, they so much prefer fresh grass, clover and such forage that they will typically have little interest in many agricultural crops; they are excellent weeders for such crops as strawberries and potatoes. They can also be used to mow lawns and are

especially valuable at trimming grass and weeds from along fence lines, ditches and other hard-to-reach areas. It is quite possible to raise geese almost entirely on a homegrown diet, provided you are not seeking high egg production. Though not noted for laying large numbers, geese can produce formidable amounts of egg protein as each egg weighs up to five ounces: Chinese geese can lay over 100 eggs per year, equivalent in weight to average production from Leghorn chickens, one of the more productive breeds.

Geese are hardy, cold-tolerant and need little care, though they must have constant access to drinking water. Like ducks, they enjoy swimming water and prefer to breed in it, but can be raised without, though fertility rates, especially among the heavier breeds, are higher when mating takes place in water. Lead them around the farm with water. Geese are often kept in pastures with simple field shelters available. They can protect themselves from small varmints, but are vulnerable to attack by dogs, foxes, coyotes and large owls. If these creatures are present in the area, be sure to pen your geese in a shed or fenced yard every night.

You should always keep both male and female geese. They are typically monogamous and mate for life, but if widowed, can usually be mated again after a period of mourning. Sometimes a gander and two geese will form a durable trio. If breeding in a flock, about one gander for four to six geese is recommended.

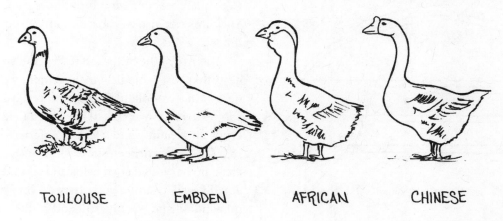

TOULOUSE   EMBDEN   AFRICAN   CHINESE

Four Major
Breeds of Geese
[Credit: Jami Scholl]

Geese as well as ducks are famed for the high fat content of their carcasses. Fat is part of the waterfowl's adaptation to cold weather and exposure to water, but the amount of fat ducks and geese put on is dependent on diet to some degree. When fed corn and other concentrated feeds, young ducklings and geese will develop fatty carcasses, but if allowed to feed on forage, they will be much more lean, taking longer to acquire a full complement of body fat.

Goose down, the fine inner feathers of the bird, is the world's premier insulating fiber, though duck down is also well-regarded. Both species of birds can be plucked live, though this is easier with ducks than geese because of the size and relative strength of the birds. If plucking a live bird, do so early enough in the season that it can restore its full feather complement before winter. If you raise waterfowl to eat, it will be worth your effort to pluck the carcasses carefully of the down — you have to get it off anyway unless you skin the bird — because it makes excellent pillows, quilts and insulated clothing.

### Turkeys Glean Woodlands and Fields

These large North American birds were first domesticated in Mexico or Central America about 400 BC but have been taken to many other parts of the world. They are nearly as common in Europe today as in the US or Canada. Heritage breeds bear the more streamlined profile of wild birds and can still fly; indeed seeing a flight of turkeys pass over the tree tops is a stunning and magnificent sight. The turkey is an omnivorous forager, ranging farther than chickens and tolerating heat and arid conditions better. Turkeys eat chiefly acorns, beechnuts, chestnuts, grubs, seeds, bulbs and insects in the wild, though they will also consume small reptiles, rodents and amphibians as well as fruits, producing a carcass with a higher proportion of meat than a chicken, and of very high quality.

Most turkeys are raised commercially in confinement for the holiday trade, but they can be raised on range. Scavenging birds are quite well-known in Mexico, wandering through the village during the day and returning to their roosts at night so long as they are provided with water. Turkeys do well on legume or grass pasture seeded to alfalfa, bluegrass, ladino clover or bromegrass. They also have been raised well on annual plantings of soybeans, sudan grass, sunflowers, rape, kale and reed canary grass (*Phragmites australis*). The birds will forage green matter from sunflowers, reed canary grass and sudan grass and can benefit from the shade provided by these tall plants or by a few rows of corn. Some shade must be provided during the hottest months. If the range is fenced against predators, open, covered field shelters with roosts provide sufficient cover; allow about two square feet per adult bird under roof. The birds can also be roosted on the ground if it is dry.

The size of the birds and their habit suggests that best ecological management would be within a woodland setting or forest garden supplemented by access to harvested field crops for gleaning and late season fattening. Turkeys have been successfully used to clean up harvested corn fields and should do equally well in sunflowers. They do very well on windfall and wormy fruit and could help

Turkey Field Shelter
[Credit: Abi Mustapha]

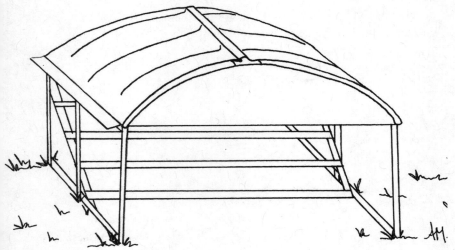

to maintain orchards free of pests. Heritage breeds are likely to be hardier and more disease-resistant, but clipping their wings may be necessary to prevent flight. Though large and not bothered by most smaller varmints, turkeys need protection from canines and large owls, so any management scheme should provide secure overnight enclosure.

Unless you hope to manage a breeding flock, raising a few turkeys needn't be overly complicated. Buying and raising poults for slaughter late in the year could be a successful part of a household food plan if enough vegetation, surplus fruit and spoilage is available. The turkeys could be confined to a fenced pen with about 25 square feet each, allowed to range in selected parts of a perennial system using movable electric fencing and returned to a secure nighttime house for protection against predators. It's possible to raise most of the concentrated protein and energy food turkeys need using a formula similar to that for chickens or ducks: corn and sunflowers, squash, sorghum, leafy greens, waste fruits and plenty of insects, grubs and snails. Where bearing fruit trees are established, the turkeys could be finished on the fallen fruits of autumn just prior to the holidays. At about 20–25 pounds the birds are easier to manage and harvest than hogs, and have some of the same diet with a lighter impact on the land.

## Guinea Fowl Guard the Flock

These African grassland natives (*Numida meleagris*) are similar in size (averaging 5.5 pounds) to larger chickens or ducks and can be run in mixed flocks with either or both. Guineas have a compact body shape and are determined foragers of insects. Their feral instincts incline them to roost in trees. But guinea fowl are also notably noisy, having a very loud and raucous call that announces the arrival of guests or any strange creature coming into the yard. While not burdened with intelligence, they are quite unintimidated by larger animals and will attack almost anything

Guinea Fowl
[Credit: Abi Mustapha]

that threatens them, including venomous snakes, which they have been known to kill. This combination of traits makes them excellent watch animals for a poultry flock, and many times a guinea has saved the lives of less suspecting chickens, ducks or turkeys.

Guinea fowl are hardy in both hot and cool climates and are relatively free of diseases. They will not, however, lay until seasonal temperatures exceed 59°F. They forage well on grain, leaves, ant eggs and other insects and may also eat carrion. When foraging, they do not scratch, and so can be let in among crops with much less damage than chickens will cause.

Guinea hens will begin laying at around 18 weeks of age and will continue laying throughout the warm season. Production depends on diet, with free-ranging hens laying about 60 eggs a year; intensive feeding can triple that number. The eggs average about 1.5 ounces, slightly smaller than chicken eggs but with a stronger shell. Laying can be delayed til about 32 weeks of age by holding the birds in windowless housing and controlling their access to light. This improves egg size and the rate of hatching and survival of the offspring. The young birds, called keets, will hatch out in about 27 days. However, the parents are almost incapable of raising them. Keets are often fostered under chickens to improve their chances of survival, and may be kept indoors for three or four weeks after hatching to protect them from predators and rain. While they live in groups, guineas are monogamous by nature and tend to form breeding pairs. However, in a flock one male can take care of four or more females.

Guineas never become tame, but neither will they leave the farm. If confined to the poultry yard until 12 weeks of age, they will continue to return throughout their lives. At first, they are unaccustomed to foraging for their food and seek the supply provided for them, but eventually they learn to scavenge. The birds range farther than chickens, preferring to lay their eggs in the bush in small hollows dug in the dirt.

Guinea fowl are hardier than chickens or ducks and well adapted to dry conditions. They prosper under semi-domestic management where they are allowed access to range, can use a great deal of green matter, scavenge for insects and seeds and are much better able to defend themselves against predators. This means their cost of production can be quite low. They also have better resistance to common poultry diseases. However, under intensive management, the birds are less efficient at converting feed to meat than broiler hens or young ducks and do not reach a marketable size until 12–14 weeks. The meat is dark and delicate, resembling that of wild game birds, and brings a premium price.

As part of a mixed poultry flock, guineas offer the opportunity to harvest wastes and wild feeds from the landscape with minimal management. At the same time they can do a great deal of pest control and provide early warning to companion birds against predators. In dry or wilder landscapes and where broader acreage is available, they should provide the small grower with more resilient poultry, though at a lower level of productivity.

### Pigeons Forage Widely

Eaters of seed and scraps, the pigeon (*Columba livia*) is a prodigious and swift flier long in domestication and trained to carry messages over great distances. The young of pigeons (squab) grow rapidly and produce large tender breasts of meat that is easily digested. It is considered a delicacy and commands a premium price. Because they require little space and thrive in urban environments, pigeons have considerable promise as city livestock.

Pigeons will range over many square miles to feed themselves. Under intensive culture, pigeons are fed whole grains and pulses, but on their own they will also eat leaves, fruits, meat scraps, bread, insects and small invertebrates. Whatever their dietary regime, pigeons need a continuous supply of fresh water to drink and for bathing at least once a week. They also need small grit for their crops and for minerals. Pigeons have little smell, and if good hygiene is practiced to absorb their manure in deep litter, they can be among the least troublesome of birds to keep.

Pigeons raise two 1- to 2-pound squab per cycle; these are harvested between 21–30 days of age, before their flight feathers develop. A breeding pair can be expected to produce 6–8 clutches per year. Though not continuous, squab production can provide a steady supply of good meat for the household or market through much of the year. Selected breeds have been developed for racing and squab production, and these should be sought out by growers with a market interest as they will outproduce ordinary birds by a wide margin.

Traditional Nepali
Mud-Brick Farmhouse
Showing Pigeonholes
[Credit: Peter Bane]

Pigeons in the wild nest in cracks of cliff faces while domesticated birds prefer small *pigeonholes* below eaves and in building walls. In traditional farmhouses in Nepal, a few bricks are omitted from the outer courses of mud-brick walls. The birds nest there, taking off over the dooryard gardens into which they drop their rich manure.

Though at home in temperate and tropical countries, pigeons do not produce well in very cold climates, and in hot humid environments are subject to more pests and diseases. They prefer nesting shelves, two for each pair. A small board can be added to the front of these to prevent eggs and squab from falling out. For farm operations, a small shed can be adapted to house a dozen or more pairs. Pigeons are prey to every imaginable predator; care should especially be taken to exclude rats from their nests.

The stories of pigeons as racers and carriers are legendary. The US employed 100,000 pigeons in the Army Signal Corps during World War II, while Britain used five times as many. Birds can travel up to 600 miles in a day and even as much as 1,000 miles in two days, often traveling long distances without food or water en route. Though subject to occasional losses from predators, hunters or other airborne hazards, pigeons are otherwise extremely reliable messengers. In Normandy, a regional hospital even uses them to deliver blood samples from outlying clinics along the coast to a central laboratory. It is estimated that they save hundreds of dollars a week that would otherwise be spent on gasoline and vehicle repairs.[7]

### Quail Fit in Small Spaces

Four subspecies of these small birds (*Coturnix coturnix*) are native to Eurasia and southeastern Africa and have been farmed for meat and eggs since ancient times. Despite their long domestication, the Japanese quail, which is the only cultivated type, will readily cross with wild quail, producing fertile hybrids that can themselves be backcrossed either to wild or domestic birds. Feeding on insects, grain and other seeds, quail need a fairly high-protein diet to flourish. Females, which are slightly larger than males, reach about one pound at maturity and may live five years, though peak egg production occurs at about six months of age. The birds can lay 200–300 eggs per year, but age rapidly if they do so. Females mature in five to six weeks and reach full egg production in seven.

Quail meat is dark and slightly gamier than chicken, but the eggs, which average about ⅓ ounce, are similar in flavor to chicken eggs. A good market exists for both meat and eggs, but the birds are not suitable for free-range management; they must be confined or they will escape. They have no homing instincts, nest on the ground and are subject to severe predation if not protected.

Quail hens make poor mothers, so the grower wishing to raise quail for meat should expect to provide artificial incubation. Quail eggs can be incubated under bantam chicken hens, but are subject to minute fractures that can allow the embryo to dehydrate and die. It may be possible to incubate quail eggs successfully under bantam ducks, which maintain more moisture in their breast feathers. Quail are resistant to Newcastle disease, common among other poultry.

Despite their vulnerability, quail are relatively easy to raise and take very little room. Many growers feed them on chick mash, which is higher in protein than ordinary poultry feed. They can be kept in a small enclosure similar to a rabbit cage at a ratio of one male to six females.

## Small Mammals

Small animals are all of potential interest to the garden farmer because they can so quickly accelerate soil building. That they can turn wastes into meat, eggs, honey, fur and other useful goods is a wonderful bonus. Among mammals, first place goes to the rabbit, a species so prolific that permaculture teacher Dan Hemenway has written that rabbits would be the perfect domestic livestock if only they laid eggs. They don't, but they do the next best thing: they make lots of bunnies.

### Rabbits: Putting "Not Meat" on the Table

Native to the Iberian peninsula, rabbits were not taken under domestic management until sometime during the Christian era, probably by monks, and then became popular after the Pope in 600 AD declared their flesh "not meat" and thus suitable for consumption on fast days. Cultivation spread quickly. They reached Britain in the 14th century where they were called *coneys* after their Spanish name, *conejo*.

Rabbits are among the most economical and productive of all livestock. A ten-pound mature female of a large breed such as a Satin, New Zealand, California or Flemish produces a litter after one month gestation and can raise eight bunnies to five pounds of gross weight each in 12 weeks. Such young rabbits dress out at 3.5 pounds for a total of 28 pounds of clean meat in less than four months. One doe can triple her body weight in live protein with each litter and can raise four litters a year without strain. A small rabbitry can easily provide an adequate supply of high-quality meat for a household.

More than 40 breeds of rabbit are recognized by the American Rabbit Breeders Association,[8] and some types have been developed for virtually all conditions, both tropical and temperate. Rabbits need protection from cold drafts — despite their fur coats — and in hot climates must be kept out of the sun in well-ventilated conditions. They die from heat and drafts, not generally from cold temperatures. Rabbits shed heat through their ears, so the longer the ear, the more tolerant the animal will be of hot conditions. There is considerable genetic diversity among rabbits — though not equally in all breeds — and almost every breed has its adherents. Though commercial rabbitries exist and the meat is used in finer restaurants, the rabbit remains the small grower's livestock par excellence. The investment to set up a rabbitry can be less than $200, and the returns are rapid. As a commercial venture, rabbit-raising may be subject to considerable market swings and whims, but as a subsistence activity, it can't be beat. Rabbit breeding (which can develop very rapidly for obvious reasons) is much in the hands of devoted amateurs and small commercial growers.

Domestic rabbits can be fed grass, leaves of herbs, trees and shrubs, crop residues, legumes and kitchen scraps. If given too much soft vegetation they can develop diarrhea, so while Peter Rabbit and his literary clan favored lettuce and cabbages, your rabbits should get only a bit of that and much more roughage. Commercial feeds are based on alfalfa, now threatened with GM contamination, but other leguminous plants such as cowpeas, clover, vetch or field peas may be substituted. I have successfully fed rabbits a mixed diet including stems and leaves of tulip poplar (*Liriodendron tulipifera*). Many other trees and shrubs are suitable fodder, but if you don't know whether your rabbits will do well on something, introduce them to it in small

A rabbitry requires separate cages for each breeding animal and extra cages for maturing young. Shade and windbreak are essential.

[Credit: Peter Bane]

amounts. If using non-commercial feeds, salt must be supplied. Lactating females and newborns require a diet of about 16% protein and 18% fiber, and a large doe with kits will eat about eight ounces of feed a day.

Rabbits are known to enjoy twigs and foliage of fruit trees and shrubs, especially of the rose family, which they eat thorns and all. Kenaf has been reported as a suitable feed, which suggests that other Malva-family plants such as hibiscus, hollyhock and okra might be appropriate. Above all, be sure that rabbits get a generous portion of fiber in their diets: this helps keep their digestion in good shape.

Rabbits begin breeding between four and six months of age and may continue for 4–6 years. Breeding adults should be housed separately, the doe always taken to the buck's cage for mating. The gestation lasts 30–31 days. A nest box large enough for the mother to avoid stepping on the bunnies should be provided at about day 26 from mating. The mother will pull fur from her belly and breast to line the box, exposing her nipples. Does can conceive again within 24 hours of giving birth and can produce a second litter within four weeks, however this is not usually done. The ideal is eight bunnies per doe. If faced with too many babies, a mother may become so stressed that she kills them all.[9]

Young open their eyes at ten days, but remain in the nest box until two to three weeks of age, then begin to move about. By three weeks they can eat on their own. The bunnies will continue to act as a litter for three to four months, but should be separated by sex if not harvested at 12 weeks. At around four months they begin to become rowdy and territorial. Maximum weight gain and feed efficiency occurs in the first eight weeks, but harvesting at 12 weeks produces a larger carcass.

Rabbits can be handled with relative ease if accustomed from an early age to human beings; they are easy to kill and clean. It is best to begin handling those selected as breeding animals or pets at about three weeks. Hold the bunny on its back and stroke down the belly toward the tail for short periods; this is very calming. Rabbits always urinate and defecate in the same place, usually in one corner of their cage. They have 18 very sharp claws and though they seldom bite, they will kick and scratch if frightened. Always protect your forearms when handling rabbits, especially when breeding them, as the female may be skittish or uncooperative.

Most of the problems rabbits have in captivity can be managed by good hygiene. If manure becomes too runny, feed the rabbits some coarse hay or other fibrous, non-watery vegetation. In North America, the animals are usually kept in wire cages that allow the droppings and urine to fall through, but under more primitive conditions they will do well in a packing crate or similar enclosure. Cages must be cleaned regularly and thoroughly to limit disease, and the animals must be kept dry. While occasional direct sunlight is helpful for sanitizing the cages, you must guard against the animals overheating during warm months. Never house them on the west side of a building or exposed to western sunlight as it is usually too hot, and also that is very often the direction of prevailing winds.

Some growers like to raise worms under rabbits; poultry can be run beneath them as well to harvest insects from the manure piles. Rabbit manure is almost perfect plant food, whether applied directly to the garden or made into a tea for houseplants or seedlings.

A minimum rabbitry would consist of a buck and two does, each housed separately. Even when rotating the breeding of only a few does, rabbit production will result in temporary surpluses of meat, so this implies the use of a freezer or other effective means of storage. Rabbit carcasses can be quickly and rapidly skinned and cleaned when warm, and the pelts make excellent garments. To monitor nutrition and herd health always check the color and condition of the liver when butchering — it should be dark red-pink and free

Rabbit Tractor
on Pasture
at Polyface Farms
[Credit: Peter Bane]

of defects or spots. If you keep dogs, the guts make a good food for them.[10]

While keeping caged rabbits is far and away the most common method for managing these animals, they have been run on pasture in tractors and some growers are experimenting with yarding. The problems with those systems involve the rabbits' tendency to burrow and thus to escape. Polyface Farms in Virginia has successfully run rabbits in open-bottom tractors by laying out netting on the ground well in advance of the animals. The grass grows up and rebounds easily above the wire when a tractor brings the rabbits over it.

Systems have also been devised to allow easy harvesting of wild rabbits and to give them no other care at all. As rabbits graze on meadows and pastures during the day and retreat to their warrens at night (often in brushy ravines), they have predictable patterns of movement. By placing a fence across the tidal zone of rabbit movement, they are forced to cross it. If provision is made in the fence for gates and those gates are live traps, the rabbits can be accustomed to entering the live traps when they are kept open on both ends, passing through on their daily rounds. Then when rabbit is wanted, the traps can be set to catch the animals. This system, which was implemented at Ragman's Lane Farm in Gloucestershire, England, is an excellent example of permaculture design, using least effort for greatest results.

### Guinea Pigs—the Indoor Livestock

Native to the central Andes mountains of South America, *Cavia porcellus* are kept and eaten in large numbers in Peru and Bolivia. They could be of much greater importance elsewhere if more widely known. Small, clean and extremely docile, these rodents are often kept indoors. They have great potential as urban livestock, even for households with little or no land. With no odor, minimal housing needs — they can be contained by a 4-inch board — and needing only kitchen scraps and weeds as food, as few as 20 females and two males could provide an adequate meat diet for a family of six.[11]

Low fencing herds rabbits through opened live traps to which they become accustomed. Setting the traps then enables easy harvest with no feeding costs.

[Credit: Jami Scholl]

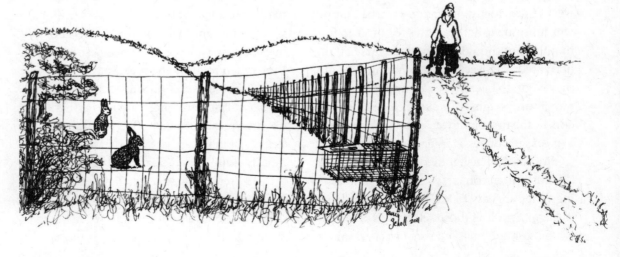

Though most guinea pigs sold in the pet trade are of smaller varieties, improved types have been developed in Peru that can exceed 4.5 lbs at maturity — the size of a medium-breed rabbit. Those with curly hair are preferred in Peru for meat production; they are quieter and stockier than animals with smooth pelts. The cavi, as it is also known, enjoys green shoots of grain and legume hay. And they need a supply of fresh greens or fruits to supply vitamin C which they cannot synthesize. They are not hardy outdoors in freezing temperatures — presumably they live in burrows or other simple shelters in the wild — and must also be protected from high heat and sunlight. They tolerate living in wooden boxes with little or no light.

Cavi, the Humble Guinea Pig
[Credit: Jami Scholl]

Guinea pigs mate at all seasons and average two to three young per litter, sometimes more. Gestation takes 65–70 days. The females come into estrus every 13–24 days and are fertile shortly after giving birth. Sexual maturity occurs at about three months, and the females may have four litters a year for up to four years. Newborns are large and are born fully developed with fur and open eyes. Resembling adults in miniature, they can start eating grass and other feeds within hours of birth. Weaning may occur as early as 21 days. The young gain weight rapidly to about six weeks, but then slow down, reaching a harvestable weight in 10–13 weeks. They dress out at about 65% of body weight. Despite small litter size and relatively long gestation, the rapid maturing of cavis makes them potentially very productive when raised for meat. Their conversion of feed ranges between 3–6:1.

Cavis communicate among themselves relentlessly by squeaks and other noises. Males are good-natured around other animals, but may fight among themselves. In some parts of Peru, they are ranged during the day and herded into adobe coops at night. In captivity, they need very little space and can be kept on a litter of wood shavings, dry corn cobs, shredded paper or straw. Their droppings are odorless, but if fed extensively on green matter, they produce quite a lot of urine, which needs extra bedding or more frequent changes.

The presence of sizeable Peruvian communities in the US means not only that there is likely a good market for cavi, but that the improved varieties will soon be available, if they are not already here. There is considerable cultural resistance in North America to eating animals that people have become accustomed to thinking of as pets, but cavis are simple to

## RECIPE FOR CAVI OR CUY

3 or 4 cuys, dehaired, gutted and cleaned
¼ cup of ground toasted corn, or cornmeal
4 lbs. of parboiled potatoes, cut in slices
8 cloves of garlic
6 fresh hot peppers, either red or yellow
½ cup oil
½ cup water
salt, pepper and cumin to taste

1. Rub the cuys with a mix of the pepper, salt and cumin and bake. You can also skewer over a barbeque.
2. Prepare a sauce with the oil, peppers, garlic and cornmeal with the water from the potatoes or broth. Cook a few minutes until the peppers are done. When tender, place the meat in a serving dish and spoon the sauce over it. Serve with the boiled potatoes.

Credit: Bonnie Hamre

care for, provide a good-quality meat in small portions easily consumed and without need for storage, and should, because of their small size and unassuming appearance, be easy to view and raise impersonally.

And unlike rabbits, which people are used to eating, and around which we have created mythologies of adventure (Bugs Bunny) and saintliness (the Easter Bunny), we have made no anthropomorphic projections on guinea pigs. Bon appetit!

### Hogs Make the Fat

Swine (*Sus scrofa*) are intelligent, adaptable and economically valuable animals, and they deserve a whole book (of which many have been written).

Pigs have gained favor in production agriculture in the past 40 years because they put on weight fast, mature rapidly, produce large litters and are thus very cheap to raise. Small farmers have always valued pigs because they can be fed on wastes, are self-reliant, friendly and loyal and are at least as smart as dogs. When people in the west country of Britain and in Ireland went visiting, they often made a point of calling on the family pig as much as on their human neighbors.[12]

Because of their temperment and characteristics, pigs make good working animals in many situations. Their keen sense of smell makes them the animal of choice for locating (and unearthing) the underground fruiting bodies of the choice truffle mushroom. They have been used to turn the bedding pack of stabled cattle. After the cattle are returned to pasture in the spring, pigs can be turned into the barn; they will root down searching for sweet morsels of malted grain that the farmer has scattered into the pack as it accumulated.[13] This tendency to root can be harnessed not only to turn compost, but to dig out perennial weeds, shrubs and even tree stumps as well as to make terraces. A little fencing and a little food, and off they go. When confined to a newly excavated pond, pigs can help to seal it (called *gleying*) by laying down their manure and wallowing to compact the bottom. The animals are, despite their common association with slop and mud, relatively neat when given the right space. Because they have no sweat glands and are naturally woodland dwellers, they rely on shade and water or cool soil to regulate their body temperatures.

Pigs are true omnivores and flourish on bulbs, roots, insects, rodents, grain, food and garden scraps, as well as bitter fruits and even ruminant manure (they seek out the undigested grain). Permaculture designer Mark Cohen raised pigs as a sideline to his main crop of winter squash seed. The prime fruits went to market, the ripe but damaged or misshapen fruits were opened for seed (sold under contract to a seed merchant) and the resulting cut squash went to feeder pigs. On less than two acres in a narrow Ohio valley, Mark made $8,000 in one season between the seed, the squash and the pigs. Unlike many other livestock animals, they put on fat in adolescence and continue to build up reserves throughout their lives. Among the pig's many fine qualities, its fat may be the most valuable. Lard is a high-quality cooking and eating fat that melts easily and stores well. The leaf lard from the back around the kidneys is considered to be the finest, but the pig's flesh is

[Credit: Peter Bane]

naturally well-marbled, fat occurring in layers throughout the animal's compact body. Pork is also among the easiest of meats to preserve, needing only salt or immersion in melted fat, though it is often smoked.

Consolidation of meatpacking and large Confined Animal Feedlot Operations (CAFOs) have driven down the price of commodity pork, but a small producer on less than three acres could make a decent living by specializing in certain niches within the industry, particularly breeding high-quality animals. There is also a well-established market in small pigs, called *feeders*, and you can buy one or a few to fatten and finish as desired, whether for the freezer or the market. Small orchardists might find feeder pigs very useful for cleaning up waste fruits and pests under the trees. A pair could convert the wastes and leftovers of a CSA or summer market garden into winter meat and fat for the farmer. Because of their small footprint, anyone with access to spoiled food and a small amount of land can raise pigs. A venturesome tenant farmer with a ¼-acre lot, a small outbuilding and bicycle or pickup access to local restaurants could finish a half dozen from the fine-dining wastes of any small city. Raise a heritage breed famed for flavor, and you could sell the slop back to the restaurants converted into gourmet ham, loin chops and spareribs. Pigs are quiet, and they need not smell. Worms and worm compost for urban gardeners could be a sideline. Anytime you had more worms than you could sell, they could become pig food.

The pig matures sexually at four to six months of age (in smaller breeds), estrus occurs throughout the year at three-week intervals, and a sow will have between 6–10 piglets. Gestation lasts "three months, three weeks and three days." The greatest cause of death in young piglets is being crushed by their mother, followed by diarrhea (called scours in pigs) and starvation. Farrowing cages are designed to prevent the sow crushing the pigs while still allowing them access to her.

Farrowing Cage
[Credit: Jami Scholl]

The sow's milk is the richest of any common barnyard animal — contributing to the rapid growth of the pigs — and once the piglets sort out dominance, they will adhere to the same teat until weaned. This can take place within a few weeks of birth, and typically animals raised for meat will be slaughtered at six to seven months, matching the annual season of growth in temperate regions. As with most animals, handling the young from an early age makes their management easier as they mature.

In many traditional cultures from Mexico and Southeast Asia to Minnesota and the Appalachian Mountains, a pig was the household's insurance against a hungry winter. Let about to scavenge in the village or the woods for its dinner and drawn back to the house by affection and offerings of scraps and leavings, it would eat whey from cheesemaking, spoiled and leftover foods as well as grubs, vermin, snakes and toads, dead poultry and whatever fruits, vegetables and carrion other animals would not eat. In our southern mountains, it would happily clean up the pecans, the chestnuts or the apples complete with worms, and such meat became a delicacy. Over the months of summer and early autumn, almost magically the pig grew fat. And with the onset of cold weather in November, it would be sacrificed in an all-day community ritual — a hog slaughter — that drew together the whole neighborhood or clan to skin, butcher, render, smoke and salt meat, and to grind and make sausage, head cheese, chitterlings and a vast array of delectables. The feet were pickled, the intestines stuffed for link sausage, and every scrap of fat not left on cuts of meat would be melted down and clarified to make lard. It has many times been said that the only part of the pig that isn't eaten is the squeal.

Ranging from 25–150 pounds, small pig breeds will be most useful for small properties. Even the tiniest of these animals — at 25 pounds, the Cuino of the Mexican Highlands — retain the virtues of large hogs: fast reproduction and weight gain, self-reliance and even temperment. An advantage of smaller pigs is their ability to withstand heat (bigger surface-to-mass ratio), lesser need to wallow and relatively clean habits (a pig will select and use one corner of its pen as a toi-

A Young Guinea Hog
[Credit: fishermansdaughter via Flickr]

let). The Guinea Hog, topping out at 130–150 pounds, is a lard breed — one included in the Slow Food Ark of Taste for its fine flavor — and has garnered considerable attention from permaculture farmers here in Indiana who have sought it out for its gentle disposition and ease of management. As a heritage breed, its fertile offspring also bring a premium price as demand for this useful animal is high. A single low strand of electric fencing can contain the herd. The sows give birth easily and will raise from four to eight pigs.

Because they are woodland animals, pigs, especially small ones, are compatible with perennial polycultures, though their movement must be contained. Pork and chestnut were the traditional meat-and-potatoes of southern Europe for many centuries.[14] Famed Austrian permaculture farmer Sepp Holzer runs Mangalitza and Turopolje hogs — two cold-hardy Central European breeds — in his alpine orchards to eat pests, harvest windfall fruit and churn up cobbles from the rocky soil. (The stones help gather extra solar radiation to ripen fruit at 4,000–5,000 feet.) Holzer leaves his animals confined among a patch of trees until they have removed all the other vegetation and are eyeing the cherry or apple trees hungrily; then he puts them in a new paddock. The hogs live in south-facing artificial caves that he has built on his mountainside terraces using trunks of blow-down spruce trees that used to cover the land. With deep straw bedding and the rest of the small herd for warmth, the pigs go comfortably through the winter with no supplemental heat.

Needing little exercise and with dominance hierarchies easily established within litters, pigs are among the easiest livestock to keep in small spaces. Many ecologically minded farmers provide small field shelters against rain and snow that allow the animals access to fresh pasture, but the traditional sty is not unkind to the pig. The animals should, however, be moved periodically to prevent a buildup of parasites in the soil.

Hog manure, which has gotten a justifiably terrible reputation due to the abusive practices of CAFOs, is valued in the Far East for fertilizing ponds, and there pigs will often be housed on a platform above the pond in order to deliver the manure directly to where it will be eaten (by plankton, which are in turn eaten by fish). At the experiment station in Longdenville, Trinidad, hogs are confined and their manure is collected in troughs that feed to a digester where bacteria produce biogas (methane). After the gas is drawn off, the residue is then used to fertilize trees.

Pigs, like poultry, can exchange diseases with humans; influenza is the most common. But as with any creature, sanitation makes the principal difference in whether health is maintained or illness fostered. In confinement operations, where hogs never see the light of day, a slow drip of antibiotics keeps acute illness at bay, at the cost of breeding more drug-resistant bacteria both in hogs and in the humans who eat them. We should object more strenuously to this dangerous practice — entirely a product of industrial agriculture — which stands ready to loose a pandemic on the world at short notice. Neither pigs, per se, nor their time-honored relationship to humans are the problem; rather, the problem stems from too-large-scale and corporate detachment from consequences.

## Goats — Milk from Wastelands

Asked about goats, famed grazing expert Joel Salatin said, "I won't keep animals who are smarter than I am." On the other hand, goats have earned the scorn of permaculture author Bill Mollison, as "maggots eating the landscape" for the role they play in preventing regeneration of forests across much of Africa and the Near East. The conflicts inherent in these comments point to the animal's potential and its management challenges. Jim Corbett, who wrote a 1991 adventure tale of a man and two goats, made clear what nomads the world around have known for eight millenia:

South-facing, earth-sheltered "caves" provide winter housing for Sepp Holzer's pigs in the Austrian Alps.
[Credit: Peter Bane]

The above-ground portion of a digester that turns hog manure into methane and fertilizer.
[Credit: Peter Bane]

goat's milk provides a near complete diet for a human being, and two of the animals will support an adult.[15] When the land is poor and even more so when you own no land, a herd of goats is a powerful form of security.

Along with pigs, goats are one of the land manager's prime biological tools for converting scrub and waste land to cultivation. Goats can be brought onto overgrown, brushy ground to turn thorny and even toxic plants into meat and milk. The animals can work their way 20 feet or more into the canopy of a tree to eat leaves. Goats do not do well on succulent and rich herbage, except in small amounts, preferring the twigs and stems of shrubs and trees to soft grasses. The range of what they will eat includes many species that are problematic in the inhabited landscape such as multiflora rose, honeysuckle and poison ivy. They can be used to open up tangled thickets and reduce woody perennials to a manageable size that can then be uprooted by pigs. For the same reason, goats are more difficult to use in orchards than sheep, poultry or even pigs, because they can climb the trees. When run in mature orchards in Germany or olive groves in Spain, they are fitted with halters that prevent their eating high on the trees.

Like any other animals, however, goats will rarely go far out of their way or work more than needed to feed themselves. So close management and providing adequate fodder are key to preventing them from damaging cultivated systems.

Goats are notorious escape artists, and can only be confined by quite substantial fences. The only hedge that can endure goats for years on end is one of blackthorn or sloe plum (*Prunus spinosa*).[16] Other hedge plants may deter passing traffic, but will not confine goats. Many keepers resort to a tether, which

Halters are used on goats ranged in orchards to prevent them climbing the trees.

[Credit: Abi Mustapha]

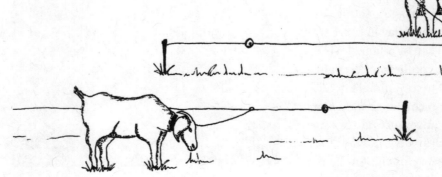

Goats tethered to a wire are unlikely to tangle themselves.

[Credit: Abi Mustapha]

can give satisfactory service if a few concerns are taken into account. Several goats should be tethered in the same area for companionship, and may be on the same wire if boards are stapled to it to divide their grazing areas. If goats are tethered out in wet or windy weather, provide them a shelter at the windward end of their range.

Despite improved breeding, goats remain little changed from their wild ancestors. Leadership of the flock is shared between the king billy, based on his strength and courage, and the old she-goat who is the queen of the flock. Both are selected (or acknowledged) by the whole group based on evidence of intelligence, strength and agility. For this purpose, the flock must be allowed some form of communal life. Separating individual goats from the flock or preventing them from selecting a queen will simply make them neurotic and dependent on their human "leader." Goat instincts for herd obedience and cooperation exceed their discernment of what foods are best, and it is through education by their mothers and the guidance of the wisest female that the flock remains in good condition.

If goats are threatened, they turn, face the threat and base their defense on superior agility. They are sure-footed and nimble from thousands of generations of climbing rocky mountains and can outstep most predators. Unlike sheep or cattle, goats cannot be driven. Across Eurasia where goats are herded, small boys from seven to twelve years of age easily lead the instinctively orderly goats, milk them and keep them out of trouble. You may discipline goats by flicking a bit of water at them (they detest getting wet), and thereafter, a flick of the wrist will express the same displeasure. But it is difficult, once you have assumed the role of leader, to step out of the role. In wooded country, the herder may give the flock the slip and they will carry on, but in open country, he will have to calmly and without rudeness throw a stick to indicate that he is stepping down from his role as the

king billy. Goats, of course, consider this to be ridiculous monkey behavior and immediately recognize that the human is not one of them.

Goats flourish under a regular and peaceful routine. If they are to forage, lead them on a series of circular routes that avoid places you want them to stay away from (the neighbor's fields, your garden), and after a few weeks they will know their beat and can then be expected to stay on it, returning at nightfall to the comforts of their shed.

In cold climates, the breeding cycle of the goat is annual. Does come into estrus when the light begins to decline at a certain rate. In southern Europe and Anatolia, latitudes equivalent to the northern US and southern Canada, this happens in mid-October and kids are born in April, six months later. In far northern regions, breeding can occur in early August and kidding takes place in the harsh conditions of January or February. In tropical regions of more steady light levels, breeding can take place at any season. Billy kids become fertile by three months and so must either be butchered or segregated from their sisters by that age if to be kept for breeding. Many goat keepers simply drown the male kids at birth rather than feed them, though a young goat can put on 20 pounds in 10 weeks and makes a very worthwhile carcass in that time. Both mother and kid will be more content and well-adjusted if the little one is permitted to suckle at least four to five days before being weaned to a bottle.

Goats are more efficient producers of milk than either cattle or sheep and can eat far more in relation to their body size. A 140-pound goat will typically be carrying around 50 pounds of browse in its rumen. Goat milk averages slightly more butterfat (3.8%) than cow's milk, and is considered more digestible because its fat particles are smaller and more emulsified. A milk cow will produce about two pounds of milk from each 100 pounds of body weight, while a goat will produce five pounds from the same weight.

A wooden box provides goat beds within and on top.
[Credit: Abi Mustapha]

But to yield that, the goat must eat a diet from which the cow could derive little value. It must also have from five to six gallons of water per day, and that should be warmed a little if possible. Cold water lowers the internal heat by slowing fermentation in the rumen.

In most areas, if goats are allowed free access to range they can meet their nutritional needs quite easily, but mineral deficiencies can occur if they are confined to pasture or fed concentrates. Goats need more salt than cattle, and an adult will consume 15–20 pounds of it per year. Iodine too is an essential nutrient. Goats, especially young ones, will sometimes eat dirt to get copper and iron. Cobalt can also be added to pastures or crop fields at two pounds to the acre by mixing cobalt sulfate with sand and broadcasting it, or a salt mix can be made for the goats from one ounce cobalt sulfate in a half-pint of water, used to wet six pounds of salt. Dry this mix and allow the goats free access. Another way to get trace minerals into the goats is to mix about six ounces of seaweed meal into their daily feed ration. Chicory in paddocks is another excellent source for all these minerals, and given access to enough of it, goats seem not to develop mineral deficiencies.

If fed on rough browse, goats can stand very inclement weather and can live in simple shelters without heat, but counterintuitively, if they are fed concentrates and soft, rich feeds, they should be kept in warm housing, as they will develop much less internal heat. The goat, it seems, was meant to live in marginal landscapes. In the wild, goats make a thick bedding pack of their own droppings, which absorb their urine and do not smell while giving off useful underheat to warm the animals at night. If providing goats a warm house, they should be protected from downdrafts, which can occur if the building is not insulated, and also from cold floors. If insulation cannot be provided on all sides, it is best to create sleeping boxes for the goats 9–24 inches above the floor.

In yards, it is recommended to provide goats individual rations as they can easily eat far more than they need, and are not especially good at sharing. For this purpose, feeding troughs can be fitted with slatted neck restraints.

Traditional dairy breeds include Saanens, Toggenbergs and various crosses. Nubians are smaller goats that give a very rich milk, but most producers find them uneconomical because the volume of milk they give is quite a bit less. If, however, you are using the milk for the household, their smaller size and the milk's high butterfat content may make them a good choice.

Sheltered sleeping boxes, a communal yard with segregated feeding racks or troughs and separate quarters for a billy if you keep one can provide satisfactory housing for your flock if range is not available or convenient. As little as 3,600 square feet can grow most of the diet for two goats, though supplements will be required and you should be prepared to hand feed them. You may also find it profitable to carry them bulk feeds or tree trimmings. Two-thirds of an acre could house and feed a small flock.

### Llamas—Fleecy Guardians and Pack Animals

The llama (*Lama glama*) is a South American camelid domesticated about 5,000 years ago. The llama's wild ancestors originated in North America as did all the world's camels, and from there spread to Asia and South America. Camelids once ranged across the southern mountains and plains of the US and northern Mexico, but became extinct at the end of the last ice age, about 13,000 years ago. They are now making a comeback in their homeland.

Llamas are three-stomached ruminants with two toes and padded feet. They are intelligent, alert animals with a calm and gentle disposition who can learn tasks after only a few repetitions. They have been used for meat and wool, and to carry burdens and pull plows and carts. About 150,000 llamas and perhaps 80,000 of the somewhat smaller alpacas are kept in the US and Canada. They have gained popularity as a pack, fiber and guard animal over the last 30 years; alpacas are kept primarily for their very high-quality fleece. Breeders are now widely distributed in almost all regions of the continent, and the animals have become a familiar if not yet common sight on farms.

Llamas require little more than forage and have very few if any veterinary needs. Adults will consume 1.5 — 2.0 pounds dry weight of forage daily per 100 pounds of body weight — about 8–10 pounds of hay — and can be fed bromegrass hay, alfalfa or corn silage along with a wide range of pasture plants. Two to four may be kept on an acre of pasture. Llamas can be contained by a fence of 4–5 foot height. They are social animals and should be kept in a herd of at least two if not used as a guard animal for a flock of sheep or goats.

As a camel, the llama is adapted to dry environments and can get most of its water from the plants it eats. As a result, the droppings are fairly dry pellets giving rise to few flies. Wherever kept, the llama has discrete toilet habits: the herd will use a communal dung pile, which makes collecting manure a fairly simple task.

Females mature at 12 months but are usually first bred at 18–24 months. Males mature at about three years of age. There is little visible difference between the sexes, though males are slightly larger. The female llama is an induced ovulator, releasing an egg shortly after mating. There is no regular estrus; the animals may be bred at any time of year. Gestation lasts 350 days and in the wild is typically timed to result in births during the warmer months. Single births are the rule; twins are very rare. The young, called *crias* (from the Spanish for baby), are born between 8 AM and noon and will be standing and nursing within an hour of birth. They nurse frequently as the mother produces only a few ounces of milk at a time. The young will begin grazing within 60 days and wean at six months.

Llamas are appealing animals and the crias especially so, but excessive handling should be avoided to prevent them treating humans as they do other llamas. Animals in the herd jostle for dominance and to maintain a hierarchy of privilege. They will spit, neck wrestle, and males will butt chests against each other.

Llama

[Credit: Tom Fogg]

If a young animal comes to regard humans as one of the herd, it will be more obstreperous as an adult and may spit. Otherwise, this typical behavior is uncommon toward humans.

While a number of llama herds are made profitable by selling breeding stock, other economic niches have arisen. There are several hundred llamas in use as guard animals in North America for flocks of sheep, goats or alpacas. A single two-year old gelding is the preferred choice and should be introduced to the flock in a separate but adjacent pen until the animals are accustomed to each other. As the largest animal, the llama will assume a dominant and protective role; it has strong instincts to repel predators. Llamas will charge, kick and have been reported to injure and even kill dogs and coyotes. At mature weights from 280–450 pounds and heights of 5.5–6 feet, they can be formidable guardians.

Llamas are increasingly used as pack animals, being able to carry from 25–30% of their body weight for 5–8 miles. Their sure gait, soft padded feet and delicate browsing habits evolved in sensitive montane environments, make them well-suited to trail work. In many western and mountain regions, they may be hired for camping expeditions. They have also been trained to pull carts, either singly or in teams of up to three. Historical accounts suggest that llamas were used in the Andes to pull plows.

Far more common is the use of llama fleece for spinning and weaving. The fiber, which comes in every common color from white through grey, brown, roan and black, is without lanolin and relatively fine. There are two strains of the animals, one with hair of the thickness of mohair, the other with slightly coarser fiber. Animals can be shorn once a year in the spring; sometimes a second shearing is done in summer.

Llamas may seem exotic, but they have a proven track record. And their ancestors were well-established in North America until just before the current geologic epoch. With their long gestation and small birth numbers, breeding llamas will require some patience and insight, but the niches they can occupy look likely to expand. Their nutritional demands are modest, so a herd could be kept on small acreage. Llamas, either as pack animals or pulling a cart, might well be an effective way to move produce to market over a short distance while creating a striking attraction.

### The Milk Cow (with a Little Beef on the Side)

Cattle are intimately bound up with the story of our civilization, in particular with the westward expansion of settlement in the US. Beef and milk are major agricultural commodities in North America, and many hobby farmers with pasture raise a few beef cattle for their own use or to sell. Because of their size and demand for feed, I don't see beef cattle as a suitable livestock for most garden farms. It is, of course, possible to raise for slaughter a few head of steers without much involvement in breeding, and thereby to meet the family's or the neighborhood's requirement for beef if a few acres of pasture are available. But I think the niche for a milk cow is much more significant.

Dairy cattle represent an important and fairly economical way to transform grass into fat and protein. And grass is a useful perennial plant that grows well in many climates, but especially in cool and rainy regions. The strong and growing interest in raw milk and the irrational hostility to it being created by some elements in government and industry point to an opportunity for the small producer. Smaller heritage breeds like the Dexter, a multipurpose animal of Irish origin, and the Kerry, also Irish, were developed for small farms with low-quality forage. Both are multipurpose breeds, having been used for milk, meat and as oxen. And they are able to thrive on very meager pasture. Both breeds produce good-quality milk high in solids and very good for making butter and cheese. Another popular and slightly larger breed for household use is the Jersey, whose milk is very high in butterfat.

As with other dairy animals such as the goat, milk production in cows is related to pregnancy and birth. For a cow to give milk, she must first have been bred, undergone 41 weeks of gestation and then birthed a calf. The calf is part of the economic package that accompanies milk production. Feed requirements for cattle are typically calculated in cow-calf units. For a dairy cow and calf, as few a 1½ and as many as 8 acres may be needed to provide sufficient feed for a year, so only the larger garden farms will find that a milk cow fits into their program.

Even a smaller breed of cow will give two to three gallons of milk per day and more prolific milkers such as Holsteins, widely used in the dairy industry, can produce five gallons per day. This is enough milk for fresh drinking and for making butter, cheese and other products. It may also be enough to share. The traditional use for milk was to make cheese, as this kept for months through the winter when the cow would be dry and fed minimal rations. Whey from cheesemaking usually went to a pig or chickens. In some cultures it is valued for human consumption and has been made into a kind of lemonade with juice and sugar.

A cow's lactation can run 300 days, but is only likely to be sustained that long if you manage to milk regularly — twice a day at 11–12 hour intervals and always at the same hour — and maintain a good level of protein in the feed. Commercially produced concentrates are readily available and therefore easy, but being composed of conventionally grown crops will likely bring GM-contamination into the food stream. Wheat bran and wheat middlings are available and not yet GM. Also linseed oil cake and peanut cake may be available in some areas. Neither crop is GM yet. You can grow some of your own corn, soybeans, sunflowers, barley, oats, sorghum or other energy and protein-rich foods for the

cow and make your own concentrates with mineral supplements — chiefly calcium and phosphorus. Be sure that you grind or soak large seeds before feeding to improve digestion. Salt licks suitable for cattle are readily available to ensure access to trace minerals and salt.

It is also possible to raise the cow primarily on pasture and hay and accept the reduction of milk and shortening of the lactation that comes with that choice. Most farmers find that supplementing with concentrates makes better use of their investment and time. Cattle need about two pounds dry weight of nutrients per 100 pounds live weight. For a 750-pound cow, this might look like 75 pounds of pasture grasses at 20% dry matter content. (Dry matter is the weight of solids in the feed after water has been removed.) Average pasture yields might amount to 10,000 pounds per acre per year, but larger yields are possible in well-watered areas on good soil. Cattle will eat many other kinds of feed as well. In winter, stored roots are a traditional staple and cows enjoy them. Consider also: kale, cabbage, corn silage, turnips, mangels, rutabagas, beets, potatoes, carrots or wet brewers grains (spent mash).

Of course, what goes in must come out, and the production of a large volume of manure is one of the distinctive yields of the cow: from 12–15 tons per year. About half of this manure will go back on the pasture as the cow drops it. The rest will accumulate near the barn or shed where the cow spends its winter months and where it is milked. This is a great resource but also a large volume of sloppy material that must be handled and stored for some months until it can be returned to soils. One approach is to provide abundant dry bedding material to absorb the urine and to pile and compact the manure and bedding during the winter, taking care to keep it protected from rain which can leach nitrogen quickly. If the bedding is very dry and the pile is not compacted, much nitrogen can be lost to fungal decomposition. If the architecture permitted it, piling all the manure and bedding and having the cows compact it and even sleep on it would be optimal (it gives off heat). The pile does a slow compost in winter that can be accelerated by turning it with pigs in the spring once the animals are back to pasture. Salatin does this with beef cattle in his barn, and has special feeders that rise on rails as the pile deepens. By larding the pile with barley and corn, which sprouts during the winter, he entices pigs to root down to the bottom of the stack. One might also introduce manure worms to begin digestion. Then it all goes out onto the pastures with the tractor and a manure spreader. A small operation might use a wheelbarrow.

Cattle are kept off of pastures in winter in most cold regions for three reasons: 1) there is little productivity as the grasses are dormant; 2) these are typically wet seasons also, and wet pastures with dormant grasses are subject to heavy damage (called pugging) from the animals' hooves and 3) cold stress causes the animal to lose weight or to consume more feed at a time of year when it must be bought or stored ahead. Cattle are hardy, but in extreme winter weather, they can freeze to death. So you will need a barn large enough to house the cow and her calf for the months between its birth and the availability of spring pasture. Since the main yield from the cow is milk, and she will be nearing the end of her lactation going into winter, you will still need to milk at least once a day and twice a day once she freshens. The barn, therefore, should at least be well sheltered against cold winds — built in a protected location no more than 100 feet from the house (because you have to walk there twice a day to milk) — and tight against drafts. You will need at least 200 square feet to provide room to store concentrates and hay. A concrete floor is easier to clean and allows the collection of valuable urine.

A cow prefers a box stall of 10 feet square in which she can lie down. You will need a great deal of bedding to keep this covered and to absorb her droppings. Wood chips are suitable and usually available cheaply or for free. Wood shavings and sawdust are fine if you have a source for them. Hay that the cow has nuzzled through and drooled on can be pulled from the feeder and put onto the bedding pack. You will also need a milking stanchion or stall. Besides its use for milking, you can confine the cow here when you need to clean out her box stall. Feed should always be stored in tight enclosures against vermin. Hay should be kept separated from the cow's area. Overhead storage is ideal if you can easily load it there, as the feed can then be tossed down to the cow as needed and she won't be able to break into it.

Keeping a dairy cow, even a small one, is a major undertaking, probably of value only to those with large households or wishing to provide for themselves and neighbors. For those who appreciate them, fresh raw milk, cream and homemade butter and cheese are wonderful foods, but in the end, you will keep a cow only if you have the resources of land and time, and if you have a feeling for the animals. The main challenges are the size of the animal and the need to keep a regular routine with her. Services and feeds for cattle, including butchering, veterinary attention and artificial insemination are widely available, so, despite the size issue, cattle are in some ways among the more accessible of livestock.

### Dogs and Cats Need to Work

And we come at last to the animals most people know best. The main role for them on the garden farm is to guard against predators and vermin. Dogs on the small farm can be defenders of the garden against deer and other pest wildlife and of the livestock against predators. To be effective, the guard animals must be well-matched to the threat. If you live in a suburban area or in the country near a town, you may have coyotes, though larger predators will be less likely. In that case, and where perimeter fencing of the entire property is impractical, two smaller dogs kept and fed outdoors make the most sense. They are cheaper to feed and easier to train and control than one large dog, but working together they have a powerful advantage over a single predator or a herd of deer.[17]

Cats are a bit more troublesome. The garden farm with a barn and stored hay or grain may need a cat for purposes of rodent control. Similarly, if you have indoor spaces such as greenhouses with perennials or sheds with awkward corners where mice can wreak havoc, a cat may be important. Unless you are in a situation where you can use multiple cats, or where owl predation is significant enough to control surplus population, I would not allow a cat to breed. Free-roaming domestic cats range over an area of about five acres and have a significant predatory impact on birds and amphibians, so you should definitely limit the farm to a single outdoor cat, less if the territory is already cat-rich. The cat should not be fed overmuch so that it will learn to provide most of its own food. If kittens are raised with chicks, they are said to leave them alone later. However you train the cat, do not tolerate predation. Cats are cheaper and easier to come by than chickens. The same goes for a dog, but even more so, as a cat will likely not bother adult chickens but only chicks,

where even a smaller dog can easily kill the adult birds. Left to their own devices, cats can develop an unpleasant habit of crapping in garden beds and may dig things up as well, so provide yours with an outdoor litter area and train the animal to use it.

Guard animals should not become house pets. Create shelter for your guardian animals in the barn or the shed where they will do much of their work. Be friendly and affectionate, but remember that they are there to work. Small farms in a rural or rugged area are more likely than those in suburban or urban zones to find dogs valuable. We fenced our property of ⅔ acre for about $700 and have been able to exclude deer successfully. Fences are not foolproof, however. They do require maintenance from time to time, and we are obliged to open and close gates every day and night. They are also not effective against smaller varmints like possums and raccoons who can be devastating to poultry or against groundhogs which can burrow under them. But our capital cost is not rising, and we have no feed bill. Feeding a dog requires 365 days of care a year, whether you want to be at home or not. And in four years of feeding a dog, we would have spent as much as for the fence. The dog goes on eating for 15 years and needs shots. On a larger property, dogs make more economic sense and may indeed be indispensable.

# CHAPTER 15

# Living with Wildlife

The farm, if it is successful, will attract many wild birds and animals. This can be a source of joy and also of frustration if their hunger overwhelms your cropping plans. Plants are rooted and stay in one place; animals move around and eat them. You have to remember that what appear in your garden as pests are simply wild animals doing their job. It is easier to respond appropriately if you don't take predation personally but understand how to counter it. Virtually all of our food crop plants are delectable to wildlife, including parts of the plants that are inedible to us, such as leaves, stems and even roots. And, while plants can defend themselves against insect predation by a variety of internal chemical means that you as the farmer can support, hungry larger animals can and often will eat almost anything that is growing well and within reach — usually at a most inconvenient time. Strategies for crop protection include fencing or caging, screening, camouflage or dispersion, repellents, pest predation (including hunting), overplanting and timing or evasion. None of these are foolproof. Even hunting is only a temporary and partial measure.

Deer are the biggest and most destructive wild herbivores at large in the North American countryside (in most places where they are not, it is their cousins the elk, moose or caribou that hold the honors). Nor are they limited to rural areas; they flourish in cities and suburbs equally well. Their ecological niche is the woodland/meadow edge just as it is for humans, and deer have catholic tastes. They are swift, alert, reproduce rapidly and can leap over 11 feet vertically if allowed enough running room. As ruminants, deer can convert almost any green matter into venison. They make good food, but the hunting season is limited by law in most places, and it's hard to feel safe shooting at deer in a crowded neighborhood.

Protecting against deer comes down to dogs or fences or both. And not just any dog, but a trained guard dog that lives outdoors. Our neighbors on three sides have dogs — at least four of them, but before we fenced, deer meandered through our lot with impunity, day and night, though mostly at night or when we weren't around. They came within a few feet of the back door, walked down the alley in broad daylight, ate from the garden and would even stand across the street in small groups at mid-afternoon. Since we fenced, I have seen them sneak in the open gate at twilight on a foggy day even though they get no ready access most of the time and so are not accustomed to browsing the property anymore. They can crawl under fences if gaps are too large.

We keep no dogs, but our fence is approximately eight feet in height — the minimum you can expect to deter deer — and, though broadly effective, it is not perfect. The fence does not have to be robust, in fact, if it has a floppy, indistinct top, that is even better than if it is visually clear and rigid. A floppy-topped fence is also effective against climbers like raccoons and cats. In places we have a six-foot, solid wooden palisade fence that works, in part, because shrubbery and buildings on the neighbor's side (and their dogs) add to its de-

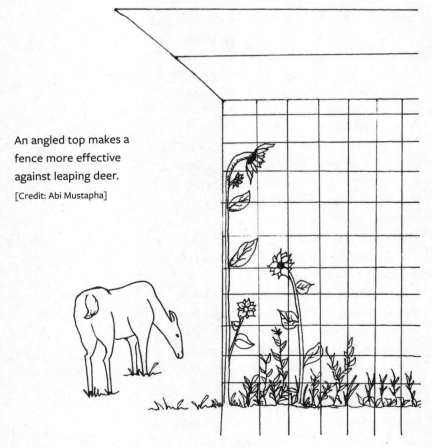

An angled top makes a fence more effective against leaping deer.

[Credit: Abi Mustapha]

terrent effect. Deer will not leap over a barrier they cannot see through. Double fences can be effective at lower heights. You could also run chickens in this "wire moat" surrounding a garden to control insect pests. Wire mesh laid horizontally above the ground near the fence can be a deterrent, though perhaps if it was inclined at a 45° angle, vegetation underneath could be more easily maintained.

Another strategy for fencing offered by permaculture designer Toby Hemenway is to plant vegetation the deer are invited to eat on the outside of your hedge while growing your fruits and other delectables on the inside.[1] We are growing a living fence within and adjacent to our steel fence. Though many of the species we selected are suitable forage crops for herbivores (plum, chokeberry, crab apple, hazel, willow) and we expect to coppice them to feed to rabbits and possibly other livestock, they are nothing on which we depend for our own food. Anything we harvest would be a bonus. Many are thorny species, and they were selected to fill out the range of height from low shrubs up to about 12 feet. Our aim is to achieve considerable density of stems and foliage, so most of the species are shrubby, and they are planted on 18-inch centers so that they can grow into each other without gaps. Plants of the same or related species can also be pleached, or in effect, grafted to one another. To pleach woody species, bind young branches together using string so that the bark touches; the stems will fuse over time.

## Carnivores on the Prowl

Foxes represent a menace to poultry, and bobcats can also endanger larger livestock though neither poses any problem for plants. Good housing and fencing are the best defenses, and a pair of dogs should be effective against either of these animals.

Larger predators are unusual threats to the farm, but not unknown: alligators in Florida or Louisiana; mountain lions in the Appalachians, Rocky Mountains and California (and increasingly elsewhere); bears in many parts of the US and Canada. Though these animals don't generally want to mess with humans, urban sprawl has encroached on their habitat, and occasionally they prove extremely dangerous. Don't be foolish or leave yourself exposed if any of these critters is on the prowl near you. Gators may be fenced out, but big cats and bears can leap and climb. A bear will sometimes tear apart a small tree in the process of harvesting fruit. Pay special attention to protecting beehives against bear predation by robust caging.

## Other Mammalian Pests

Wild rabbits are sometimes a pest of gardens, but easy to exclude with low fencing. Two-foot-high poultry netting will keep them out. Rabbits know how to burrow but won't generally take the time to go under a fence unless they are confined. (You should always be mindful of gaps, however.) Another creature that can be destructive to gardens is the groundhog (or woodchuck as it is called in some places). This large diurnal rodent is vegetarian and tasty; it has a hearty appetite for garden produce. They den underground and always have two or more entrances to their burrows. A favorite location is under a woodpile or stack of lumber where they cannot easily be gotten to. If we can find both entrances, we place a trap at one end and fill in the other.

The raccoon is famous for getting your sweet corn the night before it will be perfectly ripe. A large raccoon is extremely powerful and has articulate hands that can open latches. Be sure that your poultry coop is well secured. Wooden walls or hardware cloth are required to exclude them. We have also had success with live traps in nabbing both raccoons and possums. Thus far we have caught only one skunk, which agreed to be released without spraying (cover the cage with a sacrifice cloth). They are here but do not seem to be destructive, preferring to eat earthworms,

grubs, rodents, snails and fallen fruit. In other locations we have used their hunger for insect larvae and eggs to get rid of yellow jackets nesting in the wrong place. (These small members of the wasp tribe have a vicious sting, but they are useful in many ways, so we tolerate them if they do not threaten people.) Put a few chicken bones by the entrance to the insect nest and the skunk will be drawn to them at night. It will then dig out and ravish the yellow jackets.

Squirrels are a nuisance if you have nut trees in pots. They like to dig up the seeds. Mice and chipmunks may do this too, especially with smaller-seeded plants. Smaller rodents can be trapped and dispatched easily, but for especially high-value crops you may wish to take extra measures. We have used hardware cloth as a guard over the soil, but this can be expensive, usually warranted only for plants like ginseng or nut trees. A cat, if it were vigilant, might be a better solution. There isn't much good to be said about squirrels except that they are tasty. Since you can hunt them by shooting up in the trees, it's safer than hunting deer. Though squirrels contribute to planting nut trees in the wild, their brainless approach can make a hash of well-laid plans in the orchard or nursery. They will be a greater problem where you have large trees on or around the property.

Mice can be damaging to young trees in winter, when they may ringbark them under the cover of snow. Wild rabbits may do the same. Tree guards, whether tubes or plastic spirals, are very helpful against this kind of gnawing damage which, in severe winters, can involve deer too. Tree guards are convenient, if somewhat expensive, but have the advantage of providing a warmer microclimate for the young tree in winter, helping to reduce freeze damage and speed growth. But you can make suitable and cheaper guards from poultry netting if you are willing to take the time. Always be sure to remove or adjust the guard before it constricts the tree's growth.

A greater threat to young trees (and to many other plants) are voles, which eat plant roots. Daffodils repel them and can be planted near young specimen trees. A good shovelful of gravel in the top of the planting hole is also helpful in keeping rodents away from the trunk. Though we experience damage from voles, we have only lost a few plants altogether. Pocket gophers pose the same challenge to gardeners in California except that they are larger and more destructive, being able to kill smaller trees. They can be trapped, but underground fencing may be a more permanent solution. In very gopher-prone areas, and where underground perimeter fencing is impractical, it may be necessary to use a substantial portion of gravel in the planting hole of new trees.

## Snakes and Raptors Eat Small Pests

Snakes are effective against not only rodents but snails and slugs, so if you don't have enough ducks yet, try to attract some of the smaller snakes. They enjoy a bit of mulch cover near some stones in the sun that provide enough shelter so that they can retreat out of sight. As our farm has accumulated more mulch and the perennials have expanded to provide more cover, the population of reptiles has grown to include garter snakes and black snakes. The latter will reveal to you any gaps in your siding or soffits as they are quite comfortable in slithering up a downspout, tree or post in search of a hidey-hole or a choice baby bird. Black snakes are territorial and will chase off even venomous snakes, so we welcome them. In California, the king snake plays much the same role. It is immune to rattlesnake venom and kills and eats rattlers by constricting them. The copperhead is a hazard in some areas of the East, being venomous and also easily irritated. With the exception of deep woods and swampy or riverine areas — or in the West rocky hillsides — rattlesnakes are uncommon. If you live on the suburban fringe or farther into the countryside, they may show up. They are not inherently aggressive, and though I don't recommend attempting to handle them, I've seen people grab and bag them for relocation. A noose on the end of a longish stick is the usual instrument for capture. If held behind the jaw, the rattlesnake cannot strike. How to get your hand there is another matter! Relocation is hardly ideal, as they are fairly territorial, but better than having them loose in an area frequented by children or livestock, and it's preferable to killing them outright.

Raptors are in general so useful and important in the ecosystem that we should always attempt to protect our livestock against them and give thanks for the valuable work they do for us. Most owls are hunting small rodents or rabbits, though they can certainly take poultry that is not penned at night. Hawks are a greater danger as they hunt during the day when your birds are out. Tree cover is helpful because raptors must dive in to make best use of their size and sharp talons. In close quarters, a good rooster has a chance to chase off a hawk. Poultry netting over the yard can provide a secure defense. Unlike weasels or raccoons, a hawk will not usually take more than one bird at a time, so you have a chance to respond if this threat emerges.

## Sharing with the Birds

Birds are mostly a blessing, as they help control insects, but some are seed eaters and most will eat small fruits. We lose a few berries each year, but the loss is minimal and if we keep up with the picking as fruits ripen, the birds don't get many. There are always enough mulberries for everyone, and their abundance helps distract birds from other more choice fruits. Cherries can be netted if the trees are small. Birds will peck at peaches, but not cause much damage. Scarecrows, owl balloons, used CDs dangling in the wind, prowling cats and dogs and complex and diverse planting patterns all help reduce predation. The best single strategy against birds is overplanting. You should be able to afford to let them turn some of your fruit and seed into phosphates for you.

## Integrated Pest Management

It is better on all grounds to tolerate small pest pressures than to mobilize resources against them, so long as damage is less than 10% of a crop. This requires monitoring all the time but intervening only when the dynamic crosses this economic threshold. Past experience is valuable, but no two growing seasons will be exactly alike. Ecological conditions are constantly changing. Your aim should be to create a mosaic of diverse elements and to maintain a balanced ecosystem by regular small interventions.

Herbivores (the ones that eat your plants) always outnumber carnivores, even in the insect world. From both their numbers and impacts, it is easier to see pests than it is to notice many of the predators who keep them in check. Quite tiny wasps can kill and eat much larger tomato hornworms. Our response to the mass reproduction of pest insects should be to encourage their natural enemies which includes leaving some of the pest bugs to be eaten. We also need to disrupt their mating, minimize their habitat, confound their (programmed) expectations, overwhelm and confuse their senses (this includes spreading panic among them), lure them into traps and force them to work very hard for every bite they eat. What we don't want to do is spray poisons to which the bugs will become immune while we succumb.

Insects and other mini-pests are by and large not visually directed. They move by smells and to a considerable degree by chance. They also have no great ability to process complex patterns: "Smell food, fly or crawl to food, eat food, mate and lay eggs near food source, repeat until dead." This means that when we plant rows upon rows of exactly the same kind of plant, we are streamlining operations for pests: "Yum. Corn. Two thousand

Patch Gardens, Pattern #50 (See Chapter 6)
[Credit: Abi Mustapha]

acres of corn." The opposite approach, one that nature pursues vigorously, is to create a mosaic of diversity. This means interplanting, crop rotations, patch gardens, varied heights and times of bloom, wild edges everywhere, genetic variation, multiple colors, camouflage, livestock and arable crops interacting, a cacophony of scents, sights and forms.

### Foundations of Diversity

The translators we need to ensure a healthy garden farm are mostly small insect eaters. The list is long and includes many species we hold in high esteem: dragonflies and damselflies, frogs and toads, songbirds, poultry and turtles; some on which we look askance: snakes, skunks, possums and yellow jackets and some we barely notice: skinks, newts, lizards, preying mantises, hoverflies, syrphid flies, lacewings and tiny stingless wasps. Among this motley crew, about the only creatures we have to house and feed are poultry — but then we ask eggs and sometimes more from them and gather their rich manure. For the rest, it is mainly important that we understand how and where they live, and that we work to create conditions that support them, so they in turn can support us.

Providing small ponds in the garden helps to sustain a population of frogs and toads, which are important allies against pest insects, slugs and snails — and which in turn help to feed snakes. Tadpoles hatching from eggs laid by the amphibians will control mosquito larvae in permanent ponds, and some of them winter over, so they are ready to go in early spring. Toads too like to hide out under rocks and logs. To attract amphibians you need only a reliable supply of standing water. Ponds as small as three feet across can provide a home for them. With frogs and water it is pretty much a case of "if you build it, they will come," often within 24 hours of filling the pond, although reliable occupation may take longer to allow the pond ecosystem to develop. Amphibians are active as early as the ice melts and conditions are a little moderated.

Small ponds should be supplied with a few goldfish or similar small hardy fish. These can overwinter in moderate climates — even under ice — and are cheap enough to replace every year in colder regions.

Together with the tadpoles laid by frogs and toads, goldfish will control any mosquito larvae that appear. Goldfish are also an attractive element in the water with their bright colors. In warmer areas (zone 7 and above), tiny mosquito fish (*Gambusia* spp) will control mosquitos without showing any interest in eating frog eggs.

Small outposts of frog, toad and snake habitat scattered throughout the farm and garden are more important than one large zone. If your landscape is rolling, there will undoubtedly be opportunities for creating terraces, drains, stone retaining walls and other features that will serve nicely as habitat. These will also nurture skinks and lizards, which enjoy sunning themselves as they keep an eye out for bugs.

### Protection in the Skies

Amphibians and small reptiles will cover the ground, but you also need allies in the air, and that has two components: birds and predatory insects. Birds are always attracted to gardens; wherever soil is disturbed or exposed, there are worms and pillbugs to be eaten. Later in the season, ripe seeds attract other species. And, in between, there are airborne insects to be preyed upon. Though all small wild birds can contribute to the garden farm, we should especially support the energetic hunting species that feed on insects. These birds struggle to find live protein in winter. You can supply suet in feeders to help them through the lean months and to ensure that they will be familiar with your airspace when the insects start

Goldfish are easy to maintain and will control insects and algae growth in a small pond.

[Credit: Creighton Hofeditz]

to fly. Most small birds can shift their diet to some extent, so offering bird seed is also useful.

Birds that hunt like to have a bird's-eye view of the territory. Not only will they swoop and dive to catch bugs on the wing; as they launch themselves into flight they will deposit many small packets of fertilizer on the ground below. Birds like to be up at 5 to 10 feet above the ground for close-in hunting. Some have better eyesight and will perch 15 to 20 feet up in the trees or on power lines, but you want to offer many inviting perches at head height and just above. Eventually, perennial plantings will fulfill this function, but don't wait for the woodies to get taller. Start with whatever you have. Beanpoles and tomato cages are helpful. Tall grape and kiwi trellises are useful. But birds appreciate most a high branch or short stalk above everything surrounding it. Give them many good options, and they'll do the rest.

### Refueling the Air Force

Predatory insects are more numerous than birds, and contribute mightily to suppressing pest outbreaks in your garden if you provide them food and habitat. Among the most useful of these are small stingless wasps that parasitize aphids, cabbage loopers and a host of problem bugs. Ichneumon, chalcid and braconid wasps are the smallest and most numerous of these: over 6,000 species inhabit North America. These wasps also feed on flower nectar, and they are most attracted to plants with many small flowers. These include Umbels — plants of the *Apiaceae* or Carrot family — as well as Composites (the Daisy/Aster/Sunflower family, by their name compound or multiple flowers) and Crucifers (mustards and brassicas, which have many small, cross-like yellow or white flowers close together). Umbels offer a wide range of bloom times, so use as many different ones as you can to provide nectaries throughout the growing season. Most are annuals, and you can ensure their convenient return by flinging a few

A chain and cone can reduce squirrel thieving of bird seed.
[Credit: Jami Scholl]

seeds around from the ripe flower heads. Dill and fennel are naturalized in our garden, and cilantro (coriander) is close to a weed. Lovage, on the other hand, is perennial and among the largest of this family. Tiny wasps also use the nectar of mint-family plants, which include not only the familiar spearmint, peppermint and their many cousins but such culinary favorites as sage, thyme, basil and oregano. Non-edible members of this family that support predatory insects include pennyroyal (useful as a mosquito repellent), lavender and ajuga.

All wasps and bees are useful, however, and not just the myriad smaller ones. Yellow jackets and paper wasps tear right into large caterpillars and plant pests many times their size. They effectively control cabbage loopers on our brassica crops. Ladybugs are ferocious eaters of aphids. Damselflies and dragonflies in their many forms, which are attracted to water, have tremendous eyesight and wing control and can catch and devour almost anything smaller than themselves. Fireflies and particularly their larvae are effective predators on many pests, and the larvae overwinter, being active through fall and in the early spring months.

### Nectary Sources

Some plants we consider to be weeds are valuable as nectar sources for beneficial insects. These include stinging nettles, yarrow and burdock (all mineral accumulators as well), Queen Anne's lace (wild carrot), tansy and fleabane. Some wild plants deserve special attention, such as milkweed, a favored food source for monarch butterflies, ox-eye daisy, New England aster and goldenrod, the last two providing important nectar flows for honeybees in the late summer and autumn. Most of these plants play such important ecological roles that we need to find places in the landscape where they can grow and flower. Burdock makes a good ground cover beneath trees, as it can tolerate shade. Yarrow is a beautiful ornamental that comes in several colors and can easily grace a flowerbed or become part of a mixed planting of perennials. Goldenrod is allelopathic toward many plants and can grow in pure stands, so you may find that you want to restrict it to the edge of the farm or to strips between other land uses.

Among the pest control concepts developed by permaculture designer Jerome Osentowski is the idea of *bio-islands*, which he used for a Colorado golf course that was constrained from applying biocides to maintain its fairways. Jerome designed guilds of plants to attract pest predators, pollinators and other beneficial insects, then planted these guilds throughout the property as patches of habitat for micro-wildlife. They ranged from blocks the size of a living room or strips the width of a driveway to more expansive patches half the size of a front yard.

If we scale that idea down to the garden farm, it translates quite well. Your bio-islands might be as small as a three-foot-square endcap on a bed of vegetable plantings, anchored by a shrub and some native herbs or nectary

The Roaring Fork Golf Club in Basalt, Colorado, maintains ecological balance using bio-islands that host pest predators.

[Credit: Patrick Brunner]

plants. Even on very tiny properties, perhaps especially so, there are always scraps of land and niches that can be planted to natives and other ecological supporters. Everywhere land is divided by property ownership, there are fences and edges — usually already wild or scraggly. In urban areas, many of these are suspect for soil contamination, but this doesn't preclude their providing wildlife forage and nectar for insects. Make use of these marginal areas to give natural pest predators a boost.

### Welcome the Natives

It is also important to keep some pockets of native vegetation in your landscape. You may actually have to plant some to ensure their presence. Many disturbed landscapes of settlement are filled with ornamentals or agricultural weeds which have a competitive advantage over many natives when soil and light conditions have been modified by humans. Most of our horticultural and food crops are not native to North America. And many of the most useful economic plants are not either. Annual and perennial native flowers have well-established relationships with the insects, birds and other small animals of the region. They provide food, pollen and habitat of value to the larger ecosystem. In a permaculture it is easy to include native shrubs and small trees, many of which give economic yields: willow, elder, salal (*Gaultheria shallon*), chinquapin, hazel, plum, sumac (*Rhus typhina* and *R. glabra*), saskatoon, hawthorn, crab apple, manzanita (*Arctostaphylos* spp), dogwood, mulberry, buffaloberry (*Shepherdia argentea*), persimmon, Oregon grape (*Mahonia aquifolium*), blueberry, aronia, viburnum, spicebush (*Lindera benzoin*), pawpaw. Some native legumes and nitrogen-fixers are helpful: indigo (*Baptisia* spp), indigobush (*Amorpha fruticosa*), small locusts (*Robinia* spp), redbud (*Cercis canadensis*), Illinois bundleflower (*Desmanthus illinoensis*), alders. Most of the sunflowers too, both annual and perennial, are native to temperate North America. Milkweed (*Asclepias syriaca*) provides food for monarch butterflies, and the buds can be eaten. Other useful and pleasant native herbs include wild ginger (*Asarum canadense*), columbines (*Aquilegia* spp), rudbeckia, echinacea, bloodroot (*Sanguinaria canadensis*), penstemons, coltsfoot (*Tussilago farfara*), some poppies, irises and yucca. There are hundreds more.

Native and nectary plants, wild edges and water, patches and perches all increase pest predation on the farm. If you distribute these to form a mosaic or network across the garden, you have laid the foundations for garden and farm health. The patchwork design strategy works on many levels: by scattering predators throughout, pests are forced to run a dangerous gauntlet on their way to lunch. Variations in color, scent, texture and height of plants confuse and screen against insects moving quickly through the territory. Since many if not most pests are specialist eaters, distributing rather than concentrating similar plants forces the bugs to travel farther and to spend more energy seeking their food. This leaves them less energy and time to reproduce. All of these qualities arise from the layout and patterning of the garden, but the most fundamental level of crop protection comes from good mineral nutrition and soil health, which we have discussed in Chapter 11. From these, plants derive the chemistry they need to repel most insect pests.

### Spot Strategies

We can employ other strategies when a zone defense isn't sufficient or when plants are stressed by climatic or other factors. Basic predatory responses can be enhanced by spraying beneficial nematodes — tiny invisible organisms that parasitize many pest insects — over the affected area. These are purchased inputs, but simple to use and carrying no unwanted side effects (they come on a sponge that you wring out in a bucket of water that is then applied to the plants). Insect diseases can be released with more or less collateral damage. Milky spore is a fungus that attacks many pests. In Tennessee, Permaculture designer Adam Turtle used Japanese beetle pheromone traps to lure these pesky insects toward a pond where they became feed for fish. He attached the traps to short sections of PVC pipe and hung them over the water. As the beetles homed in on what they thought would be a hot date, they hit the trap and did what Japanese beetles do instinctively when confused or attacked — they dropped to the ground, or in this case, through the pipe right into the pond. The fish were quick to learn where lunch was being catered.

Shallow dishes of cheap beer are attractive to slugs when offered in the garden; they happily drown themselves. It is sometimes effective simply to collect pest organisms yourself. Good friends of mine used to spend some evenings in the garden during slug season, picking and drowning. I usually spend a few minutes a day during the onset of Japanese beetles each summer to gather them in a

The Karmic Converter—a pheromone trap suspended above a pond attracts Japanese beetles which fall through the tube to become fish food.

[Credit: Peter Bane]

small bucket of soapy water. Without exhausting yourself it is possible to make a significant dent in the local population. This is also a good occasion to observe what they are eating and how many there are, a basic requirement for pest management.

Some broad-spectrum and harmless sprays can be effective for localized infestations. Safer's soap — a liquid compound that acts as a surfactant and carrier — mixed with garlic and cayenne can repel many organisms. Don't get it in your eyes. Kaolin clay slurries under various brand names such as Surround™ are available to spray on fruit to prevent damage from insects and fungus. Where the pest is visible and can be collected, you can also gather a good sampling, liquify them with water in a blender and spray the resulting gory mess back onto the crop under attack. This both spreads diseases specific to the pest and also distributes pheromones signaling predation and death, causing panic and terror among the remainder of the population.

Modifying pest habitat is helpful — if you can figure out where it is. In orchards, fallen fruit harbors many pest organisms, so cleaning it up using chickens, ducks, turkeys, sheep or pigs makes sense from several points of view: it's cheap and nourishing feed, and the animal foraging breaks pest cycles. Some organisms such as cutworms, which overwinter in soil as larvae, are subject to predation by moles, small sightless rodents that tunnel near the soil surface. Many gardeners and even more lawn caretakers resent moles and attempt to exterminate them, but unlike voles and gophers which eat plant roots and can kill even fair-sized trees, moles are primarily insect eaters. They are on your side. If they burrow inconveniently, just walk carefully along the trail they leave and flatten the mounds. They will go somewhere else. They will not eat your bulbs or your plants.

Animals are such an essential component of balanced ecosystems that it's fair to say we cannot farm without them. Our domestic stock and larger wild birds, amphibians and reptiles give us enormous leverage on soil fertility and the ability to limit damage from the millions of insects, slugs, snails and vermin that would otherwise get most of our crops. Animals are insistent and immediate, but they are creatures with habits and instincts that can be learned and managed, and this is true whether they are wild or domesticated. Keeping animals primarily means finding economical ways to keep them comfortable while they work for our aims as well as their own needs.

Remember that animals belong naturally in polycultures. Whether you are using poultry or hogs to clean up after cattle, or sheep to complement their grazing; running geese with hogs in a wet landscape; using poultry or pigs to manure a pond for the benefit of fish or guarding your flocks with a llama, your animals will be happier, healthier and more productive if they are in relationship to others. Whatever animals you live with, shape your design to allow their life processes to enrich and support your efforts and the vitality of all elements in the farm landscape.

# CHAPTER 16

# Trees and Shrubs, Orchards, Woodlands and Forest Gardens

There are trees alive today that were old before the Industrial Revolution began. They stand witness to the ephemeral nature of fossil fuels, which will be gone before some of them drop their last leaves. Reminding us that another way is possible, trees and forests have also endured the long centuries of ecological vandalism that humans and their agriculture have visited on the world.

Forests store some two trillion tons of carbon, more than twice the amount present in the atmosphere, yet forest cover on the Earth has diminished by ⅓ since the post-glacial maximum of 8,000 years ago and continues to wane at a quickening pace. Between 1945 and 2005, world forest cover was reduced about ⅙.[1] Forest destruction and degradation continues. Though a few areas in industrial countries have recovered a bit since their 19th-century low, tropical forests are being devastated around the world, and industrial consumption has turned fiercely on the boreal forests of Siberia and Canada.

It may not be possible to prevent the world's climate from entering a wholly new regime (one unlikely to be hospitable to humans or to our agriculture), but if it is even remotely possible, we must attempt it by every humane means at our disposal. In writing about the possibility of large-scale carbon sequestration — returning carbon from the atmosphere and the oceans to stable terrestrial storage as soil humus and tree biomass — Albert Bates has made very clear that

1. Reduction of industrial fuel use is essential and
2. No serious reduction of atmospheric carbon can take place without large-scale tree planting in addition to forest preservation.[2]

Garden farms take part in many solutions that address this and other pressing needs of the present and the foreseeable future. Theorists of energy systems have speculated that nuclear fission and perhaps ultimately fusion would be humanity's final energy transition in a sequence that began with wood many ages ago, turned to coal in the 1700s, to oil from 1900 on and along the way embraced methane gas. Given what geologists know about the state of the world's fossil fuel reserves, the next energy transition is almost certainly back to wood.

## Trees for Food, Fiber and Energy

Growing trees for wood fiber became a worldwide industry in the last century, though the large-scale wild harvest of native forests that preceded it continues wherever it is not restrained by law and public oversight. Neither wildcat logging nor plantation forestry is kind to soil, wildlife or forest-dwelling peoples and cannot even be depended on to result in net carbon sequestration.[3] There are, on the other hand, excellent and long-lived examples of sustained-yield forestry at all scales from the artisanal to the industrial, supplying many products while also conserving ecological health.[4] The principles they have demonstrated should inform those aspects of woodland management that affect the garden farm.

The challenge that tree crops pose for us is twofold: forests produce much less human food per acre than do the fields of our present agriculture, and woody crops take several years or even decades to bear. There is a third concern, and one that has tended to weight the course of a bloody history toward peoples of the grass: when vandals strike (and they

have been striking for as long as there have been grain-based surpluses), it is possible to grab your seed corn and flee over the horizon to live and perhaps to plant another day. Tree cultivators live by the virtues of their ancestors and work for the benefit of their descendants. Fire and the sword can undo in one season many generations of care, and recovery is slow.

To embrace permanent agriculture based on woody perennials, we must enter into uncharted territory. Tree cropping is not wholly unknown: humans have been cultivating tree fruits for nearly 7,000 years, and until perhaps 1920 the intellectual resources applied to cereal and vegetable cultivation on the one hand and to woody perennials on the other were not dramatically divergent. The past century, however, has seen an enormous explosion of scientific and agronomic research into the intensive cropping of grains, legumes and oil seeds, and secondarily into the mass production of a small number of livestock species. These crops have become the basis of industrial food.

Adopting tree crops is part of a broad-based citizen initiative to correct the imbalance of research effort in our food system — away from large-scale, chemical-intensive, arable and livestock cropping for market — and toward small-scale organic and perennial systems for household consumption. This return to earth from the industrial to the domestic economy and the great leap skyward from two-dimensional to three-dimensional farming entails more than new information. It involves new attitudes.

And that requires us to delve a bit into culture and psychology before we take up our hoedads and our pruning shears.

## Living in the Garden

Anthropologists have shown strong inverse correlation between levels of violence and gender and sexual oppression in human cultures on the one hand and relative tree

North American farms are characterized by vast terrains of monoculture — the growing of a single crop only.

[Credit: fishhawk via Flickr]

cover in the landscapes they inhabit.[5] While few societies can be said to be completely free of violence, the powerful coincidence of war and environmental destruction over the center of the Northern Hemisphere's ancient civilizations — from Gibraltar across North Africa to Baluchistan and the Gobi, an area of increasing desertification over the past 10,000 years — should strike any elementary student of geography and history as too great to overlook. At the deepest level, humans know our security to be anchored in the treetops whence we descended to stand on two feet, and our opportunity to unfold on the plains and grasslands where we learned to run with the herds. We invariably create the edge between these environments wherever we settle and resources permit. In prairie regions, we surround our homes with trees, and in wooded areas, we carve out lawns and open vistas over meadows and pastures to secure the long view. Frederick Law Olmsted cast these elements together in his designs for Prospect Park in Brooklyn, Jackson Park in Chicago and other beloved urban landscapes in North America.

North Americans in particular suffer from the illusion that trees must be either looming and dominant (as the forest appeared during the early years of European settlement) or completely absent (as early settlers were determined to make them — and very nearly did by the early 20th century).

What our ancestors saw in North America (when they arrived from a Europe already seriously deforested) was a landscape once extensively cultivated that had reverted to an overgrown condition. The myth of a forest primeval, implanted during our civilization's childhood and reified in the years of our own credulous youth, lives on in uninspected attitudes. But archaeology has since revealed the existence of a very different sort of pre-Columbian landscape that was vanishing along with its cultivators just as Europeans stepped off their boats. In the Americas, both

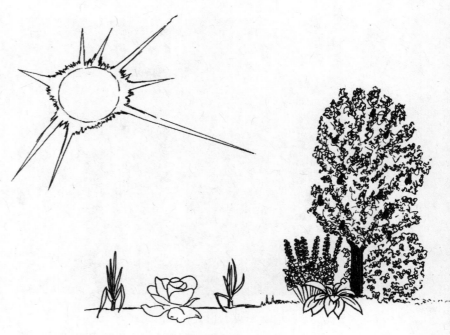

South and North, vast areas were maintained, largely through the use of fire, as a woodland mosaic in which native peoples grew crops, harvested game and gathered wild foods.[6] This was a productive landscape, call it a garden, that fed millions, perhaps a hundred million throughout the Hemisphere. In North America in the 16th century, DeSoto, Coronado and Ponce de Leon delivered Old World diseases — smallpox, measles, typhoid fever, perhaps the plague — against which the native peoples had no immunity.[7] Their gardens fell into neglect from want of labor.

Two-dimensional vs. three-dimensional agriculture [Credit: Jami Scholl]

This image of Central Park in Manhattan reveals the remarkable quality of intimacy with nature that Olmsted brought to his urban landscape designs.
[Credit: Nicola since 1972, Flickr]

**An Example of Coppice**
[Credit: Dave Fincham]

can while still feeding ourselves. We shall do well to keep the vision of a woodland garden in our minds as we proceed.

## Coppice—the Art of Cut-and-Come-Again

What our European ancestors left behind (or abandoned in the shock of encountering vast and towering forests) was a well-developed tradition of growing trees for a myriad of uses. Though industrial forestry and the harsh demands of two world wars in the 20th century nearly killed it, coppice forest management is making a comeback in Britain and elsewhere in Europe. Coppice is the art of cutting trees and allowing them to regrow from the stump. It has much to offer the garden farmer.

Traditional products of coppice forestry included all manner of building materials, timber and non-timber bounty.

While we are not quite in the post-petroleum era, it is likely that our culture's

Our 21st-century challenge is not to restore some imagined pre-Columbian paradise in North America, but to do as well or better than aboriginal people by reclaiming and managing a productive landscape and at the same time, for the sake of our very survival, by growing as many trees in it as we possibly

Coppice products include wattle fencing, furniture, building materials, fuelwood and forage for animals, as well as non-timber products.

[Credit: Peter Bane]

love affair with plastic is winding down — certainly the romance has dulled — and as fossil fuels and even many of the metals they make possible can no longer economically supply the routine goods of ordinary life, something must take their place. That something will be wood.

Instead of allowing trees to grow to a mature height and girth, under coppice systems they are grown only to the dimension that meets the need of the products for which they are cultivated. If you need stove wood of three-inch diameter, it makes no sense to fell and split a two-foot-diameter tree. Better to cut the stems when they are the right dimension for the job. Coppice originates from a time before fossil energy and machines made relatively easy the felling and sectioning of giant trees. It is a system suitable for handwork and hand tools and a necessary part of the skill set for garden farmers.

When a flowering tree is coppiced (most conifers do not respond well to coppice), some of the roots die back, but not all at once. The tree sprouts many new stems vigorously, putting up much more growth and more quickly than it did as a young seedling. It can do this because its large root network holds a great reserve of energy. By freeing the tree of the need to cover its existing trunk, limbs and branches with new cambium cells, coppicing resets the tree's life clock, enabling it to go on growing indefinitely. The coppice worker takes advantage of this, not only to harvest a wide range of wood product from beanpoles to bedposts, but to achieve high rates of growth from trees in a smaller area than would be possible if they were not cut. Regular coppicing results in a tree that remains always in a juvenile condition — young and vigorous — and incidentally small and useful. The part of the tree that endures is the stump (or *stool* as it's called in Britain) with its large root network and ability to resprout vigorously. Some coppice stools there have been worked for over 1,000 years, several times the

lifespan the same tree might enjoy growing unimpeded.

Trees do not need to be cut completely to the ground to yield benefit by this form of management. Where they are grown in pastures or in areas grazed by animals, trees are often cut at six feet above the ground so that the new growth will not be entirely eaten by livestock or wildlife. This is called *pollarding*, and remnants of this tradition, now without productive purpose, can be seen across much of the southern US, where front yard trees (and not only those under power lines) are routinely *lollipopped*, often at a very inconvenient height, for cosmetic or forgotten reasons.

## The Hidden Life of Trees

Much else goes on in the life of trees that we know too little about. Though they can grow to great heights above the ground, trees seldom mirror this growth under the soil. (In drylands, some trees can send roots down hundreds of feet in search of water, though this is uncommon in humid climates.)

Tree Functions and Structure
[Credit: Jami Scholl]

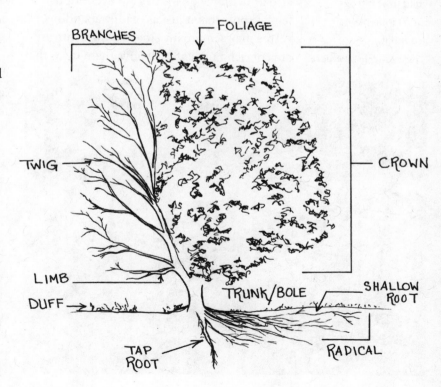

A tree's root mass may be as great as its top growth, but will tend to spread widely rather than dive deep for very practical reasons. Deep soil layers have very little biological life: most nutrient is exchanged in the top 12 inches where oxygen is readily available. Deep soil layers tend to be compacted, dense, even rocky and difficult to penetrate. Most tree roots stay in the upper layers of the soil, but they don't stop at the drip line of the tree. Roots can spread up to three times the diameter of a tree's crown.

Tree roots are large, so it is possible to observe their development as they sometimes break above the soil surface, heave nearby objects or emerge as a result of erosion. Many plants, including trees, grow in spirals, and in addition, the exchange of water and minerals for sugars from the crown reveals a complex interaction: the branches on one side of a tree are often fed by roots on the opposite.

Trees have different types of roots. The ones we notice cracking sidewalks are large structural roots that anchor the tree to the ground and stabilize it against wind. The tree also grows large quantities of adventitious feeder roots that it just as readily abandons. A tree may send many small rootlets into an area of rich nutrient, then give them up as that material is consumed. Most feeder roots are sloughed off at the end of the growing season when photosynthates cease to be available and the tree begins to draw on its reserves. The growing and dying of these many small feeder roots is, along with the growing and dropping of leaves and twigs, one of the chief ways that trees build organic matter in soils.

## Trees for Fertility

The young growth of woody plants (up to about finger size) is very rich in enzymes, vitamins, minerals and other nutrients. With many species this makes the material an excellent fodder for ruminant animals and rabbits, but with all species, when this *ramial wood* (or what English coppice workers call *brash*) is cut and applied as mulch, it breaks down into humus very rapidly, in most cases within a year. Rates of decay are, of course, influenced by the amount of moisture and the degree of soil contact, but we should see woody plants as a prime source of fertilizer and mulch to maintain fertility in our gardens and farm fields. This is especially true of those species that fix nitrogen, and also of those that are cut in leaf, where the green matter adds extra fertility.

Shrubs and small trees are the woody workhorses of the garden farm. They are adapted to heavy browsing pressure from animals — because they grow at mouth height. Browsing or pruning stimulates heavy re-growth. Particularly valuable are those species without thorns and with soft or brittle wood that can be snapped off by hand. Among natives, *Acer negundo* (box elder or Manitoba maple) makes excellent seed-free mulch and can be kept quite small, though it grows rapidly. Similar in ease of hand-pruning and rate of growth are the exotic *Ailanthus altissima* (tree-of-heaven) and *Paulownia tomentosa* (princess tree), both Chinese in origin. These trees are widespread, generally reviled or overlooked, yet can easily be used to improve the fertility of ecosystems. That is their

Autumn olive in this guild fixes nitrogen for fruiting plants around it.

[Credit: Creighton Hofeditz]

role. Our role is to recognize how they can be useful. Trees and shrubs we might choose to plant include legumes such as caragana in cold regions, tagasaste (*Chamaecytisis palmensis*) (in California and the warmer parts of the Southwest), amorpha (in temperate regions) or pigeon pea (*Cajanus cajan*) and sesbania in hot climates. Willows are prolific wherever water is available. Poplar and aspen are fast-growing and respond to coppice. *Elaeagnus* species, especially the widespread autumn olive (*Elaeagnus umbellata*) and Russian olive (*E. angustifolia*), are often seen as dispersive problem plants: make them work for you rather than spend energy trying to exterminate them. We had two autumn olives show up unbidden in a neglected section of our garden. We cut them back every year for mulch and nitrogen, and they have become fertility anchors for the crop species around them. They will eventually provide edible berries. Appendix 3 provides a more extensive list of woody nitrogen-fixers and other useful pioneers.

To manage trees and shrubs for mulch, cut them more heavily in the spring when growth is vigorous, moisture is adequate, lengthening days will help the plant recover and seeds have not yet been formed. If you cut perennials for fodder, you are more likely to need the material in late summer when cool season grasses are dormant. The presence of seed will then be a plus (extra protein). Whatever the purpose, take no more than ⅓ of the plant at any one time unless you cut it to the ground to grow new stems.

For the release of subsoil nitrogen, trees and shrubs need to be adjacent to the arable zone — thus alley and strip cropping provide the greatest edge for nutrient diffusion on the broadscale. For more intimate settings, nitrogen-fixers can be alternated in rows of crop trees or planted in a hub-and-spokes pattern amidst gardens.

Every farm needs an architecture of trees and shrubs: to give it a three-dimensional

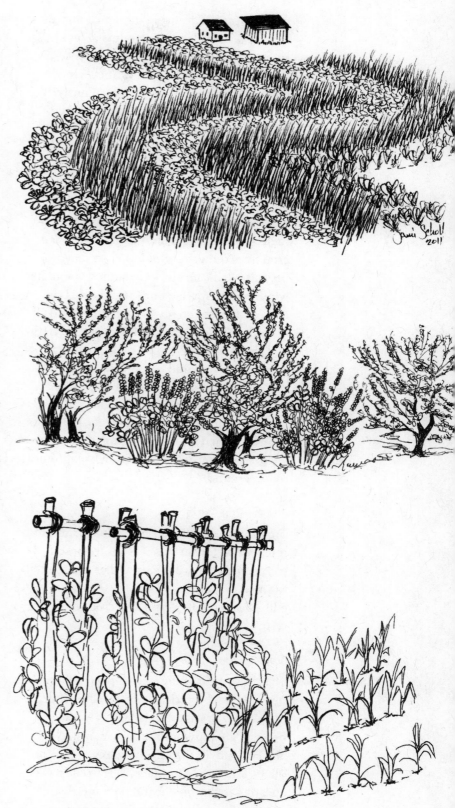

Nitrogen-fixing plants can be interplanted with crops in a variety of patterns.

[Credit: Jami Scholl]

form, to exploit available sunlight comprehensively, to cycle large amounts of biomass and to mediate environmental influences. Woody plants may be seen as anchoring plant guilds or as dividing more intensively cultivated areas. Such divisions should be designed to double as windbreaks, visual screens and as part of an IPM strategy to confuse pests, attract birds and harbor beneficial insects. The way the perennial architecture develops depends on the scale of the system and the size of the trees. Small properties may use perimeter plantings and specimen crop trees centering guilds. Larger landscapes may also dedicate blocks of land to woodlots, windbreaks, riparian buffers and field borders, and these can involve tall trees. And all home landscapes can incorporate small trees and shrubs among garden beds, surrounding buildings and in odd corners.

Trees for fertility should always serve as many functions as possible and may include edible products: leaves, nuts, fruits, flowers or saps. But when food production is the main yield, trees or shrubs move from background to foreground.

## Trees for Food

Traditional tree food crops include fruits and nuts, which for various reasons we need to treat somewhat differently. Other trees suited to small properties include those that bear edible leaves or flowers, salad crops or useful resins, gums or saps.

### Nuts to You

The chief nut-bearing trees that grow in temperate North America in decreasing order of cold-hardiness are hazel, pine, walnut, hickory, gingko, chestnut, pecan, pistachio and almond. In the warm regions of California to Texas and Florida, plus Hawaii, araucarias, including the monkey-puzzle tree (*A. araucana*), may be hardy (zones 9–11). They are native to the Southern Hemisphere and bear large starchy nuts. There are native

as well as exotic nut-bearing hazels, pines, walnuts and chestnuts. Hickories and pecans are all from North America. Gingko is Chinese in origin, though long established here, and almond and pistachio are from Western Asia. Each of the main groups save almond and gingko consists of several species. There are ten species of hazels (*Corylus* spp) worldwide; most are shrubs. The Turkish hazel (*C. colurna*) is a tall tree. Edible pines come from Korea, China, central Europe and the American Southwest (*Pinus edulis, P. coulteri* and others). Walnuts include the English, Carpathian or Persian (all varieties of *Juglans regia*), the black (*J. nigra*), butternut (*J. cinerea*), heartnut (*J. ailantifolia var. cordiformis*) and crosses such as buartnut (*J. cinerea x ailantifolia var. cordiformis*). Pecan is a single species (*Carya illinoinensis*) related to walnut, but has been crossed with its close cousin the hickory to yield edible "hicans." Hickories include several species of which some have better flavor than others; most are thick-shelled. *Carya ovata*, the shagbark, is among the best-flavored. Black walnuts have been selected and also bred for improved size of nuts, flavor and cracking ability (they are notoriously tough nuts). Chestnuts are native to the eastern US (*Castanea dentata*), Europe (*C. sativa*), China (*C. mollisima*) and other parts of East Asia (various). The related shrubby or small tree chinquapins are grouped under the genera *Castanea* (deciduous) and *Castanopsis* (evergreen). They also bear edible nuts, some of good quality.

With the exception of most hazels, almonds (*Prunus dulcis*), pistachios (*Pistacia vera* etc.) and chinquapins (*Castanea pumila*), all of these are large trees that reach 40–150 feet in height and take from 10–20 years to begin bearing nuts. In the last 15 years, Phil Rutter of Badgersett Enterprises in Minnesota has done some selection of precocious chestnuts that bear within three years of planting.[8] Ken Asmus of Oikos Tree Crops in Michigan has been selecting precocious hazels.[9] Korean nut pine is said to bear at a small size under some circumstances. For the most part, however, planting a nut orchard is an investment in the next generation and requires several acres of land devoted to it. Walnuts and the like may be planted on 20-foot centers (about 100 trees per acre) with the expectation that they would be thinned to at least half that density by maturity, the thinnings to provide some timber harvest. Purdue University has demonstrated that wide alley cropping of black walnut with corn can be economically profitable. The trees take little away from the grain and may indeed improve environmental stability, yet provide a secondary long-term yield.

### Bread from Trees

Chestnuts are a valuable source of carbohydrates that traditional cultures of southern Europe and the Appalachian Mountains raised for "bread." Unlike oaks, beeches and some hickories, the chestnut is an annual bearer, making it a suitable staple food. In both cultural settings, swine were also foraged on the nuts to provide meat and fat for the diet. Where space and time permit, chestnuts should be considered for a main crop. Rotation of poultry through the orchards would help to limit damage from chestnut worms, which infest most growing areas. (The worms are actually quite delectable when sauteed in butter, but they do detract from the nut's commercial value.) The limitations of chestnut are, besides the regrettable loss of American chestnut as a supremely multifunctional tree, the relatively long interval between planting and bearing — up to 13 years — and the spiny burs from which the nuts must be extracted. Clonal propagation of precocious seedlings shows promise of reducing the fruiting age to three years and of making it possible to crop smaller forms of the plant. With Chinese chestnut (*Castanea mollissima*) — the species typically grown as a replacement for the functionally extinct American chestnut — the large nuts make harvest relatively easy. Smaller

Black walnut alley cropped with corn in northern Indiana. The trees are shown at three stages of growth from young seedlings to 20-foot saplings. Corn is cropped throughout this period, but as the trees approach early harvest, corn yields may decline. [Credit: Hugh and Judy Pence]

chinquapins have good-flavored nuts also and are worth a try on smaller farms.

Extensive breeding efforts have been underway for over 30 years to hybridize American with Chinese chestnuts, and cultivars have been developed that show some blight resistance while retaining most of the desirable characteristics of the American species (*C. dentata*) — tall form, high-quality timber and excellent flavor in the nuts. My study of the blight and its development suggests that restoration of the American chestnut, with or without breeding efforts, will require mimicking the unusual conditions that provided this shallow-rooted tree of mountain glades with high levels of nutrient.[10]

### A Few Special Cases

Almonds are a valuable crop worldwide, but require intense concentrations of honeybees to pollinate. Almonds will grow from Oklahoma to Virginia and southwards, but are not reliable bearers due chiefly to humidity, fungus and spring frosts. Some regions in Texas, Colorado, New Mexico, Arizona, Utah and Nevada are suitable for almonds. In selected microclimates of Oregon, Washington and BC, home-scale cultivation may be possible. California is, of course, the center of commercial almond growing in North America.

Pistachios are grown in California though they are native to Persia. With a similar range but just slightly hardier than almonds, they may bear as far north as southern Oregon but are not economic outside the Central Valley of California and parts of Arizona. Pistachios are dioecious: one male tree is needed to pollinate about nine females.

Gingkos are also dioecious, and male trees are frequently planted as street trees because the species is tolerant of air pollution. The females are usually avoided because, though they produce an edible nut, the flowers are rank. The main market for gingko nuts is among ethnic East Asian communities familiar with it, so only small amounts enter commerce, most of them imported. Gingko leaves are reputed to contain compounds supportive of circulation and stimulating to memory. The market for these, harvested when yellow in the fall, is considerably larger than that for the nuts at present. Gingkos can reach 100 feet in height, but they can also be managed as shrubs if trained to a small growth habit and pruned annually. Coppicing male trees in this manner would be the way to manage for leaf production.

Some 15 species of pines bear fine and valuable edible nuts ranging up to half an inch in size. They are undemanding and tolerant of a wide range of conditions. However, like most other nut species, pines take more than a decade to begin bearing and will grow large relative to available moisture. Their seeds are also tedious to collect and process. Because they have few problems, live long and can yield valuable timber, nut pines should be considered for larger properties where their evergreen growth habit would be an advantage. Korean pine (*Pinus koraiensis*) is the most hardy of the lot (zone 3); Italian stone pine (*P. pinea*) the most tender (zone 8). Those growing in warmer environments appear to produce the most protein (34% in *P. pinea*, 30% in *P. sabiniana*, zone 7); the piñons of the Rocky Mountains yield relatively high-fat nuts (60–70% in *P. edulis*). The various pines of western North America are all much more drought tolerant than Old World nut pines. But harsh growing conditions result in sporadic and low yields.

### Queen of the Nuts

The North American walnut tribe (*Juglans*) are all large trees with an affinity for water. Pecan is the most tolerant of wet roots, but black walnut and butternut both grow well alongside streams. Butternut (*J. cinerea*) which, as its name suggests, has a fine-flavored meat, is suffering from a widespread canker that is killing many trees. It is the most cold-tolerant walnut. Black walnut (*J. nigra*) is the most common of the group and a valuable timber species, nearly as hardy. In addition to its tough-shelled though flavorful nuts, the tree exudes juglone from its roots while alive, an allelopathic chemical toxic to many other cultivated dicots. Polycultures with black walnut are challenging, but a number of approaches have been suggested. Mulberry is said to buffer other tree fruits from this allelopathy.[11]

Pecan is without a doubt the finest flavored of the nuts and the highest in oil content (63%); pecans will burn if lit. They crack more easily, but are less hardy than their walnut cousins. Though commercial production ranges from Georgia to Missouri to New Mexico, hardy varieties can be grown to zone 5, southern Michigan and Ontario. The nuts are smaller and commercial prospects are poor, but in a well-designed home system, this would matter less.

Buartnut, a heartnut-butternut hybrid, deserves attention from growers with a spare acre or three. It grows well to zone 4, with better flavor and crackability than its heartnut parent and better disease-resistance than its butternut parent. It is still a very large and vigorous tree. Some varieties are said to bear within five years.

Though black walnuts are used in flavorings and candies, the main walnut of commerce is *J. regia*, the English or Persian walnut, hardy to zone 6a. Most are grown in California or the Old World. The species will grow and bear across North America and has much less allelopathic effect than *J. nigra*. Carpathian cultivars, originating in the mountains of the Czech Republic and Slovakia, are much hardier (to zone 3b).

All the large walnuts, pecans and hickories are suitable for silvopasture in a modified savanna environment, with livestock maintaining a clean sward beneath the trees, thereby facilitating nut harvest and garnering an economic yield during the early years of tree establishment. In warmer regions, the dappled and short-season shade from the trees can increase pasture growth.

## A Small Nut for All Growers

For smaller garden farms across most of the US and Canada, only the hazel (or filbert) justifies much attention or space. American and California hazels (*Corylus americana* and *C. cornuta*), European species (*C. avellana*) and the many crosses between them (hazelberts, filhazels) show a compact, shrubby form and can be coppiced. Hazels will begin bearing after about four years and may live for half a century. They may reach 20 feet at maturity but can readily be maintained at 8–10 feet for cropping purposes. Agronomic research by Badgersett and others suggests that plantations or intercrops of hazels can be mown periodically and will begin yielding again in the second autumn after cutting. On a farm scale, this minimizes pruning labor while allowing the plants to be kept small and still productive. Hazels, though deciduous, make a fine dense screen with their many stems; they tolerate a wide range of soil conditions, growing best on deep loam. They can be interplanted with tree or cane fruits or alley cropped with vegetables. Like other tree

This 400-acre Missouri pecan orchard functions as a savanna ecosystem by providing pasture for livestock beneath the open tree canopy.

[Credit: Peter Bane]

crops, they would benefit by periodic foraging rotations of poultry or even pigs. Squirrels are a significant pest of hazels, so where the crop is economically important, measures should be taken to hunt, guard or net against them.

Hazels fruit on one-year-old wood, so pruning should emphasize an open center to maximize light to the fruiting wood. Otherwise, little pruning is necessary, though removing root suckers can improve yields. As a screen or low windbreak, hazel's tendency to sucker is valuable. The bushes can be regenerated by cutting back to the ground and allowing new shoots to grow. Hazel shoots or spars were traditionally used as crosspieces for hurdles or movable fence sections. These were harvested from the understory of coppice systems, cut on 6 or 7 year rotations and grown among taller timber trees.

Hazelnuts grow in fibrous husks that are readily removed. The shells are easily cracked at home, but whole nuts would not be suitable as a fodder for animals other than hogs and perhaps goats or turkeys. Their high-quality oil is chemically indistinguishable from olive oil, so they represent a promising vegetable source of dietary fat.

### Fruiting Shrubs and Vines

Though not trees, fruiting shrubs and vines are commonly grown with them, and as we consider forest gardens, we will see how these smaller perennial plants become essential to our tree-centered assemblies. Shrubs are rarely going to be in full sun, so they compensate by creating very dense or evergreen foliage, leafing out earlier than most trees and keeping their fruits small so that birds can eat and spread them into sunny niches. Included in this category are the *Ribes* clan — currants, gooseberries and hybrids — which do well in partial shade, bearing crops in as little as half light. Other shrubby fruits of note include many serviceberries (*Amelanchier* spp — there are also tree forms), some bush cherries and bush plums (both *Prunus* spp), some crab

apples (*Malus* spp), some quinces (*Chaenomeles* spp), blueberries and cranberries (both *Vaccinium* spp). Figs (*Ficus carica*) will grow as shrubs in marginal climates where they die back to the roots and resprout each year (zone 6), but are basically trees. For that matter, apples at the extreme northern edge of their range grow as shrubs too.

Some fruits are borne on woody vines; grapes (*Vitis* spp) and kiwis (*Actinidia* spp) are the most common of these. Akebia is an East Asian vine bearing sweet edible blue pods. By their nature, vines are expansive: they depend on fast growth to climb the trunks and branches of trees to reach sunlight in the canopy. The trade-off is that they use much less energy making wood. Old vines do, of course, get thick and woody, but this takes a number of years during which they remain pliable. Vines are usually trained to grow horizontally on a trellis of wires strung between posts. The wires are suspended at a convenient height for pruning and picking fruit, but in some commercial orchards, high trellises are being used to allow perennial crops such as asparagus and rhubarb to grow underneath.

As a consequence of their vigor and in order to obtain good fruit production, it is necessary to prune vines heavily. About 90% of new growth should be removed from grapes during the dormant season — usually just after leaf fall. The typical pattern encourages a central stem with four and perhaps later six side shoots running out horizontally on the wires, two in opposite directions at two or three levels.

This is the basic architecture of the grape as it starts the season. New growth is then trained along the wires. Where winters are severe, grapes are sometimes grown as fans. In this case, branching occurs low to the ground with main leads spreading up and out in a plane. This allows the branches to be pruned back at the end of the season and the main growth points covered in heavy mulch to pro-

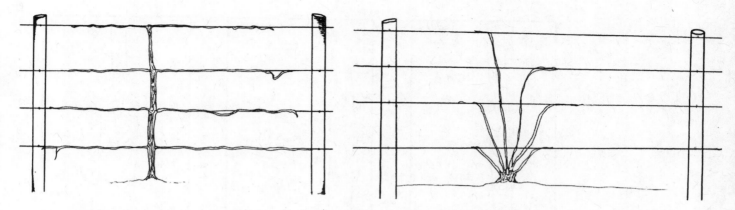

Two Systems of
Grape Pruning

[Credit: Abi Mustapha]

tect against the cold. Kiwis are vigorous but somewhat slower to bear and can be heavily pruned at two- to five-year intervals, depending on need. If not pruned they can become massive plants that will completely cover a large area such as a patio. If they are grown on a tree that is coppiced for leaf production, for example linden, the vine and the tree can be cut back at the same time. This pairing of vine and tree is a forest mimic, as kiwis in the wild know nothing about wires. Grapes are not so suitable for this type of management because they are one of only a few woody plants that need major pruning every year.

The logic of pruning vines is similar to that of coppicing. If the vine has to expend energy to cover a long "trunk" and branches of woody material with new cells, it will have less energy to produce fruit. Replacing green material is rather easier than growing wood. Each year it grows, the trunk, main branches and the root system of a grapevine get bigger, so its new growth comes out faster and more vigorously. Vines can live a hundred years or more if maintained in this way.

Grapes produce large amounts of sugar, which is why they are so often turned into alcohol. Also, crushing them and screening the juice eliminates a vast amount of fiddly labor in removing seeds from small fruits. (Seedless varieties solve this problem in another way, but there is considerable value in grape seed as a source of high-quality oil and antioxidant compounds.) Because of their potential to produce sugar, grapes need nearly full sun-

light to bear heavily. So it is worth dedicating distinct spaces to them, another reason that high trellising can be worthwhile: it leaves more space on the ground for vegetables and smaller plants. As long as they have sufficient water, grapes love the heat of sunny spaces too. Dry conditions intensify their flavor. They struggle in humid or rainy conditions which stimulate the growth of fungus eager to eat all that lovely sugar. Kiwis are not quite as sweet and will tolerate partial shade.

### Growing Tree Fruits

Better than money, most of our main fruit crops do grow on trees. Some of these are naturally small like peach, plum and medlar. Others are forest giants such as apple, pear and sweet cherry. Many of our common fruits belong to the rose family and consist of two large tribes — the pome fruits (apple, pear, quince, medlar and hawthorn) and the stone fruits (plum, apricot, cherry, peach, almond — see nuts above — and their crosses such as nectarine and pluot). For a variety of reasons, some climatic and environmental, some related to size and vigor, farmers have learned, by the art of grafting, to combine hardy rootstocks that limit the size of the tree with top growth from cultivars with choice fruiting qualities.

Grafting consists of creating a living bond between a *scion* (a piece of the top growth of a desirable plant, usually a tree fruit or nut) and the root and lower stem of another of the same or a closely related species. To

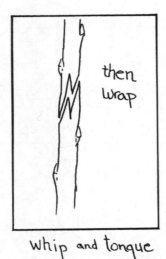

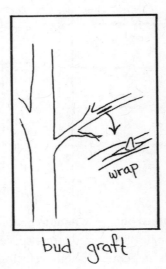

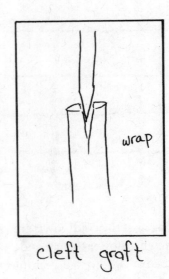

Methods of Grafting
[Credit: Jami Scholl]

whip and tongue    bud graft    cleft graft

accomplish this, the cambium layers of scion and rootstock must be aligned very carefully, held together and kept moist for several weeks while they fuse. This usually requires clean cuts with a sharp knife and a strong but flexible wrapping of the joint with elastic grafting tape or rubber bands. Single buds of scionwood can also be inserted into the bark of a rootstock, typically in late summer.

Grafting is usually done in spring or early summer when the energies for growth are abundant. A graft can also be used to repair damaged plants by bridging above and below a major wound, enabling the tree to continue circulating nutrients. When the graft is successful, the cambium layers of the two parts become one that grows out and around the break in the woody parts, eventually building a thick callus that supports top growth for years or decades into the future.

With pome fruits — apples and pears almost universally — pollination is by other varieties of the same or closely related species. This sexual reproduction results in genetically diverse offspring, which is good for the species but bad for the grower. Take two fine apples, cross them and you are likely to get daughter trees that produce tasteless or mealy junk. Occasionally new varieties are superior, and it is important that apples and other fruits that cross-pollinate be allowed to

do so freely at the wild edges of cultivation, so new varieties may be created. Forests of wild apples, for example, still grow over tens of thousands of acres in Kazakhstan. Long may they live! Grafting, which is necessary for reliable reproduction of type among pome fruits, presents an opportunity to choose the size of the mature tree and to give it cold or drought hardiness, resistance to blight and disease or tolerance for difficult soil conditions. Besides its obvious advantages, grafting can put several scions on a single rootstock, allowing growers with very limited space to achieve cross-pollination and wide variety from a few trees.

Stone fruits pollinate themselves fairly well; they are said to be self-fertile. You can plant seeds from peaches, apricots, cherries or plums and about ⅔ of the time will get a tree that closely resembles its parent. This percentage isn't good enough for commerce, but it is helpful to the small grower who can afford to grow cheaply from seed and select with good odds among the resulting plants.

### Choosing a Size

A standard apple grown in grass lawn or other urban conditions may be expected to reach 25 feet at maturity. Apples in forest conditions grow taller because they are competing for light. The cultivated apple may grow taller than 25 feet under fortunate circumstances, but this should guide your choice of root-stocks. Harsher conditions (cold winters, heavy soils) will limit this potential. A standard rootstock will allow the tree to reach 90–100% of full size. Trees on semi-standard rootstocks will grow 70–90% of their potential. Semi-dwarf stocks come in a couple of ranges (40–55%) and (55–70%). True dwarfing stocks limit the tree to 20–40% of its full size. Rootstocks affect not only the tree's age of bearing, but its total yield. A standard tree, though slow to start, will greatly outproduce anything smaller, yielding over 400 pounds of fruit in a single season. Yields from semi-dwarf and smaller trees are proportionately less. Yields are also affected by soil fertility, rainfall, pollination and other factors, but the smaller the rootstock, the smaller the potential harvest. This can sometimes be an advantage.

Apples may be grafted to apple rootstocks and also to crab apples and hawthorns, of which there are many species, some edible and even choice. Apples may also be pollinated by some crab apples, which belong to the same genus and have the same number of chromosomes.

Pears are another pome fruit, closely related to apples, but with a more columnar habit of growth. They are long-lived, hardy, more tolerant of damp or heavy soils and slightly faster to bear than apples on comparable roots. Pears too are grafted, but the rootstocks are somewhat less satisfactory than those used for apples; pear scions are often mismatched to the quince rootstocks used to dwarf them, for example. Hybrid rootstocks of Old Home and Farmingdale varieties, called OHF, yield trees about 70% of full size and have perhaps the widest utility. Bees seem uninterested in pear flowers, so for best pollination and fruit set it is recommended to plant three or more varieties of pears together. More than apples, pears suffer from fireblight, a bacterial disease that causes blackening of the leaves and branch tips. Asian pears are more resistant and in humid climates should be considered a first choice for the home orchard.

With the exception of sweet cherries, all the stone fruits are relatively small trees, and most are grafted to rootstocks that allow them to reach full size, approximately 15 feet in height and spread. Plums are sometimes put on St. Julien, producing a 60% tree, or may be dwarfed on Pixy. Recently, a variety of roots developed in Russia at the Krymsk experiment station have come into use that allow sweet cherries to be held to between 50–70% of full size, making them far easier to pick and to include in smaller landscapes. The vigor of the scion still affects the size of the mature tree, and this is true of all grafted fruits.

*Selecting Trees for Your Climate*
I have tried to assess the center of the genome for each crop in placing it on this scale. Some cultivars of plants listed below may be hardier than indicated and would be well worth a try. Some individual species, like elderberries, are widely adapted. Others are members of semi-continous tribes, for example various amelanchiers (juneberry, saskatoon, serviceberry) of different species but broadly similar fruits, that range from the subarctic nearly to the Gulf Coast.

Gardeners who live in the coldest sub-arctic regions of North America will find their choice of cultivated fruit species limited. Saskatoons (*Amelanchier alnifolia*) will grow right to the edge of the tundra, as will the wild pin cherry (*Prunus pensylvanica*) and chokecherry (*P. virginiana*), showy or northern mountain ash (*Sorbus decora*), *Rosa acicularis* and highbush cranberry (*Viburnum trilobum*). This is zone 0. In zone 1 add mountain ash (*Sorbus americana*), northern mountain cranberry (*Vaccinium vitis-idaea*), some crab apples and one or two small apple varieties. Zone 2 brings a much wider range of cultivated apples and crab apples, as well as apricots (*Prunus mandshurica*), Canada plum (*P. nigra*), currants, gooseberries, low-bush blueberries, cranberries (*Vaccinium macrocarpon*), elderberries (*Sambucus nigra*), hawthorns, *Rosa blanda*, some varieties of sand cherry (*Prunus pumila* var. *besseyi*) and nannyberry (*Viburnum lentago*). None of these zones occur in the lower 48 states, but only in the Canadian North and interior Alaska.

Zone 3 includes some parts of the Adirondacks and Maine, northern Minnesota and upper Michigan, and high elevation parts of the mountain West, as well as large swaths of the Canadian prairies and northern Ontario and Quebec. In these regions add to the above lists tart cherry, American plums, a few Japanese plums, more varieties of bush cherry, other large-fruited roses such as *Rosa canina*, *R. rugosa* and *R. villosa*, European mountain ash or rowan (*Sorbus aucuparia* var. *edulis*), American hazel (*Corylus americana*), blackhaw (*Viburnum prunifolium*), more apples and crab apples and a small handful of pear cultivars.

At USDA zone 4, we enter a broadly agricultural region and have many more choices of crops. Some of the most cold-hardy wild fruits will not grow in regions much warmer than zone 4, but others stretch all the way to zone 9. I will not remark on those that should be removed from the list as we proceed southward, except to say that when you can grow figs, you become much less interested in cranberries, even if they tolerate your climate. Judge accordingly.

In zone 4, apples flourish and pears are widely hardy. Butternut and black walnut may be grown. Many grapes are possible as well as hardy kiwis (*Actinidia arguta* and *A. kolomikta*). More apricots join the crowd (*Prunus armeniaca*) as well as some European plums (greengages) and white mulberry (*Morus alba*).

Zone 5 brings access to hardy varieties of peach, a few northern pecans, buartnut,

gingko, highbush blueberries, sweet cherry, Damson and prune plums, red mulberries, American persimmon (*Diospyros virginiana*), *Cornus kousa*, pawpaw (*Asimina triloba*) and medlar (*Mespilus germanica*).

Zone 6 allows us to add more pecans, peaches, heartnut, Persian walnut, most Japanese plums and figs under cover.

Zone 7 is the territory for figs to grow as trees and not just as annual shrubs. Nectarines do well. Black mulberries begin here and go all the way to the subtropics. Jujube or Chinese date (*Ziziphus* spp) is hardy. Asian persimmons may be attempted. Pineapple guava (*Feijoa sellowiana*) is barely possible in protected locations.

Zone 8 brings us to the cusp of temperate and subtropical climates. Rose family trees (apples, pears, cherries, apricots, some plums) have difficulty because of mild winters and erratic springs. Low-chill varieties should be selected and careful attention paid to microclimates and good air drainage. Peaches are commercially successful. Figs are in their glory. In drier regions, pistachios will bear. Meyer lemons are hardy but not highly productive, Asian persimmons come into their own, fuzzy kiwis (*Actinidia deliciosa*) flourish, feijoas yield and a few bananas can be nurtured under special conditions.

Zone 9 is the realm of loquat and satsumas or mandarin oranges, among the hardiest of the citrus family. Bananas will bear more reliably but still require protection in winter.

Zone 10 allows the growing of most citrus, date palms in particular conditions, starfruit and the tree tomato or tamarillo (*Cyphomandra betacea*).

Zone 11 is the edge of the tropics. Papaya (*Carica papaya*) can be grown, along with avocado and mango in protected locations.

### Cultivating Tree Crops

If you didn't plant your fruit trees ten years ago, the next best time to do so is today, provided it is late winter, spring or a moist

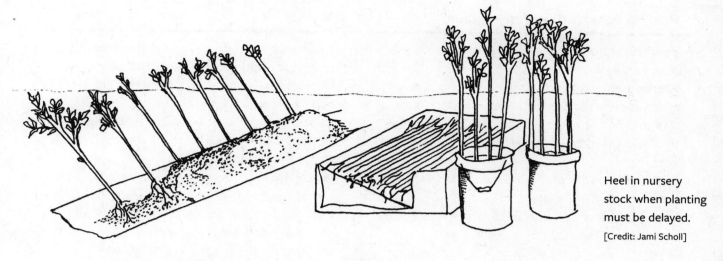

Heel in nursery stock when planting must be delayed.
[Credit: Jami Scholl]

autumn. Growers in subtropical regions have much more latitude but should plant during times of mild temperatures and when rain is expected within weeks. Temperate zone gardeners must work with the annual cycle of growth. The window for planting is especially narrow in very cold climates.

Transplanting is always a shock to the plant, but this can be minimized when it is dormant. Consequently, most trees are shipped from nurseries as bareroot stock at the end of the dormant season. Some nurseries also ship in the autumn after plants have again become dormant but before the ground freezes. After mid-May, most nurseries hope to have moved out the bulk of their inventory. They typically will pot up remaining plants and sell them over the counter.

You shouldn't assume that plants you can buy down the road were grown in a climate like your own — they may have come from hundreds of miles to the south, or even across the continent. When you buy from a small catalog retailer or local grower, you are likely to get a well-tested product, but you should still be alert to the possibility (always present in dealing with nursery stock) that plants may be mislabelled. A reputable nursery will replace mislabelled material, but only if you can spot it! Do your research, and be familiar with what you expect to get.[12]

After making your design based on the amount of land you have, your production aims and an assessment of your climate and soils, select trees from a reputable nursery and order early — by mid-February at the latest — in order to get the plants you want. If you don't receive everything you ordered, or if you have some last-minute change of plans or a new idea, it should still be possible to get additional stock to fill gaps from local retailers. And, there is always next year.

Order only as many plants as you can plant within a week, and begin preparing your planting holes in advance. Bareroot stock is very vulnerable and must either be planted immediately, or *heeled in* by placing the roots in a shallow trench and covering them with soil and mulch.

Always keep new plants well watered, out of direct sun, and never expose the roots more than the very brief time needed to remove the plant from its container and put it in the ground. If you are expecting a delivery, be at home, meet the shipment and deal with it immediately. Trees and shrubs that are heeled into a shallow trench will leaf out and begin growing roots, so when the time comes to plant them into a permanent location, handle with great care. Be sure that your temporary holding location is shaded from direct sun, protected from vermin and is someplace that you will not overlook!

# HOW TO PLANT A TREE (OR A SHRUB)

For plants larger than 12 inches and more valuable stock, careful preparation is warranted.

1. Prepare a large bucket of water into which you have dissolved some sea minerals or fish emulsion. Place your trees to be planted into this bucket and hold them there with the roots immersed for 30 minutes or more prior to planting.

2. Cut a hole 18″ in diameter with a spade and remove the sod, setting it aside for later.

3. Using a large sheet of cardboard or a scrap piece of plywood, paneling or other waste sheet (even plastic) to hold the soil, remove the earth from the hole down 18″, keeping the sides vertical. Set the "A" horizon soil in one pile and subsoil in another.

4. Using a sharpshooter, mattock or garden spade, make several stabbing cuts into the sides of the hole. This will aid the tree roots in spreading. If the bottom of the hole is heavily compacted, use a spud bar or mattock to break it open a bit.

5. Take a small portion of the subsoil and build a mound in the bottom of the hole. The tree will be planted on this mound, so it should be high enough that the graft union of the tree will be several inches above final soil grade.

6. Separate and spread the roots of the tree carefully in all directions and place it on the mound in the hole. If the tree has a prominent side branch, turn this toward the southwest or into the prevailing wind to provide the tree and the soil a little extra protection from that quarter. Holding the tree upright, begin to return the subsoil layers to the bottom of the hole.

7. If you have mineral amendments (limestone, greensand, rock phosphate or sea mineral supplements), mix them into the soil as you return it to the hole. If the subsoil is very heavy, you may wish to mix in a portion of small gravel. Under no circumstances should you put any kind of manure into the hole. Even compost is not a good idea, as it may be insufficiently broken down and could cause rot.

8. Tamp the soil firmly around the tree roots as you replace it. Don't wait until you have filled the hole. You can use your feet. The roots need to be well anchored and in good contact with the soil. Large air pockets will not help. After all the subsoil is replaced, add the topsoil, continuing to press it down firmly.

9. If your subsoil is rocky or very dry, your hole may refill about as it emptied, but in most cases, digging will increase the soil volume temporarily, resulting in soil that overfills the hole. Use the extra soil to create a small berm around the lower side of the tree about 18″ out from the stem. This will help harvest and retain runoff to water the tree.

10. Reuse your sod elsewhere that you may need grass, or compost it. If your soil level in the hole is too low and you need the volume of the sod, shake as much of the soil off the roots as possible, then invert it to smother the grass. Grass is the enemy of young trees.

11. You can finish the top of the planting with a shallow layer of gravel, or apply well-finished compost or aged manure (only at the surface). Or you can combine these treatments.

12. Water the new tree well, giving it at least two gallons. Some of the sea mineral or fish emulsion would be helpful.

13. Cut a six-inch keyhole into a large sheet of cardboard; lay it down as a collar around the young tree. Do not bring the cardboard directly up to the tree stem as you will create habitat for mice. Cover the cardboard with wood chips, leaves or straw and wet it all down. The cardboard mulch suppresses weed and grass competition for about a year. You will still need to weed right around the stem for a while.

14. Be sure that the tree gets water every week for a month unless it rains substantially, and every two weeks for the remainder of the growing season. If the autumn is dry, water well twice a month until the ground freezes.

## After you plant

After planting the tree, trim the main stem to about 24 inches in height, cutting just above a strong and healthy bud. Make the top cut at a slight angle to shed water. Cut back side branches to no more than 10 inches, and limit the number of these to 3 or 4, each pointing in a different direction. If two branches come off directly opposite each other at the same height, remove the weaker one. Be sure to remove any side growth below the graft union, whenever it may occur.

Training a young fruit tree is an important part of achieving good yields over time. Your best hope of managing the crop as the tree matures is to establish an economical and sound structure that it can support. This is achieved by early training and maintained by yearly pruning of excess growth. Two systems are used for most temperate species. Apples, pears and cherries tend to grow tall and are usually pruned with a central leader and a series of scaffold or side branches. Peaches, apricots and plums tend naturally to open into a vase or goblet shape, so pruning of these emphasizes an open center.

Once a young tree is set on the right course, it should only be pruned lightly until it begins to bear fruit, allowing it to develop sufficient leafy surface to grow a strong root network. Remove any dead or damaged branches and any that cross or grow toward the tree's interior. If two branches align with one closely above the other, remove the weaker, usually the lower, but use your judgment. Pruning cuts to remove a branch should always be made at the outer edge of the growth collar, that slight swelling where a branch angles out from a larger stem: do not leave stubs (which lead rot into the heartwood) and do not cut too deeply toward the crotch. Only the growth collar has the tissues that allow the tree to properly close its wounds.

### Pruning Bearing Trees

All tree fruits, after they begin to blossom heavily, will tend to set more fruit than they can ripen effectively. There is a natural self-pruning process that occurs early in the season, but with young trees, pruning out some of the immature fruits helps the remaining ones to fill out well, relieves stress on the tree and evens out the tendency to bear every other year.

A young apple, pear or peach needs about three dozen leaves per fruit, or six to eight inches of stem per fruit. A 3- to 4-year-old tree may do well with about 50 fruits, a number that will increase from year to year. About four to six weeks after flowering, you can see which blossoms have been pollinated, as these will be showing small fruits. Remove all but one or two in each flower cluster, favoring those that are largest and have no blemishes or insect damage. If you cannot tell which are going to fill out, wait another week or two to allow them to swell. With young cherries, plums and apricots, there is less concern about thinning the crop. The tree will thin itself to some extent, and the weight of small fruits is much less.

### What to Expect

Dwarf trees may begin to bear in their third year, semi-dwarfs in years four to five, semi-standards in years five to six, and standards from year seven on. Local conditions and stresses may advance or retard these times. A tree will typically put out a few blossoms a year or two before it begins to bloom heavily and may even set a few fruits. Some trees are biennial bearers.

Pay particular attention to the time of blooming and record this each year for each tree or major block of trees. Notice also what wild plants are then blooming, what birds are migrating through, what frogs are mating at the same time and other climate-sensitive phenomena. Each variety has a predictable period from bloom to ripening, so by keeping good records, you will learn when you can expect a harvest.

## Designing for Perennial Crops

With this basic understanding of tree biology, we can turn our attention to trees as components of the landscape and players in the ecosystem. Good design requires that we consider all the possible functions and roles that each element may serve. Trees can provide all of the following:

- shade and protection for people, livestock, buildings and smaller plants

- windbreak
- cooling through transpiration
- lowering a high water table and reducing salting from irrigation
- soil protection and soil building
- visual screening
- living fences
- mulch from leaf and twig drop
- wood products and fuel from coppice, pruning or felling
- fruits, nuts, saps, resins, flowers and leaves for food, medicine and utility
- fertilizer in the form of nitrogen fixation, canopy drip and leaf harvest
- fodder and forage for animals (including bees and wildlife)
- habitat for wildlife (including children)

With all these possibilities, we should never plant a tree that cannot serve at least three functions. Beauty comes in the bargain. And since trees live for many years and can be quite large, we need to think about the changes they will undergo and will create around them over time.

Trees also need certain things to support their health. Among these are:

- protection when young from predation by herbivores and rodents
- protection at all times from very large animals such as cattle, horses, donkeys
- protection from mechanical damage to bark and root damage by vehicles
- a thick layer of mulch over their roots
- adequate moisture
- mineral fertilization to improve soil (trees can provide some of this themselves)
- periodic inputs of nutrient, mostly as mulch, but also animal manure and soil-based nitrogen
- synergistic companion plants, soil fungi, birds and insects
- suppression of allelopathic plants such as grass or black walnuts

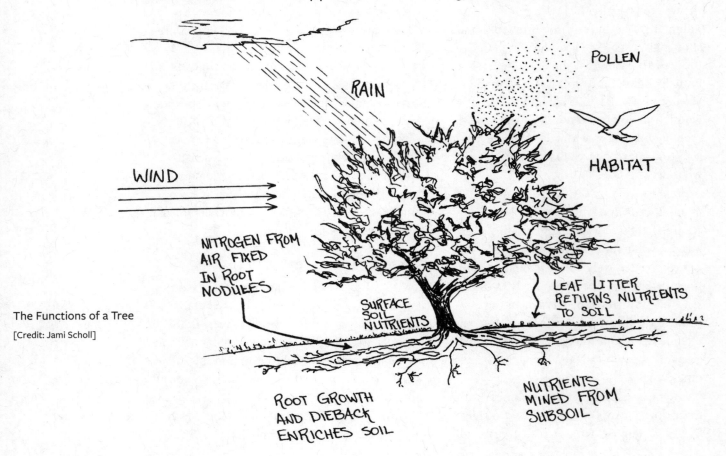

The Functions of a Tree

[Credit: Jami Scholl]

POLLEN

RAIN

WIND

HABITAT

NITROGEN FROM AIR FIXED IN ROOT NODULES

LEAF LITTER RETURNS NUTRIENTS TO SOIL

SURFACE SOIL NUTRIENTS

ROOT GROWTH AND DIEBACK ENRICHES SOIL

NUTRIENTS MINED FROM SUBSOIL

- if fruiting, they may need pollinators (insects or small animals) or pollenizers (nearby companions of the same or similar species)
- if fruiting, adequate air movement to reduce fungus
- sanitation, including the cleanup of excessive fruits, nuts or pods on the ground
- clean and competent pruning (no bad cuts, no spreading of diseases)

As they grow, trees create habitat around them, not only for birds, animals and insects, but for other plants and fungi. In nature, we call this habitat the forest, but in cultivated systems, we can create analogs that are *food forests* or *forest gardens*. The literature on these is considerable and growing[13] and builds on a large body of academic knowledge and practical discipline in agroforestry and ecology.[14]

### Forest Gardens

The term for these vertical and edible landscapes comes to us from Robert Hart, whose 1991 book introduced an ancient idea from tropical cultures to the English-speaking countries of the temperate world.[15] Hart's own garden was very small, perhaps ⅛ of an acre, and as authors Dave Jacke and Eric Toensmeier point out, it was seriously overgrown and chaotic. Nevertheless, Dave Jacke writes of coming to tears on seeing its astonishing beauty, which for him and for many of us, was reflected not only in the plants and the homeliness, but by the sheer intellectual splendor and emotional power of something truly creative, fully human and deeply connected to nature.[16]

Hart's models were tropical systems from Java, the Philippines, East Africa, Ecuador and Central America, and in those settings with high light intensity, seven vertical layers would be likely.[17] In the lower light levels of temperate latitudes, a more open canopy (from 40–75% coverage) and fewer layers permit adequate energy to reach the ground and make everything more productive.

The structure of a temperate forest garden is based on the canopy elements — the tall trees — which might be apples, pears, cherries, walnuts or chestnuts. Tall elements might also be nitrogen-fixing trees for fertility such as black locust, alders or even mimosa. Provision must be made for adequate spacing of these for their ultimate density, but since this can take up to two decades (depending on the species), short-lived perennials, even small trees, can be grown and removed before the design fills out. It's also possible to overplant deliberately as one might for a windbreak, and expect to remove every other tree after six, eight or ten years.

A compatible ground cover layer needs to be established early as well, in order that grasses and agricultural weeds not become a problem while light levels on the ground are still high. Though orchards are often planted to grasses for ease of maintenance by mechanical mowing, better choices would be other flowering plants that are expansive but more easily controlled than grasses. Martin Crawford of the Agroforestry Research Trust in Britain uses mints as a main ground cover in his forest garden.[18] Masanobu Fukuoka, in writing about orchard plantings, suggested cruciferous vegetables such as turnips and radishes, as well as alliums and nasturtiums.[19] Like most permaculture gardeners, Robert Hart grew comfrey. We use all these as well as burdock, rhubarb and horseradish, sedum, cup plant, fennel, lovage and yarrow; we also plant many annuals in open patches. Perennial clovers and vetches are effective. We tolerate henbit, which is an inconspicuous spreading weed that is easily removed, and wild carrot, both of which bees love.

With high and low layers in place, it is easier to fill in the middle bit by bit. Small trees and shrubs spread the cover. Shade tolerant *Ribes* plants have a prime place, as do serviceberries. Elderberry flourishes in wet soils. Chinquapins and hazels do well in the understory. And there are numerous nitrogen-fixers that prosper as shrubs (see Appendix 3) and can be cut for mulch and fodder. Many trees that would be taller if left to grow can be pollarded to manage them for leaf growth or mulch. If these are indigobush, Siberian pea shrub or autumn olive, you get nitrogen with each cutting. If linden, then you get flowers and leaves for tea and salad.

### A Different Character

What distinguishes forest gardens from conventional orchards of tree crops? In a word: polyculture. Forest gardens mix many species. Among these are always included functional supportive plants for mulch and fertility, for pest control and pollination, for fiber, fodder and medicine. One of the best controls for plum curculio in orchards turns out to be poultry.[20] And if your tree crops need birds for weed and pest control and for fertility, then you had best include plants among the trees that can feed the birds. Simple, no?

Scale is another differentiating factor. Forest gardens range from tiny plots of a few thousand square feet (such as Robert Hart's) on up to several acres. Martin Crawford's excellent forest garden in Devon is two acres, though he grows tree crops and has a nursery on a separate eight-acre plot. Conventional orchards, on the other hand, are rarely less than ten acres and often reach into the hundreds. They are designed for commodity cropping and compete on price by mechanized operations at large scale. The forest garden is meant to provide many kinds of products for household use and local trade, as well as a welcoming environment for its cultivators. It is a garden, and meant to be lived in. Joe Hollis, of Celo, North Carolina, calls his a *paradise garden*, which may be a little redundant but is certainly appealing. His garden is 1.5 acres and contains over 1,000 species of economic plants. The forest garden, literally surrounded by forest, provides Joe his living as an herbalist, apothecary and teacher.[21]

### Getting Started

With even a small forest garden, garden farm or permaculture system, many hundreds of species will ultimately be cultivated or introduced. And the density too, will be greater than that of conventional gardens and farms, so many more plants per acre will be needed. It therefore pays to learn to grow plants and to establish a small home nursery. This can also be a livelihood, a cash crop or a source of material for barter. The skills required for propagating plants are useful regardless of the design into which they are placed, so you can begin growing plants whatever your circumstances, as long as you have a little sunlight, access to water and the ability to make soil.

To support plant propagation, a small greenhouse or hoop house is very useful, not only because it gives tender plants a boost early in the season, but because it provides a

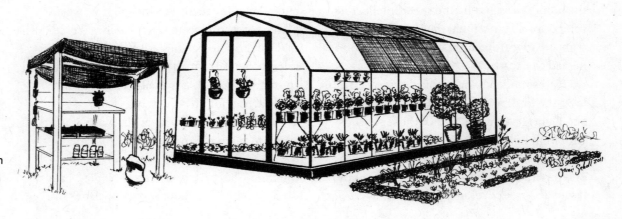

Greenhouse, Pattern #38 (See Chapter 6)

[Credit: Jami Scholl]

somewhat dry space for working when rain drives you out of the garden. It also provides a measure of protection from predators for the young plants. The greenhouse, of course, needs a potting table, consisting of a large surface (15–30 square feet) where soil can be mixed and flats and pots filled. Add a few bins and buckets, put up a rack of shelves, salvage an old metal cabinet, repaint it as a blackboard to post reminders, and you are in business! Oh, and don't forget — you need a hose connected to a water source and a supply of compost, so start gathering biomass.

Now, go looking for seeds and plants. You can search catalogs, collect in the wild and peruse public plantings. Get yourself a good pair of hand pruners (secateurs) and keep them with you all the time. And remember to have some collecting bags and envelopes for seeds. Don't forget to swap with friends.

## Designing Polycultures

When it comes time to put plants in the ground, especially long-lived ones, you need to consider the environmental influences on the site such as sun, wind, soil and water flow — and also the cultural influences such as proximity to the center of your farm, connections to other functional buildings (barn, shed, root cellar, summer kitchen), water supply, noise and pollution, views by and of the neighbors and privacy gradients. If you're unclear on these, it would be a good idea to review Chapters 5 and 9, which cover design practices and microclimates. For tree planting, a niche analysis (needs-and-yields) is very helpful, as is a view to the changes that will come as the tree matures.

Design can be made easier when it is understood as a language: elements of the landscape (plants, structures, animals, cultural influences) are words that can be put together in a variety of ways. Complete fluency allows you to express anything, but none of us starts out speaking language in full sentences with large vocabularies. We learn words and then string them together into memorable phrases. An archetypal phrase in permaculture design is based around the word *fruit tree*. A fruit tree could be any familiar or exotic fruit that will grow in your area: a pear, a mulberry or a lemon. The fruit tree word becomes a phrase when it collects its companions and becomes a *fruit tree guild*. This phrase occurs in the garden farming pattern language (see Pattern #53, Chapter 6) and is a widely recognized expression of permaculture.[22]

A fruit tree guild is a group of plants and their associates that centers around a fruit tree as its companions and supporters. The fruit tree is the organizing element, the star player, while the smaller plants, along with insects, fungi and animals, are the chorus, supporting actors and the orchestra — maybe even the ticket takers. Without them, the star won't succeed, but, no matter how excellent the other players may be, buttery and succulent

The Greenhouse
Potting Table

[Credit: Creighton Hofeditz]

pears or juicy lemons are going to steal the show, hands down.

### The Hazards of Creating Habitat

An important shortcut to successful forest gardening is to learn about and use a good range of friendly weeds. These are expansive or dispersive plants that are easy to use or to live with. They don't have thorns. If they show up where you don't want them, they're easy to move or to get rid of. They're pretty or edible or make good compost or medicine, so that you don't mind having lots of them. Lettuce is a friendly weed in our garden. We let some of it seed, and it shows up everywhere. Sometimes we reorganize the mob. Chickweed, mache and lambsquarters are friendly weeds. Dandelions are a bit stubborn but basically OK — they provide the first substantial nectar flow of the spring for honeybees. The reason we get many of these edible, medicinal and soil-nourishing plants is that they make lots of seeds, and we let them do it. And we create the conditions they like. In order to get friendly weeds, you have to grow a lot of useful plants and you have to develop your soil. And, I might add, you also have to suppress unfriendly weeds.

There is no getting around the fact that most domesticated landscapes are full of tenacious and often unfriendly weeds: creeping euonymous, poison ivy, japanese honeysuckle, multiflora rose. Every region has its candidates. Your goat would enjoy these, but you probably don't. They've had the run of the place for years without anyone saying otherwise. You can think of them as the neighborhood bullies. They'll have their way until you put your foot (or your hoe or weeding trowel) down. It's also true that as you create a woodland garden, build soil by accumulating biomass and mulch and change existing plant regimes, you'll get all the plants that love disturbance and do their best to cover bare ground. Our own weeds du jour are

hackberry trees, silver maple seedlings, black cherry and tap-rooted ashes, plus perennial grasses — especially Sudan grass — and bindweed, a morning glory relative. The latter two are stoloniferous plants that send runners under our sheet mulch. They can be vexing, but they can also be suppressed. They're only flourishing because we don't yet have enough things planted nor enough grazing animals eating the weeds. You can expect these kinds of challenges as you convert a system to perennials — don't let them discourage you. These opportunists dominate in the new beds filled with coarse mulch and not yet wearing their ground covers. As you fill the empty niches with plants you choose, unwanted weeds will fade into the background.

Forest gardening is part of a larger strategy to create productive woodland mosaics in our inhabited landscapes. We need the trees to capture carbon and store energy, mitigate climatic extremes, and we need them to help feed us as well. We also know that forests are healthier when they are connected to more parts of themselves. Forest gardens in our backyards and on our garden farms should be the most productive vertical landscapes we can design, and they should connect through hedges, fencerow plantings, street trees, neighborhood woodlots, community edible parks and regional forest reserves to a broad fabric of forests throughout the bioregion. These in turn should stretch out corridors to connect the islands of our national and state or provincial forests and parks to cast a net of healthy ecosystems over the whole countryside.[23] Only with a vision of this breadth and determined efforts to implement it will we be able to restore some semblance of the grandeur and natural wealth that was present on this continent when the first Europeans arrived. We owe ourselves, our children and the native peoples of America whose mantle of stewardship we have usurped no less than this.

# CHAPTER 17

# Productive Trees and Where to Grow Them

I view trees and shrubs as valuable elements to include in the food garden, whether it looks like a woodland or merely has small vertical elements dotted about. But there are many other settings in which trees are absolutely essential or can be greatly helpful. Trees can be used to enhance the productivity of pastures or of row crops. They can benefit aquatic systems, including ponds, riparian zones, wetlands and aquacultures. Trees are the main show in orchards of fruits and nuts, but polycultures of perennials can increase yields, reduce costs and work even there. Trees are critical components of windbreaks, hedgerows and living fences. And even in the tiniest landscapes, trees can add amenity, beauty and offer economic yields. In this chapter, we'll examine what elements to use and how to organize them in each of these situations.

## Living Fences, Hedgerows and Windbreaks

These strip forests play important roles in confining or excluding livestock and wild animals, buffering against wind and storms and in providing valuable habitat and reserves of fodder, timber, biomass and even food against times of dearth. Many small strips of woodland define boundaries, filter pollutants, screen unwelcome views and buffer noise, and they do this work on millions of properties.

Trees and shrubs are often planted along fence and property lines, and even when they are not planted by people, they often grow there, planted by birds that alight on wires and small animals migrating along the brushy, unmanaged strips of grass and weeds growing under fences. Fencerow trees are often neglected and weedy themselves, and just as often they are a maintenance headache, throwing limbs, buckling fences, harboring unwanted vines and sheltering garden pests. In defense of these little bits of wild land that weave the human landscape into nature's fabric, I must say that they also shelter native pollinators and a wide range of creatures who have too little habitat in our cities, suburbs and devastated farm fields. But we could maintain this service to wildlife without the headaches and with much greater yield of products for ourselves by turning some intelligent attention toward these linear enterprise zones.

### Harvesting Weedy Trees

If your city lot or garden farm is bordered by unmanaged trees, the first opportunity is to take a harvest of timber for poles, firewood and even sawlogs. This need not involve removing everything, but it does require a willingness to prune heavily. As with any forestry operation, the guiding rule is to "leave the best and take the rest." If the ownership of the trees is in question, and even if everything is entirely on your side of the property line, a decent regard for neighbors would involve a little discussion about impending changes and plans for replacement. People are often invested in the landscape around them even when they don't pay taxes on it. Regrettably many fencerow trees are misshapen, but any of them is likely to be worthwhile as fuelwood, logs for mushroom inoculation or sundry poles, stakes, trellis parts and of course woody mulch from small branches, twigs and leaves. Many smaller and more succulent parts may be fed to goats, cattle, sheep or rabbits (using good judgment and discretion in feeding, of course). Anything not processed by animals can be chopped up by machete or ground in a chipper to feed the soil of garden beds. Heavier limbs still too small to be worth stacking as firewood make fine durable mulch around trees.

When I take a tree apart, I assess the potential harvest. If the trunk or any large upper limbs have straight sections of ten-inch diameter or greater over lengths of at least eight feet, there is some potential for saw timber. Smaller diameter but straight sections may be split or used round as poles, either in building or, if rot resistant, for fencing. Except for use in crafts or boatbuilding, crooked sections are better cut into smaller lengths. Sound hardwood with undamaged bark of diameter from four to seven inches — often upper limbs of larger trees but also stems of smaller ones — has potential for mushroom inoculation. Mushroom log sections should be about three to four feet, depending on thickness. This is primarily a practical matter of ease in handling.

After sawlogs, poles and mushroom logs, the next crop to be taken from a felled tree is firewood. I cut for fuel everything not used for better purposes and larger than the thickness of my wrist. Smaller pieces are rarely worth the effort to carry and stack — it just takes too much time for the small heat value returned. If you don't have a use for wood fuel, there is likely to be someone in your neighborhood or network of friends who would trade labor for it, and in many areas firewood is a worthwhile crop, especially if packaged for retail sale. Buck the wood into lengths that can easily be picked up by one person. I find that it is easier to pick up and handle fewer pieces as long as they do not exceed a threshold weight that strains the body. Picking up 50 bolts of 25 pounds each is actually much less strain than bending to the ground to pick up 150 pieces of 8 pounds each. The latter will tire your hands and wrists as well as your back because you'll try to grab each piece in one hand and won't think much

Hedges harbor and feed pollinators, pest predators and domestic livestock. They also buffer against noise, storm and pollution and can provide a reserve of food and timber.

[Credit: Jami Scholl]

Albert Bates displays inoculated mushroom logs in a Tennessee woodland. Inset: Shiitake fruiting on hardwood.

[Credit: Peter Bane, inset photo by Kathie T. Hodge]

about bending your knees, and you'll bend over more times. You can cut wood more accurately and efficiently to length if you use a sawbuck.

And if, as we do, you produce solar electricity, you'll be able to cut some of the smaller wood on a chop saw in your wood-yard with green energy rather than running your noisy and polluting chain saw on blood-drenched gasoline. This will also keep more of the sawdust, wood chips and shavings where they can easily be diverted to a garden, a compost pile or a dry toilet instead of leaving them in the woods or on a parking lot.

## CULTIVATING SHIITAKE ON LOGS

Select undamaged log sections weighing less than 30 pounds from trees that are alive when cut. Reject material that has split ends, branch stubs, large gashes or evidence of canker or rot on any side; use it instead for firewood. Harvest for mushroom cultivation is best done when trees are not in leaf or are going into dormancy (roughly September to March in the mid-northern latitudes). Inoculate the logs within 30 days of when they are cut. Plugs or sawdust spawn may be used; place the spawn in drilled holes or in chainsaw cuts if you are working away from a power source. The holes or cuts need be only ⅝″–¾″ deep. Cover the spawn with melted cheese wax and rack the logs in a damp location out of direct sunlight. Water occasion-ally if there is no precipitation for several weeks at a time. Within three weeks shiitake spawn will have "run" through the log, and within 9–12 months the log will begin to fruit when heavy rain, seasonal temperature change or immer-sion triggers a flush of mushrooms. After the first flush, additional fruiting can be stimulated at about five-week intervals by soaking the logs for 24–48 hours. Mushrooms will emerge within eight days. A heavy-fruiting log can bear three pounds of mushrooms in a single flush, though smaller harvests are more common. Logs can yield well for up to five years and at a much re-duced rate for twice that. Shiitake have natural-ized in mushroom yards where cultivation has been continous for over a decade.

A sawbuck made from salvaged lumber holds 3–7 logs at a time, reducing the number of saw cuts required to produce stovewood.
[Credit: Creighton Hofeditz]

Hügelkultur beds are built up deeply with coarse woody material and topped with soil.
[Credit: Jami Scholl]

Anything not suitable for firewood might, if long enough, still be used for beanpoles or similar trellis, but the remainder winds up as mulch or animal fodder. We chop ours into 8- to 10-inch lengths and pile it up to make *hügelkultur* beds — raised garden beds built on a base of woody debris.

We also mulch tree crops with this coarse carbon. It's a valuable way to increase the fungal content of your soils.

Once you have removed unwanted trees from the fence line, you can clear out the troublesome vines and shrubs and begin considering what you want to having growing there.

*Designing Boundary Woodlands*

The design phase of a living fence or hedgerow is similar whether you have to remove unwanted vegetation or are starting with no competition. You first determine the height you want, and then select species that can be managed at that height and that give yields you want. For a fence of say 8 to 10 feet, figure on one tree or shrub every 18 to 24 inches initially. This forms a single row that can be kept three to four feet thick. You may eventually thin out a few things, or even add in more where plants don't do well, but this will allow your living fence to close horizontally within three to four years. It is important to vary the height and growth habit of adjacent plants so they can nest among each other, filling in all the space in two dimensions.

For a hedgerow, you can expect to plant a thicker strip, say 10–25 feet, consisting of two or even three rows of trees and shrubs. If the hedge is narrow, you may plant in a kind of zig-zag pattern with two rows of plants offset from each other.

A hedgerow is likely not to be cut as short or as often as a living fence. Therefore, the trees selected for it should be chosen for their ultimate height. If you want the hedge to be a dense visual screen, sound buffer or filter against pollutants, you need it to hold leaves at all levels, so plant shrubs among trees and leave side branches low to the ground.

In both hedgerows and living fences, thorny shrubs and trees are welcome. While you might not want the 12-inch thorns of honey locust except at the far edge of a pasture, many other species such as blackthorn

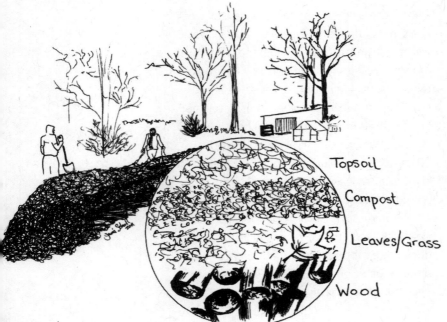

Topsoil

Compost

Leaves/Grass

Wood

Two-Row Planting Pattern for a Hedge or Living Fence [Credit: Peter Bane]

or sloe plum (*Prunus spinosa*), hawthorns, roses, quince or hedgeapple (*Maclura pomifera*) could be very helpful. You want to manage for dense, shrubby growth, so the tendency to sucker is also beneficial. Not only does this provide you with a source of new nursery stock, but it fills in gaps in the planting with many small stems, ultimately making the hedge or fence almost impenetrable. Trees in the fence can also be cut partly through the stem and bent over to the ground while still connected to their roots by cambium. These will then send up vertical shoots and this new material can be woven. This technique is called *laying hedge*, and it provides a thick, sturdy barrier against animals. The laid stems continue to grow; they may even root where they touch the ground, and the whole assembly becomes much stronger. Even where you do not cut through stems to bend them over, you can weave smaller branches together to form continuous horizontals — getting your trees and shrubs to link arms.[1]

Besides their value as barriers, these strip forests become an important reservoir of genetic diversity and habitat for birds, small animals, amphibians and reptiles. They dramatically increase the ecological diversity of small properties without taking anything away from main cropping areas. Beneficial insects and pest predators abound and from their habitat on the edge will have excellent purchase on garden pests. A profusion of flowers provides bee forage through much of the year, and you have a steady source of canes, poles, wattles and eventually stovewood from trimmings. If you like, hawthorns can be grafted with apple scions; Peking cotoneasters (*Cotoneaster acutifolius*), quinces (*Cydonia oblonga*) and mountain ash (*Sorbus* spp) can take pear scions. There's room to play and experiment.

### Multi-Species Windbreaks

Here diversity remains important, but with generally larger areas to cover, it may become somewhat more formulaic. A small living

fence might contain as many as 40 species, while a long windbreak might be limited to 5–10 for simplicity of planting. Attention should be paid to filling in gaps, and this is at least as important with a windbreak as with a fence or hedgerow. If a windbreak is needed, it's worth designing it to succeed, and that means it needs from three to five rows of trees and shrubs, and these should include several species at different heights.

The most difficult aspect of establishing a windbreak is often simply getting trees to take hold in a tough situation. In the harshest conditions — on the edge of the desert, the ocean or the tundra — plant anything that will grow, no matter how otherwise useless or undesirable.

In the lee of these sacrifice plants, you can start those that are slightly more tender or useful but will persist and grow well once rooted. Another challenge with windbreaks under strong windy conditions is the loss of some plants, creating gaps. Narrow openings channel and accelerate the wind, increasing its destructive force. This is an important reason (though not the only one) for using multiple rows and multiple species. Both of these kinds of redundancy ensure that diseases,

This tender bromeliad (inset) was flourishing in the lee of these severely flagged trees on the coast of Chile, facing constant salt winds off the Pacific.

[Credit: Peter Bane]

accidents and natural stresses will not render the windbreak ineffective. In the same way, the windbreak should also extend beyond the area it must protect because wind whips around the ends of any barrier with increased force.

## Alley Cropping Trees

There is one other important kind of strip forest or linear planting of trees, and that's *alley cropping*, where arable crops of grain, oilseeds or vegetables are grown in broad alleys between lines of trees, called production hedges. This is a form of stacking yields in time: the annual crops bring in cash while the trees grow to a harvestable size. It can also be a way of supporting the fertility of the land and of stabilizing microclimates to improve yields of the arable alleys. From annual leaf drop and regular root pruning but also from mycorrhizae, trees will make minerals, mulch and sometimes nitrogen available to the grain or vegetables. Facing competition for water

Alley Cropping,
Pattern #15
(See Chapter 6)

[Credit: Jami Scholl]

and nutrient at the shallow soil layers, tree roots will dive deeper, making better use of the whole resource but also sharing it through leaf and twig drop and regular root turnover. The herbaceous or annual crops can also provide surplus nitrogen to the trees if the low crops include clovers, vetches or other nitrogen-fixing species. As Patrick Whitefield points out, trees can take up nutrient even in winter, so their presence in these multicrop systems helps to prevent leaching of soil nutrients at the times when this is most likely.[2] The same applies to their use of water, which can drain away or evaporate if not being actively cycled by plants.

In choosing species for the tree element of alley crops, upright canopy bearers should be selected, and interactions with the potential arable crops should be considered. Trees may be pruned or coppiced to increase the light available to the understory or to improve the value of the timber. Paulownia, which requires coppicing to produce straight stems for timber, is a good example. In China, it is grown amidst grain and vegetable crops. Rick Valley, a permaculture designer, teacher and nurseryman in Oregon, has grown black locust (*Robinia pseudoacacia*) with *Phyllostachys* bamboos to form a fencing polyculture. The locust are harvested as poles, and both when they are cut as well as when they grow, they release nitrogen harvested at their roots by colonies of rhizobia. The bamboo loves the nitrogen, and its canes provide adequate horizontal members for the fences. The horizontals need replacing more often than the posts — black locust is durable for about 50 years in the ground — but the bamboo grows faster and more copiously. Italian alder (*Alnus cordata*) is a tall and upright species that also fixes nitrogen and can support understory crops. Hybrid poplars have been grown on three-year rotations between alleys of wheat, rape and turnip in Europe. The poplar reduces crops yields about 12% but gives nearly four tons per acre of biomass from very little land. This appears to be more productive than monocropping the trees.[3]

## Pasture and Fodder Trees

*Silvopasture* is the practice of growing trees in livestock paddocks or of grazing animals in woodlands. When both trees and animals are managed for optimal interaction, the benefits can be considerable.

The most productive pasture grasses are so-called cool season species that grow well in spring and again in fall but go dormant during the hot summer months just as young ruminant livestock are growing fastest and the need for forage is peaking. This is the season when trees are in greatest leaf and are often ripening edible fruits, pods or nuts. So tree forage can complement the productivity of pasture grasses, potentially removing a limiting factor (the late summer dearth of grass) to the number of animals that can be raised on a given acreage. A light shade cover over pasture can also lower soil temperatures and thus increase grass productivity by enhancing biological action in soil.

The challenges in silvopasture are 1) to introduce and establish trees where none exist, 2) to select suitable fodder species and 3) to distribute the tree effect for most benefit to land and livestock. If we examine these in greater detail, we will see that they are interactive, so the solutions must be holistic.

When there is shade in part of the pasture, the cattle will loaf there, manuring more under the trees, and thus will concentrate nutrients from the whole pasture in a small area, leading to declines in productivity. The need is to spread the shading effects widely and evenly. On the other hand, the establishment of trees requires their protection when young, and one of the easiest ways to do this is to plant them between lines of fencing that separate paddocks. For spreading species that drop edible pods or fruits such as honey locust and apples, fencerow establishment makes sense. The problem of uneven graz-

ing can be addressed by tight rotations with portable electric fencing to move the animals. The zones along the fencerows might then be reserved for late-season grazing when grass productivity in the open areas is lowered and the trees are yielding fruit.

Tree can also provide fodder from leaves and stems. Experiments in Europe have shown promise for mulberry, black locust, *Amorpha fruticosa*, *Colutea arborescens*, *Medicago arborea* and *Coronilla emerus* (the last five of these are legumes). Willow, alder, hazel and poplar leaves and stems are all known to be edible by ruminants, and many rose family fruits are as well. Basswood (*Tilia americana*), ash and elm are also good candidates. Ruminants relish tree fodder and will "serve themselves" if the trees are within reach, thus reducing labor. Modified pollarding would keep most of the foliage within reach without risking the main stem. Where animals are confined to stalls or cages, as with rabbits for example, trees and shrubs can be trimmed or coppiced to yield "cut and carry" fodder. Though the practice today is largely confined to eastern and southern Europe, *tree hay* can be made by lopping branches and storing them for later use. As with grass or herbaceous hay, the material is best taken when green and succulent, allowed to wilt slightly and then bundled. Reducing moisture levels helps to prevent mold.

To reduce damage to young trees in pasture, there are three basic strategies:
1. Fencing the animals out of blocks until the trees are large enough to withstand contact
2. Providing individual tree protection structures during establishment
3. Spraying with repellents

In the first case, some effort must be made to manage the understory in the period prior to grazing; haying is the most common practice. Electric fencing is the usual preference for block exclusion. We have used rough-stacked cribs of small-diameter tree limbs to create temporary fencing around fruit trees with good success. Protective structures should be matched to the size of the grazing animals. Repellents consisting of animal dung, egg and abrasives such as sand have been shown affordable and effective against biting damage.

Another method of establishing trees excludes the animals for two to three years and uses fast-growing species that get tall enough in that time to eliminate most subsequent damage. Pines, eucalypts and poplars are all grown in this way. Hay is taken from the ground between the young trees in the seasons before the animals are introduced. A variation on this is to overplant and graze the land when the trees are less than two years old. This will result in some damage, but mainly to smaller and weaker trees that would be culled in any case.

## Trees in Aquatic Systems

Along flowing waterways, streamside trees provide important benefits. These riparian buffers are essential to maintaining shade to keep water cool for the health of fish and invertebrates. Strips of woodland within 50 feet of streams also dramatically reduce soil erosion and can mitigate the damage and danger from floodwaters by anchoring streambanks and by slowing and filtering high water overflows.

On the other hand, trees adjacent to very small ponds may be problematic. Their shade reduces aquatic productivity. Leaf fall can overwhelm the pond ecosystem with too much organic matter, and a reduction of air movement over the pond can lower oxygen levels. With larger ponds or where trees are sparse, none of these issues is significant.

Many trees grow in swampy land where their roots help to stabilize soils and filter water flow. The North American woody species with greatest affinity for water are the willows (*Salix* spp), sycamores (*Platanus occidentalis*) and bald cypress (*Taxodium distichum*), but alders (*Alnus* spp), birches (*Betula*

spp) and witchhazel (*Hamamelis virginiana*) are also commonly found in riparian zones. Sycamore is a huge tree, among the largest that grow in temperate latitudes, and unlikely to be suitable for small properties, but willows come in all sizes and there are more than 200 species to choose from; many are used for furniture, basketry and medicine. Cypress is a valuable rot-resistant timber species, now somewhat overharvested and confined to warmer regions. It grows in our county in southern Indiana, but is most often associated with the lower Mississippi Valley. In Louisiana it is intercropped with crawfish in a kind of aquatic silvopasture.[4]

Willows have also been used in water treatment systems that are a specialized form of wetland. Using their propensity to pleach or fuse stems and their pliant tops and branches, Richard Wade, an American permaculture designer living and working in Spain, has created large woven baskets of willow as filters for sewage systems. The baskets, which are up to 6 by 12 feet each and 6 feet deep, are filled with soil, and the willows are allowed to root into it. The effluent is then piped into the top of the baskets and allowed to filter through the contained soil. The willows take up much of the moisture and extra nutrient, and the baskets can be trimmed periodically on the outside to harvest mulch material and cuttings for more willow work.[5] Danish sources document the use of willows in closed waste treatment cells where the plants are able to transpire and filter the effluent flowing under and through a gravel medium in which the trees are rooted.[6] The ability of willows to transform waste and to grow well on marginal land is promising for short-rotation coppice management with the aim of producing biomass for energy.

## Tree Cropping in Orchards

Most conventional orchards provide poor examples for the garden farm, which may nonetheless have several acres of crop trees planted as an orchard. Where this is the case, the grass beneath the trees should be mulched out by smothering it under cardboard and covering with wood chips or coarse tree trimmings. Alternatively or as a first step, poultry or pig tractors could be used to weaken or eliminate the sod. Seedballs of daikon, turnip, mustard, nasturtium, yarrow, fennel, dill, caraway, coriander, clovers and vetches can be prepared and cast into the coarse mulch. The umbels, cruciferous plants and legumes provide habitat for pest predator insects or fix nitrogen to feed the trees. Coarse woody mulch should be used to build up the fungal component of the soils and assist with the growth and development of mycorrhizae.

Some excellent heritage varieties may be discovered in older orchards: don't get rid of anything until you know what it is. Poor existing stock presents an opportunity to top-work well-established roots with new, more desirable scions. If major prunings are to take place, the space thus opened up between trees can be planted to supportive shrubs such as autumn olive, caragana, indigobush or other climatically appropriate nitrogen-fixers that should then be managed for coppice and fertility enhancement.

When regenerating an older orchard planting, attention should be paid to mineral supplementation of the soil. In humid areas, lime is almost always useful, and trace minerals in the form of seaweeds, fish emulsion or crabshell should be applied in almost all cases. If the soils are contaminated with persistent organophosphates or heavy metals, the best approach is to apply a heavy mulch of woody debris and chips over cardboard, effectively isolating the contaminated soil and sequestering it in a matrix of organic matter. Any organic pesticide residues will be digested by fungi over time. Heavy metals will be immobilized, and if good nutrition is supplied to the trees they will ignore the toxins. Applied along with the woody debris, a light dusting of sugar or a molasses spray will aid fungal

growth and digestion of the mulch. Some species of mushrooms will strongly bioaccumulate heavy metals — *Gomphidius glutinosis* is a powerful concentrator of radioactive cesium, for example — and the fruiting bodies can be collected and removed to decontaminate the underlying soil.[7]

This Danish plantation of fast-growing willows supplies wood chip fuel for a biomass power generator.

[Credit: Peter Bane]

## Trees for Small Situations

Dedicated gardeners have grown citrus and figs in pots for centuries, bringing them indoors to conservatories and greenhouses during winter months, so it is quite possible to have productive trees even on balconies — you simply have to keep them small. Where growing space is on the ground but limited, fruit trees can be grown on wires — this is called *cordon*. Trees on dwarf rootstocks are planted with the main stem at a 45-degree angle; side branches are trained onto the wires running in either direction, much like grapes. Growth on either side of the wire is pruned back to keep the trees in a plane. Adjacent trees in a cordon may be pleached to provide additional root support. Similar to this but taking advantage of the thermal mass and microclimate of masonry are *espalier* trees, also grown in a

Espalier trees are staked against a masonry wall to take advantage of the special microclimate it provides.

[Credit: Cynthia Edwards]

plane but pegged to a south- or west-facing brick, concrete or stone wall. The anchors must be solidly in the wall but must open for adjustments and to allow the tree branches and stem to slip and grow. The extra heat of these special zones enables one to push the climatic edge of any location, while using a tiny space to good effect. The leafy growth of the trees also provides cooling for any building to which they are attached.

All these kinds of trees, whether in pots or rooted in the garden, must be given special care and attention to ensure that their roots are not trampled, bound or waterlogged — and that the plants have adequate nutrition. Foliar feeding of micronutrients, which benefits all fruit trees, is very important for these vulnerable specimens. Also, for espalier trees, irrigation may be more often needed because the tree can only draw water from the ground on one side.

## Trees for Fuel

The ethics of burning wood are first and foremost that you must plant trees. Secondly, cut no living tree for wood unless you are doing a necessary job of pruning or removing a tree that is dying, diseased, damaged or in the wrong place. Thinning woodlands to improve the health, vigor and productivity of the remaining trees (resulting in a net increase in standing wood) counts as virtuous action. Thirdly, respect all wood. When wood products are worn out or broken, remove their incombustable parts and put them in the stove or on the floor of the forest garden to decompose. That carbon either belongs in the soil or it should be keeping you warm. There is so much wood used for pallets, packaging, construction and furniture that goes to waste every year, that hundreds of thousands if not millions of North American households could be heated if it were properly salvaged. Put in landfills, that wood waste creates methane. Left on the ground, it turns into carbon dioxide fairly fast. Neither gas is helping the planet to stay cool.

The economics of using wood for fuel require that it be harvested nearby. It's too bulky to be worth hauling more than a few miles. We get virtually all of our fuel from within a half mile of our home. Healthy forests can grow about a cord of wood per acre per year.[8] In the days of inefficient stoves and open fireplaces, great value was placed on hardwood, and it is true that it burns longer, but for heating a well-insulated building, lighter-weight wood is often ideal, especially in spring and fall, or at other times when a quick, short-lived fire is all you want to take the chill off or dry out the room. The highest heat value comes from the hardest woods: black locust, dogwood, osage orange, hickory, all of them denser than oak.[9] Black locust burns so hot that it can warp cheap stoves. These are the woods that will burn all night and leave a good bed of embers glowing for the next morning. But for most situations, far less select fuel will do just fine.

In a small permaculture system, wood for fuel is much like meat for food. It's a by-product — in this case, of growing trees for other purposes. You can still choose species with fuel value in mind, and if you heat with wood, you'll come to appreciate the varying qualities of your local forest species and urban woodlands. Where you have the room to grow trees for fuel alone, weedy or disparaged species are often a good choice for a coppice woodlot — precisely because they are vigorous and hardy, e.g., Chinese and Siberian elms in the Great Plains and Rocky Mountains, silver maples and sassafras in the Midwest or ailanthus in Virginia and other parts of the East.

If you accept the obligation to use whatever wood comes your way, and not to waste anything, you wind up with plenty of odd shapes. Perfectly fine fuelwood can be grown in a coppice woodland, and so the living fence, hedgerow or windbreak that you planted ten years ago will have quite a bit of fuel in it for the harvesting. In the meantime, there are plenty of trees that need to be trimmed, felled or salvaged where they have fallen. Empty lots, rental yards and alleyways are always getting overgrown, and quite a few people would be delighted to have someone trim out the weedy fence line that shades their garden. If you are determined to heat with wood, you'll find there's no real shortage — unless you are living in the desert. This will change in a decade or less; for now, take advantage of it.

## A Tree's Unseen Companions

No consideration of forestry or tree cropping would be complete without a view of the fungi that make tree growth possible. All manner of fungi are essential to tree growth and decomposition.

Fungi are valuable crops in their own right and can be cultivated. In some settings, they may be the most valuable crops that can be taken from forest land, exceeding the timber in potential value by two or more orders of magnitude. Among the best known and most widely grown species on wood is shiitake (*Lentinula edodes*), a choice and dense Asian gilled species that is rich in nutrients and immune support compounds.

Using about 180 logs with a weight of about 2,000 pounds, we were able to generate in one six-month season $1,600 in wholesale value of shiitake mushrooms. The mushroom yard occupied an area of 250 square feet (little more than a large room) under the dense shade of hemlock trees. After inoculation costs, which ran about $100, the only input besides our labor was rainwater diverted from a nearby roof. The logs represented a by-product of necessary tree removal, but could have been harvested from about a half-acre of hardwood forest on a sustained basis if needed. For financial contrast, the value of that much oak firewood is about $60; an equivalent weight of sawlogs (and these small-diameter logs were not suitable for the mill) would have been about $200. Shiitake are considered choice edibles and have nutritional properties, but the return on cultivation of highly valued medicinal species like Reishi or Maitake would be much higher, with prices exceeding $100 per pound for top-quality material.

Woodlands provide good conditions for cultivating mushrooms because they are not only filled with food for fungi, but they are shady and moist. Fungi do no photosynthesis, but feed on the breakdown of plant matter. Like all cultivated mushrooms, shiitake logs benefit from being kept in deep shade, helping them conserve moisture. Direct exposure to sunlight for many hours a day can cause the mycelium to dry out and lose vitality. Where tree shade is not available, artificial shading must be created.

Oyster mushrooms (*Pleurotus* spp) also grow on logs, and may be cultivated, but the fungus is aggressive and they are common in woodlands so can often be wild harvested. More delicate than shiitakes, they must be gathered quickly before insects infest them. Oyster mushrooms are one of the best for breaking down complex hydrocarbons such as petroleum or pesticides. Reishi (*Ganoderma lucidum*) is one of many polypore or woody-shelf mushrooms valued for their immune stimulating and even anti-cancer properties. They too can be grown, usually on stumps or short, thick logs. Other stump-sprouting polypores of note include sulphur tuft or chicken-of-the-woods (*Laetiporus sulphureus*) and maitake (*Grifola frondosa*), also known as hen-of-the-woods. Both of these are eaten and can be cultivated. Maitake and Reishi have shown significant anti-cancer, antiviral and antibiotic properties in clinical tests.[10]

Many other mushrooms can be cultivated, some on wood, some on straw and other substrates. And choice edible and medicinal varieties also may be wildcrafted. When mushroom season comes round with the rains of April and May, many Midwesterners go looking for morels (*Morchella* spp), a delicious edible fungus with a sponge-like cap that commands nearly $40 per pound in local shops. No one quite understands how morels grow: they are saprophytes (wood-eating fungi) but seem to exhibit mutualism with trees as well. We have had them simply appear in our garden beds mulched with coarse shredded wood and sticks. You don't have to be religious to consider that a blessing!

Whether for their fruits, nuts and leaves, for their flowering beauty or for their beneficent and valuable fungal companions, trees and small woodlands provide the fundamental ecological structure of the garden farm. They should be planted, grown and harvested in as many ways as imaginative design and good management can sustain.

Having sketched out a framework for the cultivated ecosystem of a garden farm, we must now look more closely at the human-made structures, both visible and invisible, that allow it to function smoothly for the people whose lives its supports.

# CHAPTER 18

# Structures, Energy and Technology

A farm is a working landscape. Even though I address people who live on lots of 10,000 square feet or even smaller (as well as those with up to 25 acres), I expect that any home with some land around it can be made more productive than it is today, and that you, dear reader, want to do just that. The primary design aim of garden farms is sustenance for their inhabitants; secondary aims include support for regional ecosystems and societies and cash income for the operators. In order to achieve sustenance, we have to know what sustains us. We need, besides clean air and the ability to regulate our body temperature, redundant sources of clean water, shelter from rain, cold, excessive heat and predatory organisms. We also need adequate amounts of nourishing food throughout the year, day in and day out. And we have many other emotional and psychological needs — needs for other people, for affection and inclusion, for creative expression and for freedom to learn, exploring and understanding ourselves and the mystery of the world around us. We also have special needs that occur when we are very young, very old, ill, infirm, pregnant, nursing or disabled. These non-material needs have, in some cases, material components or supports, and I will touch on a few of those in this chapter.

The structures of the home system are an extension of the human and communal bodies. The house itself is a third skin, beyond our clothing, within the environment of the planet, helping us to regulate our temperature and to store valuable materials. You can farm a small property without living on it, but it's harder. You will still have to live somewhere (else), and so your time, energies and attention will be divided. Where you sleep and eat is where you have the greatest influence. It's where you see what's going on, where you know the inhabitants, where you are welcome and empowered. If these things are not true of where you live, then you are living in the wrong place. And if you dream of providing much of the food, water, energy and livelihood for yourself and your household, then you need a home with land around it, or very nearby.

The structures of the garden farm begin with the home: a building of some kind in which people can sleep and store valuables against weather and predators. Most homes have a kitchen (a place where food can be prepared and eaten) and also a bath. This is enormously helpful and common, but not essential. In many cultures, the kitchen and the bath are separated not only from each other, but from sleeping quarters and from meeting spaces. All these functions may be grouped close together in a family compound, village or camp, but for the homeless they are often scattered over an entire cityscape. In North America we put kitchens and baths close by each other in the same building to save the cost of plumbing water, and because people expect to be able to eat, bathe, sleep and visit with each other in their homes — and to do all these things without putting on galoshes.

Most conventional North American homes built within the last 100 years embody the assumptions of the middle class: that food comes from the grocery but is usually prepared at home (frequently for individuals, often for the family and sometimes for guests); that life needs can be met by the marketplace; that adults will be employed outside the home with the limited exceptions of old people, the infirm and disabled and mothers of very young children; that children will be educated in school; that machines will perform or aid as many repetitive types of work as possible (sanitation, laundry, cooking, cleaning, landscape maintenance, transport); that people will live as couples (households of two adults) sometimes with children and sometimes as singles and that specialized forms of work needed at home (carpentry, plumbing, electrical and mechanical repairs, tree trimming) will be jobbed out to tradespeople. Besides wastes and pollution, the typical North American home produces almost nothing more than some meals and does only clothes cleaning, minor repairs and perhaps grass-cutting for self-reliance.

The garden farm, in contrast, grows, harvests, preserves and prepares food; builds soil by composting and other means; captures, stores and distributes water; cuts and processes plant and animal matter to make other things and to yield food; recycles nearly all of its own biological wastes; provides a good measure of its own energy for heating and other purposes; makes and repairs common household goods and even specialized tools and machinery; may provide medical and veterinary care to people or livestock and supports its residents to work gainfully at home at least part of the time and as much as possible. And of course, it also supports bathing, clothes washing and house cleaning, storage of valuables, plant maintenance, socializing, sleeping and everything worthwhile that also goes on in conventional homes.

To support these other functions, structures not commonly found in the conventional middle-class house are needed. Some of these can be adapted from existing structures, and other will have to be built. To do the work of a garden farm, the labor of more than two adults will be needed, at least from time to time. And to facilitate that labor, meeting over food is important, even essential. So the dining capacities of the home will be put to good use capturing energy in the form of voluntary or traded labor. Most homes have a kitchen; the successful garden farm needs an ample one with room for several people to work at the same time. It should be intimately connected to spaces for socializing and eating: a dining room, a covered porch, a deck with a picnic table. Most people already stage some form of social gathering in their homes now and again, and most people could organize a party. Now imagine doing it every week and even several times a week, perhaps working up to every day or at least five or six days a week. How would you streamline available spaces? Do you have enough chairs, tables, flatware, plates, bowls and cups? Paper cups are fine for the big summer barbeque, but everyday? These challenges are very modest but real.

## Structures for Food Storage

Most houses are limited in their food storage capacity. There will be one refrigerator with a small freezer unit built in. There will be a few cabinets in the kitchen for canned and dry goods, perhaps a few shelves in a small closet as well. Large homes have basements or garages where shelving units can hold extra canned food, and there may be a large freezer or an extra refrigerator sometimes used for drinks. In the event of a prolonged power failure (the sort that is predictable every 5–10 years), refrigerated and frozen food may be usable for only a few days. Lessening dependence on the industrial food system means moving toward a condition in which the household stores several thousand pounds of food, more in the months of late summer and autumn when food is naturally abundant.

A resilient household, at the center of a garden farm, will begin increasing its food storage. This means creating a pantry where preserved foods in stable storage can be kept without electric power beyond the minimal maintenance of baseline building conditions (dry, not frozen, not excessively hot, free of vermin). The pantry should be near the kitchen, next to it if possible. Stored food should be kept cool, dry and largely in the dark. Some food can be stored raw for months without cooking, salting, drying or smoking it. But where this is the case, as with some fruits and vegetables, grains, nuts and pulses, special storage conditions are needed. A cool dry pantry is suitable for onions and garlic, winter squashes, sweet potatoes, grains and pulses in airtight containers and for nuts in the shell and free of worms. Such a space is also useful for the storage of seeds in jars and tins. A cool damp room or root cellar is needed for apples, pears, root vegetables and potatoes.

A pantry should have shelves and room for a few bins of bulk items. Air circulation is important for vegetable storage. A sure way to get winter squash to rot is to pile them up and leave them in a closed room for a few months. Sweet potatoes can be binned, but need to be wrapped individually in newspaper. Garlic is often braided; this allows the bulbs to be handled in bunches but keeps them from spoiling. Onions can be treated in the same way, but their tops are often clipped, so storage in open trays is a good alternative. Many items that are needed in the pantry will be put up in cans and bottles, cooked, dried or pickled. Shelving helps make them visible for ease of use. A spare bedroom, an insulated porch or part of a dry basement can be converted to pantry use without completely foregoing its previous functions. We have a guest bed in our pantry, and sometimes people stay there. But remember that the pantry will become one of the most important rooms in the house, and it should be convenient to the kitchen.

Converting part of a basement into a root cellar is one of the best options for houses with below-ground space. Sometimes large

Interior of the above-ground root cellar at Renaissance Farm

[Credit: Creighton Hofeditz]

volumes of produce will need to be brought there, so a direct entrance from the garden or driveway is desirable. A root cellar must be vermin-proof and should be connected to a large thermal mass. Traditionally, this was the earth itself, but we have built a cellar attached to our aboveground cistern.

This has an insulated roof and outer wall that is further isolated thermally by earthen backfill and a green roof. We chose not to dig a cellar underground because our water table is very high at least part of the year, and we didn't want to have to pump the cellar to keep it dry. Permaculture designer Doug Clayton built a cold room in his New Hampshire house that uses small fans and thermostats to draw cold air into the room from outdoors whenever the ambient air temperature is colder than the storeroom itself. We achieve a similar effect crudely by leaving the door open with a screen closure overnight on cold nights in spring and fall. Ventilation is needed at other times, so design the building to draw air down into the earth in a cool location and vent it up into the root cellar and out through a small chimney that can be closed off as needed.

An earth-sheltered cellar can be expected to maintain earth temperature in summer (about 55–57°F), but can be gotten cooler

A root cellar needs positive ventilation to limit the growth of mold.

[Credit: Abi Mustapha]

in winter by drawing in outside air. Under ideal conditions, it can maintain 34–36°F for months on end, effectively becoming a refrigerator. This is workable because the greatest food storage need is during the cold months of autumn, winter and early spring. As temperatures rise in late spring, new crops begin to ripen for fresh eating and stored food will be consumed but not replaced until late summer or early autumn. A root cellar is a good location for doing some kinds of ferments: beer, wine, cider and sauerkraut to name the more obvious ones. A cellar is also good storage for hardwood cuttings being held over winter and for tender bulbs that must be pulled from the garden. In our situation, its location attached to the water tank and covered with earth and stone makes it fireproof and a good tornado shelter, which we don't have anywhere in or under our house.

## Other Food Storage Structures

Food that can't be stored raw, whether damp or dry, must either be smoked, salted, dried, fermented, pickled, immersed in fat, canned or frozen. The choice of method and the actual embodied energy of the preserved food will depend on your climate, the quantity and nature of the crop to be preserved and the weather at the time of year when it must be processed. Canning and freezing are strictly industrial technologies, the former somewhat older and more resilient.

Smoking and salting (along with drying) were traditional preservative treatments for meat. Before refrigeration was common, most meat was consumed shortly after being butchered. Storage was "on the hoof." Large livestock, whose meat would be eaten for a prolonged period, were slaughtered in the autumn when feed became scarce and cool temperatures permitted the meat to be more easily preserved. Even when cold autumn air made slaughter safer, there was a need to retard spoilage by other means. Salting produced a reliable but not very palatable prod-

A Simple Smokehouse [Credit: Jami Scholl]

uct, while smoking could in some regards improve the quality of the food. Evidence for this is still with us in the popularity of "smoked" bacon, ham, fish and turkey. A smokehouse is not uncommon today in regions where wild game and fish make up an important part of the diet. This is a somewhat specialized structure that might best be shared in a community, just as the tasks surrounding the slaughter of large animals have traditionally been a community endeavor. Despite what might be implied by its name, smoking is a cool process, so provision must be made to burn a small fire to create smouldering embers that are isolated from the meat by some distance. The smokehouse serves to channel the cooled smoke up and past the hanging meat, often after the smoke has passed horizontally through channels buried in the earth.

We take salt for granted, but in the preindustrial era, salt production was a signicant enterprise and salt commanded a high price, so important was it to food preservation, flavor and health. At some point in the late Middle Ages, it was discovered that mixing gunpowder into salt for preserving meat increased its storage life. The active constituent is saltpeter, or potassium nitrate. But

## Vessels for Fermenting

Preparing food for fermentation requires various forms of slicing, juicing or maceration. The chief tools and structures needed for fermentation itself are non-reactive containers that can be partly sealed, sometimes with an airlock, and this applies to ferments of whatever type. Kettles, jars, bottles, carboys, vats, crocks and barrels are used. Many of these items can be found in a kitchen, and most of them can be used there. Vessels for fermenting must be of a size that can be handled and washed, or of a type that is fixed in place in a room with the right temperature and where the vessel is provided with a water supply and a drain for cleaning. Large vats for wine or beer are not often moved, but they are washed out. Perfectly excellent sauerkraut can be made in gallon jars and prepared with a knife or a grater and a cutting board. We use specialized crockery made in Germany with a collar that creates an airlock using a moat of water.

The traditional North American crock, a simple, open-topped ceramic cylinder, was covered with a wooden lid that was then weighted with stones or bricks.

German-made ceramic crocks for ferments

[Credit: Creighton Hofeditz]

the preservative effect comes from nitrites, derivative compounds of nitrate ions. Today, meat is preserved commercially with sodium nitrite. Although used in far smaller quantities today than in past centuries, nitrites are implicated in causing cancer. Their use is thus a calculated trade-off between lowering food losses and reducing acute digestive ailments versus some small increment of long-term disease. Salt or brine can be applied to fresh vegetables and herbs as well as to meat. It preserves food by rupturing the cells of bacteria and other digesting organisms that would otherwise attack the food.

Drying is suitable for any place that has sunny months with moderate humidity. In the US, western regions enjoy these conditions more frequently than those east of the Mississippi, but with the right architecture, a food dryer can work almost anywhere the sun shines or electricity can be supplied. Food to be dried should not be left in full sun, but hung or spread in a dry place with good air movement and protection from animals. Wet foods like fruits, vegetables and meat tend to draw insects and are better dried in a vermin-free enclosure. This is one of the purposes of a food dryer. Electrically powered units are available commercially — Excalibur is a well-known and reliable brand. They usually combine a rack of plastic trays with a small fan and a heating element.

A thermostat keeps the box temperature constant. There are also many good plans for solar food dryers which are bigger and more articulated structures designed to draw a large volume of warm air over the food.

Solar dryers must, of course, be located in the sun. Most models are portable though bulky. They use solar heating to create a strong draft of warm air, but don't deliberately expose the drying food to sunlight. Eben Fodor has written a fine book on the subject, complete with detailed plans for making your own.[1] An electric fan attached to a solar food dryer makes sense, because in humid climates it may be rare to get three days of dry sunny weather when you need it. We have dried tomatoes in southern Indiana with an electric food dryer, but it gets expensive. We find it more economical to dry herbs electrically, as this takes little energy but gives a fine finished product. Tomatoes are easy enough to can in a hot water bath. For most of our herb drying, we use ambient air in the house or under porch eaves, and we find the conditions of early autumn ideal. The air is already dry, in October the sun is still bright and we may have an occasional stove fire that provides that extra bit of energy to dry things to the right level of humidity for long-term storage.

Freezing offers convenience because it's quicker than any other method of preservation, but the energy cost is high and frozen food remains vulnerable to prolonged power outages. Freezing is unsuited to most off-grid locations, though I have used freezers successfully where constant hydropower provided the renewable energy. If you use a freezer, keep it full; they are more efficient that way. It's better to replace food with plastic jugs of water than to leave the freezer half empty. If you have frozen jugs of water, you can easily use them to chill coolers for transporting food or for temporary storage.

Electric food dryer showing trays

[Credit: Creighton Hofeditz]

They will also give you several days of backup cooling for the food in the freezer when the power fails.

Date and label all items going into the freezer. Don't freeze liquid in glass as it can shatter. Don't fill plastic containers more than 90% as they can bulge and split. And if you store food in plastic bags, either use heavy-grade freezer bags with sealable closures, or double bag. The environment of the freezer is extremely dry, and food can acquire off flavors and "freezer burn," a kind of dessication. Freezers should be inspected and reorganized thoroughly about every three months; if you find anything that is more than a year old, get rid of it.

All food preservation efforts except direct cellaring involve kitchen work, but the most involved of all, and the most durable form of preservation, is canning. It's the method best suited to preserving large amounts of food for months and years with no additional energy inputs. And the energy used in canning is heat energy, available by lots of low-tech processes: burning wood or biogas, or concentrating sunshine in a chamber.

The kind of large surpluses that canning allows the household to preserve easily and safely are most often available at the end of summer and in early autumn. Usually, this begins in the month of August, which in most of North America is still a hot month — sometimes the hottest. Thus arises the dilemma of working over a hot stove and steaming kettle at the hottest time of year, and doing so in the kitchen where all that heat and steam goes into the house, making it even more uncomfortable than it already is, or doubling the cooling bill. Another problem with canning is that it's labor-intensive, benefits from more than two hands working in concert and takes over the center of the household for many hours. Canning also requires a set of specialized equipment: one or more large kettles, a large pressure cooker or two and scores to hundreds of glass canning jars (ordinary glass

food jars are not reinforced at the neck and bottom, and shouldn't be used as they are more likely to shatter). Plus, there are useful specialized tools for processing foods into pastes, purees and chunks, and for handling hot jars.

## The Summer Kitchen

The long-established answer to these problems is to create a canning kitchen. Whether this is a summer kitchen used by the household for canning and also summer meal preparation, a converted neighborhood garage or a community kitchen dedicated to canning, it stands apart from the dwelling and offers the facilities needed for processing large amounts of food into stable storage.

The summer kitchen was common to all large households in the southern US well into the 20th century. It was an entirely sensible adaptation to life without air conditioning, and it still makes more sense than heating up the kitchen and then pumping that heat outside by the use of fossil-fuel-generated electricity. Few people have summer kitchens today, but many people grill or barbeque on the deck or the patio.

A proper summer kitchen had a roof and some semblance of four walls. It also had a sink and some counters or tables as well as a heat source for cooking and could be supplied with water, if only by a hose. The sink often drained into a bucket or into the yard. Where space is limited, a temporary summer kitchen can be assembled on a deck or patio, or even in the backyard near the kitchen door. Access to a water source and the space to set up and operate gas- or wood-fired cooking surfaces safely is needed. There should be a sweepable floor — one can roll out an old carpet over lawn. Folding tables can supply the working surfaces. A salvaged double kitchen sink can be set in a wooden frame and drained into buckets.

Besides moving the heat of canning out-of-doors, it may be possible to move it later on

Summer Kitchen,
Pattern #39
(See Chapter 6)
[Credit: Jami Scholl]

these rare circumstances, a wall-mounted gas burner may provide all the extra energy required. For the millions of us with older houses that face the street and not the sun and who cannot afford to thicken walls to 12 inches, no reasonable amount of caulk, extra insulation and better windows can eliminate the need for a substantial heat source (whether a furnace, heat pump or wood stove) except in the mildest climates.

The consideration of how to use energy in the home is dictated on the supply side by which fuels can be used for which purposes. On the demand side, design should be driven by two concerns: 1) the cost, stability and flexibility of various fuels and 2) the resilience needed in each area of energy use. Solar water heating is practical in most regions of the country — easy in the South and West — but it isn't economical yet, and it remains very convenient to have hot water systems that work on demand. So unless you get hot water as a by-product of geothermal heating and cooling or live in a sunny climate, your economical choices for hot water heating are resistance electrical, pipeline gas or propane. If you heat or cook with wood, then you can also heat water incidentally. This may be inconvenient, but it is possible and can be made much easier with the right equipment. You can also back up your bathing functions with a wood-fired sauna or hot tub. There is a considerable value to having a resilient cooking system. Wood or gas provides this; gas is easier to live with in warm climates. Under emergency conditions, you can live quite a while without hot water or even much lighting if you can cook, especially if you have a reserve of drinking water and can stay warm. Where there are reasonable and economic choices, I think gas — either pipeline if you have it or propane if you don't — is the superior fuel for cooking and water heating, wood the best choice for space heating. These are resilient fuels, though only partly renewable. Pipeline gas is unlikely to be shut off, though prices may fluctuate.

the calendar. Friends with a small household in eastern Canada use a freezer to hold food for a month or two, from the end of summer when it is abundant until early autumn when the weather is better suited to lengthy indoor cooking. They then pull the frozen produce out and can it for long-term storage. In that way, the heat and moisture of canning are a boon rather than a burden. For larger households or warmer climates this might not be as convenient.

## Heating the House

The house needs to be made as energy efficient as possible, and that includes adapting it to the capture of solar energy. Placing and building the house for best solar gain is an option for only a few people. Rebuilding, repairing, remodeling and repurposing will be required of almost everyone else. Increased energy efficiency may involve reglazing, adding a room or extension in a way that intercepts sunlight or attaching a greenhouse or solar collectors. But in most regions, supplemental heat is needed often. If the house is superinsulated and very tight, this may involve only tiny amounts of heat, and in

Propane is more expensive and more volatile, but since you store it onsite, you won't be cut off in an emergency if you manage your supply well. Methane gas can be produced from waste biomass. And wood, of course, is widely available. You can always burn the furniture to keep warm.

Geothermal systems circulate water deep into the ground to draw earth heat up for boosting to room temperature with the use of heat pumps. They are relatively efficient but remain dependent on a steady source of electricity—in almost every case, the grid. From studies done by members of our Peak Oil Task Force in Bloomington, Indiana, a region where the state subsidizes geothermal installations and virtually all electricity comes from coal, we have learned that switching home heat from a propane furnace to a geothermal heat pump reduces the dollar cost of heating, but make little to no impact on the carbon dioxide footprint of the home.[2] Just as much $CO_2$ from coal goes into the atmosphere as would have previously done from gas burnt in the furnace. If electric rates rise, some of the savings from geothermal erode. If the grid goes down, you sit and shiver in the dark.

I certainly don't want anyone to be cold, and I know that millions remain dependent on nuclear electricity, coal and methane gas increasingly sourced by dirty and destructive hydraulic fracturing. I also understand that not everyone could switch immediately to using renewable energies, but I encourage you to reduce your need for heating and cooling energies by improving the thermal performance of your house, and I recommend that where wood is abundant, you consider using a wood-burning stove to provide space and water heating if you are now dependent on heating oil, electricity, propane or even pipeline gas. All these fuels are imperilled to one degree or another; they may not go away suddenly, but they will rise in price. Hydrocarbon fuels are subject to extreme price volatility. The problem with electricity will be grid

failure and blackouts. Even if you maintain systems for using industrial fuels to heat, keep them primarily for backup and supplemental use. Wood heats more comfortably; it also makes you a more conservative energy user because you don't just push buttons to get warm. Another important aspect of a wood heating system is its use to dispose of scrap wood that has no further value for construction. This material is common, bulky and a nuisance to have around. It is obscene to bury it in landfills and often illegal to burn in the open. If you have enough room, you can let it rot in the woods, but many of those living on small properties don't have that option.

## Harvesting Energy from Wood

If you heat with wood, you must gather it in advance of when you burn, store it in a way that's convenient to your house and stove (or stoves) and, because it's a bulky, solid fuel, stack it in an orderly way. Woodsheds need some room adjacent to them that is paved or planted in grass and can take relatively heavy foot traffic and occasional wheelbarrow and even vehicle traffic during dry weather. These are your woodyards, where you will keep and use your sawbuck and splitting blocks. The yard can overlap with other functions like drying laundry, occasional camping, play space for the kids or even animal grazing.

Wood gives most heat when it has aged at least two years in dry storage. It does not need to be indoors, but should be under roof or at least under cover with good airflow all around it. This means that you need a woodshed or two in which you can store enough wood to provide heat for two winters. In our case, that means we must store 5–6 cords of wood to meet our annual need of 2–3 cords.

After experimenting with ad hoc stacks on the ground and with two woodshed designs, I feel confident in saying the ideal woodshed is long and thin and can be filled entirely from one side if necessary. This shape works well along a fence line or as a divider between

In this self-supporting design, longer logs are placed parallel to the length of the shed in between the posts to form "walls" between the bays. Then all other wood is stacked perpendicular to these walls. [Credit: Edward Carter]

These 13-year-old all-steel wheelbarrows have a narrow profile and a center of gravity over the wheel. The "nose" does not protrude unduly, and the curved front struts will never get caught on rough ground. [Credit: Creighton Hofeditz]

outdoor spaces. Our linear woodshed measures 27 by 5 feet with a 30 by 7-foot roof providing 12-inch overhangs on both long sides and 18-inch overhangs on the ends. The shed has four sets of vertical posts forming three bays. By filling a nine-foot bay three feet high, I can store one generous cord; by stacking to six feet, two cords. The whole shed can hold a fat six cords when full. Stacking consists of racking the longest pieces of wood between the posts so the logs are running parallel to the length of the shed. This eliminates making corners, the fussiest and most dangerous part of stacking wood.

In the space above the racked wood and below the roof, you get a kind of open attic that is very convenient for storing odd bits of lumber, pipe, planking and other bulky, useful materials out of the weather and not underfoot. This forms part of your Resource Inventory (Pattern #58) of Salvaged Materials (Pattern #57).

## Tools Used for Work

We have lots of hand tools and not very many machines. If our farm were larger, we might appreciate some additional machinery, but for 30,000 square feet, we get by with a pickup truck, two gas chainsaws, a chop saw, a table saw, a circular saw, a couple of battery-driven screw guns, a jigsaw, a router and a small hand-held electric planer. We've borrowed a power washer, and we've rented a log splitter, a trenching machine, a floor sander, a cement mixer and a backhoe. I like the idea of cutting grass with a scythe, but most of the areas we need to cut are small, irregular and cluttered with stone, shrubs or flowers we don't want to cut, so a scythe isn't very useful here. Instead, we use a small battery-powered weed whacker to trim out odd bits of tall grass and weeds in areas that haven't yet come under cultivation, using our surplus solar kilowatt-hours to recharge the batteries: Goodbye lawn mower! With more than five acres, some farmers might want a tractor and a cart or a three- or four-wheel ATV for hauling electric fence, dead livestock, salt blocks or water tanks around. We use wheelbarrows for everything on property, and we have four good ones.[3]

Tools divide into three broad but overlapping categories. The first consists of garden tools and some small machines such as tillers, chippers and chainsaws. These are on the same general scale and should be stored under roof in a location central to the garden but not too far from the house. Machinery that contains fuel and oil should be stored securely out of the weather but not in a building where people live. All these could be kept in a barn, a garden shed or a garage. Where the property is relatively secure, a convenient arrangement is to rack garden tools on the outside of a shed or barn under a wide eave. Broadly speaking, garden tools include axes, mattocks, pry bars, peaveys, billhooks and post hole diggers as well as the obvious shovels, rakes, hoes, spades, forks and brooms. Outlines for the various tools can be painted or marked on the wall and long pegs put in to enable several shovels, rakes, hoes or mattocks to be nested together.

The second category consists of hand- and electrically powered tools for repair and construction, including plumbing and electrical work. This is everything from hammers, levels, drill guns and wrecking bars to pliers, wrenches, pipecutters and voltage testers. These items are smaller, more valuable per pound than garden tools and more easily lost. Some of them have batteries that need to be charged. They need to be stored in tool boxes and drawers or in cabinets, organized by general type: woodworking tools here, wrenches in that box, levels hanging on the wall or standing in the corner. These tools are used inside as often as out, so keeping them in the house or in a garage near the house makes a lot of sense. They could also be kept in a well-equipped workshop.

The third category consists of supplies that facilitate repair and construction: hardware, caulk, lumber crayons, wire nuts, plumbing teflon, string, duct tape, lubricants, pipe glue, paint. They aren't exactly tools, but there are many jobs you can't do without them. Some, like paint, must be kept from freezing. Others, like pipe glue, are volatile and highly toxic and shouldn't be kept in the house if another location is available. And most of these items are small; many are very small. They need to be in cans, drawers, bins and specialized cabinets.

## From Tool Room to Workshop

Sorting out all these items — and a garden farm needs a large inventory of them — is much easier if there are appropriate buildings or parts of buildings dedicated to their storage and use. Some older homes have garden sheds, and these can be bought ready-made for a few hundred dollars and dragged into place. This is a place to keep the garden tools and the lawn mower, perhaps the patio chairs during the winter. Most conventional homes today have a garage, though many older and especially city homes do not. A garage may be your first best hope for a workshop, a tool room, a resource inventory and even for housing a few small animals through the winter.

Many houses also have a utility room or large utility closet where the furnace, water heater or laundry machines may be located, but, unfortunately, these are often buried in the interior of the building with awkward access. If so, they are unlikely to meet your need for a tool room or shed. We are lucky in having a utility room in one corner of our house, and we were able to convert a window there into an outside door giving the room access directly to the laundry lines, woodshed and a small working yard where things can be cut, drilled or pounded. We outfitted the walls of the utility room with metal cabinets, one of them a specialized cabinet with bins for hardware. We built in shelving, hung things on the wall, used every nook and cranny. The room is only 7 × 13 feet, but we've been able to concentrate most of our small tools there. A back entryway or covered enclosed porch might work in a different situation.

The point of having and organizing tools is to work with them to do useful stuff. The workshop function is little activated in the conventional consumer household, but garden farmers find themselves drawing on it almost daily. While writing this page, as if to punctuate the point, I had to repair a broken vent register in the ceiling of my office. Fences and gates need repair, door hinges squeak and must be oiled, handles break and have to be replaced, things need to be deconstructed or attached or…well, it goes on and on. A lot of this happens out-of-doors or in connection with parts of the farm that are not the residence.

The higher purpose for a workshop is to create things of value. Eventually, those might become craft items for sale. Officially, the workshop might have to be a two- or three-car garage, depending on your local regulations, or a carriage house with an upstairs studio. It could be attached to the house, but it's probably easier if it's free-standing but close enough for a covered breezeway connec-

tion. Build the garage deeper than the length of the longest van or truck you might imagine parking there — make it 25 to 30 feet deep — and you'll have room for a workshop in the back, even if cars are parked in it. The cars are just an excuse at the moment. Not too many years in the future, we won't be driving them much, but we'll really need to make and fix things at home and to have a place for a small repair shop, woodworking studio or neighborhood store. Think ahead and get ready.

## The Return of the Barn

Farms need barns, and garden farms need small barns with multiple functions. A barn was always a multifunctional structure in the past. It housed animals, stored tools, machinery and feed, often including large amounts of hay, and besides holding these agricultural inputs and implements, it was a resource inventory, filled with extra doors, stacks of planks or plywood, spools of rope and cable, a canoe for getting out on the pond. It could be converted to living quarters, a dance hall or even a classroom for a day or a decade. I've taught in quite a few barns. The garden farm might not have horses who need to be shod,

Storage Barn, Pattern #22 (See Chapter 6)

[Credit: Jami Scholl]

but it might have goats whose hooves need to be trimmed or who need to be milked. There has to be a place to store bales of straw, barrels of grain, extra animal gear, temporary growing structures like hoop houses and cloches (taken apart) and other bulky consumable and seasonal items. There might need to be a freezer or two in the barn, perhaps a place to pack vegetables for customers. Maybe you need a place for business inventory? The bottom line is this: If you're going to have a home-based economy, then it has to have covered space that can serve many purposes, and not just in the house. The barn is the center of your economy.

One of the important reasons for having a barn, a shed or more is so that you can organize your resource inventory. The industrial economy is beginning a long, slow spiral of devolution, and it's scattering its parts across seven continents. The garden farm has to sift through some of this detritus and collect the more useful bits. A resource inventory is your storage of industrial energy. This consists of leftovers, salvaged materials, great finds and white elephants. You can't do anything useful unless you have a stock of spare parts. That means things like scrap lumber (all sizes and dimensions), paneling, fence posts and fencing, wire of every gauge and description, hardware of course, but also salvaged window sash, an extra sink, a zillion buckets, cardboard by the bale, straw, compost, wood chips, poles, siding, pipes and fittings, irrigation line, insulation, spare gutters, sheet metal, bicycle parts, glass jars and tin cans and plastic tubs, live traps and cages and harnesses and straps and hinges and twine and rope. You can't afford to have endless amounts of these things, but you need a little of each. In short, you need your own hardware store. Sometimes you work on a big project for a few years, and stuff to build it accumulates slowly: doors, windows, a toilet, a countertop, concrete block. Sometimes you can find wonderful things for free or trade that just can't

be bought at any price. It pays to make room. And every so often, you have to purge some of it to a bonfire, the River Styx recycling center or the scrapyard. Don't get attached.

If you have a large outbuilding already, then make it work for you as long as the placement and structure aren't a disaster (in a flood plain, blocking a road, riddled with termites or toxins). Place a new barn where it will best serve the whole farm. You need vehicle access to it. If it might someday house animals, will there be an easy connection to paddocks or a strawyard? Could you move manure in and out? Make sure the doors are generous and easy to use. Give it deep eaves, and you can store things on the outside of it as well as on the inside.

There are advantages to placing any building along a north property boundary so that open sunny space lies to the south of it. This is true of greenhouses, barns, animal sheds and studios. Of course, topography matters. Many barns were built so that they could be directly accessed both at ground level and at an upper level. This meant building into a hill. The barn doesn't have to be solar heated (though you might want to later), so it could be on the south side of a narrow valley against a hill or tucked into a side valley in the shade as long as these locations have good access to the working parts of the farm. Use the new structure to mask an ugly view or block sound, but don't let it block light to growing zones or other buildings that need it. Think about the prevailing wind and where odors will move. Consider any new building to be part of a complex of buildings; this is especially important on small properties where every element counts many times over. Place new construction so that it creates courtyards and outdoor rooms between buildings and improves the utility of existing structures.

If you build a barn, a shed or a workshop, consider installing a composting toilet. There's no reason to remove your flush toilets, but you want to be able to close the nutrient loop between your land and the food you eat. You won't always be able to obtain free mulch, spoiled hay, bags of leaves from foolish town dwellers and other riches, but if you can process your own humanure into compost and capture your own urine, you'll have the nutrients you need to grow your own food. Add in a few animals, capture a little wild bird action, grow some trees that mine deep soil nutrients, and you can be in surplus, which is where you want to be.

There are commercial composting toilets made by a number of companies, and there are plans for building your own. I favor the two-chamber mouldering type of toilet, as I've used these for many years and know how they work. If you live in an arid and sunny climate, John Cruickshank's Sunny John is an elegant design. The simplest kind of system is one practiced and described by author Joe Jenkins in *The Humanure Handbook*: a bucket toilet and a proper set of composting bins.[4] In most parts of the US and Canada, as long as you have a flush toilet in your house connected to an approved septic tank and field or sewer system, there's nothing illegal about doing whatever you want in the bathroom, nor is composting illegal. People I know have been using bucket toilet systems at home in cities

and suburbs across the country for years, decades even.

## The Greenhouse

In almost all temperate climates, a greenhouse is invaluable for supporting the small farm. You want its special microclimate to extend your growing season, which requires a quick start-up in spring and benefits from a prolonged autumn. A greenhouse can give you valuable fresh food in winter when little is growing outdoors and can make winter marketing possible. And it can also serve valuable living functions as a place for doing projects or getting exercise in inclement weather. A greenhouse can be a simple set of hoops covered in plastic film, or it can have a much more sophisticated structure. It can be freestanding or attached to another building.

The greenhouse produces seedlings for the garden, and it may become a production center for salad, vegetables or bedding plants to be sold. It may also house some perennial fruits — grapes, peaches, figs, feijoa, bananas — whatever is marginal in your climate zone. It needs excellent access to sunlight and should be placed where trees cannot throw limbs down onto it. The site need not be completely flat — it can slope — but installation will be

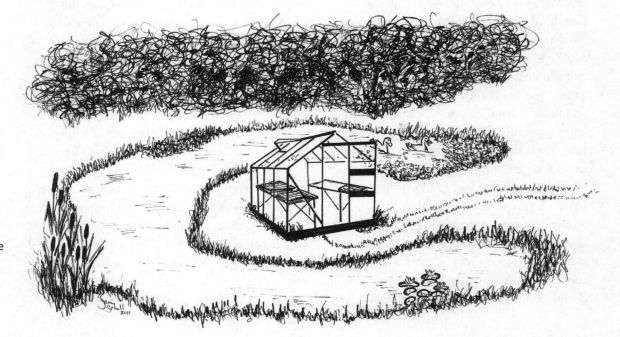

Place a greenhouse near or surrounded by ponds to increase light and moderate temperature.

[Credit: Jami Scholl]

easier if the site is level or the slope falls from north to south. Place the structure where it will not flood or intercept much surface run-off. A greenhouse requires attention almost every day, so it can't be more than 100 feet from the farmer's residence without imposing undue burdens. Give the greenhouse as much topographic shelter as you can: place it against a hill to the north and west or place it slightly in front of a treeline or against a tall solid fence. If possible surround the greenhouse with water. A greenhouse on a south-pointing peninsula in a lake or pond would be an ideal location — with water on three sides and extra reflected light.

Much was made of passive solar greenhouses as a way to heat homes and grow food during the first US energy crises of the 1970s. But, while there are exceptions to every rule, attached greenhouses seem to work best in arid climates where moisture buildup is less likely. In humid climates, I recommend greenhouses not be attached to residences. A classic setting for a small greenhouse is on the sunny side of the chicken coop: the birds' heat and $CO_2$ support plant growth, and the solar gain from the greenhouse can be vented into the coop to warm the birds in winter.

Good ventilation is needed for summer, but the greenhouse can actually draw in cool air from the north side of the combined structure, helping to keep the birds cool.

Most commercial greenhouses use a metal structure. Footers are always masonry and should be insulated to below the frost line. Good ventilation is critical, and that means the capacity both to exhaust air out of the greenhouse through upper vents and to pull warm, moist air underground into the soil of the beds where the moisture can condense and thereby release its heat to the soil. This is the physical principle underlying what Jerome Osentowski and John Cruickshank have dubbed the *climate battery*.[5] You do not have to build a greenhouse in this manner, but it is the most energy-efficient way to do

so and achieves the most stable and capacious ecosystem inside. The climate battery becomes economic and even essential in climate zones 5 and colder.

Whatever system you use — fans and pipes under the beds, racks of water barrels painted black or fish tanks — you need a large amount of thermal mass inside a greenhouse. The soil is the largest and cheapest form of this, but it's most effective if you can circulate warm air through it as well as absorb sunlight falling directly on it. Greenhouse insulation can be improved by using double layers of treated polyethylene with air blown between them. Polycarbonate in multiple layers with insulating gas between them is an excellent choice, though expensive. Glass makes a poor greenhouse; even when it doesn't leak or fail, glass cuts out a considerable part of the UV spectrum that plants use. The quality of light under plastic or polycarbonate is far superior, and this use of oil and gas is one of the very best we could possibly make.

A greenhouse should, of course, be aligned with its long axis east and west to maximize solar gain, but it doesn't need light exposure

This elegant solar greenhouse is combined with a chicken coop.

[Credit: Denny Henke]

on the north side. You can put extra insulation on the north side or bury it into a hill or a soil bank, use the walls there and the corners especially for fish tanks or perennials such as tender trees, or build in a connection to a wood-fired sauna as a backup heater for very cold nights. End walls can also be partly solid; they don't need to be fully glazed. Any interior wood walls or stuctural members should be painted white to maximize light distribution within the house. If you use water tanks or masonry walls, paint them a dark color to absorb light fully and radiate heat. Black is best for heat collection, and dark red captures heat almost as well as black, while still giving a useful frequency of light for plants.

If a greenhouse is to be a production center for garden seedlings, it needs a potting bench and an area for making soil mixes. This should be near the main entrance and convenient for the delivery of materials. Storage for pots, flats and other paraphernalia should be nearby. Be sure that you have a water tap and hoses that can reach throughout the building, or install irrigation lines. You may want some bottom heat for flats of seedlings. Radical Roots Farm uses an old refrigerator and some electric heating pads for a *hot box*. Flats stay in the hot box until they germinate, then move to the growing benches. The seeds don't need light until they break the surface, and by then they're ready to grow and don't need extra heat.

Remember that the soil in your greenhouse needs as much feeding and attention as the soil in your garden, maybe more. Plant some perennials so that you develop normal soil-building processes. Because it's under a roof, greenhouse soil won't get watered by the rain; you may have to irrigate even when it's wet outside. Greenhouses that can be opened widely admit the mobile elements of the broader ecosystem: you'll get cabbage moths, slugs, chipmunks and Japanese beetles, but you're also likely to get toads, the neighborhood cat and maybe the odd garter snake.

It's better to take part in the larger balance of nature than to try to keep it out.

## Shelter for Animals

Greenhouses are invaluable for supporting gardens, but animals need housing too, and if you don't have a barn or you're working out of your garage, you may need a shed or two for livestock. Consider whether you can consolidate several types of animals under one roof to save costs and make feeding and care more convenient. Rabbits and poultry, for example, go together fairly well. Chickens or ducks can range under caged rabbits and harvest the worms that will naturally flourish in rabbit dung. Hogs are probably best kept separately or in a structure with other large animals. As omnivores, they may take an unseemly interest in your baby chicks or ducklings. In providing shelter for animals, you need to know their requirements for space, warmth and water. Design water catchment into the roofing so that you have a good supply on hand, and make sure there is space for feed storage that is protected from vermin and moisture. Consider too, how you will clean out stalls, harvest manure and eggs or milk dairy animals. Make sure that the animal sheds are downwind of the house and not upwind of your neighbors. Good management should be able to limit odors, but no one is ever perfect: you might run out of straw during a bad wet period, or the heat might spike just before you can clean out the bedding.

## Outdoor Rooms

A greater proportion of the area of a small farm is covered with buildings and infrastructure than on a large farm. But no one can afford to provide four walls and a roof for every function. Many things have to take place out-of-doors, but they can be made easier or more difficult by containment, screens, surfacing, shade or the lack thereof.

A porch, covered or otherwise, creates a transition between indoors and out. However, a porch, unlike a walkway, should be an ample place where many things can happen: eating, socializing, working.

Porches (and decks) step down onto patios, which are almost the same kind of space, though less often covered. It's not very hard to put a pergola or trellis over a patio and grow some shade, making it a more appealing venue in hot weather. These outdoor rooms come more alive when their boundaries are marked by architecture or vegetation. Even a low hedge of shrubbery along the street side of a porch or patio makes it more private and more comfortable.

We are fortunate in having two buildings of similar size on our farm, and we live in both of them. In the 30 feet between them we have a patio and several walkways. This is really the heart of the farm, though, of course, in winter we don't use it much. But it's like our own Times Square, a place where all the paths meet and cross. Next to it is a grassy yard under large trees. This is almost as busy, but a little off to one side. It's bounded by both houses to the east and north, the open face of a woodshed to the west and the laundry lines to the south. We use the grassy yard for

drying clothes, for camping space, for moving and cutting wood, for picnics. On the other side of the patio is the driveway, part of which is paved with asphalt. It's bounded on two sides by our main houses and on a third side by a temporary storage building. Next to it is

Awnings and Porches, Pattern #42 (See Chapter 6)

[Credit: Abi Mustapha]

The patio is flanked by walkways and on either end by two other outdoor rooms, a paved driveway and a grassy enclosed yard.

[Credit: Keith Johnson]

a little strip of garden and some fruit trees. Of course we park vehicles on the asphalt, and they come and go, but it also serves as a working space for big projects. We can open out large tarps or sheets of plastic to dry, then fold them up. We can sweep it clean and find small things that fall on it. We can set up sawhorses and paint or varnish wood. We can sort junk, bend metal, store piles of sand or gravel temporarily or set up the table saw in good weather. It's a convenient space for a yard sale or a plant sale.

Both the driveway and the woodyard/laundry yard are outdoor rooms, and they're both expressions of the Garden Farm pattern I call Drying Yard (Pattern #40, see Chapter 6). I think this important enough that it needs to be included in an expanded understanding of the permaculture zone 1, which is classically the garden and specialized structures like a greenhouse or a toolshed, perhaps a children's play area.

## Fencing

If you have existing fences, find out if they are on your land or the neighbor's. Whichever the case, if a fence exists on the line, it should be possible to repair it, improve it or raise its top level without causing a stink. If you modify existing fences, I would strongly suggest that you consider installing gates to connect to your neighbors' yards — all of them — wherever these don't already exist. You may not use the gate often, but it's insane not to be able to walk into your neighbor's yard for a conversation, a visit, to retrieve a lost frisbee or, if nothing else, to warn them of the danger of fire or storm. There's also the matter of chasing an errant chicken or duck that gets over the fence. Do you want to have to drive around the block to pursue?

Putting up a fence at the front of your property is shoving something in your neighbors' faces. It needs to be done thoughtfully. There are, I think, two preferred approaches.

Well-placed gates should connect you to your neighbors while keeping your livestock in and wildlife out.

[Credit: Creighton Hofeditz]

The first of these is the one we elected to follow: use a nearly invisible fence. We are outside city boundaries and regulated little or not at all by township government. Yet our neighborhood feels suburban, and the street carries a fair bit of local traffic. Plus, we are in a very prominent location and highly visible. Our thin galvanized welded wire fence is held up by green T-posts. You can tell it's there, but it doesn't really obscure the view in or out of the property. The top two feet of it really are invisible from more than 20 feet away. We're still connected to the neighborhood. To our surprise when we put it up (with some trepidation), the neighbors from up and down the street were strongly supportive, even outspoken. The other approach, which we considered but rejected on grounds of cost, is to erect an elegant fence. You can do masonry, wrought iron or stained or painted wood.

If you decide to put up a fence, consider how a few small additions might gain neighborhood approval. If there's a fair bit of pedestrian traffic past your property, you could create a bench or some seats under a tree out by the front of the lot; these make an offering to the public space and invite people to enjoy views of the garden while they catch their breath or wait for the bus. You can also use art: put up a kiosk at one of your corners with a place for neighbors to post notices or a chalkboard for leaving notes or even graffiti. Make a small shrine, or mount birdhouses on the fence. Make one panel solid and paint a mural. Use whimsical bits of metal sculpture: mount masks or birds or leaf shapes.

Plant flowers thickly along the fence; use lots of color and vary the heights. Make them perennial so they'll come back stronger each year, and keep the flower beds weeded and trimmed. Or you could just get a 15-foot pink fiberglass T-Rex and a couple of inflated palm trees, and no one will even notice the fence.

This crenellated brick wall is in a downtown location, but it demonstrates an economical and elegant way to erect an attractive fence that can be combined with vegetation.
[Credit: Peter Bane]

# PART III

# Outcomes

## Radical Roots Farm, Harrisonburg, Virginia, USA

[*Color Insert #2 includes photos of this property.*]

**LOCATION:** A rural neighborhood of small farms and former farms in the middle Shenandoah Valley at about 1,450-foot elevation. Rolling hills with small patches of woodland give expansive views of the Blue Ridge Mountains to the south and east. Deeply eroded clay soils under former pasture reveal rock outcrops. Climate zone 6b trending toward 7a, with 36 inches of precipitation per year.

**DISTANCE TO URBAN CENTER:** 4.6 miles over gravel and asphalt roads.

**ECOSYSTEM:** Agricultural savanna, amidst a naturally forested region of mixed hardwoods. Deer, raccoon, groundhog, skunk, squirrel, possum, turkey and bear inhabit the slopes of nearby Massanutten Mountain, which rises about 1,200 feet above the surrounding valley.

**SIZE OF PROPERTY:** 4.7 acres.

**OPERATORS:** Wife and husband Lee Sturgis and Dave O'Neill have two children under the age of eight. The farm hosts between three and five interns who work the growing season from March through October.

**ESTABLISHED:** 2003.

**PRODUCTS SOLD:** Annual vegetables, salad, tomatoes, nursery plants, eggs, chicken. Dave teaches permaculture design and consults on properties. They provide their paid interns with education besides on-the-job training.

**MARKETS:** Weekly farmers markets in Harrisonburg, Charlottesville, and Staunton (April-October). On-farm sale of nursery stock, once a year in the spring. CSA of 50 shares (2009) with pick-ups in Harrisonburg and on the farm.

I visited Radical Roots Farm in early March of 2009. Lee was seeding salad greens in the greenhouse; Dave was on kid-duty with Isaiah (4) and Miranda (2) in the couple's tiny apartment in their Farm Center Building, the first permanent structure they erected. The season's first interns—one of them returning from the year before—had been due at 8 AM, but were delayed by a winter storm that had blown up the East Coast the night before and dumped four inches of powder snow on the rolling hill pastures of the Valley. Winds were still blowing it into drifts with temperatures hovering in the lower teens when the overnight clouds parted to reveal a brilliant and dramatic day.

"I love being self-employed," said Dave. "On a day like this, when weather throws us a curve, I can just switch gears and talk about the farm." With a big grin he acknowledged a little relief from the obligation to bang away on trim and kitchen counters in the family's new (almost finished) house next-door. I had spent a warm night in that tight, energy-efficient solar home, amidst construction debris and sawdust.

After graduating from university, Dave visited a permaculture farm in Guatemala near Lake Atitlan, a highland lake ringed by volcanoes that has been a tourist mecca since the 1970s. "There were these wild gardens, with herb spirals and chicken tractors, and in the farmhouse were stacks of old *Permaculture International*

*Journals* which I devoured," Dave recalls. "I was amazed and enthralled. I didn't know what anything was, but after several weeks, I walked out with a new mission. I wanted to grow things in this way, but I had no idea how to do it."

It was 1998 and the couple took off for Oregon, where Dave's sister lived. Dave wanted to work on an organic farm to learn market gardening. He landed a job at Herb Pharm, a botanical grower in Williams, though it didn't pay much. But the internship work there involved field trips, which took him to places like Lost Valley Educational Center (LVEC), where he encountered more evidence of permaculture, including Mollison's book, *Permaculture: A Designers Manual*, which laid out the design system in great detail.[1] By the end of that year, both Lee and Dave had taken the Permaculture Design Course at LVEC.

After the farm apprenticeship ended, the couple went down to Baja California where Dave worked for a time with Gabriel Howearth, one of the co-founders of Seeds of Change and a student of mad botanist Alan Kapuler. Kapuler took a global view of plant diversity and advocated assembling gardens that brought together representatives of as many plant families as possible. Howearth was growing kinship gardens based on Kapuler's work. "The legume family is very old," Dave told me, "and these gardens showed plants from the oldest layers of the family in a continuum to the most recently evolved. You could stand there in the middle of millions of years of evolutionary biology and read the story with your eyes, seeing the emergence of new structures, new flower forms. It was a remarkable education."

Returning to Virginia in late 1999, Lee and Dave felt they were ready to farm, so they rented an acre on the banks of the Shenandoah River and began Radical Roots. Living in a tiny woodshed on the property, leasing equipment and trying out what they had learned over the past two years, they grew a market garden. "It was my first own garden," Dave admitted. "We spent $2,000 and made $8,000 that year." His father and grandmother were both avid backyard gardeners, and the family told stories of an uncle who farmed and raised chickens in the middle of the Massachusetts town of Whitinsville. "We all heard about Uncle Vin's adventure with chickens and the butchering, and how the birds all ran around the yard with their heads off," he laughed. But at that point, he and Lee had become vegans, and the notion of keeping animals seemed like Uncle Vin, a story from another time.

After the first season, the couple got married, asked for a tandem bike as a wedding gift and took off on their honeymoon for Australia and New Zealand, intending to WWOOF and travel the countryside. "We would WWOOF a week, then travel a week on our 'bicycle built for two'," Dave reminisced. "That was where I really saw a lot of examples of permaculture."

"After many months, we ended our New Zealand travels with an afternoon tour of Rainbow Valley Farm, the home of Trish and Joe Polaischer," said Dave. Rainbow Valley Farm was at that time widely acknowledged as one of the premier permaculture demonstrations in the world, and for Dave it was a revelation. "At a casual glance you couldn't see what was going on," he noted. "The productivity was huge, but not obvious. All around you had a young food forest filled with animal systems. It was incredibly sophisticated. I would never have seen all the layered connections without Joe's explanations, but it was beautiful and delightful to be in. From the multicolored tapestry of the living roof, to the cool-sink shaded patio in the back, to the complex perennial polycultures that provided food, forage and fuel, it was a work of art, a real statement of the possibilities of permaculture design."

Returning to the US after nine months abroad, the couple spent the next year looking for land and farming their rented acre as Radical Roots again; they saved money from odd jobs, Lee doing social work and Dave landscaping. In 2003 they located their present farm, a 4.7-acre parcel that had been a cow pasture about five miles east of Harrisonburg near the village of Keezletown.

With $20,000 in savings, they obtained a $40,000 mortgage from the Farm Service Agency to purchase the property.[2] At $13,000 an acre, it seemed both a good value and a huge amount of money. Their parents lent them an additional $10,000 with which they purchased a camper trailer to live in and their 3,000-square-foot steel-rib and plastic-covered greenhouse. The farm was launched. Shortly after moving onto the land, Dave landed a half-time job as Director of the James Madison University Arboretum. It was an important financial anchor that helped the couple gain purchase on their mortgage and buffered the vagaries of selling market crops for income. Dave finally gave up the position in early 2007 after second child Miranda was born and the farm had proven its income potential. He still gets invited back to teach and consult on campus.

Dave described the beginning of their design for the land, "This place was an open slate—grass and a few walnut trees along the road. We began by doing a one-foot contour map with an urgent need to determine how to place the greenhouse. I wanted to put it up here initially," he said, gesturing out the window to the new house across the drive. "This hill had great sun and commanding views, but friends pointed out that this would be a much better home site, so we put the greenhouse further to the west where it is now. ... It would have been a strategic blunder to put the greenhouse here. I'm glad we took our time to figure it out."

The second problem to be solved was the location of the Farm Center Building. This 16-by-28-foot straw bale and pole structure is divided into three distinct sections; at the south end is a small apartment where their family have lived since 2004. The north end contains "the packout," an open area under roof where they process harvested vegetables for delivery to local markets and CSA customers. In between is "the barn," containing tool storage and a small loft for open-air sleeping. This multifunctional struc-

ture was a key to their early success. It provided needed comfort and utility so they could sink their roots into the land and at the same time optimize market garden operations. "I struggled with the orientation of the building," Dave said. "From permaculture analysis of energy and from contact with natural builders, I was inculcated with the importance of solar orientation and initially couldn't get past the idea that the building had to run long east and west, with a broad south side for solar gain, but it didn't work. We needed a cool space for the packout, and a long orientation this way would have compromised our use of the best building space on the land. Then I started just playing around with other possibilities, and I swung the axis of the building around to the north. It was one of those 'Aha!' moments when the lightbulb went on. By turning the building so that it ran long north and south, we got a solar apartment on the south end and a cool, shaded open porch for the packout on the north end. The straw bale internal wall between the apartment and the barn, plus the structure itself, protects us from the cold north winds, and the actual living space is wider east-west than it is deep, so it works as a solar room." He pointed out that they deliberately made no door between the apartment and the barn. "The psychological separation between home and work is important even on the farm. Maybe especially on the farm."

Dave and Lee built the structure themselves based on their experience of natural buildings in Oregon and down under. Besides the straw bale components, it is framed in local lumber and carries a coat of earthen plaster. The enameled metal panel roof enables them to collect clean rainwater. The apartment has big windows to the south, east and west, a one-room sitting space with a kitchen at one end, a tight set of built-in storage cupboards, including bookshelves, a central counter and a tiny bathroom. Radiant heating pipes under the earthen floor keep the place comfy when the sun isn't out. Stairs at the back lead to a sleeping loft, which the growing family still shared in the winter of 2009. The

transitional structure was designed from the beginning not as a permanent family residence, but
as limited term shelter for Dave and Lee while
they stabilized their income situation and got
the farm running.

After getting advice from me and my partner Keith Johnson in 2006 about layout of the
driveway, placement of a pond and the spacing
of swales, Dave called in the bulldozer operator
that summer and set to work. The following winter, he got advice from Dave Jacke about forest
gardens and planting productive trees, over 200
of which went into the ground in the fall of 2007.
Sitting at the kitchen table, we could look out on
the terraced rows of young apple trees descending to the pond between the house and the road.
"The perennials are planted in contour strips
about 40 feet apart. At our 7% slope that means
about a 4-foot drop in most places between the
terraces. We're experimenting now with berries on either side of the trees. The apples are
interplanted with goumi, Siberian pea shrub and
Baptisia, which are all nitrogen-fixers that will
support the fruits. We can coppice the fertility
plants to pump nitrogen into the soil. I am all
about design for management," he explained.
"For example, those apples over there are next
to alders (a nitrogen-fixing tree). We can mow
the alders and rake the resulting mulch onto the
apples. That releases nitrogen from the roots of
the alders to fertilize the apples, and at the same
time gives us a mulch that breaks down in about
a year to make very black soil."

"We do the same thing with the comfrey," he
added. "We mow it and rake it to the chickens. In
between the rows of trees and shrubs, we have
wide alleys with good sunlight where we can
raise market vegetables, grow cover crops or
rotate animals, depending on our needs."

"I love the size of this business and of the
farm. It's human-scale. We walk everywhere," he
enthused. "And that scale applies to our tools.
Of course we have a pickup truck and use it
for trips to town and to move large volumes of
material around the farm sometimes, but our
main power source is a BCS walk-behind tractor." Dave attributes his choice of power systems
to Eliot Coleman's influence. Coleman farms,
largely in the winter in coastal Maine, growing
salads and other vegetables under double hoop
houses. "The BCS is a beautiful machine, rugged and well-built, and it comes with a range of
implements including a rototiller (which we use
only sparingly), a rotary plow and even a sub-
soiler. The rotary plow will fold in a waist-high
cover crop, and I have discovered that it works
like a dream to prepare terrace beds in between
the swales. The machine even has a pull-behind
cart that will carry half a yard of material (about
three wheelbarrows worth), and you can ride it
around."

Getting animals working was a key to channelling fertility in the system. "It closed the loop
for us," as Dave explained. "We came a long way
on this, from being vegans and sort of looking
askance at all these little farms in Australia that
had chickens and goats, to getting in a few
chickens of our own when we moved here, to
last year (2008) when we finally figured out how
to integrate the animals so everything worked
better.

"Rotating animals is key to using their services without damaging the land," he explained.
"I found a design online for chicken housing that
involves the use of 16-foot cattle panels (welded,
galvanized semi-rigid steel grates). We make a
kind of Quonset hut out of them by combining a
number of panels into a modular chicken coop—
as long or as short as we need. Then we hang a
nest box and a waterer from the center of the
hoop. It's high enough to walk into comfortably
to harvest eggs. And we surround this unit with
electric fencing to allow about 5,000 square feet
of grazing area. We move the house every day,
and the electric fencing that surrounds it about
every week. That gives us an area of intense
impact (under the coop) and lets us be flexible
about the shape and space of the lighter grazed
area within the fence. It becomes quite an interesting exercise to 'drive' the chicken tractor
around the landscape."

"This year (2009) we bought in 250 pullets

# Case Studies C and D

## Case Study C: Old 99 Farm

*See text description following Chapter 13.*

1. The grey-sided barn and outbuildings reveal Old 99 Farm's rural heritage.

[Credit: Ian Graham]

2. An accomplished businessman, proprietor Ian Graham spent lots of time planning his moves. Here we plot out major land use divisions for the former soybean field.

[Credit: Keith Johnson]

3. One of Ian's earliest efforts was to develop a market garden.

[Credit: Ian Graham]

4. These terraces converted the front paddock and yard into garden beds on contour, announcing to the neighborhood that something new was underway.

[Credit: Ian Graham]

5. Occupying the top of the slope behind the new gardens, seven older apples were brought back into production by careful pruning.

[Credit: Ian Graham]

1. Hoops for cloches extend the growing season. Ian has focused on cold-season vegetables as a market niche.
[Credit: Ian Graham]

2. One of the first new constructions was this strawbale lean-to greenhouse against the south side of the barn.
[Credit: Ian Graham]

3. The barnyard is becoming the new productive center of the farm.
[Credit: Keith Johnson]

4. Inside the temporary greenhouse.
[Credit: Ian Graham]

5. Taking the crop to a local shop.
[Credit: Ian Graham]

1. Preparing the site for two large greenhouses.
[Credit: Ian Graham]

2. Installing the climate battery.
[Credit: Ian Graham]

3. Erecting the greenhouse frames.
[Credit: Keith Johnson]

4. Some 20 acres of former soybean field are now seeded to permanent pasture.
[Credit: Ian Graham]

5. Ian raises Dorset sheep, which can lamb at any time of the year.
[Credit: Keith Johnson]

1. Canadiennes are one of two heritage Canadian cattle breeds, seen here rotating through new pasture.

[Credit: Keith Johnson]

2. An operator cuts a swale for new orchard terraces on the largest north-facing slope of the farm.

[Credit: Ian Graham]

3. An existing pond was deepened and reshaped to increase its catchment and improve water holding. Rip rap (coarse stone) covers the drains.

[Credit: Keith Johnson]

4. Rich edge vege-tation and floating plants indicate good water quality and a healthy aquatic community.

[Credit: Keith Johnson]

5. Proudly proclaiming the harvest.

[Credit: Ian Graham]

OLD 99 FARM
PERMACULTURE
SITE

FOR SALE
BEANS
BOK CHOI
CHARD
GARLIC
LETTUCE

## Case Study D: Radical Roots Farm

*See text description at the beginning of Part III.*

1. This 3,000-square-foot greenhouse was the first permanent structure built. It has been the hub of vegetable production at Radical Roots Farm.
[Credit: Keith Johnson]

2. The greenhouse provides seedlings for the market garden and for sale and provides the family with winter produce.
[Credit: Keith Johnson]

3. Thousands of seedlings move outdoors to harden off in large cloches as the spring operation ramps up.
[Credit: Keith Johnson]

4. The farm center building provided storage, a pack-out space to prepare vegetables for market and a small apartment for Dave and Lee in the early years, allowing them to start their family in a cozy solar home.
[Credit: Keith Johnson]

5. The pack-out space provides cover and facility for boxing vegetables for market and for the CSA.
[Credit: Keith Johnson]

1. Dave has developed very exacting procedures for the vegetable operation.
[Credit: Keith Johnson]

2. Three to five interns are employed each year for the season. In addition to pay and garden training, they are given market responsibilities and permaculture-based education.
[Credit: Keith Johnson]

3. Though production oriented, Radical Roots pays close attention to ecological balance. Here intercropping allows more yield per square foot while disrupting pests.
[Credit: Keith Johnson]

4. Lee oversees vegetable packing in preparation for one of the Saturday markets.
[Credit: Keith Johnson]

5. A BCS rear-tine rototiller has provided much of the heavy lifting in the early years, allowing green manure crops to be readily folded into topsoil to build fertility.
[Credit: Peter Bane]

1. These contour planting beds are being converted to no-till.
[Credit: Keith Johnson]

2. A wide array of flowering plants helps support the organic air force of pest predators.
[Credit: Keith Johnson]

3. Long seasons, three public markets and a 50-member CSA place heavy demands on the single cultivated acre, requiring careful intercropping and tight rotations to maintain adequate fertility.
[Credit: Keith Johnson]

4. Third-phase developments on the farm include a new driveway approach over the dam of this pond, plus extensive terracing and many perennials. A liner was required to deal with the intrusion of limestone boulders, one of which can be seen in the middle of the pond.
[Credit: Keith Johnson]

5. After six years, Dave and Lee began construction of their permanent home, building during the winter months.
[Credit: Keith Johnson]

6. These innovative chicken tractors operate within a perimeter of movable electric fencing. The fencing is adjusted daily and the sheds move once or twice a week.
[Credit: Keith Johnson]

1. Two hundred birds share these two shelters, which have adequate headroom for the farmer to maintain feed and water and to collect eggs. [Credit: Keith Johnson]

2. A detail of the new house upstairs. Solar design and natural materials make this space a delight to be in, but only the children know all the magical nooks and crannies. [Credit: Keith Johnson]

3. Commercial success rests on dependability, personal service, abundant production of popular crops, and marketing with a flair, including the good use of showy elements like these sunflowers. [Credit: Keith Johnson]

4. The farm is not just about production, but is also a home and at least sometimes a show-case for customers and visitors. Small graceful touches make a difference. [Credit: Keith Johnson]

5. Thoughtful patterning that follows landform is revealed in this dramatic summer morning aerial shot. Buildings cluster at the crest of the high ground; vegetable production occupies the flat top, and contour beds of crops cascade down toward the public road. A young shelterbelt is emerging around the perimeter. [Credit: Norm Shafer]

at $5 a piece in April. All year long we sold eggs in the market at $4/dozen and offered an egg-share to our CSA customers at a little less than that. Then in October, we began selling the birds as laying hens. We got rid of about 150 of them that way at $8 apiece—people just calling up, wanting four or five hens for a home flock. So really, it was like renting the birds for a year," he laughed. "For the remainder, we rented a portable chicken processing plant, and a friend and I slaughtered and butchered 100 birds in a day. The unit had four killing cones, a scalding vat and a centrifugal plucker. That particular processing unit wasn't very portable, but that's what we need [Agricultural] Extension to own and rent to small farmers: mobile abbatoirs so we can process animals on the farm, keep the wastes here and claim the retail value of the meat."

"We start the chicken flock in the market garden in April on winter cover crops. Then as the vegetable operation gets cranked up, we move them into the forest garden. Later we can bring them back into the market garden after harvest to glean, weed and fertilize. We used the chickens to start preparing our terrace alleys too. We ran them in to weaken the old pasture, then seeded daikon radishes. Then we ran two pigs into the daikon, which they loved. After the pigs harvested the daikon—and they do it by rooting, so the soil is nicely tilled up—we broadcast winter squash, and then followed that with a fall planting of garlic to overwinter. We got four crops in one growing season."

"Our aim is to shrink our zone 3—the cash cropping section of the farm," Dave explained. "I think this is the measure of success in permaculture. We don't want to convert the whole world to agriculture. We should be aiming to use the smallest area possible to grow crops for ourselves and for cash income, so that we export from the farm the minimum volume of material necessary to pay for purchased inputs, taxes, investments and other cash outlays. The rest of the land needs to be building up the ecosystem, making long-term deposits in the carbon bank, enhancing species diversity and habitat for wildlife and for us. All of that stuff takes less management time and energy to produce, because it's mostly perennials."

I asked Dave about the continuing evolution of Radical Roots. "One of our next moves is to transition the market garden into permanent 'synergistic,' deep-mulch raised beds, the way Emilia Hazelip demonstrated.[3] We've begun next to the greenhouse by moving the chickens into that area to weaken the old pasture grasses, then used the rotary plow to form the deep beds, which are separated by paths mulched in wood chips. Then we introduce perennials on an open pattern, leaving plenty of space for annuals which are seeded or transplanted in around them. We don't think we will ever till these beds again.

"I put a new stone facing on the public end of the greenhouse because we're growing more and more nursery crops and bedding plants, and we want to do more on-farm sales. That's very high-value horticulture," he explained. "We raised basil in small pots last year. In an area of the greenhouse about 4×10 feet, we had about 700 pots which sold for $2,000. With that kind of return, we can really shrink our farming footprint.

"We'll be favoring the CSA over the markets—we presently travel to Harrisonburg, Staunton and Charlottesville. I have a love-hate relationship with the Charlottesville market. It's our most lucrative, but it's a 1¼-hour drive each way over the Blue Ridge Mountains, and I spend too much time on that highway in rough conditions—snow, fog, high winds and rain, not to mention crazy drivers.

"We also want to get into more value-added products. Keezletown has a cooperative cannery that was set up in the 1930s and still operates. They have an industrial kitchen, and Lee's been down there to put up applesauce. We want to try sun-dried tomatoes, salsa, sauerkraut and some other foods we can easily make from our produce. We think that's going to help us stay closer to home and get more return on our efforts."

# CHAPTER 19

# Diet and Food

We live in an age of fusion cuisine, world music and seemingly endless consumer choice. This is also an era of growing hunger and runaway obesity, of food riots, butter mountains and empty calories. Here I want to explore food as a product of the farm and the kitchen, as a commodity and as a support for the body's metabolic processes.

## Food Is Energy

Humans need an average of 2,700 calories[1] a day (a million calories a year) to remain healthy and active. Hard-working adults consume more, children and seniors less, men eat more than women, pregnancy adds extra demands. The UN Food and Agriculture Organization (FAO) says 2,300 calories is enough, but North Americans are larger than the world average, even without obesity. Also, many of us live in cold winter climates, and I'm writing about a way of life that involves physical work. The body constantly burns up this energy. It keeps our temperature at 98.6°F, it digests our food, repairs our cells, keeps our organs pulsing, our brain neurons firing and our muscles contracting. The calories we need to eat as food in one day are equivalent to the energy in ten ounces of gasoline or a pound of coal, but of course we can't eat the fossil fuels directly. We do consume seven times this amount of fossil fuel each day when we eat a conventional diet, because it takes 7.3 calories of oil, coal and gas to produce and deliver a calorie of food energy to the table.[2]

We can convert about 30% of our metabolic energy into work at any one time, making humans more efficient than the cars we drive, albeit many times less powerful. But, of course, we must sleep and rest, so on average about 12% of the total energy a person eats as food becomes work. The rest is given off as heat, about three kilowatt-hours or 9,000 BTUs each day.[3] This is roughly the amount of heat given off by a single gas or electric stove burner left on high for one hour.

All this energy comes from food, and all of our food starts out as living plants, animals, fungi or microbes, powered by the sun.

## Fast-Food Continent

North Americans actually eat on average nearly 4,000 calories a day, some 40% more than they need for health.[4] Our obesity grows directly out of the energy-squandering, toxic and unhealthy industrial food system, which includes bought-and-sold politicians who dish out massive subsidies for a handful of commodity crops, giant mechanized farms and feedlots (complete with antibiotic-doped animals and antibiotic-resistant microbes), rail lines, barges, trucks and a few hundred airplanes, factories that butcher, extract, distill, emulsify, blend and preserve thousands of products from a tiny handful of basic crops, a vast amount of packaging and refrigeration and advertising agencies — this last element part of an industry worthy $650 billion a year by itself alone.

The scale of this system ensures that when mistakes happen, the consequences are magnified and they spread far and wide. There is no easy way to control quality and safety when the sources are distant and not under modern regulatory control — even close to home these centralized systems fail. The dangers of acute illness and death from food contamination are real and regularly cause scandals. There is an illusion that they can be controlled; laws and money are sometimes thrown at these problems, almost always ineffectually. Even if we could control the outbreaks of salmonella, E. coli 157 and aflatoxin, ban the launching of rockets over the California lettuce fields and stop the Chinese from putting melamine and lead in baby food — and we can't prevent them all — we would still face insuperable systemic problems that undermine our health.

The basic foodstuffs that provide raw inputs for this system and for our conventional diet are deteriorating in quality because our farming has depleted the soils and polluted the waters in which they grow. Adding a few vitamins to "enriched" white bread, cornmeal or white rice doesn't replace the minerals that have eroded from our once-rich prairie and woodland soils. The body needs calories, but it also needs proteins, fats, essential amino acids, vitamins, minerals, enzymes and a host of nutritional factors that are synergistic and that science understands poorly. Frankly, we also need bacteria to digest what we eat, not just in and on the food, but in our guts as well. By fits and starts over nine decades, we've isolated and named about 20 vitamins and perhaps two dozen minerals that are implicated in metabolic processes in the human body and most higher animals. But we are far from having turned this fragmented knowledge into a robust and resilient culture.

Another negative consequence of eroding mineral nutrition in soils (more than 70% in 70 years)[5] and in crops is a steady loss of flavor; in the orderly world of nature, food that's bad for you tastes bad too, and vice versa. This loss of flavor is compounded by the industrial demand for commodities that can be sold like bricks — in other words *shelf life*. Even green vegetables have been shipped and refrigerated for days before they reach the supermarket. No wonder they taste like day-old shrubbery. Food technologists (not cooks) have reliably determined that people respond to brightly colored labels and to high concentrations of sugar, salt and fat. These stimuli, all of which can be manufactured cheaply and are, in fact, hard-wired into our brains in regard to food, overwhelm the subtle details of nutrition. But, of course, they can't fool the body's cells.

So North Americans, and increasing numbers of people from around the world eating the industrial diet and imported food, are walking around fat and starving at the same time. Craving real nutrient, their bodies tell them to eat more. Downing empty calories, they blow up like balloons and get sicker by the day.

The best argument I can make to you for garden farming is that you need to do it to save your life and health and the lives and health of your family and those around you. Never mind having enough food to eat; you won't be getting food worth eating if you don't start growing some of it yourself and doing what it takes to grow healthy soils and ecosystems to support it. This is not about saving the Earth — that would be a nice by-product — it's about saving our own asses or, more accurately, our brains, our hearts, our guts and our glands. Gross hunger is a real issue for millions of Americans and nearly a billion other people, but empty belly hunger is an issue of poverty and injustice, not a lack of food. One-third of all the food grown in the world each year is wasted before being consumed by humans. The day may come when we cannot grow enough, but for now the greater threat is from food not worth eating, food that poisons and corrupts our bodies and farming that is ruining the land.

## Soil Remediation

Nutrition begins with the soil. Plants that are overfertilized and undernourished — just like hungry fat people — will draw all kinds of critters to attack them. Poor soil nutrition induces diseases and pests to flourish; they are working to eliminate deficient and inferior plants and animals. If your fruit isn't sweet, if your vegetables are riddled with holes or attacked by fungus, it's almost certainly a lack of minerals. Soil degradation takes time and so does soil repair. A garden started on a lawn will be three years before it pops. This is the beginning of significant integration and accommodation by microbes, fungi and plant roots in the soil. Seriously damaged soils can take more than a decade to restore. You can put rock powders down until you're blue in the face and some of that will help, but the key to holding and making soil minerals available to plants is organic matter, especially soil humus, which is durable. Remember: A.O.M. (Add Organic Matter). And the key to transforming organic matter is soil life — from earthworms to microbes. Fortunately, you only have to build it and they will come. Organic matter is the daily bread of any soil. Mineral supplements are the salt: apply them lightly and frequently.

## Use Your Senses

Assuming that you are doing right by the soil, you will be growing plants in it, and some of these are to feed your household, not just the soil. Planting and cultivating is necessary but not sufficient. You must also harvest. And this becomes a spiritual act as well as a practical one. Teach yourself about the cycles of the garden. Keep notes in your garden journal about what is blooming, what is fruiting and what you are eating. You are getting to know a lover — taking him or her into your body. You need to learn the rhythms of arousal and response. Yes, there is work and waiting, but there must also be excitement, pleasure and satisfaction. The day will come when you will eat yourself silly in the berry patch. With a garden there is a long courtship and the prospect of a lasting marriage. Harvest is consummation.

## Learning Garden Cycles

Harvest is also work and has its requirements. Ripe food doesn't wait for long. If you don't get it, somebody else will — the birds, the squirrels, the raccoons, the ants, the mold. You must be in the garden every day doing something: taking note of what has changed and what is needed, seeing what is ripe and what is ready. When you have learned the garden cycles, you'll be able to anticipate the harvests and the births and maturations well enough to time the planting and the breeding. That knowledge forms the basis of a home-grown food system.

One way to train yourself in these arts is to plan a harvest calendar and work backwards in time. I suggest starting with the perennials, especially the fruits. Figure out the seasons of ripening for your area and the sequence of fruiting. In most of temperate North America, strawberries are the first fruits. As I write on May 12, our southern Indiana town is having its annual strawberry festival, but in central Florida the strawberries ripen in February, so you can't depend on someone else's knowledge. Rhubarb (which isn't a fruit but is often eaten as one) is ready now too, and that means cobbler and pies are possible. Mulberries will follow in about 10–15 days, and juneberries will ripen in, well, June. Then it begins to be a free-for-all with currants and gooseberries, red raspberries and black. Early plums come next and blackberries by July, early peaches and blueberries sweeten in the heat of summer, as do apricots if you get them, and cherries. Then it is mid-season with many things continuing and new varieties joining the mix. August brings pears and summer apples, more plums and peaches, and late in the month perhaps figs that will begin to ripen and carry on into the fall. September swells the harvest of apples and pears as plums, peaches and blueberries decline. Pawpaws make a short appearance — don't miss 'em. October brings the red raspberries back for a second crop and sees persimmons start to lose their tannins; maypops ripen. Pome fruits hang on into November as cranberries float into view, but frost finishes the figs and the berries. In milder climates, harvest continues into December with loquats, feijoas, Asian persimmons and the first citrus — tangerines marking the beginning of the holiday season and citrus running through to April when mangos from the South awaken our sweet tooth. And the year turns again to small fruits. As you plan your garden, lay out this calendar for your area and fill in something for every week that you can.

The process is similar for other crops, but the time frame is determined by your planting schedule and not run so tightly by the waxing and waning of light and heat. Of course, cool-season crops must be planted to grow in spring and fall, and heat-loving vegetables and fruits set out to ripen through the fullness of summer. But within those broad confines, lettuce or beans or summer squash will be ready 60 days or so after planting. So, plant and plant again. Don't let two weeks go by without putting in more — you are filling a pipeline. Stop for too long, and there will be gaps in the harvest.

Some crops do indeed require the whole season to ripen, or such a large part of it that you will only plant once or over a short window in spring: potatoes and sweet potatoes are this way, the latter taking nearly five months. Potatoes ripen in three months, but can stay underground for some time. It's better to harvest them later in the year because they will store better into the winter that way. Basil runs its course from May til frost, and you can clip and dry it or make pesto for months on end. Some vegetables, such as cabbage, have to complete their whole growth cycle during spring or else wait til fall. Plan on one crop for slaw and a second for kraut and keeping. Our crazy bipolar climate is to blame. Few of our garden vegetables are native and fully adapted to this cold-hot-cold rhythm. Sunflowers, which are North American, get going after the weather is settled and don't stop til frost; squash can do the same, though they usually succumb to borers or exhaust the nutrient in their little patch of ground before the season ends. With richer soil, look out! Some things like collards are successful in both winter and summer, but that's rare. Onions are day-length sensitive, so they need to launch in early spring and ride the tide of lengthening days to form big bulbs. Garlics, of course, are an annual crop, planted in the fall, harvested at midsummer. There are perennial vegetables too, each with their season: early asparagus and late artichokes and cardoons if your climate is mild enough for them.

### Eating in Spring

Animals also respond to the changes of day length and heat. If you have a poultry flock, you may already know that egg laying surges in spring. What might have been three or four eggs a day out of a dozen birds in the winter becomes 6, 8, 10 and 12 by late March. Suddenly, omelets are on the menu: asparagus and morels, spinach/cilantro or kale and shiitakes. Scallions are abundant in spring and grace every dish. Each day's salad is a little different with lettuce, endive, beet greens, mizuna, arugula and early radishes mixing it up. Cut the bitterness with sweet berry jam and rich oils in the dressing. Later in the season, lambsquarters with their brilliant magenta tops enliven the bowl, and along the way redbud flowers, violets and pansies adorn the greens.

The crops of spring are meant to awaken us from the lethargy and depression of winter. They don't fully relieve hunger, but begin to replace minerals lost during the months of eating stored food and battling cold. Bitter greens are tonic: the endives and first dandelions bring a powerful infusion of mineral nutrients to clean the blood and stimulate the liver. As dairy animals get onto green pasture, their milk fattens up and gains large amounts of vitamin A, and this plus the extra eggs from hens make up some of the caloric deficit from cold spring days and extra work. Steam a pot of buckwheat and open a jar of applesauce to bulk up the menu.

Soon enough, the heat of May (or June if you are farther north) brings fruit and then new potatoes, peas and more carbohydrates, which need the extra light and heat to ripen sugar and starch. There may be some early beets, but the main excitement comes from the burgeoning berry crop. I love the seemingly endless fruits of this season but care less for the steamy days that sometimes accompany them, so I get up early and spend an hour picking while morning shadows still stretch over the bushes. By the time the sun has burnt off the dew, I've got a big bowl full and head for the kitchen with some fresh mint and basil in hand to blend up cold fruit soup for breakfast. Whizzing homemade yogurt and berries du jour, fresh herbs, a bit of kelp with sea salt and a dollop of honey, I have a luxurious breakfast that can go with me in a quart jar and leaves behind no heat in the kitchen. When early peaches start to ripen, they go into the soup as well, making a mellower blend.

### The Beginning of Surplus

As the first fruits ripen, you can eat them all standing in the garden. Then, after a few days you are carrying around bowls and baskets, headed for the kitchen. But as the harvest stretches out into the second, third and fourth weeks and currants give way to raspberries and blackberries, you have to do something else or let birds take the rest. You can freeze berries, or you can make jam or pie or both. Since the jam needs cooking, and you're using heat anyway, plan to can it. Acid fruits and jams can be canned in a hot water bath. Vegetables, meats and low-acid fruits should be canned in a pressure cooker. Most of the foods that need pressure cooking for long storage ripen late in the summer or in early autumn. If your springs are long and drawn out, by June you might have a surplus of peas. You might also have a surplus of beets in early summer, and those might be worth pickling or canning. Later in the season, I would cellar them straightaway, but your cellar might not be cool enough in June to take them through to March or April. New potatoes can be dug in June, but they are for eating fresh and not for storing. There's little point in preserving green leafy vegetables at this time of the year (unless you live in the far north) because many months of fresh harvest lie ahead. Instead, focus on capturing and enjoying as much of the fruit harvest as possible, and plant waves of starchy crops for later in the summer: beans, corn, squash, tomatoes, eggplant, cucumbers, melons.

As the dog days of late July and early August roll around, it's time to prepare for the sudden change into autumn that is only a few weeks ahead. Storage crops need to be started and transplanted. This is the season to plant the bulk of your root crops for harvest in the fall and storage through the winter. Cabbage, kohlrabi, fall broccoli, kale and brussels sprouts need to go in for harvest through the end of the year. Prepare the greenhouse for a new round of plantings. Salad for cool weather can be planted at this time; it will be ready as the temperatures moderate from September onward. Sweet fruits continue to ripen: plums, peaches, early apples, even some grapes and pears. Eat what you can, preserve, trade or sell the surplus. The tomato harvest is in full swing. Much of this needs to be canned in a hot water bath. There should be enough heat to dry some for special uses. Tomatoes are a major crop that becomes an ingredient in many cooked meals throughout the winter. At the end of the tomato harvest (first frost), pull any green fruits from the vine and take them inside. Set on a sunny windowsill, they will continue ripening for several months. We have eaten fresh tomatoes as late as early January.

Leafy salads are scarce in hot weather, but sliced tomatoes and cucumbers with salt and oil, basil and vinegar, make a splendid fresh dish. With some fresh mozzarella cheese — delicious! As much as possible during the heat of summer, we try to minimize cooking. We shift to roasting main dishes on the grill, make hummus from garbanzo beans and serve it cold with salad and bread or chips and turn eggplant into baba ghanoush. Eggs cook quickly; quiche with vegetables makes a simple main dish that can be refrigerated and warmed in a few minutes for a meal. Despite the abundance of food in summer, we find ourselves eating less. The body's not burning as many calories to stay warm, and activity levels drop in the heat — you just want to stay cool. So meals can be made simpler and with little hot food. Cold savory soups like gazpacho are a hit. Meat wants to be lighter and simpler too: poultry or rabbit grilled with a sauce or kebabs mixing little bits of meat with chunks of onion, tomato and pepper.

Summer is also the time to have lots of cold tea on hand. Japanese quince makes a good refreshing drink if sweetened a little bit in water. There are also mint and elder flowers for making drinks.

### Storing Sunlight for Winter

It's easy to be imaginative in summer when the garden is abundant, but you also need to think ahead for the cold months to come when this panoply of flavors will have to come out of the cellar or the pantry. Our diet this past winter was greatly enriched by four simple things: a good store of apples and winter squash that kept us fat with pie through all the dark days. These were from neighborhood trees and local farms, not yet our own. Blackberry jam from our own garden made winter salad dressings sparkle with summer sunshine, and fresh salad from the greenhouse turned every meal into a special occasion, not least when it included a few pansy petals. For the main dishes we had a small rotation of sausages, meat loaf and poultry, leavened with millet cakes, roasted root vegetables, noodle dishes, rice and fermented cabbage, both sauerkraut and kimchi. No flu.

Food Storage, Pattern #44 (See Chapter 6)

[Credit: Abi Mustapha]

The point of this parade of food flowing from the garden into the kitchen and onto the table is to enjoy life. That's much easier when the household is healthy. Health arises from a matrix of environmental and dietary factors, some physical and some metaphysical, laid over a framework of genetic competence. Sleep and sunlight and fresh air and moderate exercise form a foundation. It helps to have a positive attitude, to maintain the psychological state of flow and to keep a regular rhythm. I find this easier to do when I can set and keep my own routines and when those around me do the same. Working at home is an important element of this because it reduces interruptions and allows one to move fluidly from one project to another based on what is most important next. Food, of course, is central, as it both comes from the environment — a healthy garden soil, ideally — and becomes a focus of group activity and sharing, both of which are important psychological supports for health. We need to be included with those around us and to make useful contributions to collective well-being. Gardening, along with its companion art of cooking, is a synergistic satisfier of many of our fundamental human needs.

## The ABCs of Nutrition

And as to the biochemistry of it all, fresh foods contain vitamin C which most cooked foods have lost. This nutrient is critical to immune support and cell repair and, since it's water soluble, must be replaced frequently. Lacto-fermented cabbage dishes such as sauerkraut and kimchi are rich in ascorbic acid and of course in probiotic factors that aid digestion and promote health. Leafy greens provide not only beta-carotene that the body transforms into vitamin A when they are served with butter and other fats, but deliver minerals and B vitamins needed to regulate metabolism. We feel much better when we eat a good amount of protein every day, and limit carbohydrates to what satisfies.

### Feed your Brain

Besides mineral-rich and fresh foods, ample protein and eating with others, one of the most important elements in a healthy diet is fat. Fats feed the brain, and if you don't get enough of the right kind you'll lose your mind. Ever heard that fats are bad for you? Well, trans fats are bad for you. Hydrogenated fats are bad for you. And rancid fats are bad for you. Guess what? All of these are the fats routinely served up by the industrial food system. But butter is a health food — which means that cream and cheese are as well. Properly handled and stored saturated and monounsaturated fats from healthy animals and plants are good for you and should make up 30% of dietary calories.

Fats are difficult to produce from plants — lipids take a lot of solar energy to synthesize from sugars. It's not an accident that the best plant-based oils (palm, coconut, olive, peanut) are, for the most part, grown in tropical or other hot-climate regions. The sunlight there allows plants to store up more fat and more stable fat. Saturated and monounsaturated fats (lard, butter, coconut, olive, nut and palm oils, and to a lesser extent peanut oil) provide the better forms of fat for the body. Polyunsaturated vegetable oils of the sort common in industrial food are far too often rancid; their many open bonds make them easy targets for oxidation. This is why industrial processing so often hydrogenates them — to prevent rancidity becoming obvious.

In colder regions, nut trees are the largest source of fatty seeds, but they take a long time to begin bearing; fat is more readily obtained from animals. And in the realm of fat, the pig is king. Butterfat from milk (both cow and goat) is very important, and goosefat is also of high quality. You may not raise any of these animals, but you need to know a farmer who does. Most other animals don't produce large amounts of fat. Poultry have enough for their meat to be savory, and of the poultry, ducks have more fat than chickens or turkeys. Rabbits produce excellent meat but haven't much fat, so if you eat them, you need another source of fat to cook them with. We use lard. Fish are very lean, though the omnivorous and carnivorous fish put on more fat. If you grow and eat vegetarian fish such as tilapia, you'll need to supplement their flesh with other fats in your diet.

### Matching Proteins

The adult body needs from two to four ounces of protein a day, more when dealing with stress. If many people in Africa get too little protein, most North Americans get too much, eating in excess of 250 pounds of meat a year per person. Asians average about ¼ of that amount, a far more sustainable level and one toward which we will inevitably move. Animals are an indispensable component of a healthy landscape and of healthy farms and gardens, but the numbers of them that may be kept ecologically and fed on grass and feeds inedible to humans are far fewer than are presently raised in the US and Canada — perhaps no more than ½. We need them, but we and the land would be healthier with fewer of them.[6]

 **ESSENTIAL AMINO ACIDS**

Amino acids are components of protein. The nine listed here are considered "essential" because the human body cannot synthesize them from other amino acids, of which there are 22.

| | |
|---|---|
| Phenylalanine | Methionine |
| Valine | Histidine |
| Threonine | Leucine |
| Tryptophan | Lysine |
| Isoleucine | |

A mnemonic for remembering them is PVT TIM HaLL. Specific populations (infants, individuals with unusual metabolic conditions) may require other amino acids not listed here.

Credit: Peter Bane

Protein in the diet can be obtained from a wider variety of sources than fat. Nuts, seeds and legumes are relatively easy to grow at the garden scale. Eggs are a good source of both fat and protein, and small animals are not difficult to raise. Most green vegetables have a good percentage of protein in them, especially peas and beans, but humans typically don't eat large volumes of kale, collards or Swiss chard, despite their being rich in amino acids. Protein is more useful for the body if it matches the profile of amino acids we use metabolically. Eight of these cannot be synthesized by humans and must be eaten.

Some foods, such as eggs, have a nearly perfect profile of essential amino acids, but most foods are imbalanced in one way or another. Grains have been combined with beans by traditional cultures in many ways because their protein profiles are complementary. Francis Moore Lappé made the science of complementarity very clear in her excellent *Diet for a Small Planet* in the early 1970s.[7] It's still a valuable reference on the subject. You can get more protein from less food by mixing sources in the diet. They don't have to be eaten at the same meal, but it's best if you complement plant proteins and supplement them with animal proteins regularly.

### Eat What You Kill

Small animals can and will be butchered throughout the year, but there are pulses in reproduction that affect the availability of meat. If you follow seasonal cycles, you will have more poultry in the summer months than in the winter, and as each clutch of chicks or ducklings is hatched, so it will move toward maturity. Inevitably, if you raise your own, there will be surplus cockerels and drakes to be eaten. Rabbits can breed throughout the year so, if you live in a mild climate where plant growth continues in winter or are willing to feed concentrates, a doe can raise a litter of bunnies just about any time. They too must be butchered within about 12 weeks of birth, excepting only those few individuals you select for continued breeding or sale. In this respect, animals support the diet during times that might otherwise be lean. Remember too that the organs contain the highest concentration of minerals and vitamins in the body. Neglect them and you have sharply diminished the nutritional potential of a slaughtered animal.

Larger animals have longer gestations and often breed and reproduce on an annual schedule. Feed for them is much more dependent on pasture growth, and it's uneconomic to carry more than a few breeding or dairy animals through the winter when large amounts of hay or concentrates must be stored and supplied. Butchering of hogs, steers, geese and turkeys therefore most often takes place in autumn with the onset of cold weather and the advent of the holiday season when large carcasses and cuts of meat are in demand for feasting. Rural societies until about half a century ago had the skills, tools and facilities for butchering large animals, though capacities were dwindling even then. They had holding pens and barns and big kitchens, tractors and guns, and they knew how to use them. Our rural ecovillage in North Carolina organized a hog slaughter a few years ago, and the neighbors, some of them in their 70s and 80s, showed up by the dozens to make sure the newcomers got it right. Even they were impressed and moved because no one had locally held such an event in four decades, though they all knew vividly how it was to be done. Today, few communities make the attempt, which requires a coordinated collective ritual far more meaningful and complex than stringing holiday lights or putting on a prom. Apart from the widespread home butchering of poultry that marked small towns and even urban backyards before World War II (my great-grandmother did her share) and the clandestine butchering of rabbits that continues to this day, 20th-century urban societies left the killing and carving of livestock to butchers and commercial meat-packers. To bring home butchering back to our suburban and city farms will require determination and finesse.

## In the Kitchen

Most households have an excess of small kitchen items and relatively few implements that would be practical for self-reliant food processing. A turkey roaster might be the outer limit for most. To turn garden abundance into real self-reliance requires a good collection of kettles, large stainless bowls, roasters, trays, skillets, pressure cookers, food processors, food mills, cutting boards, long-handled spoons, ladles, spatulas and the counter space and stove burners to animate it all. Gourmet cooking has become enough of a consumer fetish that quite impressive equipment can now be found in retail shops catering to cooks, although much absurd, flashy, poorly designed and completely unnecessary nonsense is on offer as well. The 12-quart stainless kettle is a basic item that bridges between the space of the home range and communal quantities of soup and sauce. A hot water canner or large pressure cooker can be even larger, but these can sit on more than

The basic implements of canning: a pressure cooker, canning kettle, jars and tools for handling them. (Inset) A 12-quart stainless kettle.

[Credit: Creighton Hofeditz]

one element of a stove top if not being used out-of-doors on a free-standing burner. If you are going to can, you need at least one of each.

One of the consequences of garden farming is that you spend more time at home. You also often work with more people so meals become a central feature of the day. In order to get things done on the farm, eat well and to enjoy the things you grow, it becomes necessary to shift the scale of cooking somewhat — and also to simplify menus. Working housewives and kitchen men have long understood that it pays to make large batches of things and put half or more of them in the freezer for the week to come. We make good use of our pressure cooker to prepare hummus, chili and a variety of soups, usually combining beans and meat. We always make enough to feed 12, so we get three or more meals out of every dish. Getting a good result and keeping costs down is relatively simple: starting 24 hours ahead, you soak any dried beans that will be involved. Drain them the next morning and put all the ingredients of the soup, stew or spread in the pressure cooker with enough liquid, and heat it up. Within 30–60 minutes the hard work is done. In the meantime, you can be eating breakfast or doing the laundry. You can salt these dishes before cooking, but seasoning should be done after the pressure cooker cools (when the pressure seal releases). Then, when the soup is still steaming hot, you can add fresh raw garlic, dried herbs and other delicate flavors that don't need to go to 250°F for an hour. Eat the soup fresh and for several days to come, or freeze a couple of quarts (in plastic tubs) for later in the month. A meat loaf, a nut loaf or a paté (made with lots of inexpensive but flavorful ingredients) works the same way: bake two, freeze one.

When hot weather graces us with armloads of basil, we make pesto (you can make spring pesto with cilantro, which pops up in the garden from overwintered seed). The classic recipe calls for pine nuts, but who has 'em? We use peanuts, sunflower seeds and sesame, and I imagine we'll use hazelnuts when our trees begin bearing. We blend herbs, seed, olive oil (no, we don't grow our own) and salt. You can add grated hard cheese if you like. Pack this in small tubs, label, date and freeze. Pesto makes an easy all-purpose spread for bread or veggies, a salad dressing or an ingredient in omelets, and a little goes a long way.

If you have your own eggs, or even if you don't yet, it's easy enough to make mayonnaise at home in the blender in a few minutes. We do this often, adding herbs if we like, or varying the flavor mix. Pie, quiche and casseroles are all sure to please and easy to serve. They can use a variety of homegrown ingredients, stretch a little meat over many servings and will keep for several days.

It's wonderful to work with your own homegrown ingredients, but it takes a while before you have enough of everything you need for the kitchen and the diet to roll out a full, rich, nourishing and abundant cuisine. Inevitably, store-bought foods will continue to be part of the household supply. In many communities there are good local food co-ops run by people who are committed to healthy and organic foods. If there is one near you, join and support it. If not, consider forming or joining a buyers club with a group of other households. You can usually open an account with a natural foods wholesaler and get access to high-quality bulk foods as long as you can buy once a month. You and your co-op members will have to meet the truck and divvy up the goods amongst yourselves. Six or more households (perhaps a House Cluster) buying together can typically meet the minimum purchase threshold.

## Stocking the Pantry

At Renaissance Farm, we have focused on key foods (see Chapter 13) in our gardening up to now, adding in staples a little at a time. This is both an economic and a nutritional strategy to save money, support health and grow what the garden can easily supply during the years when its soils are being built. To round out our diet, we obtain our meat from local growers and purchase staples from the local food co-op. We always seek organic foods unless we know the source and can be confident that we are not being unduly exposed to toxins from chemical production methods. As a practice, we keep tubs of several kinds of beans and grains on hand at all times, using them for soups, spreads and main dishes. We have salvaged plastic olive tubs with square sides and tight-fitting screw-on lids that will hold more than 25 pounds of grains or beans. I put a small cloth bag (reusable tea bag) filled with pennyroyal and bay leaves in each tub to repel bugs, and keep the tubs in a cool place. For flavor, protein, versatility and ease of digestion, I prefer black turtle beans, garbanzos, lentils, rice, millet and buckwheat. We also have some homegrown mongrels. Choose your own favorites and develop recipes to use them all. Then just rotate your stock.

Beyond a base level, more ingredients do not make for better food. You don't have to reproduce at home the variety available in the grocery store. A hundred pounds of dried staples per person is a good target to have on hand. It's about what we can expect to go through in a year. A vegetarian household would probably need more. Remember that dry foods cook up into several times as much food on the plate. Our garden provides most of the onions and garlic we need and can often supply us with tomatoes for sauce and salsa. But in years when the summer is too rainy or cool, we will buy cases of canned tomatoes. And we can usually find local apples in bulk in the fall markets. We also supplement our own cabbage crop with purchased heads as needed and keep a dozen or more on hand in the root cellar during winter. This goes a long way, even when we make fermented vegetables. Those too become a staple.

Your pantry needs a stock of salt and sweeteners, vinegar and oils and any prepared condiments that are important to your diet —

we use mustard in salad dressings, for example. For winter use and convenience, we keep large quantities of dried culinary herbs and herbs for tea on hand. Because we often make medicinal extracts of plant leaves, flowers and roots, we also keep a stock of neutral alcohol for making tinctures.

## Making Medicines

I am not a trained herbalist, but I have learned bits and pieces from various teachers and by my own studies. We have some favorite herbal

Every pantry should hold stores of dried beans and grains.

[Credit: Creighton Hofeditz]

Dried herbs are one of the easiest and most valuable storage crops of the small garden.

[Credit: Edward Carter]

Echinacea (flowering) is a perennial herb of the Composite family. Its roots produce immune-stimulating compounds.

[Credit: Edward Carter]

remedies for home use, and there are many hundreds of others worth knowing about. Botanicals can be incorporated into wines, cordials or teas, used in cooking and made into tinctures or salves. Some are applied topically as a poultice. Some plants we dry and use as teas or decoctions. Nettles are an excellent source of minerals; I drink them periodically as a tea. We also use spearmint and peppermint in this way and with food. We have grown stevia, which is a natural non-sugar sweetener that commands a high price in shops. We always have lots of comfrey around for use as a mulch and fertilizer — it's usually employed medicinally either in salves or as a poultice. Plaintain too is helpful for bee stings and such irritations, but there's never a shortage of it during the warm months (in winter you don't suffer insect bites), so I don't usually collect it. The seeds are a demulcent (used to cleanse the bowels); psyllium seed comes from a plantain relative.

Some plants I find useful to take internally. We grow a small East Indian herb called spilanthes (*Spilanthes oleracea*) that has many medicinal uses. It's also known as toothache plant because some of the active constituents cause an analgesic tingling of the gums that reduces tooth pain. I have found it effective against bee stings too. Internally, it works as an antimicrobial and immune stimulant. I tincture the flower heads and sometimes leaves as well. You can chew a little of the leaf for a toothache, but the flowers are much too strong for the mouth. Echinacea is well-known as an immune-supporting plant. It is perennial, and the roots need to be three years old before they are harvested, at which time an alcohol extract can be taken from them. Wash and chop the roots finely, pack them into a pint or quart jar and cover with grain alcohol (I use 100 proof or 50% pure). Shake the jar daily for six weeks, storing in a dark place and decant after six weeks, using a strainer or filter if necessary. Label and date as the tincture will keep for many years.

Elderberry is considered a good cold medicine; if you can keep them from the birds, the berries are the part to use, either in wine or simply in an alcohol tincture. Catnip and fennel combined in a tincture can provide relief from intestinal discomfort due to gas or indigestion. You can also chew the fennel seed directly. We tincture dandelion roots as a liver cleansing medicine. Turkey tail mushrooms (*Trametes versicolor*) are common in our area and have strong demonstrated anti-cancer properties. In traditional use they have been consumed in soups and teas.[8] The active ingredients are also capable of being drawn out by alcohol.

The important thing to know is that most botanicals can be processed easily at home. If making a tincture, it helps to dry the leaves, flowers or root material a little before soaking them in alcohol. This simply removes some of the water and leaves a more concentrated medicine. Salves are not complicated, though a little practice may be needed to get the right consistency. Most are made with a blend of olive or coconut oil and a small amount of beeswax to thicken. Warm the oils to liquify

## A Fourth World Pharmacy—Common Weeds that Make Useful Medicines

**Plantain (*Plantago major*)**
First aid for stings or bites as a poultice (chew and apply); internal use for sore throat, dry cough, stomach ulcer—anywhere a surface is dry or irritated. Helps wounds heal faster; prevents scarring. The seeds are a mild laxative as well as nutritive.

**Chickweed (*Stellaria media*)**
Demulcent and salve for chafed or irritated skin. Make a poultice or juice for fresh use, or wait a day after making a salve to allow it to dry slightly. Also edible.

**Yarrow (*Achillea millefolium*)**
Decoction induces healing sweat to break a fever. Leaves applied to a wound will staunch bleeding. A poultice for cuts and abrasions. Tincture of flowers effective for minor urinary tract infections. Flowers mildly immune stimulating and antiseptic. Can be used for stomach upset. Stimulates the liver, digestion and blood flow; a cardiovascular tonic. A bitter used in brewing beer.

**Dandelion (*Taraxacum officinale*)**
The root is a cleanser for both liver and blood; improves digestion. Aids pancreas to balance blood sugar. Also a kidney tonic, stimulating urination during infection. Can lower blood pressure as a diuretic, but helps supply potassium to avoid loss of electrolytes.

**Burdock (*Arctium lappa*)**
A mild immune tonic and like dandelion but more subtle in supporting the liver and the blood. Effective over time against dry skin, eczema and acne. The root is used.

**Yellow Dock (*Rumex crispus*)**
A liver tonic stronger than either dandelion or burdock. Good for moist skin conditions. The root is a mild laxative and astringent to the large intestine. A tea helps the body absorb more iron, combatting anemia. Strongly flavored so suitable for blending; use in small amounts.

**Nettles (*Urtica dioica*)**
A herbal multivitamin that stimulates milk flow in nursing mothers. Alternative, or blood-cleansing, helping rid the body of excess uric acid. Effective for allergies by healing the mucosa of nose and throat. It can also be used for joint problems as it is anti-allergic and detoxifying.

**Ragweed (*Ambrosia artemisiifolia*)**
The leaves supply antihistamine effective against allergies when used as a tea or tincture, both preventively and as a remedy.

**Red Clover (*Trifolium pratense*)**
Nourishing and gently cleansing. Supports the body during cancer treatment, cleaning accumulated toxins. The flowers are edible.

**Mullein (*Verbascum thapsus*)**
Tea or tincture of the leaves is soothing to the lungs, effective for cough. Stimulates the lymph system to purge toxins. An oil infused with the flowers is specific to ear infections.

**Chamomile (*Matricaria recutita*)**
The flowers and leaves are rich in calcium and calming to the nerves, it is effective against irritable bowel, mild diarrhea and related painful conditions.

**Comfrey (*Symphytum officinale*)**
Extremely mineral-rich, comfrey makes a fine poultice for all wounds, even broken bones. Especially suitable for mixed wounds of bruising and cuts, it supercedes arnica in more severe cases. Excellent in cases of broken fingers, toes, ribs, or collarbones which cannot be set. Also useful for sprained and swollen joints. It should be drunk as tea or eaten only with caution.

[Credit: Peter Bane, referencing Corey-Pine Shane. "Ten Good Friends: Using Weeds as Medicine." *Permaculture Activist #45* (March 2001); Anna Newton. *Herbs for Home Treatment: A Guide to Using Herbs for First Aid and Common Health Problems.* Green, 2009.]

them. Infuse dried herbs by packing tightly into a small jar. Cover with oil, seal and stand the jar in a warm place for a month, shaking periodically. Then strain and use the oil directly or thicken to make a salve.

## Rhythms of Eating

There is scarcely a cultural norm anymore around even so basic an element of life as food, but, as with so many things in our era, this is largely a function of the excessive energy in use in our society. Among many things, this enables a proliferation of (apparent) choice, and so contributes to the breakdown of formal structures. When half of all meals are eaten away from either home or the workplace, there is no center anymore. Food is eaten on the street, in cars, at desks, and much of this scattering is unwelcome even by the people engaged in it. If you are going to farm, you will build your workday around meals, and they will largely be meals taken with others who will be working with you.

For now, it might be enough to remind ourselves of the wisdom of people who have examined human life in great and loving detail, from scientists to religious leaders. The old adage "breakfast is gold, lunch is silver and supper is lead" flies against the practices of my youth and contravenes many common habits of eating in the US. Nevertheless, I think it sound. I'm hardly immune from the influence of the world around me, but my choice (when I have it) is to eat a hearty breakfast, a moderate and leisurely dinner at midday with friends and to sup on salad and light fare before dark, filling in with drinks and snacks when I feel the need.

I spent a short time in Norway as a youth where my brief exposure to farm life revealed to me that meals can be quite plastic in different cultures. But even there, the emphasis was on eating more heavily early in the day. My Norwegian farm hosts served five meals in the 20 hours of daylight that prevailed at midsummer. Breakfast at 7 AM was followed by second breakfast at 9:30, virtually a clone of the first, heavy on meat and starch. Dinner was the midday meal, and it too was substantial and served by 1 PM. At 5 PM there was a soup, bread and savory items (little bits of fish, cheese). And about 8:30 PM we would have what the English might call tea: pastries, jam, tea, cakes. This was a diet designed to support heavy and prolonged physical labor. What we know of nutritional science confirms much of this as well. A good breakfast is important for sustaining energy throughout the day, and eating smaller meals more often is easier on the digestive system. Cultures of hot regions often take part of the midday for rest and rumination. We sometimes have to do the same during very hot periods in our Midwestern summers. For different climates and cultures, the patterns may vary, but for all of us, regularity and deliberate rhythm in eating are valuable supports for a healthy life.

# CHAPTER 20

# Culture and Community

Each expansion of the human sphere has diminished the capacities of nature to sustain other species (extinctions have increased above the background rate observable in the fossil record) and to sustain further human growth. From about 1989, humanity has been in *overshoot*, which means that each year we use up more natural capital than the remaining systems of nature can regenerate. And that ecological deficit has grown steadily. In 2010, the human economy exceeded the sustainable limit of the planet's ecosystems on August 22, almost a month earlier than the year before. It is arguable that we are headed into default and foreclosure of the only sort that actually matters.[1]

Energy descent has been underway for more than 30 years (the peak of per capita world energy extraction occurred in 1979), but in the US the contraction has been masked by public borrowing, various investment bubbles and the exploitation of cheap Chinese labor. That game collapsed in 2008 as George W. Bush slipped out the back door of the White House. His administration had gambled that by invading Iraq it could seize control of the largest unexploited oil reserves on the planet, but the bet failed. The US now faces decades of economic decline, a dire condition that our public officials refuse to address honestly. A further wild card is in play: the massive release of carbon from fossil fuels over the last two centuries has destabilized climate such that it will increasingly disrupt agriculture and ecosystems.

We can imagine the contours of an energy descent future and infer some possible cultural responses to them. Since oil is fundamental to transportation, mobility will be constrained. Industrial agriculture, too, is energy-intensive, so food will become expensive and sometimes locally scarce. Between population growth and climate stress, water will become a limiting factor to economic and agricultural activity. Drought will not be confined to the West.

Centralized systems will deteriorate, contract or both, leaving large subsets of the world population and even many wealthy societies outside the social contract. There will be growing gaps in public services: street lighting, public safety, fire control and

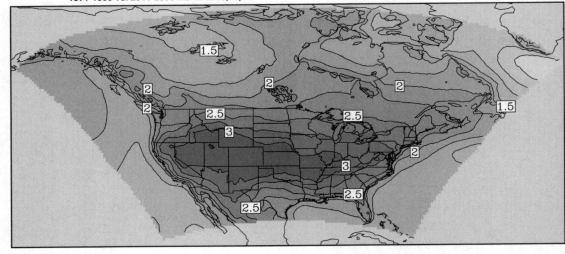

## ENSEMBLE MEAN CHANGE: 2 m Air Temperature
### 1971-1999 vs. 2041-2069 Months: 06,07,08

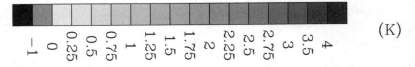

Projected North American Climate Shift from Late-20th to Mid-21st Century Shown in Degrees Celsius for Summer (June—August) Based on a Composite of Models

[Credit: North American Regional Climate Change Assessment Program]

education are already on the block in US states and cities. Public health and sanitation, including waste disposal, are coming under pressure. And pensions and benefits for the lower classes will be cut in order to preserve profit for the elites and funding for the military. These privileged elements of society will eventually break down into smaller and poorer units but not, I think, without the impetus from blowback, revolution and imperial overreach.

With jobs scarce, food prices rising and government spending slashed, millions will be left to fend for themselves. Earlier and more primitive levels of society are emerging to fill the gap left by the shrinkage of the corporate welfare economy: families, bands, villages and tribes. The Rainbow tribe have been ritually enacting the end of the world as we know it in nature settings for more than 30 years while continuing to draw on diverted wastes and surplus from a very wealthy society around them. Urban youth have organized gangs to carve up economic territories

from the bodies of our decaying cities. Adult children are moving in with their parents and vice versa. The best of these forms will evolve into groups that can generate resources for their members in ways that reduce conflict and help to stabilize society around them. My hope is that garden farms will provide a bulwark against both the rapid deterioration of society and economy and the violent and authoritarian upheaval that such change could engender.

## Adapting to Hard Times

Cultures are fashioned from resources, the technologies that can exploit them, the social and economic structures that mobilize work and wealth and the attitudes, beliefs and rituals that propagate these systems and lend them meaning. The culture that needs to evolve now across North America must be, for practical reasons, based in some familiar and accessible elements, but must also be much less dependent on fossil energy, distant resources and high levels of mobility. It must adapt to lower levels of energy and the wealth that has come from it. We can already see this process of change underway.

### Thrift

The first adaptive response to economic contraction is thrift, which reduces the consumption rate of existing resources. Call it *making stone soup*. It means doing more with less — and requires us to change our habits. A Depression-era ditty captures the attitude underlying this change, "Use it up or wear it out. Make it do or do without." Today, politicians are making hay by claiming that "We're broke, and we have to tighten our belt." But while this hard-nosed if hypocritical advice must become a permanent shift for most, it will not be enough to address the predicament we face because, by following this strategy alone, you starve.

### Social Capital

The next level of response, to substitute social for financial capital, is also essential but will be more challenging. Essentially, it involves much more sharing, surrendering a measure of autonomy for collective security. Call it *taking the bus*. This affects social status more acutely and involves close cooperation and coordination with others; it is inherently political. Culture wars and other manufactured antagonisms, including racism and ethnic conflict, make this useful approach more difficult. Sharing provides a measure of material surplus by changing social and economic structures. And, while this runs contrary to the ideology of the consumer society, it has strong roots in our religious traditions.

### New Resources

The third level of response is to find or generate new resources — and to fashion or adopt technologies appropriate to them. This approach involves the redesign of our productive systems and ways of life; it is here that permaculture has the most to offer. Call it

*make a new plan, Stan.* These are the most difficult changes because they involve the built environment, tools, physical plant and new investments of external energy. All these approaches, if they are to succeed, must be reinforced and stabilized with attitudes, beliefs and rituals that can imbue them with meaning.

## A Culture of Garden Farming

The new resources needed by the culture of energy descent are primarily biological, local and rooted in land. The energies of the future are also those of the past: sunshine (and wind and rain), biomass, animals and human labor. These will be supplemented by salvage from the industrial economy: some tools, some machinery, some parts. We won't give up fossil fuels immediately, but we must stop leaning on them so heavily, reduce our rates of consumption and make what remains serve the highest-possible uses. Long before liquid fuels become unavailable, they are likely to become very expensive, and so we will learn to what uses they can most profitably be put. If you are not now planning to get by on ¼ of the fossil energy used by the average household, you haven't set your sights low enough yet.

Massed Solar Arrays in the Desert
[Credit: Carl Berger, Sr.]

Biological and land-based resources are neither guaranteed nor readily exploited. They are thin in drylands and in very cold regions. I can't offer a lot of hope to most people living in the desert southwest of the US, which is seriously overpopulated and vulnerable. Minerals and solar energy are the primary resources there and the only hope for a long-term economic future in those regions. As the archaeological record clearly shows, even riparian and oasis agriculture — a mainstay of deserts the world around — may fail if climate shift proceeds far enough.

The availability of biological resources also varies from season to season and year to year. Wildland foraging makes sense where wild lands surround our settlements, but hunting and gathering are no longer a viable way of life for more than a part of the rural population. Urban foraging consists largely of scavenging; pickings will get leaner as times get harder and more eyes scan the territory. The common thread underlying all these limitations means that biological resources will be difficult to extract from afar: they will belong to those on the ground.

## Living in the Garden

I call the worldview that makes it possible to thrive from resources of the land Living in the Garden, Pattern #8 from the Garden Farming language (see Chapter 6). Many traditional peoples saw the world around them as a garden, even a pantry — a place where they were at home and that fed them — a place they cultivated. However much land we own or control, this attitude can help us enrich our own lives and those of the community around us. The Kayapo people of the Amazon were nomadic, but they followed well-established trails, and along those trails they planted seeds of trees and shrubs that would feed them and provide their medicines. They moved about a large territory in accord with the ripening of fruits and the availability of game.[2] Guerrilla gardening uses land we do not own or control to create food and other resources unbeknownst to authorities, landlords and absentee owners. With maintenance budgets stretched, the opportunities to push succession toward edible and economic species on publically accessible ground increase.

Living in the garden requires us to see land through the eyes of a gardener and always to be asking the questions "What could be growing here?" and "What else could I plant or graze?" Gardeners move about in their territory and notice changes; they also intervene to push ecosystems in the direction of greater productivity for themselves. By adopting the attitudes of a gardener, you notice much more.

Seeing with new eyes is part of the necessary shift in culture that we must undertake. Changes in the structure of work are also needed. I don't expect many people to give up high-paid employment in the formal economy to take up garden farming, though some may do so by choice. But I think those who have become redundant, unemployed or underpaid, as well as for those young people without an obvious future in the formal economy, it may hold a greater appeal. In order to create a new resource base, one less dependent on fossil fuels and paid employment, it's necessary to spend time at it. This isn't just a matter of learning a new career as a farmer, but of actually creating the farm out of a landscape that may not have been productive for two generations or more. It may be possible to enjoy dabbling in the garden evenings and weekends as a hobby, but to create a productive system at home requires thousands of hours of work over many seasons. You've got to reorganize your life. You have to be available to work when the seasonal energies and the weather are going in your direction. You also have to create continuity of attention in order to keep the complexities and potentials of the system fresh and alive in your mind. The land won't wait for you to give it your attention. If the system isn't already on a trajectory reflecting your aims and intentions, then it's going to follow the default path: weeds, trees, varmints and random growth in response to the weather. Even if it's going in a direction you have already given it, change happens, new conditions arise and you need to stay on top of it. So you have to carve out dedicated time when the farm gets your full and best attention. And you've got to get help.

## Getting the Work Done

Writing or computer programming may be solitary work, but farming has always been a group enterprise. Even garden farming in the suburbs is a group activity.

Our present average household of 2.6 persons is too small. We will have to grope and stumble our way toward better situations. This will require us to learn about sharing, to expand the number of adults and teens in the household labor force and to make do and bear the stresses of not having enough help all the time. It will give us a new appreciation for people, even if at first they seem poorly suited to the tasks that need to be done. The culture of garden farming will be one based on learning what we need to know.

Two people are almost always more efficient at garden tasks than one alone because they benefit from division of labor. Some tasks can only be done by two (or more) people, and even when both are doing the same one-person operation (such as planting seeds or weeding), working near someone else you like is more fun. You'll need other members of your household to pitch in, and not just once in a while, but every Tuesday and Saturday (or whatever days you can manage). Give the garden farm a focus at certain times of the year and on a regular rhythm during the week. This builds expectations and allows you to direct flows better.

Shared work provides an opportunity for a fluid division of labor. Two people can be digging holes for trees while a third person is soaking their roots and pruning them in

Friends help lay out a sheet-mulch garden.

[Credit: Creighton Hofeditz]

preparation. Or the third person may be collecting and moving the cardboard and chips to mulch them, or setting up the hose or the watering buckets. When a task is done, having more hands means the tools get cleaned and put away so everything can be finished up briskly. Some tasks benefit from many hands — haymaking, large-scale mulching, fruit picking or even pruning an orchard — but work teams of two or three are the most efficient for small settings. When we have four people available, we often set up two teams of two each. It's also possible to mix it up during the day — put everyone together for a big job, then break it up for smaller tasks.

## Each One Teach One

Some people come to the farm explicitly to learn — students, interns, apprentices — but everyone comes out of curiosity. They will be more supportive and appreciative if that impulse is recognized. We always give first-time volunteers a short tour, and we make time with our long-term apprentices for tutoring, field trips and design charettes. To flourish, your new culture has to be centered on learning.

Working with people who don't have basic garden skills can be challenging, but even people who are familiar with tools or know some basics may not know how to use tools well. We find that working with novices and volunteers means that a certain amount of tool breakage is inevitable, though one should try to coach people to avoid it. It's easy to forget that people not familiar with garden work won't have these elementary lessons: Put cardboard down when you are digging a hole so the dirt isn't lost in the grass or wasted on the driveway (or use a wheelbarrow). Keep the layers of the soil separated (topsoil apart from subsoil); don't mix them. Don't leave the rake lying on the ground with the tines up! Don't overload the wheelbarrow or you may lose control of it. And in your complex polycultures, you'll have to advise people which plants are not weeds and where to step to avoid compacting the soil.

In evaluating people for long-term engagements, look closely at their basic work skills and ethic. Do they respect tools? Do they know how to self-start? How familiar are they with construction, landscaping, plants or animals? Can they work alone on a proj-ect and stay focused? When they work with others, do they know how to anticipate other people's moves? Are they physically fit, and are they graceful or clumsy? When things get confused, or a task is winding down, does the person move on to completion and the next job, seek direction or simply stand around? Consider spiritual qualities such as humor, generosity, self-knowledge and the ability to listen. In a healthy group situation, everyone makes a contribution, and everyone works to include all the others. That means stepping into leadership when it's appropriate, and yielding to others frequently.

## Work, Play and Relaxation

We can give ourselves to work more fully when it also becomes play and when, at the end of the day, we know that the body can relax. Play is experimentation, and garden farming and permaculture are both full of that. Much of what needs to be done also has to be discovered. And as we figure out what to do and how to do it, there's a third dimension, which is how to sustain it. Good work must grow out of design. There are clever devices, well-placed elements and calculated balances, but farming is never finished, perfected, systematic. It is an endless dance with nature, always changing. The challenge is to learn some steps and then to improvise in graceful ways. The calendar of the year has its stresses and pauses: spring planting and fall harvest, summer expansion and winter focus.

To incorporate these fully into a way of life particular to your own place takes a lot of practice, arguably it takes generations, but we can begin now. Just as it takes three years to accumulate an adequate supply of dry firewood or five years to get pear trees to bear amply, so it takes us a number of seasons to learn our role in the farm ecosystem: when to pull out the seed packets and when to pull out the weeds.

When you work at home, relaxation can be elusive or only available in sleep or in

Fruit picking is an activity that calls for many hands.
[Credit: Chewonki Semester School]

drink, but there are other ways to punctuate the day with relaxation. Water takes away our cares and our pains, and when it's hot, that's almost bliss. Elements of the bath need to be brought into the rhythm of the week and the day. There's nothing to do on a steamy summer afternoon but take refuge in the swimming hole. And when autumn throws its robe of cold night over everything, to seek the fire and sweat in the dark lodge or sauna is to surrender to a primal force of the universe. If you work together, then you are entitled to relax together, and nothing leads quicker to relaxation than radiant heat, steam and cold water, usually in quick succession. Make the communal bath a part of the work rhythm. Let it be impulsive as a quick dash for the pond, throwing clothes and caution to the wind, or let it be deliberate and intense with the building of the fire and the quick plunge or drench under a bucket of water. True mastery of craft means knowing when to stop.

## The Centrality of Food

When we organize workdays at Renaissance Farm, we often eat both breakfast and dinner together, dinner being the large midday meal. At breakfast, which anyone can cook, we talk about what's on the agenda for the day. At midday meal we carry on the conversation and cooperation that started in the garden. Usually one of the household members prepares the dinner or the elements of the meal in advance so that it can be heated and served in a short time. Then, at the last few minutes with several people pitching in, the food gets heated and served, the salad is prepared and a dressing is made, drinks are poured, the table is set, and later dishes are washed and put away and the floor is swept. The foods are simple: casseroles, soups, stews, chili, roasts or stir-fry, sandwiches, steamed veggies, a pot of grain or grilled items. On the side there is sauerkraut and kimchi for relish, sometimes toasted seeds. The salads are always different but always fresh from garden or greenhouse.

A shared meal is the center of the workday.

[Credit: Creighton Hofeditz]

The outside work may be useful and productive, but the shared meal sacralizes the day, giving the entire experience more meaning than it would have otherwise: we are in this together. We are working and we are feasting. All are giving and all are receiving. All are of one body and one flesh, at least for a few hours.

We have made a small ritual of eating pie with each workday meal. Pie is a humble dish, easily thrown together in a few minutes once ingredients have been organized, yet it is festive and special too. "Easy as pie" is true, but only if you practice. Made of flour and fat, salt, sugar, spices and usually fruit, perhaps with eggs or milk, pie can be almost entirely from local and homegrown foods. It is an everyday extravagance that delights and reminds us of the earth's abundance. Pie is our potlatch, an affirmation that the gift must always move.

## Closing the Nutrient Cycle

No agriculture can be truly regenerative until it recaptures the nutrient in food consumed by the whole society. The problem with one-way eating is that soil minerals are progressively removed from our farmlands and dumped into waterways that lead to the ocean. Once in the sea, these minerals do not return to land for tens of millions of years. The most critical of the many soil nutrients thus lost is phosphorus. As landscapes age geologically, their phosphorus tends to become depleted anyway, but in North America we are accelerating the aging of landscapes through our agriculture. Conventional agriculture applies phosphorus as superphosphate, an acidulated compound of rock phosphate that is mined in only a few areas of the world. The extractable reserves of phosphorus in Morocco, Florida, North Carolina and a few other locales are dwindling and will be exhausted within two generations at present rates of use. The energy required to mine and ship these minerals will also become prohibitively expensive over time. The only foreseeable solution is also the simplest one: we must recycle the phosphorus (and other nutrients) in human and livestock waste toward agriculture.

So as we open the discussion of human waste recycling, we need to emphasize the two-fold nature of the dilemma. If we return untreated humanure to agricultural soils, as many traditional cultures did and still do, we risk spreading bacterial infections or parasites, some of which can be deadly. If we fail to return humanure to soils that are growing our food, those soils will continue to degrade and lose mineral content, making our food less healthy and our bodies subject to degenerative disease. Typhus or cancer? Cholera or diabetes? Surely, there must be another way that resolves this unwelcome choice.

John Todd, formerly of the New Alchemy Institute and now at the University of Vermont, has demonstrated that drinking water can be derived from municipal sewage entirely by biological methods. These consist not only of the familiar settling, skimming and aeration that go on at treatment plants, but of further filtration through closed cells containing sophisticated communities of aquatic plants and animals. Essentially, these "living machines" eat all the waste and turn it into living organisms. All that's left then is clean water.

Some of John Todd's living machines have been installed at municipal sewage treatment plants and in other institutional settings.[3] Though this is a good thing, we shouldn't expect a wholesale conversion of public sewage systems anytime soon. But, as garden farmers, we have other options.

### Treating Human Waste for Reuse

There are four low-tech pathways toward the destruction of pathogens in human waste. One is hot composting. A combination of heat, bacterial digestion and oxygenation results in the destruction of pathogenic organisms. When compost reaches 149°F for one hour or more, pathogens are killed. Longer periods of time at lower temperatures also work, for example, 122°F for 24 hours, 115°F for one week or 108° for one month.[4] A second method is a variant of this, called *mouldering*, which may or may not achieve critical temperatures as in thermophilic compost piles, but that uses time and microbial action to achieve breakdown of organic matter and pathogens. A third method involves *dessication*, accompanied by fairly high heat. Toilets have been designed to focus solar energy on the collected excreta, turning it effectively into ash. Some commercial incinerating toilets use electric heating elements to dry the waste. The fourth method involves *aerobic digestion in water*. This is essentially what many municipal treatment plants accomplish by pumping air into their pools of sewage and churning it all around. Aerobic bacteria eat the organic matter, and that includes the pathogenic bacteria and parasites. Something similar goes on in healthy soils, though much more slowly.

If you want to close the nutrient loop between the food you eat and the food you grow, your choices in ascending order of cost and complexity are:

1. a bucket toilet
2. a commercial composting toilet
3. a custom-built, dual-chamber composting or mouldering toilet system
4. an artificial wetland that receives the effluent from your sewer or septic tank

Option 1 — the bucket toilet — is suitable for virtually any household where all the adults are in agreement to respect it. I know of many elegant urban homes where humanure composting has gone on for decades without a whiff of controversy.

For the other three systems, you must own your home or have extremely friendly relations with your landlord, as these will require some modifications to structure or new construction.

Commercial composting toilets run upwards of $1,000, and the simplest will require at least a vent connection to the outside of the house. Some use electricity to evaporate moisture from the waste, either with fans or heaters. Some are quite large and require major construction to install in the home. However, they do not require engineering, and most are approved for use in a wide range of circumstances. Some commercial models are adapted to work with very low flow flush toilets.

A dual-chamber composting or mouldering toilet sits above a sealed chamber into which the wastes fall and are held. Two chambers are used alternately, so that one can be full and digesting for up to a year while the other is in active use. After a year of breakdown, the waste material is converted into odorless compost very like forest floor duff. These systems can be built into homes or other buildings or may stand free as small structures of their own.

Mouldering toilets require good ventilation, which can be passive (no fans), and if properly designed, they have no more smell than any other bathroom, sometimes less. Air is drawn down through the toilet seat,

circulates around the compost pile and leaves via the stack into the air above the building. Screening excludes insects. Allowance must be made to drain moisture from the pile — this may include diverting urine — and to evaporate it. As with any good compost pile, the material in the toilet must be moist but not wet.

A bucket toilet needs a good lid and good habits. Every deposit must be covered by shredded paper, sawdust, wood shavings or similar cellulosic material. [Credit: Asphyxia—fixiefoo.typepad.com]

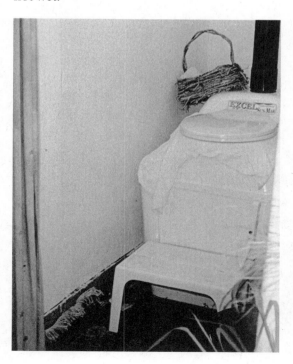

A Sun-Mar toilet installed in a California home. Note the black vent pipe. [Credit: Peter Bane]

This free-standing mouldering toilet has two chambers built into a hillside. The south-facing glass front assists venting through the solar chimney.

[Credit: Peter Bane]

This wetland cell is populated by cattails and cannas. They grow out of the gravel medium with their roots in the subsurface wastewater. [Credit: Creighton Hofeditz]

The artificial wetland takes liquid effluent that has been settled and skimmed through a conventional septic tank or equivalent system and passes it through an open-topped, lined tank into which gravel or sand and gravel have been placed. This requires either a dosing chamber and a pump to lift the wastewater from below ground to a location from which it can flow by gravity, or a significant slope. The gravel in the open tank provides pore space for flow and also supports emergent plants, called *macrophytes*. The water level in the tank or cell is designed to stay below the surface of the medium. The plants rise out of it.

Constructed wetlands are stocked with plants such as cattails, water iris, bulrush, arrowhead, jewelweed and others that grow in swampy conditions. Many of these have hollow stems and pump oxygen to their roots to support colonies of aerobic bacteria. These bacteria digest pathogens and organic matter remaining in the effluent, while the plant roots take up the nutrient and convert it into biomass. By periodically harvesting the tops of the plants growing in these wetland cells, the gardener can recycle all the nutrient in the sewage safely to gardens of any kind. The cell may discharge into a second similar cell (depending on the volume of flow) or into a series of ponds in which aquatic organisms (plants, algae, tadpoles, fish) feed on any remaining nutrient. The effluent will ultimately be discharged to trenches in the soil surface. These can be filled with mulch to produce compost.

Our society will be forced to grapple with the reuse of human waste for land fertility over the coming generation, and it will be easier to do this through public policy if some of us have gained experience with managing private systems that do this safely and effectively. The cultural shift that is implied means recognizing that the most despised material in our world is actually extremely valuable (although potentially still messy and dangerous), and that it can pay very handsome dividends to learn to transform it.

### Dealing with Other Kinds of Waste

Our household recognizes the great nutrient flow represented by urine, so we capture ours in jugs to apply to the garden. Urine, as discussed in Chapter 10, is typically free of bacteria as it leaves the body and makes a near-perfect plant food. Whatever you do or don't do with your plumbing, you should consider this cultural change. Urine can burn plants if applied directly to leaves, so we pour it onto mulch or dilute it, often by applying it during wet weather. A side benefit of collecting urine is that it becomes much easier to maintain clean toilets, as we do not pee into them.

The money economy generates enormous amounts of waste, not just by throwing away agricultural nutrients in sewage, but in every form of packaging, disposable consumer goods and through sheer excess consumption. Not a few people live almost entirely on the flow of throwaway food from grocery store dumpsters. What isn't salvaged in an edible condition still represents a massive amount of potential compost for soil improvement. The construction and landscaping industries are major generators of bulky waste, much of which is extremely useful. Lumber, plywood, tile, bricks and block suitable for reuse, parts of buildings, trim and furnishings, fencing and vast amounts of wood and brush are thrown away every day. Printed vinyl billboard tarpaulins suitable for roofing are a small item of commerce locally; I suspect they can be found in every region. Sheet metal and roofing panels can often be gotten for free or cheap. Pallets, boxes and thousands of acres of cardboard are pushed out the back of the global economy daily. Buckets, barrels, pails, tubs, jars and bottles can be had in virtually every size, even without buying their contents.

The most useful forms of salvage are materials that have multiple uses: cardboard, lumber, doors, brick, block, buckets, tubs and tarps. Rope and cable fall into this category. All organic materials — wood, natural fiber cloth, paper, cardboard — can either be mulched or burnt as fuel if you have a stove. Manufactured items are tricky; the more specialized they are, the less likely it is that you will be able to use them. Furniture is similar: tables are the most flexible form, bookshelves and cupboards next, then chairs, chests and beds. The market in these is fairly well established, but it's also often glutted.

If you have the room for one, a semi-trailer body can make a practical storage shed. They are usually surfaced in aluminum and are therefore highly durable if the envelope has not been compromised. Consider placing such a structure on a block foundation to avoid the floor rotting out. Mobile homes (single-wide) and travel trailers are similar but much less robust and desirable. However, you might inherit one on a farm or in a rural or suburban area. I don't recommend living in them, but if the roof is sound they can, once gutted, make an acceptable storage shed or chicken coop. If you are going to use one for the long term, and especially for the storage of valuable items, it pays to put a second roof over it.

### Foraging

Scavenging, dumpster diving and garage saleing are all forms of foraging in the industrial mode — tapping into the abundant resources of the territory. Over time, you gain familiarity with hot spots and can develop a regular route. There are also seasonal pulses, especially in communities with college students where the month of May always brings a raft of discards.

If you heat with wood, you are likely to have ample opportunities to forage your fuel supply. Let it be known to the neighbors that you are looking for hardwood fuel, and you will be offered dead and dying trees, fallen branches, blowdowns and other windfalls.

Sometimes fuelwood can be found in the public domain: along highway and railroad rights of way, fallen across streets in the winter or dragged into the ditch, dead and down in the no-man's-land between properties or on vacant lots belonging to absentee owners. If you notice a fallen tree in your neighborhood, approach the owners. You can serve your own needs and also generate goodwill since you are likely to save them money on clean up. You might even be able to get paid to collect your firewood.

Food foraging is also a worthwhile activity, especially in the early years when your own trees may not yet be bearing. Many fruit trees are not tended; an even larger portion of nut trees are neglected. Paying attention to fruiting species in your area is a good way to learn the calendar of harvest. With fruit as with other foods, the early bird gets the worm, so when you spot something that is ripening, move on it. As with firewood, many wild and neglected food crops grow on the boundary between properties or on public land, where neglect is a near certain sign that you can harvest without interference. If you know the owner of a property with ripe fruit on it, ask for permission to pick. If you don't, just approach whomever may be living in the house to inquire. Offering to share the harvest is a near-sure way to get permission, and you may not have to offer much. Most people don't yet want to be bothered to process surplus food. They might be happier with a jar or two of applesauce rather than a bushel of apples. If fruit turns out to be of good quality but the tree is in poor condition, a nice trade is to offer to prune it. In this way you can also build a relationship with the owners or occupants of the property and perhaps gain future access to the crop.

Where wildlands are accessible, learn to look for edible and medicinal mushrooms, wild foods and herbs. You can also collect seed and sometimes shoots, bulbs or suckers to transplant onto the farm. Be mindful that wild plant populations may be threatened locally, so never take more than 10% of any plant population, whether you are harvesting plant parts for medicine or digging a root or bulb to propagate. Seeds and fruits are usually more abundant, but again, you should leave a portion to propagate naturally. In some areas of the country, collecting seed in wild areas can be a sustainable economic activity if limits are observed to avoid overharvesting. This can include tree seed as well as the seed of wildflowers, shrubs and native perennials. For your own use, look to plants that have potential value as nitrogen-fixers, nectary sources or biomass producers.

## Neighbors and Strangers

Garden farmers work at home, so more than most people they see what goes on in the neighborhood. When changes occur, you may be among the first to know: who's moving in or out, who's building, who's doing repairs, what services are being hired in, where the children live (where the school bus stops). You will also draw your own stream of visitors to the farm, whether volunteer helpers, customers or curiosity seekers. Sorting out these many kinds of flows and relationships is part of the pattern I call Neighbors and Strangers.

Neighbors are the people you see frequently, even if only in passing, and who live nearby. Neighbors share fences, hedges, wildlife, care and concern over pets and other animals, sound- and lightscapes, traffic, safety and utility concerns. You also share and safeguard each other's privacy. With the people nearest you, make a point to offer things and to ask favors in return, building obligations both ways. Always try to give full measure plus a little more and, wherever possible, be proactive. If you are lucky, the neighbors will be great people whom you enjoy living near and who will feel the same about you, but this doesn't always happen. Some people may want to party, and others will be happy to chat

Neighbors and
Strangers,
Pattern #19
(see Chapter 6)
[Credit: Jami Scholl]

over the fence every few months. Others you will scarcely see. But in every case, you would do well to explore the nature of the boundary: Is it hard, soft, permeable, prickly or quirky?

### A Face on the Street

In a stretched-out suburban region, it can take years to meet everyone unless you go out of your way. In rural areas, people tend to take a greater interest in knowing their neighbors, if only to bolster their own security. Urban neighborhoods can go either way. You see people more often because houses are closer, and if there is a core of long-time residents, you may get drawn into it. There may be block parties, neighborhood association meetings, a local church or some other social anchor. However, in transient areas, there may be few people with a social map of the whole, so it's harder to figure it out. Many residential neighborhoods have no public environments other than the street, so getting to know people requires hanging out within view of the road, walking, bicycling, waving and establishing your public persona by the appearance of your property, gates, mailbox, entrance and your habitual activities. You may partly make your mark by the way you paint

or decorate your house. And, as a gardener, you can make showy displays of flowers or tall plants where everyone can see them.

If you're going to be in a place for the long term, plan to create public space so that you and your neighbors can get to know each other. If you are growing things, you'll certainly sometimes have surplus that can be sold or traded. Put up a sign for the farm. Choose a name, make a logo and post it out by the road. When you have strawberries or lettuce or pumpkins for sale, put out a chalkboard or some other prominent sign to let your neighbors and passersby know. You can do this even in the first season, whether or not you have regular hours or any large amount of anything. Maybe you grow extra seedlings and want to sell a few. Be prepared to give advice. People may ask you about anything related to the garden or even about building or repairs. Try to be helpful. The process of building your market goes on for years, but it starts with simple acts of opening your gate to the neighbors and getting them interested in what you are doing.

The appearance of street frontage is an important part of your efforts to make a public space. You can deliberately offer joggers and

Whether plain or fancy, make the face of the farm a generous one.

[Credit: David Barrie]

walkers on the street a bench to sit on by the sidewalk or the road.

You can put up a kiosk for neighborhood notices or a produce stand where you offer things for sale. Your plantings and your signage contribute to a feeling of public ownership, of being addressed. So too does the way you handle visitors, in particular parking cars. Some neighborhoods have little extra parking, some have lots, but using it may be somewhat taboo. Whatever the ethic, be mindful of the impact your visitors and customers will have on the neighborhood.

### Boundaries

It's said that good fences make good neighbors, but I think it's the gates that make good neighbors, and these are often missing. The biggest sources of conflict between neighbors arise over noise, animals, trash handling and smells, parking of cars and, in neighborhoods where owner occupancy prevails, appearances (which can include vegetation manage-

ment or *lawn order*). Fire is a legitimate but less frequent concern. Keep in mind that all of these problems can cut both ways. As a farmer, you may be running noisy machines from time to time, your vegetation management will be non-standard, you may have recycled or "junk" materials stockpiled and your animals have the potential to intrude on your neighbors. The reverse may, of course, also be true. Your poultry are vulnerable to predation by neighborhood dogs or cats not kept in control.

Fences can keep your livestock in and domestic predators out, though they are an imperfect barrier against cats and wildlife. Visual screening helps to reduce the impact of clutter. Hedges can filter smells. Buildings, hills and wooden fences can partially block noise. Perhaps the most difficult situation to face would be a neighbor committed to using biocides on the lawn. Your best approach is honestly to convey your concerns about the chemicals to your neighbor. Seek information: who is doing the spraying, how often is it done, what chemicals are used? Try to establish a dialog by respecting the person's desire for an "attractive" and low-maintenance landscape. Present information regarding the risks associated with lawn chemicals, in particular those in use locally. Be very clear that you intend to protect yourself and your plants, animals, children and household members from the dangers that chemicals pose and ask your neighbor for help in this. Research alternatives, ask for a reduction in frequency of spraying, request notice when spraying will occur and especially insist that no spraying take place on windy days. This is in accord with industry-recognized best practices and is not controversial. Indicate that if the neighbor will not stop the spraying, you must create physical barriers, including fencing and hedging. If the neighbor's land drains onto yours, you will also want to intercept and treat the runoff using swales filled with woody debris or other substrate on which appropriate mushroom species have been inoculated. If you have already established a farm and obtained organic certification, a new neighbor who begins spraying may be subject to legal injunction or civil suit based on damage to your business. All these measures are unproductive and take time and money, and you would be better not having to fight such a battle, so try to avoid falling into this condition blindly. Before you buy a property, investigate neighboring land use practices, especially see the landscape during warm weather if at all possible. Get to know and regularly talk to your neighbors and use every opportunity to let them know that you grow clean food without poisons, and would like them to benefit from those same practices.

### Receiving Visitors

Not all visitors will be customers, but all are potential customers. They deserve to have a good experience when they come, and you will be well served if they do. The advantages of selling from the farm are obvious: there's no transportation and little need for packaging. You pick only what you get paid for: no waste. The biggest problem with mixing production and sales in the same setting is the conflict between catering to customers and getting productive work done. As important as a good fence is the invisible boundary you create by your policy on visiting. You have to reserve some days and some hours for private activity, including work. Work can sometimes be opened up to volunteer and visitor help, but this is public outreach, education and market building and only incidentally about production. Neighbors pose a special kind of problem because they will be your first customers and will tend to feel that they should have, if not unlimited access to you, then at least random and spontaneous access during the day. You can encourage people to call first before dropping by; when they do, suggest days that are best for your schedule. Perhaps you'll have one open day a week when visitors

are welcome. It's best if you have extra help those days or can work on projects that are easily interruptible.

If you have enough room along your driveway, people will typically park there; it it's long, it also needs to have both a turnaround and places along its edge for parking that don't block traffic flow in and out. If you don't have a large driveway, or if it gets too easily congested, put up a clear sign at the gate that parking is on the street, or in the church parking lot or wherever nearby may be appropriate. This may still not be enough. If parking protocol is not completely obvious, at least one in ten drivers will do something stupid, so be prepared to direct traffic.

The same lessons apply to internal boundaries. A clear sign or a table with potted plants or other products for sale will be seen as "the front desk," and people will gravitate there. Without a focal point, they may wander into private areas of the farm, come knocking at the house door or otherwise confound your expectations. There is also a fine line between sales and tours. You can't afford to tour everyone around who comes to buy a head of lettuce or a four-pack of cabbage. It's fine to let customers know that you are at work and also minding the store: "Just potting up some strawberry plants. And how can I help you?" If your produce or plants are on display, the shop mentality will help to limit your exposure, but you may not always want to pick vegetables in advance. Then you have to take charge of the conversation, find out what people want and give them choices, as a waiter might in a restaurant. They don't have to see everything if they are buying produce.

## A Culture of Learning

Much of what draws people to the farm is curiosity and the hope to see or learn something new. We are in a period of cultural change, much of which may seem unwelcome; garden farms are a positive element in the vanguard of that shift. Lots of people have heard of

CSAs.[5] Maybe your farm will be one of them. If you understand that larger picture, you'll be able to put all of your public encounters into a deeper perspective. Visitors, students and volunteers will come to peer into the future: Will it be scary or icky? Might it be fun? You have the opportunity, not only by what you do but by how you report what you are learning, to shape people's view of the shift that is underway. If you believe in what you are doing, be prepared to give it a positive spin. You don't have to paper over the difficulties, but let people know that it's worthwhile.

People love stories, so you'll have to develop many ways to tell yours. Why did you decide to farm? How did you come to be here? How long did it take to do all this? What is permaculture? How could things be better? What's coming into season next? You need a strong one-liner that can get a laugh, a 30-second version that tells the big story—and beyond that you're giving the tour. The main thing is to find out what people are interested in and answer those questions if you can.

Out of the flow of visitors and volunteers, some people, maybe younger, maybe older, may want more than a taste. Many are looking to learn the whole way of life that goes with farming, and particularly permaculture, which is both a way of living and a design system. They are not drawn to conventional farming for all the reasons that I laid out in Chapter 2, but they may be drawn to a self-directed life close to the earth and to people. They are thinking of family and community and clean food and skilled work that pays better than flipping burgers. Artisanal farming and ecological design can be a career path for millions. We do not have too many practitioners in these fields. The problems of the day when all your neighbors are permaculture garden farmers will be fun to solve. Then we'll form co-ops and specialize in crops across the spectrum from meat to botanicals and nursery stock to exotic fruits, selling directly to consumers across the county. We'll make wine, cider, perry and ale. We'll run science-based charter schools and teach the kids to farm. We'll do agritourism, training courses and dinner theater and grow it all within walking distance. We'll form ecological SWAT teams for hard-core restoration of the desert and of Superfund sites. And we'll launch exchange programs with African and Andean farmers to learn the best things we don't know yet about farming in really tough conditions. Til then, we need more opportunities for adults to learn the basics.

## Building Community

Community forms around an apprehension of mystery, the beating heart of life and death. People seldom have words for these feelings, but they form the field of attraction that draws folk together. The community that will make garden farms the new face of our suburbs and towns and the countryside around them will emerge from people who know that nature is alive and that we are inescapably and gloriously part of it. These people, whether they know it fully or not, will have the wherewithal to make our familiar and empty landscapes bloom and drip with food, beauty and laughter. They will dare to play at being farmers until they discover one day that they are farmers, and the world has become a different and better place because of it. We have many steps to take along this path, and every one needs to be a step into greater joy. The essential work of permaculture activism is to understand and see abundance in the world around us, often before others do, and then to help others to see it also, to bring it into being.

Opportunities for building community are all around us and multiplying by the day. We can awaken a new dream in the hearts of millions, a dream that lies sleeping there, a dream that's already stretching and yawning and rising. We have to make that dream a real place where people can live and work and love and play—and that tastes delicious. That dream is not in one place only or of one size or shape but arises with a unique face and flavor everywhere from the same mystery. The keeping and the giving away of this mystery are the culture that we must knit together now. Community is the manifestation of abundance that comes from this pulse.

No community is alive that does not have a view to the generations lying ahead. Children and youth are the present reminder of that future, and they are its elders. Whether we have children of our own, and whether our own are drawn to the community that will renew our connection to the earth, there are children, youth and young adults who will be drawn. And there are parents and elders too. The work of garden farming is household work, and for most of human history, households have been comprised mainly of family. And so it may be for many if not most now. But also families will form around this vision, and households will gather and make families for the joy and satisfaction that working with the land can bring.

So stay home and garden. Hang out your shingle. Run your flag up the pole. Identify your allies. Gather some friends. Roll up your sleeves. Throw a party and get to work. Learn what you don't know and what you need to know and then practice, practice, practice — but do it with others. Share the work and the learning. Plant trees, store water, build soil, save seed. Harvest and enjoy. Take time to play, get in the water and gather round the fire. Make food together and eat it. Cut the pie, and we'll have some more. Breathe. Talk. Sing. Dance. Tell stories about what you did and dream of what you will do. Include everyone. Compost your shit and put it back into the soil. Put up a fence and be sure it has good gates. Assert your rights when you have to, but don't pick unnecessary fights. Talk to your neighbors, relentlessly. Partner with younger people if you are old or older ones if you are young. Teach the children. We're in for the long haul. Get comfortable.

# CHAPTER 21

# Markets and Outreach

To write of markets is necessary but not enough, for food is never just a commodity. That's the point of all this. Food is a cultural value, it's a gift, it's a sacrament. Still, we do trade it, buy and sell it. Or perhaps we should say more accurately that we buy, sell and trade our acumen, our timing and skill, our preparation, our understanding of the rhythms and reasons/seasons of nature. And in this field, none can corner the market, but all can market on the corner.

The garden farm is sustenance agriculture, a way of life first and only as much of a living as it needs to be. The genius of the permaculture approach is that it enables you to move many of your needs out of the money economy. This gives you resilience, flexibility and choice. Without making a dollar from your farm, your farm can still make you wealthy. Once you know that, marketing becomes a more relaxed process, part of the play and not the end game.

You can view your efforts on a scale of profit and loss, and I do recommend that you keep track of what you spend and what you earn. But accepted systems of money accounting don't know how to value clean water, wholesome food, good friends around the table, ripe fruit on the tree or a full root cellar. You are a natural capitalist: build up your assets in soil, water, diversity, information and biomass. Money will come. I approach garden farming as an education. By learning to grow soil and raise healthy plants and animals from it, I harness free energies from nature, apply my own intelligence and wind up with a harvest that meets my needs. If I need some money, I set up structures and processes that let people give me money. But fundamentally, the main business is learning to raise healthy food, grow soil and build an ecosystem based on the home.

## Design Redundant Systems

In a balanced food forest ecosystem, garden farm or permaculture, there are many products, no one of which is the main crop. Perhaps there are five or eight main crops: highly productive and making up the bulk of our diet, but 40–60 foods that we eat throughout the year, plus many minor harvests for flavor, fiber and general utility. Some years, cherries

Necessary fall pruning of figs results in many cuttings available for propagation or sale.

[Credit: Creighton Hofeditz]

will be abundant and other years they will fail. That's why we have mulberries, plums and juneberries too. The first objective is to fill the larder and grace the table. For that we need to have redundant systems: multiple sources of water, heat, income and food crops. Plant potatoes and sweet potatoes and squash and kohlrabi; something will succeed. If your activities are multifunctional, you can eat your figs and sell them too. We like to eat figs, but as it turns out, we have to prune the fig tree every autumn, and so we wind up with 50 new fig tree cuttings to carry over in the root cellar. The next spring they are available for sale. Was that maintaining the system for self-reliance? Or was it a cash crop?

We raise flats of spring vegetables in the greenhouse. Inevitably, there are extras. Potted up appropriately they are easy to sell to visitors, neighbors or in the open market if we must. We give advice about landscapes, sometimes helping people to design and plant them. Dividing perennials, selling off surplus trees and shrubs, we add value to the work. If your echinaceas are crowding something nearby and someone wants echinacea plants — boom, you have a crop. If you don't sell it, you tincture it and make medicine.

## Recognize Surplus

You can sell anything you have in surplus, whether that be information, seed, plants, herbs, fruits, vegetables, meat, animals, crafts or bedroom space. But you will make the most profit on the first three or the last three. Basically, this means that you want to sell small things of high value and to which you have added your services and your intelligence. Better to sell breeding livestock than meat, and better meat than vegetables. Animals are a natural surplus of any ecosystem, part of nature's way of balancing swings in weather from year to year. They are more profitable if you don't have to slaughter them. So too is seed a surplus. If you sell basic foods, it is better to sell fruits, which are often in great abundance and in great demand. Leaves too, are easy to raise. We sell a lot of salad, which looks and tastes good, but doesn't weigh much and doesn't feed heavily from the soil. It doesn't travel too well, however, so we prefer to have it sold before we cut it. Herbs can be very profitable, but only if you have a good restaurant or shop connection to move dozens of pounds.

If your market disappears or you find yourself with surplus herbs, dry them. With bulkier basil or cilantro, make pesto. Six ounces for four dollars with a good label. Throw in a few pine nuts for the ingredient list and bulk it with sunflower seeds; get the salt right. Offer free tastes on crackers. You can make vegan versions, and to other formulas add the traditional grated hard cheese to pique the flavor. In the US, Michigan is leading a movement to liberalize food packing laws.[1] Other jurisdictions will follow. Many foods not considered at risk for contamination (think fruits and vegetables) can now be packed at home and sold in farmers markets. This opens up *pastabilities* for sauce, salsa, homemade noodles, sauerkraut, hummus and pie. This long-overdue reform is an obvious response to the need for home-based enterprise in an economy that has shed millions of once good-paying jobs. Other barriers to trade will have to drop as cottage industries take off again.

Markets work best for you as a seller when the goods you bring are scarce and the demand for them high. Fresh vegetables in winter are more profitable than the same in summer. Eliot Coleman has proven this, and will tell you how to do it.[2] Please note that he has applied no small amount of art and engineering to his work, so do not assume that he is benefitting only from a seasonal premium.

If you raise meat, do not feed it on grain; let it fatten on grass or bugs. These latter are inedible wastes, while the former costs you money and fattens corporate oligarchs. Meat is more highly regulated than vegetable produce because it can more easily be contaminated if not handled cleanly, but if you contract directly with your customers, you can sidestep much regulation.[3]

There can be a place for wholesale trade. If you have a surplus of something that will not keep, by all means sell at any reasonable price or you will be mulching it. But, by and large, you want to sell at retail and into niche markets wherever possible: organic, ethnic, gourmet, raw, health and wellness, local.

## Keep It At Home

The best market is the simplest that has some custom: your front gate. The next best is a wealthy market with lots of customers. For that it may be worth traveling. Remember that the farmers market is essentially delivery without prepayment, the worst terms you might accept. There has to be a significant premium in the form of good cash flow to make it worthwhile. If you run a fruit stand at the road in front of the farm, you can stock just enough to make it look attractive, offer a list of other items available, and you don't even have to sit at the booth if you don't mind using a cash jar. Many farmers do, and it pays off in the long run. It does pay to check the till at intervals through the day to avoid let-

Occasional markets are easy to make with a chalkboard sign at the road. These capture lots of attention when they aren't up all the time or when they change frequently.

[Credit: Creighton Hofeditz]

ting too much money accumulate. Leave just enough cash for making change. Package your fruits and vegetables in lots for even dollar pricing: $1, $2 or $3. Mark your prices well and list them on a chalkboard too. Be sure to provide a few bags if you display items in boxes or baskets.

After your own farm gate, consider what your options may be for selling farm produce. Are there specialty retailers or a local food co-op committed to supporting local farms? What about restaurants that want to buy high-quality local food? Is there a local growers association committed to opening markets? Institutions are often difficult to reach because they are bound by regulations and need large volumes of food which a single small grower can rarely provide. However, you may be able to band together with other growers to meet larger needs for local schools, hospitals or a college food service. Even these staid operations are changing under public pressure and demand from students and parents. More immediately, and perhaps with greater effect, look to groups within the community who already support *slow food*, food that is local, artisanal, of high quality and fine flavor. You shouldn't be looking to compete on price, but

CSAs provide their customers with a basket of food each week during a prescribed season in return for upfront payment which finances yearly start-up costs.

[Credit: Christine Hennessey]

to get a fair price for quality goods. Contributing to benefit dinners can be a way to get your name before people who support good causes and like fine food.

## Subscription Farming

An alternative to the farmers market — and with 5,000 markets across the continent you are likely to have one near you — is subscription sales, also known as box schemes or Community Supported Agriculture (CSA). The variations here are endless, and books have been written about the subject.[4]

What started with vegetables has gradually evolved into a more complex and comprehensive market. Farmers have added eggs, honey, fruits, milk and bread — sometimes on a mix-and-match basis — as their farm systems matured. Others have gone to a near-grocery store model where reach-in refrigerators are stocked with the week's harvest, bins are available and customers take what they want. Greed is not an issue; one of the bigger complaints from CSA customers over the years has been "too much food."

A common problem was shares that were too large, so farmers began offering ½ and ¼ shares for smaller households. Even though

the main point was to help the farmer prosper and get good-quality food in the bargain, the traditional value of not wasting food is deeply set in the culture, and customers often felt uncomfortable in not being able to consume everything they got. No doubt some figured out that they could give the surplus to neighbors, in-laws or the food bank, but the pushback did lead to greater flexibility. The challenge lies in finding the line where accommodating customers becomes too costly and complicated. This will vary among individual farms and farmers.

While it's possible to run a very successful CSA that provides a full-time living for two people with no more than one acre in cultivation, such an operation requires two to four people working throughout the season.[5] Most smaller garden farms will find the form somewhat rigid and may instead prefer the flexibility of farm stand sales, occasional farmers market participation and distributing to neighbors, family and visitors. Direct Web-based selling is also possible, and many user-friendly brokerage sites emphasizing local food have been set up if you don't want to create your own.[6]

## Self-Reliance First

The garden farm should first meet household needs. If you and six other households in your neighborhood decide to do this all at once, there's no good reason you couldn't divide the cropping tasks and gain some efficiencies of scale and management — you'd more quickly get to scale with individual crops and would sooner have surpluses for sale outside the group. You can also do better seed saving that way, at least of corn and squash. Whole farming neighborhoods may emerge again, but I think this is still a few years away. Any steps you can take in that direction are helpful. If, for example, your neighbors raise chickens and have surplus eggs, by all means barter your lettuce or your blackberries. For most of us, self-reliance will mean growing many of the foods we need as well as harvesting our own water, energy and other products. With a mosaic of different crops in various stages of maturity (perennials taking several years to bear) and soils taking time to develop consistent fertility, you are likely to have little bits of this and that for quite a few seasons. Marketing small bits requires flexibility. You also don't know at the outset which crops will be particularly successful in your soils and microclimates and which will languish or fail. These results can vary a lot with weather too. Thus, it can take several years to plot the profile of your agroecosystem. It's best to remain loose about markets during this period and focus instead on lowering costs of living, getting used to home-based work, investing energy in systems that will sustain you for the long haul and learning to farm. Unless you're prepared to live on savings, don't give up all paid employment.

One of the best uses of your early surplus is to support farm labor. You can organize volunteer workdays and feed everyone who comes, or you can take on apprentices or interns, whether live-in or day workers.

Treat everyone to a good time whenever they are working, and build up system fertility and diversity. Use whatever the farm is yielding to reward and encourage those who can help you. You have plants, seed, food, even livestock, and you have knowledge. Share them. Ask for help to streamline the harvest. Get projects started and finished. Make friends and allies, and train people in what you've learned. They become your advocates and promoters.

## Shifting Enterprise

Many market niches will appear along the way to creating a garden farm. In the early years, you are setting up systems, some engineered and some planted. When we built our first cistern, we organized it so as to offer a workshop in ferrocement construction. The primary return on that effort was the labor that helped us to build the walls of the tank. Incidentally, it also helped to defray some but not all of the material costs. In the past we have made bulk or wholesale purchases of garden tools and sold them on at a small markup. You can do the same thing with nursery stock. Larger wholesalers typically have very good prices but also significant minimums. If you need 200 trees or shrubs for the farm, you can easily meet such a minimum where most home gardeners cannot. Buy a little extra of useful species and sell them on. If you pick well, whether of tools or plants, you can help others by increasing their choices, making available items that wouldn't otherwise be in the local markets, and you can make a little money. Buy seed in larger quantities and repackage it in retail lots. If you have a fruit stand or a stream of visitors interested in what you do, selling seed or plants on the side is an easy way to let people show appreciation.

It can be helpful to have a small business with links to garden farming. This can be something you carry with you to the farm, an enterprise you may have started while still living in rented quarters. Ideally it is something that either provides cash at all seasons or can be done in the months of the year

Getting volunteer help consists of organizing tasks, putting the word out, being willing to teach and share and treating people well.

[Credit: Keith Johnson]

when farming duties are at a low ebb. A nursery is at the top of such a list, not because it's countercyclical with farm work, but because it piggybacks on things you'll already be doing. You'll obviously need to plant thousands of plants, and some of these can be raised from seed or cuttings in your own backyard. It's easy to create more than you need, and thus to have a few hundred plants of various types on hand. These can be useful to fill in gaps in the farm landscape and can just as easily be sold to farm visitors. To make soil mixes and pot plants is a basic skill for horticulture. Beyond this most obvious planting angle, you might be involved in landscaping or design consulting; you might have specialized equipment such as a tractor, rototiller or dump truck with which you can provide services. Once you have a greenhouse, you may be able to raise *koi* (decorative carp) which command a good price but are not hardy in many northern locations.

Keepers of goats may find that soapmaking can provide a year-round income stream. Beekeepers can easily store and sell any amount of honey they can raise, and this is not a demanding occupation, though the risks of failure are greater than in the past. Anything you need to do for yourself is potentially something that you can sell to others, and the risks of getting into such a "business" are very low.

If you don't have a horticultural business, the next best thing is something that can occupy your time in the winter or on rainy days: furniture making or repair, upholstery, sewing, knitting, shoemaking, ceramics and other forms of art or craft. Anything you can do at home, whether that's in your living room or a garage or the barn, has a built-in advantage for the would-be farmer. It enables you to mind your flock, tend your garden and work on the place while still making an income. Your attention can remain focused on the farm system, and you don't have to commute.

The farm will have different yields in the course of its growth as an ecosystem. In the early years, annual vegetables and herbs will predominate, and so those might be the focus of your initial marketing. But within three years you should have a good supply of small fruits, which means that you can make jam and jelly, flavored vinegars, salad dressings, fruit leathers and other value-added products. Animals grow to a marketable size more quickly than perennials come into bearing, but the feed and forage to supply livestock may take some time to develop. One strategy is to buy in hay and use the manure thus created to build soil fertility, while selling the milk, meat or wool from ruminants.

If growing for market is an important part of the economic plan for your home system, take heart. There are lots of people working on improving market access for small farms. The movement for organic and sustainable agriculture is more than 30 years old, and though it certainly is a big tent that includes large farms, many of the growers who have pioneered in this field work with small acreage and have solved or addressed the needs of the small grower. Almost every region of North America has an organic or sustainable farm organization.[7] If household provision and self-reliance are your principal aims, then you may not be concerned about marketing, and that's perfectly legitimate. As food becomes more valuable relative to other commodities, anyone who grows food will be able to sell whatever they have.

Our objective as a farming household is to demonstrate effective and appropriate systems and to inspire others to take up permaculture. In support of this, we work with local groups promoting urban and regional agriculture, we have advised city and county officials about policy and legislation and we contribute to conferences, workshops and public meetings as speakers and panelists. We also host groups, both formal and informal, with an interest in garden agriculture and related endeavors such as renewable energy.

Fruit Stand, Pattern #68 (See Chapter 6)

[Credit: Abi Mustapha]

This approach is not primarily intended to bring us customers, but it does incidentally. By establishing our reputation as a center of information and action for permaculture and urban farming, we draw a wide range of people looking for seed, plants, advice or simply hoping to glimpse an edible landscape. There's a strong society-wide movement toward local food and food security (two linked but not identical efforts), so we haven't had to try hard to get attention. I suspect that our community may be somewhat ahead of the curve, but opportunities exist in most urban areas and even some rural ones for outreach along these lines.[8] If you put yourself forward, you will find allies and attract supporters. This may include volunteers, customers and partners at every level.

## Tasting Another Way

I said at the beginning of this chapter that food is never just a commodity, but part of a way of life. Your fundamental marketing approach, whatever products and services and at whatever level you want to sell, must recognize that people are looking for an experience when they come to you. This is true in the farmers market, it's true at your farm stand and it's even more the case when they come to see the farm itself. As with food in a restaurant, it must taste good, but it must also appeal to the nose and the eyes. Presentation matters. If the farm is attractive, people will more readily pay money for whatever you offer. Especially if they come for your advice, your *bona fides* are in your landscape. They come to see something of beauty and magic, whatever other agendas they bring. When you are in the public market or on the side of the road, you haven't as many props as you may have on the farm, but you must still create an atmosphere that is festive and delicious, colorful and abundant. Just as supermarkets have learned to use brightly colored labels and the appearance of infinite choice, so the garden farmer who wants to excite customers will develop a landscape or a peddler's cart that is colorful and interesting as well as

Make the farm look good.

[Credit: Keith Johnson]

functional. Straight lines are boring, and they are aesthetically difficult. If you use them — as with rows in a bed or field — they have to achieve machine-line perfection to look good, but then they must not be of an overwhelming scale so they don't seem monolithic. Small areas of straight lines can be attractive if the colors and heights of produce vary and the beds and fields are small, like a patchwork quilt. Curves are actually much easier to make attractive, because the eye is accustomed to curving lines in nature.

If you follow the pattern Patchwork Gardens Pattern (#50, see Chapter 6) and use contour lines to manage water and pathways, the natural beauty of a mosaic of plants will rise to meet the visitor's eye (and your own as well).

## A Little Something Extra

There's another valuable tip for marketers that good salespeople know: everyone wants a good deal. As a grower, you are in a fortunate position to put this into practice because there are always small bits of extra goodness that can be added at the end of any transaction to sweeten the package. The Cajun French word for this little something extra is *lagniappe*. It's the 13th donut in a baker's dozen. It's two plants in a pot instead of one (how much does an extra seed cost you?). It's an extra copy of a farming magazine or a couple of strawberries on the way out for a customer who just bought nursery stock. It's a fresh fig for someone who didn't know they could be grown in your area and may never have had one. *Lagniappe* may be as simple as telling a small story or offering a recommendation about plants, varieties or technique. It doesn't need to be complicated; in fact, it shouldn't be. The essence of *lagniappe* is grace — a delicate note that enlivens the musical theme but doesn't take anything away from it.

I don't believe it possible to overemphasize the importance of building a broad base of support and acceptance for garden farms and for permaculture. Our society needs these innovations to be widely adopted. They cannot be implemented from the top down but only by inspiring individuals to act. No matter that you loathe public speaking and never want to sit on a committee. You can make a difference by everything you do and all that you say to people who come to your farm. Your success as a garden farmer comes not from standing above the crowd — but from attracting a crowd who want to do the same thing you are doing. Your customers are not your competitors, even when they take up farming. They are your future allies. You will need support when the county land use plan is revised. You will need others by your side when the time comes to advocate for another market in your end of town or to argue before council that goats should be permitted within the city limits. Every new garden farm is part of the solution to climate change, energy and resource depletion and economic contraction. Each new farmer is floating a life raft for his or her neighbors, friends, family and community. So, though you may need to earn revenue from the farm, and though it may feed you and keep you warm, perhaps the most important yield of your efforts will be in showing that it is possible for ordinary people in our rich, indulgent and terribly distressed societies to provide a good way of life for themselves and to make the world a better place at the same time.

# CHAPTER 22

# Making the Change

If you own your own home and it sits on land you also control, then the prospect of a garden farm lies all around you. If the lot is small, it may still provide enough space for the nucleus of a system that can include your neighbors' lots or empty parcels in the neighborhood on a sharecropping basis. You might have the greenhouse and your nursery, toolshed and workshop on your own property, while the orchard, garden beds and even paddocks for livestock might be through the block or next-door. The only limitations are those of your imagination and your ability to make a convincing case for garden farming to your neighbors. If you don't own your own home yet, you may still be able to start down the path toward a farm of your own. Are you in the right region, a place you would be content to live for a long time? Or are you biding your time in school or in a job that you'll be leaving. Even if you are going to move across the country, you can begin making preparations now for a new livelihood on the land.

You can make no better preparation for building a self-reliant home economy than to get training in permaculture design. Courses

are offered in all regions and in many formats. Make it a resolution to enroll in one within a year and put aside $25 a week starting now to pay the tuition. If you can afford it, take the course with your partner or spouse; you may get a small discount for taking it together, and you'll get a huge bonus from learning in stereo. Reading this book is certainly a good beginning to a new career in garden farming, but it's no substitute for guided experience and practice in whole systems thinking and design, the kind that you can get from a good teacher in permaculture. Regrettably, our education system does little to prepare people to draw on both sides of the brain, to solve real-world problems or to understand the intimate interconnections of living systems, and all these are the very skills garden farmers need most.

Farming involves land, which is always located somewhere. If you are in the right community, ask yourself if you are in the right neighborhood. Maybe you're in an apartment or living in a homeowners' association with restrictive covenants. Perhaps you have a nice house in an upscale neighborhood where front yard vegetables and chickens would create a backlash. Maybe you'd be better off on the edge of a city or town rather than inside its limits. Or perhaps your cosmopolitan and funky neighborhood already has room for goats on the roof.

How much do you need? How much can you afford? Property is expensive, though in the US there are bargains to be had right now. You don't want to saddle yourself with a large burden of debt, though; it may be difficult to pay off over the coming years. If you already have ¼ acre or more, the possibilities are considerable. With ½ acre you are in the next class up with room for market production. At one acre or more, almost anything is possible. Review Chapter 7 on land scales and strategies. Consider your income at present and include the whole household or any partners who might join you in a land venture. You should be able to finance 2.5 times your annual income. If your income is spotty, irregular or in jeopardy, dial back your ambitions. Can you locate near other people who would like to do this too? Maybe they have land, and you can buy the house next-door, adding to the size of the farm. Consider your parents, your siblings, your in-laws or your children. Would you want to farm with them? Maybe they already have the farm and need young partners to make it sing. Or maybe you can help them buy the place and make a good retirement life for yourselves in the bargain.

If you are going to work with others, be sure that your visions align. Spend a good bit of time talking, and don't be afraid to ask hard questions.[1] The people who are likeliest to succeed at this will love to cook or garden or both. You need people who care about animals and spend time out-of-doors, and who don't have hang-ups about killing things (vermin, livestock, young plants) or too many other dogmas. Sometimes you'll need machinery, and sometimes you'll have to cut trees down. Get clear about pets and kids and what will work and what won't. Figure out

how you'll handle money, labor and time. And develop an exit mechanism for dissolving the partnership. If there's a clear way out, you're more likely to succeed at staying in. Spend some time working on physical projects together and see who has leadership skills and who's a klutz. It doesn't so much matter what people's levels of skill are as being honest and accepting about them. Look for partners with a sense of humor and playfulness. There's going to be plenty of work — you'll need the balance. Think about the externals in other people's lives too — parents, well or aging, non-custodial kids and ex-spouses, debts, careers, marital status and stability. And ask how well your potential partners know themselves. Ask yourself the same question.

## When You Have Land

You need a vision. Maybe it starts with a dream or something you've carried since childhood, but it has to become clearer than that. It's got to burn brightly in your mind's eye — so much so that you can see it and describe it to others in great detail. It's worth spending whatever months and seasons it may take for that clarity to come forward. You don't have to know all the details yet. But you need to see the shape, feel the texture and taste the qualities of the place you are going to make.

Develop a concise statement of your aims. This will be your touchstone. You'll know if you're on track when you can say that you are meeting your aims. Don't confuse means and ends. You might want to have chickens, but unless your aim in life is to be a chicken farmer, chickens are only a tool (an entertaining one, admittedly). Maybe you want a more grounded way of life. Maybe you want to work as a family. Maybe you want to eat better food. Maybe you want to make your town a better place. Maybe you want to solve interesting problems. Maybe you want to work with your hands. Describe the quality of life you seek — and this has little if anything to do with income levels.

Make a map or drawing of the property showing everything that already exists and will surely remain. Include anything underground that you know about (utilities, septic field, well).

Review Chapter 5 on Design. Analyze the sectors for summer and winter sun, winds, wildlife, views, pollution, noise, fire risk and whatever else will influence the site from outside. Think about how you can enhance the beneficial influences and temper, block or mitigate the hostile ones by how you place and build things on the farm.

List all the functions that you want the farm to perform: housing, food production, energy supply, water, waste treatment, privacy. Be thorough. Think about how these can be met synergistically. Can doing one thing accomplish several purposes? If so, you'll be on the right track. List all the structures you think you'll need. Can any of them be folded into existing buildings? How? Are there overlaps or commonalities that might let you combine some structures or functions?

## Make a Design

Review the Pattern Language outlined in Chapter 6. Pick from it things that make sense to you, and see how your land can take shape using its logic. You may find it useful to transcribe the patterns onto index cards so that you can mix and match and link them fluidly. You can do something similar with a big sheet of paper by making a bubble diagram where each big idea, function, structure or pattern sits in a bubble and you draw arrows showing relationships between them. Out of this work you begin to get a picture of the whole farm as a system. Play freely at this stage. Move things around. Add in whatever you need, drop things that seem to clash, look for missing pieces.

When you have a system that seems to make sense, where every function is met by one or several elements and all the outputs are used somewhere, then begin to look at the pieces. Start at the center of the system (probably the house), and make a list of needs to be met and problems to be solved. Make another list of all the functions and yields of the house. Do the entrances go where they need to? Does the house connect to outdoor rooms? What is the flow of materials through the building? Could it be made more efficient? What's out of place? Could the house serve as the center of the farm? Is there enough space for a crowd of people to eat and meet? How would you store food? Does the kitchen work for more than one person? How does the laundry move outdoors to hang? What energy systems does the house use and could they be more economical, more resilient?

Are there backups? In what condition are the major elements of the building (the roof, the foundation, the windows, the appliances, the floors)? If anything big will need upgrading, consider how long you have before this is needed and how you might use that replacement or repair to modify the design, upgrade the quality or improve the materials and functions.

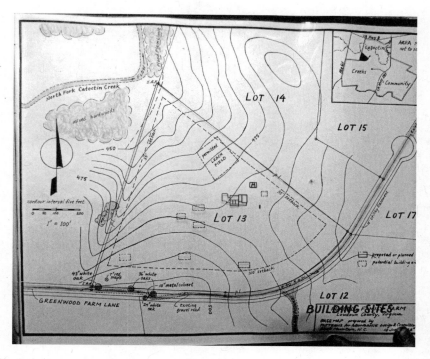

This base map shows permanent fixtures of the suburban property, including legal boundaries, watercourses, buildings and utilities. An overlay also indicates potential building sites.

[Credit: drawing by Keith Johnson and Peter Bane, photo by Creighton Hofeditz]

A resilient and economical home heating system supplied by local, carbon-neutral fuel.

[Credit: Edward Carter]

This ramshackle shed was given several years of new life by the addition of a $38 billboard tarp over a failing roof.

[Credit: Creighton Hofeditz]

Work your way through water, access, vegetation, outbuildings, energy systems, fencing, soil and any other important features of the property in the same manner. Don't be afraid to accumulate a list of many pages. This is your design prospect. You won't do it all at once, but if you don't consider everything at the outset, you're likely to overlook some important opportunities as you begin to make changes.

## Evaluate your Situation

Step away from the physical infrastructure for a bit and consider your finances and attitudes. Are you content with your present job? How long could you keep it? How about your partner or spouse? Are you still carrying debt? How much and for how long? Which one of you is more inclined to farm, and which parts of the household work will each of you do? Whose schedule can be freed up more easily, and are you willing to have one doing unpaid home-based work while the other brings in the cash? Can you switch or trade off, and how will you each feel about these arrangements in six months? A year? Longer? Do you need to bolster your savings for a while before you make any major investments? Do you have other assets that need to be converted? A second home? A 401-K or pension that looks dicey? What kind of money can you borrow on safe and easy terms? Is there family money to be had? Got any good credit card offers? Maybe you need to unload that third car or sell off the motorcycle or the plasma TV.

## Setting Priorities

Assuming you don't have any major problems with the driveway or the road in, your first efforts need to go into necessary repairs of major building systems (if any), control of erosion and water management. Make a plan for water storage and roofwater collection. How soon will you be able to create a tank or a pond or both? Divert any water that's causing problems or potential problems (running toward buildings, pooling, cutting into a road or other structure). Try to put that water some place on the property where it can be absorbed. Remove any toxic or dangerous debris, irredeemable rubbish, fences that impede movement and dead or unstable trees and shrubs. If any structures are at risk (leaky roof, failing foundation) and they warrant preservation, intervene to stabilize the situation.

Plan staging areas for materials. You'll need to bring in mulch, compost, perhaps lumber, stone and other construction materials.

Where can you create a depot for these and for any materials that may be salvaged from repairs, reconstruction or demolition? What will you do with timber if you take down limbs or trees? Will it become fuel, and if so, can you make a dry storage area for it? You may need to build a woodshed. Place any new outbuildings where they can serve other functions (collect roofwater, block ugly views, provide windbreak, complete an outdoor room). Try to make any new construction help to repair a damaged site. If an area has been graveled, scraped, devegetated or is covered with rough vegetation or debris, it's a good candidate for remediation. Consider whether an outbuilding might be located there and still serve its functions. This will prevent you covering over good growing land and leaving behind infertile areas that must then still be repaired.[2]

Consider which parts of the property would be best suited to what kinds of cropping. These choices are less certain and less critical; you may change your mind after seeing how things work. But you have to begin somewhere, so use your best understanding. Plan a garden within 50 feet of the house. Lay out contour lines and create a path system.

Dig out the paths and pile the topsoil from them onto the beds. Make the beds about three to four feet wide, or if you must make them wider, plan for keyholes to access the interior of the beds from the upper side. If parts of the lawn or future pasture need repair, conserve and transplant the sod from any places you dig. Put in a few crops appropriate to the season or plant cover crops to build organic matter.

Get a berry patch going: strawberries will be vulnerable to deer predation, so if you don't have a fence to exclude them, hold off and focus on cane fruits. Also plant some

bush cherries, currant and gooseberry bushes. Do your best to locate them well, but don't worry. They can be moved without too much grief. Plan your orchard or forest garden area and try to order some trees to arrive in the first spring after you begin.

Start collecting useful plants and set up a nursery.

Locate a place for the greenhouse convenient to the garden and the home, and if you don't have enough money for a proper structure initially, consider making an inexpensive hoop house for season extension. Concentrate

You can never have too much mulch. Other building materials are also handy.

[Credit: Edward Carter]

Contour garden beds are easier to work and are substantially self-watering.

[Credit: Keith Johnson]

A small plant nursery is absolutely essential to a successful garden farm. This one began life in a rented backyard under walnut trees.

[Credit: Peter Bane]

your propagating efforts in this area. Be sure it has a water supply.

If you have to do major exterior work on the house right away, refrain from planting too much immediately adjacent to it; leave yourself room to replace windows, paint, tuck-point the foundation or repair the roof without damaging new plantings.

If your water plan calls for major underground piping, consider how you can accomplish this with the least disruption of other infrastructure and plantings. It helps to lay out major pathways and any internal roads early, even if you don't develop them immediately. Underground plumbing should follow these pathways so that pipes can be found and repaired without undue effort and so that trees will not grow over them.

If you have sloping land that must be terraced, make a plan to begin at the top of the slope and work down, stabilizing as you go. Make sure any water held behind each terrace can be safely spilled into the next terrace below. You can begin retaining soil for terraces by using logs, rubble and even willow cuttings if you are willing to manage them.

Gather your tools and repair any that are broken or in poor condition. Add the things you need most, buying good-quality items. If handles are broken, they can be replaced if the tool head is worth keeping.

## Managing Transformation

You are unlikely to start your adventure with a functioning farm or even part of one. So any landscape and house you begin with will undergo a major transformation before the farm is fully functional. You will therefore have to live with some upheaval and manage changes to infrastructure and landscape even as you live amidst them and use existing systems. This can be daunting logistically and psychologically, never mind the financial challenges that may compound your work. It can be very difficult at the outset, but it is vitally important nonetheless to be able to pace yourself and set realistic expectations. You will not be done in one year or two. In three years, you may be well begun and perhaps will have reached the first plateau, a stable situation but one that you must move beyond. Five years might be enough time to get things looking pretty good, but don't imagine all your big projects will be complete by then. You may find yourself redoing things you've already done, either because they failed, you want to improve them or you understand better how to accomplish what you intended and the original design has proved inadequate. After five years, we are ripping out four-foot fruiting shrubs, moving our berry patch and still anticipate two to three years of major construction of outbuildings and cultivation systems. Rebuilding the greenhouse might take place in that time or happen even further out.

The upshot of this encounter with incompleteness forces you to find satisfaction in each step of the way, even as you hold a bright vision of what is yet to come — not a bad way of living life. *Unfinished* is not a sign of failure, but of growth potential. Just as important as holding a good attitude about your accomplishments is learning from them (as well as from your mistakes). Some things you cannot understand until you have lived with them for a while. Plants or crops you thought would work may prove lacksadaisical in their habit or poorly adapted. Some things take a variety of seasons to gain momentum — they might languish for years before taking off under the right circumstances. The synergies that you hope to achieve may not arise until you have crossed critical thresholds, some of which may not be apparent or measurable in advance. Absent functional connections between elements, you will have extra work to do. It can be difficult to get enough density in your plantings, and when too much ground is open, you face weed pressure and maintenance labor that can drag you down.

## Your Right and Left Hands

There are two key design methodologies involved in transforming neglected land into a garden farm:

1. Start small and work outward from a controlled front
2. Remove limiting factors

The first reflects geography; it places importance on the central infrastructure of the system: the home and any significant centers of work. The value of these elements is large at the beginning, and though it may ultimately be dwarfed by the productive value of the whole farm, nothing can take shape without these elements being in good order. The second method reflects the logic of work: everything exists on a continuum. One

Make rough terraces laying tree limbs and branches, chunks of removed sod, gravel rubble and fill dirt on contour; improve the soil later by mulching and fertilizing.
[Credit: Peter Bane]

project makes possible another. Not everything in farming or in the home system will be dependent on something else, but many — maybe most — things are. To accomplish projects, you need to line up materials, labor, time, tools, weather, soil conditions, information, perhaps even permission or at least agreements. All these must come together simultaneously or in a carefully orchestrated window of time. Anything missing or out of sequence can derail the work.

An underlying purpose to both of these work and design methods is to limit *entropy*, the dissipation and chaos inherent to material order. Anything presently existing — your house, your garden, the lawn, the old laundry line, the shed, the driveway — is in relatively stable condition. To be sure, insects and fungi are eating parts of these things, paint is peeling, iron is oxidizing into rust, sunlight is breaking down plastic cord, you lawn is getting shaggy and the garden is turning into a forest. But all that is happening very slowly.

From day to day it's difficult to see much change. In other words, if you do nothing, things will mostly go on as they are. If, on the other hand, you dig up the sod of the lawn to make a new garden bed, you've introduced energy and destabilized the situation. Soil is now loose and may wash away when it rains, weed seeds are exposed to light and will germinate with vigor and you have a hole or a lump in the yard where someone might trip. Creating a garden bed creates the need for a pathway that integrates it with the rest of the landscape and allows it to be worked without stepping on the bed. It also creates the need for weed control, erosion control and other things like fertility and water management. Work creates work. But it doesn't have to get out of hand.

Working out on a controlled front means going to the edge of those parts of the system that you have under reasonable control and making your change incrementally there. At the simplest, this means when you are ready to make new garden beds, start at the back door (or the front door) and push the garden out into the lawn.

The porch or step is still in control; weeds won't invade the new garden from that direction. The grass won't try to grow into the garden on four sides as it would if you plunked it down in the middle of the lawn, but rather on only one, two or three. When that section is mulched, weeded and planted and the drainage is controlled, expand outward. When you've managed to keep that bed full of vegetables and herbs, and you have time and energy for more, then you can expand the garden while still keeping it in control. If you keep pushing out and grow only annuals, you'll reach a point where their care requires you to:

1. Mechanize
2. Get more help
3. Convert some of the space to perennials (which require less labor)
4. Lose control — which means weeds, dead plants, unplanted areas, unharvested food

Notice that a limit is implied here. Crossing that edge reveals the limits of your attention, labor, time or skill. Adding more help or mechanizing might solve some kinds of

Plants just outside the back door get the most attention: start here.

[Credit: Edward Carter]

problems, but both of these measures involve greater complexity. They will require you to spend more time, money and energy managing that complexity. This means you will have less time and attention for the garden, and you'll become more of a landscape contractor than a farmer. Please note also that there is not a dogma here about machines or other people's help, only the advice to recognize the wisdom of limits. You should only take on those things that you can do well.

With time and with well-designed systems in place, you can increase complexity and not go crazy. Ecological design is the thoughtful effort to replicate nature's complexity and productivity. It's not unlike juggling. Once you have three balls in the air and are comfortable, that's when you can consider adding a fourth. Eventually the effort required to sustain four balls in the air is little more than that required to sustain three. And five and six and seven are possible in turn. But garden farming is potentially much more synergistic than juggling. Systems can generate outputs that become inputs for other parts of the farm. If I have a living fence planted to edible forage crops, then by pruning and managing it (though this takes time) I can also provide a yield of fodder for my rabbits or goats. Since I'd have to feed them anyway, I'm doing two jobs in the same action. That's *synergy*.

Each change you make must build on existing systems — or replace their functions altogether. Each change must be complete in itself (even if later you extend it) and, as much as possible, limit the length of the perimeter that you are obliged to patrol. Don't put delicate plants in the ground 100 feet from the house out of sight in the corner of the lot where you never visit. Don't dig a trench under the driveway until you first back the vehicles out and park them on the street, and secondly do not start until you are prepared to complete whatever work the trench was for (laying a pipe, wire, drain). Then, make sure that the work is done quickly, the trench

filled in and that gravel is on hand to cover the mud, which otherwise will stick to everyone's shoes, track onto the walkways, the porch and the kitchen. If pipes are laid but not fully connected, have caps to cover the open ends so that critters don't crawl in and water stays out. Each of these small efforts, stimulated by forethought, helps to limit entropy and control the front of disturbance. You tore up the driveway, left the cars parked in the garage and then in the middle of digging the trench, someone gashes a leg and has to be rushed to the emergency room. You forgot to charge your cellphone, and the neighbor's not home to call 911. Who can afford an ambulance anyway? The chances of such a scenario may be remote, but remember Murphy. He's in charge of those situations: if something can go wrong, it likely will. Your job is to make sure Mr. Murphy remains unemployed.

## Keep a Resource Inventory

Plan your work to match the time, labor and money that you have available. Whichever of these is in shortest supply relative to need is the limiting factor. Limiting factors can also be critical parts. That's one of the reasons you need a resource inventory at the farm. This avoids the need to drive to the hardware store across town (or in town) for a handful of screws, a one-inch pipe coupling or a few feet of electric cable.

You cannot always anticipate what will arise in the course of a project, but you can cover a lot of bases and make your life easier by being prepared. Weather is often the limiting factor in working on farm projects. You wouldn't dream of replacing the house roof during a period of incessant daily thunderstorms. If you have to dig footers for a building or a trench for a pipe, you'd rather the soil was a bit on the dry side rather than saturated as it might be in the spring. To work concrete, you need to avoid overnight frost. Experienced builders know these things, but we aren't all experienced in the construction

Resource Inventory, Pattern #58 (See Chapter 6)

[Credit: Abi Mustapha]

trades. Farmers, however, eventually have to do it all, even if only a little bit now and then.

## The Front Is Always Moving

As systems develop on the farm, new needs arise and new capacities emerge. After we put up our large greenhouse (from a kit) and used it for a while, it became apparent that we needed extra layers of internal plastic and row cover to carry some crops successfully through the winter. So we added them. It later became apparent that we needed an entrance air lock and a quick way to roll up the ends of the greenhouse to better regulate its internal temperature. You don't realize everything when you are looking through a catalog. The greenhouse did provide a special microclimate, but the zipper panel entrances that it came with wore out in a few months. Each of those incremental changes came from recognizing and mitigating a limiting factor. On the output side, the success of our efforts made the greenhouse productive enough that it fed salad every day of the winter to the whole household, and this energized us so much that we got a lot more work done through the dark months.

To design permaculture systems, you must be able to think holistically or multidimensionally. You must be able to think through the logical consequences of the choices you make. You must be able to see what inputs will be required for something to function and also what outputs must be used or dealt with as a result of success. Anticipating what failure might look like helps you to avoid it and also to be ready to ameliorate an unwelcome outcome. At the simplest level, this means having a rag in your pocket when you are painting, refueling the chain saw on the driveway rather than on the picnic table or being ready to use an extra half yard of concrete at the end of a pour.

## Be Well Prepared for Livestock

An important threshold to note is that involving the care and feeding of animals. While animals greatly enhance the power and fertility of the farm, they require daily attention (at least the smaller ones and dairy animals do), and animals are potentially very messy. The wisdom of starting small asserts itself here too: start with one type of animal and get just as few as make sense. When you have settled into the routines of care for rabbits, for instance (including the special efforts of breeding, birth and slaughter), after a year you may find that you can also manage ducks or hens. Don't start with animals until you have adequate food, water and secure housing for them. Be prepared to handle manure. Line up advice and be ready to play vet bills or be the vet yourself. Be sure that your system can handle foraging animals if you intend to let them out of confinement, and that you have adequate fencing and gates.

## Hodgepodge Growth

A consequence of following these two important directives in developing the farm is that you will experience or create the pattern of Hodgepodge Growth. One area will get going and then run into a limit — perhaps the season changes and you can't plant the trees there yet. Then you have to shift attention to a building project. Or, an opportunity arises to capture a large downed tree or a truck full of wood chips, so these are put on the driveway and then must quickly be relocated. This can look or even feel scattered, but it may be just the opposite — the most focused way to go about the whole project. You will find that things are connected functionally that don't seem to

have anything to do with each other. An example from our own farm may help to make this clear. When we decided to put up solar electric panels, we needed a south-facing roof which we didn't have. This required (but also fit with) the construction of a porch that we wanted anyway. The panels needed a sunny location in a landscape with quite a few trees. The porch happened to be located in the sunniest part of the property, and it was very near the house's electric system, making a grid intertie easy. Eureka! But the old porch wasn't gone, and worse, it was made of concrete. So, two years ago over the course of several months, with the help of a young intern's strong back and while doing lots of other things, we busted up the concrete stoop. Then we had to remove the urbanite rubble (broken concrete). With a future barn footer in mind, we stockpiled it against a fence near the site of the future barn.

One negative consequence of this was the return of poison ivy in that area where we had removed it. It grew happily up beneath and around the rubble heap. No easy way to remove that yet. We watched and waited but kept focused on the main program.

The porch gone, we built a new deck with a roof over it. (Measure, dig post holes,

Hodgepodge Growth, Pattern #59 (See Chapter 6)

[Credit: Jami Scholl ]

We hired an arborist to trim two large maples but had to find a way to get his truck into the backyard.

[Credit: Keith Johnson]

pour concrete, cut and peel cedar posts, buy lumber, hire carpenter, mobilize volunteers, juggle with autumn rains.) Now we had the structure to support the solar panels. Then we needed the installer (check), the money (check), the permit (check, almost…), the contract with the power company (check) and voilá, meter running backwards. But now the backyard maple tree was leaning into the solar window, so we needed an arborist (too big for our tools). And he brought his cherry-picker truck, so we had to remove the laundry pole that prevented him driving into the backyard to trim the tree.

And unfortunately, we'd already reset that ancient post in new concrete long before we knew that it was the key to installing solar panels. (Yikes!) We did a very good job. So we had to dig it up, smash the concrete off of it (more urbanite rubble) and set it in a new location. Truck in and out, tree trimmed, now we have firewood and brush to cut and mulch and stack, plus we have ruts in the

yard, requiring fill and sod to be transplanted. About two years after starting — just a week ago now — I think we've recovered (except for the poison ivy). We're at a new level of stability: electrons flowing into the grid, new porch useful for a hundred things including keeping the house cooler in summer, firewood drying in the woodshed, brush mulch decomposing into soil, laundry yard working better than ever, sideyard grassed, level and open for more deliveries.

The point of all this is that if you tried to understand what was going on as we were smashing the concrete front steps, digging up the laundry post, transplanting sod from various parts of the yard, allowing poison ivy to grow back into the yard or splitting and stacking silver maple logs, it wouldn't have been obvious. That's the nature of Hodgepodge Growth. Only by following the functional links between elements in the system can you determine what needs to happen first, second and so on. And only by doing that design thinking in advance — backcasting from desired outcomes to necessary starting points — can you avoid double work, backtracking, lost opportunities, compromised quality and worse. If you keep your focus on the important elements and pick your battles, you'll eventually prevail. With the new porch built and solar energy powering the house, a year and half later, we finally got around to preparing the barn foundation, using up the concrete rubble in the footer trench and driving the poison ivy back behind the neighbor's fence with a carpet barrier and clean cultivation to make sure that we can keep it out. And we were able to build the barn using free solar kilowatts. All of these complicated but ultimately fruitful actions grew out of our vision and our values: because of that they provide deep satisfactions even amidst the setbacks and the frustrations. So accept the truth that everything is connected and celebrate the fact that you have limits, that not everything can be done all at once. That sometimes little things take two years to finish. You might just be able to change your mind or your plan before it's too late!

As permaculture designers often say to each other, "The opposite is also true." Life is paradoxical, and everything works both ways. So the other side of the little lesson of Hodgepodge Growth is the common wisdom not to let the perfect become the enemy of the good. When we didn't have steel T-posts or locust poles for trellising our young grapes, we made do with bamboo. Well, bamboo breaks after a while, and then you can replace it with something better. Before we fenced the whole property, we protected our most valuable young fruit trees from the deer with makeshift cages of maple branches harvested when we trimmed the backyard trees.

They worked. They rotted and became an eyesore, and then we mulched them, but not until they'd done their job. If we'd waited until we had a fence before planting our fruit trees, we'd have lost two years of growth and wouldn't now be looking at a good crop of Asian pears and freestone peaches. But let interim and imperfect solutions be biodegradable or cheap. And think twice before you pour concrete.

## Start at the Center

The place to focus your early efforts is in zones 0 and 1, the house and its immediate environs. This would include the nursery, any small animal housing, toolshed and workshop, outdoor rooms such as patios and porches, a greenhouse and beds for basic fresh foods. Foods for storage will take longer to bring into consistent production and will require larger spaces. When you have productive trees, shrubs and ground covers growing in the nursery, then you'll be able to plant fruit tree guilds, whether as part of the garden expansion, for an orchard or in a forest garden. You can raise the stock to plant a woodlot, a windbreak or a hedgerow, any of which might cover large areas on the edge of the farm. But you can get started a few steps from your back door and have all the convenience, ease, security and efficiency that entails. Contrast that with planting out all these trees and shrubs along several thousand feet of property line or 300 feet from the back door. What do you imagine their odds of surviving would be? Could you irrigate in a drought?

Instead you could wait until the trees you've raised are healthy, vigorous and the season is propitious. You could put them into an area that had been contoured to intercept runoff, thus increasing the available water to improve their rates of survival. You can save labor and get better results by holding plants in the nursery until conditions are right; the plants will quickly make up for lost time once they go in the ground. By the same token, until you have excluded deer from the property (whether by use of dogs or a fence), you'll find it frustrating to grow certain crops and

Tree limbs too large to chop by machete and too small to be worth burning made temporary tree guards before new fencing excluded the deer.

[Credit: Keith Johnson]

ultimately difficult to sustain a large garden. But you can start a garden without a fence. Perhaps you'll use temporary fencing or repellents. You'll need to build soil anyway, and it's much easier to exert control over a small area near the house.

## Begin with the Basics

Make the house energy efficient, comfortable, and be sure that basic systems are working. If you don't do these things early, you'll be working harder than you need to under difficult conditions — not a formula for success. Can you cook and store food, do laundry, keep the house warm or cool as needed, get a good bath, put up guests? Do you have lights for the primary outdoor spaces? Do you have a place to change a tire on the truck or trickle-charge the car battery? Can you keep the raccoons out of the trash and store the recycling out of the weather? Is there a dry, safe place for your tools that's convenient to the garden? Do you have a covered work space? These are all elementary supports for your home economy, and especially for building that economy.

## Recreating the Commons

The premise of garden farming and of permaculture on which it rests is that this form of subsistence landcare is well-suited to a large swath of the population (in North America as in other places) and that self-provision can help us ride through the coming economic storms that will accompany energy descent. Indeed, little else can hope to do so. We have already begun the transition to a low-energy economy, and the time to act is now. It is true though discouraging and not very helpful at this point to admit that we in the US are 30 years behind the curve and should have taken Jimmy Carter's glum advice rather than Ronald Reagan's empty promises and fatuous rhetoric. I prefer to take an optimistic view and validate the research and testing of small-scale permaculture systems that has gone on

since 1980 in 100 countries. We know how to meet the basic needs of human beings from low-energy, high-intelligence natural systems, and our decaying industrial economy is still full of embedded energy to be harvested, transformed and put to use in sustaining households and local communities. The primary resources of the future are biological wealth and social capital. Though we can each claim our fair share of them, these treasures are not owned by anyone. They can only be held in trust for future generations and used now to meet our genuine needs. We must get together with our friends, families, neighbors and community members to build and strengthen these arks.

Climate change is making economic contraction and resource shortages more difficult and more imperative; pulling ourselves back from dependence on the use of fossil fuels while planting trees and building soil is a direct and potent response to the compound crises we face. The effects of these virtuous actions can be amplified manifold by our success in making them appealing and accessible to others, by asking for help and offering what we have learned. We have no better place to do this than where we live. Making a good life at home is or should be available to almost everyone, and it provides the most secure basis for a regenerative and healthy future. It closes off no options for humanity, indeed it preserves the greatest range of choice. If the deflating industrial economy is making many of us unemployable, then we must respond by employing ourselves to do what government, industry and academia have collectively failed to do: provide a decent way of life for everyone while limiting the imminent risks and dangers that we face — hunger, destitution, displacement, war, an inhospitable climate and social upheaval.

What I am doing and what I am calling for in this *Permaculture Handbook* has underwritten the promise of North America from its earliest days. While millions came here from the Old World and come here still to build a secure place for themselves and their families, we have been waylaid by the seductions and the horrors of empire. That empire is ending; it is not gone and it may not leave quietly, but its days are numbered. This is something much to be welcomed, but it also means the end of the something-for-nothing way of life. North Americans are morally deserving of no more material prosperity than the poorest African villagers; we have our privileges because of some extraordinary good luck in historical and geopolitical lotteries. In only 400 years we have probably lost more soil than the Chinese had when their civilization began. If we continue unthinkingly on our present economic course, we in North America will get exactly the destitution that we see among refugees of war and drought around the world, if not worse. Viewed through the other end of the telescope, the undeserved but still widely available energy and material abundance of North America is a remarkable privilege that should inspire us to roll up our sleeves and get to work building a way of life worthy of our rhetoric — not stolen from indigenous peoples and extracted from foreign workers but earned by the full measure of our intelligent care for the land and for each other.

Garden farming guarantees no one a living, but if enough of us undertake it, we may collectively create a new commonwealth that can vouchsafe dignity and freedom from want to all of us.

# Appendix 1: Metric Conversion Table

| IF YOU KNOW: | MULTIPLY BY: | TO FIND: |
| --- | --- | --- |
| inches | 2.54 | centimeters |
| feet | 0.3048 | meters |
| yards | 0.9144 | meters |
| miles | 1.609 | kilometers |
| square feet | 0.0929 | square meters |
| acres | 0.4057 | hectares |
| square miles | 2.590 | square kilometers |
| ounces, avdp. | 28.35 | grams |
| pounds | 0.4536 | kilograms |
| short tons | 0.907 | metric tons |
| gallons (liquid) | 3.785 | liters |
| bushels (dry) | 35.24 | liters |
| cubic yards | 0.765 | cubic meters |

Units in the left-hand column are taken to be US Customary units.

## Temperature:

To convert degrees Fahrenheit to degrees Celsius: $(°F - 32) \times 1.8$

# Appendix 2: Bee Forage Species

Plants listed are those that have a wide range; they are shown in relative sequence of bloom. Significant regional variations in timing occur for wild plants, less so for cultivated ones. Bloom occurs earlier in southern regions and at lower elevations. Climate notes are relative to the sectional listings in which they occur. Consult the source for more precise subregional listings.

To use this list, identify your main region and note those species that occur in your area. Extract that subset of names in descending order, and you will have a bee forage list that follows the seasons from spring to fall. By observing and noting times of bloom, you create a calendar. This design tool can be refined from year to year, allowing you eventually to create a full-spectrum plant palette.

Key:
* indicates a major nectar source in at least some subregions.
F = Forb, S = Shrub, T = Tree, V = Vine, C = Cultivated
Shrub and tree designations may apply to the same plant in variant conditions or to different varieties of the same species. Similarly, forb and vine categories can overlap for other species in response to different conditions.

Source: Goddard Space Flight Center. *Honey Bee Net — Honey Bee Forage Map*. [online]. [cited January 11, 2012]. honeybeenet.gsfc.nasa.gov/Honeybees/Forage.htm.

## Western

Covers Alaska and British Columbia to California and east to the Rocky Mountain Front. California's flora is more complex than can be encompassed here: it has more unique plants.

| COMMON NAME | LATIN NAME | HABIT | CLIMATE NOTES |
| --- | --- | --- | --- |
| Filbert | *Corylus* spp | SC | mild |
| Alder | *Alnus* spp | ST | |
| Manzanita, bearberry | *Arctostaphylos* spp | ST | |
| Osier willow* | *Salix viminalis* or *S. purpurea* | S | |
| Skunk cabbage | *Lysichiton americanus* | F | northern |
| Dandelion* | *Taraxacum officinale* | F | |
| Cottonwood, aspen | *Populus* spp | T | |
| Redbud | *Cercis* spp | ST | |
| Oak | *Quercus* spp | ST | |
| Elm | *Ulmus* spp | T | |
| Apricot | *Prunus armeniaca* | TC | |
| Plum, cultivated | *Prunus* spp | STC | |

| COMMON NAME | LATIN NAME | HABIT | CLIMATE NOTES |
|---|---|---|---|
| Radish | *Raphanus sativus* | FC | |
| Filaree, stork's bill | *Erodium* spp | F | |
| Sumac | *Rhus* spp | S | |
| Madrone | *Arbutus menziesii* | ST | |
| Maple | *Acer* spp | T | |
| Pear | *Pyrus* spp | TC | |
| Milkvetch | *Astragalus* spp | F | |
| Balsamroot sunflower | *Balsamorhiza* spp | F | |
| Fireweed/wickiup* | *Chamerion angustifolium* | F | |
| Wild geranium | *Geranium maculatum* | F | |
| White Dutch clover* | *Trifolium repens* | FC | |
| Labrador tea | *Ledum groenlandicum* | S | northern |
| Rose | *Rosa* spp | S | |
| Hawthorn | *Crataegus* spp | ST | |
| Oregon grape | *Mahonia* spp | S | |
| Tamarisk* | *Tamarix* spp | T | |
| Box elder/Manitoba maple | *Acer negundo* | ST | |
| Ash | *Fraxinus* spp | T | |
| Apple | *Malus* spp | TC | |
| Peach | *Prunus persica* | TC | |
| Cherry | *Prunus avium* | TC | |
| Pine | *Pinus* spp | T | |
| Creosote bush | *Larrea tridentata* | S | southern |
| Citrus* | *Citrus* spp | TC | southern |
| Onion | *Allium cepa* | FC | |
| Cabbage | *Brassica oleracea* | FC | |
| Queen Anne's lace | *Daucus carota* | F | |
| Parsley | *Petroselinum crispum* | FC | |
| Mustard | *Brassica* spp | F | |
| Holly, yaupon | *Ilex vomitoria* | S | |
| Canola | *Brassica rapa* | FC | |
| Alfalfa* | *Medicago sativa* | FC | |
| Plum, wild | *Prunus* spp | ST | |
| Pea shrub | *Caragana arborescens* | S | |
| Leafy spurge | *Euphorbia esula* | F | |
| Wild radish/charlock | *Raphanus raphanistrum* | F | |
| Wild buckwheat | *Eriogonum* spp | F | |
| Lupine | *Lupinus* spp | F | |
| Sweetclover* | *Melilotus* spp | FC | |
| Groundsel | *Senecio* spp | F | southern |
| Catclaw, Texas mimosa* | *Acacia greggii* | S | southern |
| Mesquite* | *Prosopis glandulosa* | ST | southern |
| Mormon tea | *Ephedra viridis* | S | southern |
| Snowberry/wolfberry* | *Symphoricarpos occidentalis* | F | |
| Crimson Clover* | *Trifolium incarnatum* | FC | |
| Mule-ears | *Wyethia* spp | F | |
| Bee balm, wild bergamot* | *Monarda didyma, M. fistulosa* | F | southern |
| Buckwheat | *Fagopyrum esculentum* | FC | |
| Mint | *Mentha* spp | FC | |
| Romona or common sage | *Salvia officinalis* | F | |

| COMMON NAME | LATIN NAME | HABIT | CLIMATE NOTES |
|---|---|---|---|
| Wild lilac, buckbrush | *Ceanothus spp* | S | |
| Marigold, Indian blanket* | *Gaillardia pulchella* | F | southern |
| Cascara sagrada | *Frangula purshiana* | S | |
| Greasewood | *Sarcobatus vermiculatus* | S | |
| Raspberry/Blackberry* | *Rubus* spp | S | |
| Watermelon | *Citrullus lanatus* | VC | |
| Pumpkin, squash, gourd | *Cucurbita pepo* | VC | |
| Smartweed, knotweed | *Polygonum* spp | F | |
| Milkweed | *Asclepias syriaca* | F | |
| Sainfoin | *Onobrychis* spp | FC | |
| Alsike clover* | *Trifolium hybridum* | FC | |
| Red clover* | *T. pratense* | FC | |
| Vetch* | *Vicia* spp | F | |
| Cranberry | *Vaccinium macrocarpon* | SC | |
| Black locust* | *Robinia pseudoacacia* | T | southern/mild |
| Cucumber | *Cucumis sativus* | VC | |
| Shadblow | *Amelanchier canadensis* | SC | |
| Russian olive | *Elaeagnus angustifolia* | S | |
| Currant | *Ribes* spp | SC | |
| Bindweed | *Convolvulus* or *Calystegia* spp | FV | southern |
| Buckthorn | *Rhamnus* spp | S | |
| Knapweed* | *Centaurea* spp | F | |
| Thistles | *Cirsium* spp | F | |
| Salal | *Gaultheria shallon* | S | mild climate |
| Cleome | *Cleome serrulata* | F | |
| Chicory, endive | *Cichorium intybus* | FC | southern |
| Broom dalea, purple sage* | *Psorothamnus scoparius* | S | southern |
| Corn | *Zea mays* | C | |
| Russian thistle, saltwort | *Salsola* spp | S | |
| Asters* | *Aster* spp | F | southern |
| Huckleberry* | *Vaccinium* or *Gaylussacia* spp | S | |
| Rosinweed | *Grindelia* spp | F | |
| Sunflower* | *Helianthus* spp | FC | |
| Cotton | *Gossypium* spp | FC | southern |
| Matchweed, snakeweed | *Gutierrezia sarothrae* | F | southern |
| Broomweed | *G. texana* | F | southern |
| Goldenrod* | *Solidago* spp | F | |
| Rabbitbrush* | *Chrysothamnus* spp | S | |
| Birdsfoot trefoil* | *Lotus corniculatus* | F | |

## Central

Covers the Great Plains and Prairie regions from southern Alberta to Minnesota and south to central Texas. This subhumid region exhibits a gradient of increasing moisture and decreasing altitude from west to east.

| COMMON NAME | LATIN NAME | HABIT | CLIMATE NOTES |
|---|---|---|---|
| Elm | *Ulmus* spp | T | |
| Milkvetch | *Astragalus* spp | F | |
| Cottonwood | *Populus* spp | T | |

| Common name | Latin name | Habit | Climate notes |
|---|---|---|---|
| Chickweed | *Stellaria media* | F | southeastern |
| Plum, cultivated | *Prunus* spp | STC | |
| Dandelion | *Taraxacum officinale* | F | |
| Vetch* | *Vicia* spp | F | |
| Sumac | *Rhus* spp | S | |
| Maple | *Acer* spp | T | |
| Box elder | *Acer negundo* | ST | |
| Peach | *Prunus persica* | TC | |
| Pear | *Pyrus* spp | TC | |
| Winter cress | *Barbarea vulgaris* | F | |
| Strawberry | *Fragaria* spp | F | |
| Blackberry | *Rubus* spp | SC | |
| Buckeye | *Aesculus* spp | T | |
| Redbud | *Cercis canadensis* | T | |
| Apple | *Malus* spp | TC | |
| Plum, wild | *Prunus* spp | S | |
| Tulip tree | *Liriodendron tulipifera* | T | southeastern |
| Osier willow | *Salix* spp | S | |
| Apricot | *Prunus armeniaca* | TC | |
| Watermelon | *Citrullus lanatus* | VC | |
| Cantaloupe, melons | *Cucumis melo* | VC | |
| Cucumber | *Cucumis sativus* | VC | |
| Alfalfa* | *Medicago sativa* | FC | |
| Marigold, Indian blanket | *Gaillardia pulchella* | F | southern |
| Milkweed | *Asclepias syriaca* | F | |
| Prairie clover | *Dalea* spp | F | |
| Leafy spurge | *Euphorbia esula* | F | |
| Sweetclover* | *Melilotus* spp | F | |
| Dutch White Clover* | *Trifolium repens* | FC | |
| Sainfoin | *Onobrychis* spp | F | southern |
| Pea shrub | *Caragana arborescens* | S | |
| Mustard | *Brassica napus* | FC | |
| Buckthorn | *Rhamnus* spp | S | |
| Cherry | *Prunus* spp | T | |
| Bee balm, wild bergamot* | *Monarda* spp | F | |
| Mesquite | *Prosopis glandulosa* | ST | southern |
| Tamarisk, salt cedar | *Tamarix* spp | S | southern |
| Alsike clover* | *Trifolium hybridum* | FC | |
| Russian olive* | *Elaeagnus angustifolia* | S | |
| Cotton | *Gossypium* spp | FC | southern |
| Pumpkin, squash, gourd | *Cucurbita* spp | VC | |
| Thistles | *Cirsium* spp | F | |
| Raspberry | *Rubus* spp | SC | |
| Cleome | *Cleome serrulata* | F | southern |
| Rosinweed | *Grindelia* spp | F | |
| Sunflower* | *Helianthus* spp | FC | |
| Persimmon | *Diospyros virginiana* | T | southeastern |
| Honey locust | *Gleditsia triacanthos* | T | southeastern |
| Black locust | *Robinia pseudoacacia* | T | southeastern |

| COMMON NAME | LATIN NAME | HABIT | CLIMATE NOTES |
|---|---|---|---|
| Elderberry | *Sambucus nigra* | S | southeastern |
| Grape | *Vitis* spp | VC | southeastern |
| Red clover* | *Trifolium pratense* | FC | |
| Basswood* | *Tilia americana* | T | eastern |
| Catmint, catnip | *Nepeta cataria* | F | |
| Broomcorn | *Sorghum bicolor* | C | |
| Canola | *Brassica rapa* | FC | |
| Snowberry, wolfberry | *Symphoricarpos* spp | F | |
| Romona or common sage | *Salvia officinalis* | F | |
| Soybean* | *Glycine max* | FC | |
| Corn | *Zea mays* | C | |
| Ragweed | *Ambrosia* spp | F | |
| Bitterweed | *Helenium amarum* | F | |
| Buckwheat, annl. | *Fagopyrum esculentum* | FC | |
| Privet | *Ligustrum* spp | S | |
| Broomweed | *Gutierrezia texana* | F | southern |
| Tickseed, Spanish needle | *Bidens* spp | F | |
| Blue/sand vine* | *Cynanchum laeve* | F | southeastern |
| Chicory, endive | *Cichorium intybus* | FC | |
| Goldenrod* | *Solidago* spp | F | |
| Aster* | *Aster* spp | F | |
| Matchweed, snakeweed | *Gutierrezia sarothrae* | F | |
| Joe-pye weed, boneset | *Eupatorium perfoliatum* | F | |
| Smartweed, knotweed* | *Polygonum* spp | F | |
| Russian thistle, saltwort | *Salsola* spp | S | |

## Southeastern

This covers southern Missouri to east Texas and eastward to Virginia and the Gulf Coast. It includes the Ozark, Ouachita and southern Appalachian mountains where higher elevation supports more northerly species. The subtropical flora of the Gulf littoral present a special case that I have not tried to detail extensively.

| COMMON NAME | LATIN NAME | HABIT | CLIMATE NOTES |
|---|---|---|---|
| Cucumber | *Cucumis sativus* | VC | |
| Tea tree | *Melaleuca quinquenervia* | T | Florida |
| Chickweed | *Stellaria media* | F | |
| Huckleberry, blueberry | *Vaccinium & Gaylussacia* spp | SC | |
| Cantaloupe, melons | *Cucumis melo* | VC | |
| Henbit/dead nettle | *Lamium* spp | F | |
| Dandelion | *Taraxacum officinale* | F | |
| Maple | *Acer* spp | T | |
| Alder | *Alnus* spp | T | |
| Elm | *Ulmus* spp | T | |
| Peach | *Prunus persica* | TC | |
| Citrus* | *Citrus* spp | TC | Deep South |
| Plum, cultivated | *Prunus* spp | STC | |
| Vervain | *Verbena* spp | F | |

| Common name | Latin name | Habit | Climate notes |
|---|---|---|---|
| Mustard | *Brassica napus* | FC | |
| Inkberry* | *Ilex glabra* | S | Deep South |
| Oak | *Quercus* spp | T | |
| Rattan vine* | *Berchemia scandens* | V | Deep South |
| Marigold, Indian blanket | *Gaillardia pulchella* | F | |
| Pear | *Pyrus* spp | TC | |
| Sweetclover* | *Melilotus* spp | F | |
| Bee balm, wild bergamot* | *Monarda* spp | F | |
| Black-eyed Susan | *Rudbeckia* spp | F | |
| Crimson clover* | *Trifolium incarnatum* | FC | |
| Alsike clover* | *T. hybridum* | FC | |
| Red clover | *T. pratense* | FC | |
| Dutch White Clover* | *T. repens* | C | |
| Romana or common sage | *Salvia officinalis* | F | |
| Arrowleaf clover* | *Trifolium vesiculosum* | FC | |
| Persian clover* | *T. resupinatum* | FC | Deep South |
| Vetch* | *Vicia* spp | F | |
| Blackberry* | *Rubus* spp | SC | |
| Box elder | *Acer negundo* | T | |
| Hawthorn | *Crataegus* spp | ST | |
| Redbud | *Cercis canadensis* | T | |
| Pepper vine | *Ampelopsis* spp | V | Deep South |
| Flowering dogwood | *Cornus florida* | T | |
| Apple | *Malus* spp | TC | |
| Osier willow | *Salix viminalis, S. purpurea* | S | |
| Catclaw, Texas mimosa | *Acacia greggii* | S | westerly |
| Privet | *Ligustrum* spp | S | |
| Currant | *Ribes* spp | S | Appalachians |
| Ironwood* | *Cyrilla parvifolia* | S | southeast |
| Mesquite | *Prosopis glandulosa* | ST | westerly |
| Holly, yaupon* | *Ilex vomitoria* | ST | Deep South |
| Alfalfa | *Medicago sativa* | FC | |
| Tulip tree* | *Liriodendron tulipifera* | T | |
| Honey locust | *Gleditsia triacanthos* | T | |
| Tupelo* | *Nyssa sylvatica* | T | Delta |
| Cottonwood | *Populus* spp | T | |
| Cherry | *Prunus* spp | T | |
| Winter cress | *Barbarea vulgaris* | F | |
| Chicory, endive | *Cichorium intybus* | F | |
| Plum, wild | *Prunus* spp | T | |
| Black locust* | *Robinia pseudoacacia* | T | |
| Elderberry | *Sambucus nigra* | S | |
| Honeysuckle | *Lonicera* spp | V | |
| Pumpkin, squash, gourd | *Cucurbita* spp | VC | |
| Ash | *Fraxinus* spp | T | |
| Watermelon | *Citrullus lanatus* | VC | |
| Milkweed | *Asclepias syriaca* | F | |
| Bitterweed | *Helenium amarum* | F | |
| Smartweed, knotweed | *Polygonum* spp | F | |

| Common name | Latin name | Habit | Climate notes |
|---|---|---|---|
| Buckwheat, annl. | *Fagopyrum esculentum* | FC | |
| Beautyberry | *Callicarpa americana* | S | |
| Buckthorn | *Rhamnus* spp | S | |
| Sumac | *Rhus* spp | S | |
| Catalpa | *Catalpa* spp | T | |
| Persimmon | *Diospyros virginiana* | T | |
| Devil's walking stick | *Aralia spinosa* | S | |
| Basswood | *Tilia americana* | T | |
| Grape | *Vitis* spp | VC | |
| Clethra | *Clethra alnifolia* | S | |
| Virginia creeper | *Parthenocissus quinquefolia* | V | |
| Corn | *Zea mays* | C | |
| Tickseed coreopsis | *Coreopsis lanceolata* | F | |
| Aster* | *Aster* spp | F | |
| Tickseed, Spanish needle | *Bidens* spp | F | |
| Thistles | *Cirsium* spp | F | |
| Joe-pye weed, boneset | *Eupatorium perfoliatum* | F | |
| Buttonbrush | *Cephalanthus occidentalis* | S | |
| Snowberry, wolfberry | *Symphoricarpos* spp | F | |
| Chaste tree | *Vitex* spp | F | |
| Mountain mint | *Pycnanthemum* spp | F | Appalachians |
| Sourwood* | *Oxydendrum arboreum* | T | |
| Anise Hyssop | *Agastache foeniculum* | F | |
| Soybean* | *Glycine max* | FC | |
| Sunflower | *Helianthus* spp | FC | |
| Prickly Pear | *Opuntia* spp | F | |
| Mint | *Mentha* spp | F | |
| Chinese tallow tree* | *Triadica sebifera* | T | Deep South |
| Saw palmetto* | *Serenoa repens* | S | Florida |
| Palmetto* | *Sabal palmetto* | T (palm) | coastal LA-SC |
| Cotton* | *Gossypium* spp | FC | Deep South |
| Lima bean | *Phaseolus lunatus* | FVC | |
| Purple loosestrife | *Lythrum salicaria* | F | |
| Lespedeza | *Lespedeza* spp | FC | |
| Cowpea | *Vigna unguiculata* | FC | |
| Partridge pea, senna | *Chamaecrista, Cassia* spp | F | |
| Blue/sand vine | *Cynanchum laeve* | F | coastal |
| Broomweed | *Gutierrezia texana* | F | |
| Ragweed | *Ambrosia* spp | F | |
| Matchweed, snakeweed | *Gutierrezia sarothrae* | F | |
| Broomcorn | *Sorghum bicolor* | C | |
| Clematis | *Clematis virginiana* | VF | |
| Knapweed | *Centaurea* spp | F | |
| Brazilian pepper tree* | *Schinus terebinthifolius* | S | Florida |
| Prairie clover | *Dalea* spp | F | |
| Goldenrod | *Solidago* spp | F | |
| Strawberry | *Fragaria* spp | F | |
| Crown-beard | *Verbesina* spp | F | |

## Northeast/North Central

This covers western Ontario to Nova Scotia; Wisconsin and Illinois east to Delaware.

| COMMON NAME | LATIN NAME | HABIT | CLIMATE NOTES |
|---|---|---|---|
| Chickweed | *Stellaria media* | F | warmer |
| Skunk cabbage | *Symplocarpus foetidus* | F | |
| Dandelion | *Taraxacum officinale* | F | |
| Strawberry | *Fragaria* spp | F | |
| Maple | *Acer* spp | T | |
| Alder | *Alnus* spp | T | |
| Box elder/Manitoba maple | *Acer negundo* | ST | |
| Cottonwood, aspen | *Populus* spp | T | |
| Osier willow | *Salix* spp | S | |
| Cherry, cultivated | *Prunus* spp | T | |
| Peach | *Prunus persica* | TC | |
| Mustard | *Brassica napus* | FC | |
| Holly, yaupon* | *Ilex vomitoria* | ST | warmer Atlantic |
| Canola * | *Brassica rapa* | FC | |
| Plum, cultivated | *Prunus* spp | STC | |
| Winter cress | *Barbarea vulgaris* | F | |
| Pear | *Pyrus* spp | TC | |
| Elm | *Ulmus* spp | T | |
| Red clover* | *Trifolium pratense* | FC | |
| Dutch White Clover* | *T. repens* | FC | |
| Shadblow, juneberry | *Amelanchier* spp | SC | |
| Sorrel, sheep/common | *Rumex acetosella* | F | |
| Hawthorn | *Crataegus* spp | ST | |
| Raspberry * | *Rubus* spp | S | |
| Persimmon | *Diospyros virginiana* | T | |
| Apple | *Malus* spp | TC | |
| Tulip tree* | *Liriodendron tulipifera* | T | |
| Redbud | *Cercis canadensis* | T | |
| Blueberry, huckleberry | *Vaccinium, Galussacia* spp | SC | |
| Tupelo | *Nyssa sylvatica* | T | warmer |
| Oak | *Quercus* spp | T | |
| Beech | *Fagus grandifolia* | T | |
| Crab apple | *Malus baccata* | T | |
| Cherry, wild | *Prunus* spp | T | |
| Plum, wild | *Prunus* spp | T | |
| Black locust* | *Robinia pseudoacacia* | T | |
| Elderberry | *Sambucus nigra* | S | |
| Pine | *Pinus* spp | T | |
| Honeysuckle | *Lonicera* spp | VS | |
| Apricot | *Prunus armeniaca* | TC | |
| Pumpkin, squash, gourd | *Cucurbita* spp | VC | |
| Cucumber | *Cucumis sativus* | VC | |
| Watermelon | *Citrullus lanatus* | VC | |
| Alfalfa* | *Medicago sativa* | FC | |
| Dogbane | *Apocynum cannabinum* | F | |
| Milkweed* | *Asclepias syriaca* | F | |

| Common name | Latin name | Habit | Climate notes |
|---|---|---|---|
| Thistles | *Cirsium* spp | F | |
| Hawkweed | *Hieracium* spp | F | |
| Bittersweet, waxwort | *Celastrus scandens* | F | northeast |
| Viper's bugloss* | *Echium vulgare* | F | S. Appalachians |
| Jewelweed/snapweed | *Impatiens pallida* | F | |
| Birdsfoot trefoil* | *Lotus corniculatus* | F | |
| Sweetclover* | *Melilotus* spp | F | |
| Alsike clover* | *Trifolium hybridum* | FC | |
| Blackberry * | *Rubus* spp | SC | |
| Privet* | *Ligustrum* spp | S | |
| Cranberry | *Vaccinium macrocarpon* | SC | |
| Romana or common sage | *Salvia officinalis* | F | |
| Vetch | *Vicia* spp | F | |
| Smartweed, knotweed* | *Polygonum* spp | F | |
| Aster* | *Aster* spp | F | |
| Catmint, catnip | *Nepeta cataria* | F | |
| Soybean* | *Glycine max* | FC | |
| Cantaloupe, melons | *Cucumis melo* | VC | |
| Bee balm, wild bergamot* | *Monarda* spp | F | |
| Sumac* | *Rhus* spp | S | |
| Fireweed/wickiup | *Chamerion angustifolium* | F | |
| Tickseed, Spanish needle* | *Bidens* spp | F | |
| Lima bean | *Phaseolus lunatus* | FVC | Atlantic coastal |
| Joe-pye weed, boneset | *Eupatorium perfoliatum* | F | |
| Purple loosestrife* | *Lythrum salicaria* | F | |
| Blue/sand vine* | *Cynanchum laeve* | F | southwest |
| Grape | *Vitis* spp | V | |
| Chicory, endive | *Cichorium intybus* | FC | |
| Sunflower* | *Helianthus* spp | FC | |
| Anise Hyssop | *Agastache foeniculum* | F | |
| Crown vetch | *Securigera varia* | F | |
| Knapweed* | *Centaurea* spp | F | |
| Sourwood* | *Oxydendrum arboreum* | T | S. Appalachians |
| Mint | *Mentha* spp | F | |
| Ragweed | *Ambrosia* spp | F | |
| Goldenrod* | *Solidago* spp | F | |
| Basswood* | *Tilia americana* | T | |
| Vervain | *Verbena* spp | F | |
| Thyme | *Thymus vulgaris* | FC | |
| Corn | *Zea mays* | C | |
| Buckwheat, annl.* | *Fagopyrum esculentum* | FC | |
| Buttonbrush* | *Cephalanthus occidentalis* | S | warmer |
| Clethra* | *Clethra alnifolia* | S | warmer |

# Appendix 3: Nitrogen-Fixing Species and Biomass Producers

Key: * indicates non-legume N-fixers

## Woody Species by Size

Theses are all legumes except as indicated.

| | | |
|---|---|---|
| Black locust | *Robinia pseudoacacia* | large timber tree and honey crop, moderate thorns (zones 4–8) |
| Mimosa | *Mimosa* spp | spreading tree to 30', light canopy (zones 5–10) |
| Alders* | *Alnus* spp | cold-hardy shrubs and medium trees (zones 3–8) |
| Mesquite | *Prosopis juliflora* | drylands shrub/tree, pods for livestock (zones 5–9) |
| Japanese pagoda tree | *Styphnolobium japonicum* | medium tree to 40' (zones 6–9) |
| Redbud | *Cercis canadensis* | native poultry forage, to 25', fixes N in sunlight (zones 4–8) |
| Tagasaste | *Chamaecytisus palmensis* | dryland forage shrub to 20' (zone 9) |
| Scotch laburnum | *Laburnum alpinum* | small tree to 20', toxic if eaten (zones 4–8) |
| Golden rain tree | *L. anagyroides* | small tree to 15', toxic if eaten (zones 4–9) |
| Buffaloberry* | *Shepherdia argentea* | shrub of plains and western mtns. to 15' (zones 2–7) |
| Amur and relatives | *Maackia amurensis, Maackia chinensis* | small legume trees to 30', Amur (z. 3–7), Chinensis (z. 5–8) |
| Siberian pea shrub | *Caragana arborescens* | shrub of cold zones to 15' (zones 2–6) |
| Autumn olive* | *Elaeagnus umbellata* | dispersive shrub, edible berries (zones 5–8) |
| Russian olive* | *Elaeagnus angustifolia* | dispersive shrub, wildlife forage (zones 2–5) |
| Brooms | *Cytisus* and *Spartium spp.* | Eurasian shrubs, small trees, dryland legumes (Though Scotch broom, *C. scoparius*, is dispersive and problematic in the Pacific NW, other brooms are better behaved; there are more than half a dozen species, all under 20', some hardy to zone 5) |
| Mountain mahogany* | *Cercocarpus montanus* | native western shrub of drylands, 8–12' (zones 3–6) |
| Hairy locust | *Robinia hispida* | shrubby native to 10', soft thorns (zones 5–8) |
| False indigobush | *Amorpha fruticosa* | shrub to 10' (zones 5–8) |
| Bladder senna | *Colutea arborescens* | small shrub to 10' (zones 6–9) |
| Miracle plant | *Lespedeza bicolor* | small shrub to 9', stems used for fuel (zones 6–9) |
| Tree lupin | *Lupinus arboreus* | native Calif. dryland legume shrub 5–8' (zone 8b) |
| Sesbanias | *Sesbania* spp | subtropical legume shrub (some hardy to zone 7) |
| Bundleflower | *Desmanthus illinoensis* | low prairie legume shrub (zones 3–7) |

| New Jersey tea* | Ceanothus americanus | low shrub (3'), Western spp are tender (zones 3–8) |
| Bayberry* | Myrica pensylvanica | low shrub (zones 3–9) |
| Sea buckthorn* | Hippophae rhamnoides | low shrub, salt-tolerant, edible fruit (zones 3–8) |
| Sweet gale* | Myrica gale | low shrub of wet woodlands and shores (zones 1–5) |

## Perennial and Herbaceous Biomass Producers

None are legumes.

| Rose-of-Sharon | Hibiscus syriacus | dispersive, fibrous, stems are persistent but easily cut |
| Cup plant | Silphium perfoliatum | produces tough stems suitable for garden use |
| Poke weed | Phytolacca americana | edible young leaves, berries make dye |
| Jerusalem artichoke | Helianthus tuberosus | edible tubers, short windbreak |
| Comfrey | Symphytum officinale | use only seedless forms, propagate by divisions |
| Burdock | Arctium lappa | tolerates shade under trees |
| Banana | Musa spp | subtropical |
| Gunnera* | Gunnera manicata | giant leaves; subtropical, N-fixing non-legume |

# Endnotes

**Acknowledgments**

1. David Holmgren. "Retrofitting the Suburbs for Sustainability." *Newsletter of the CSIRO Sustainability Network*, Update 49 (March 31, 2005), pp 1–9. [online]. [cited January 12, 2012]. holmgren.com.au/DLFiles/PDFs/Holmgren-Suburbs-Retrofit-Update49.pdf.

**Chapter 1: Garden Farming**

1. Sharon Astyk and Aaron Newton. *A Nation of Farmers: Defeating the Food Crisis on American Soil*. New Society, 2009, pp. 58–59.
2. Michael Pilarski. "Russia's Anastasia through a Permaculture Lens." *Permaculture Activist #79* (2011).
3. Hong Kong Agriculture, Fisheries and Conservation Department. *Agriculture in HK*. [online]. [cited August 30, 2011]. afcd.gov.hk/english/agriculture/agr_hk/agr_hk.html. In 2010, Hong Kong produced 2.5% of fresh vegetables, 56.2% of live poultry and 6.4% of live pigs consumed there.
4. Fernando Funes et al. *Sustainable Agriculture and Resistance: Transforming Food Production in Cuba*. Food First, 2002.
5. Nik Bertulis. "The Enduring Chinampas of Xochimilco." *Permaculture Activist #58* (2005); Scott Horton. "Floating Remnants of a Fertile Past." *Permaculture Activist #76* (2010).
6. Phil Forsyth. "Another Side of Brooklyn." *Permaculture Activist #58* (2005). "As late as 1880, Kings County, New York (Brooklyn) ranked second nationally in production of agricultural goods…" Adjacent Queens County ranked first.
7. US Department of Agriculture. *USDA Census of Agriculture, 2007*. [online]. [cited August 30, 2011]. agcensus.usda.gov. Though concentration of land continues to increase along with the number of very large farms (over 2,000 acres), the number of farms under 50 acres increased from 736,000 in 1997 to 843,000 a decade later.
8. USDA Agricultural Marketing Service. *Farmers Markets and Local Food Marketing*. [online]. [cited August 30, 2011]. ams.usda.gov/AMSv1.0/ams.fetchTemplateData.do?template=TemplateA&navID=WholesaleandFarmersMarkets&leftNav=WholesaleandFarmersMarkets&page=WholesaleAndFarmersMarkets&acct=AMSPW. Farmers markets in the US numbered 1,755 in 1994 and 7,175 in 2011. If anything, growth has accelerated in recent years.
9. Rob Reuteman. "Life On The Farm Attracts Green-Spirited Entrepreneurs." [online]. [cited August 20, 2011]. cnbc.com/id/42572373/Life_On_The_Farm_Attracts_Green_Spirited_Entrepreneurs. The Census of Agriculture showed some 75,000 new farms between 2002 and 2007 with the average age of new farmers at 48 years compared to the average of all farmers at 57 years.
10. Tom Philpott. "The Butz Stops Here: A Reflection on the Lasting Legacy of 1970s USDA Secretary Earl Butz." *Grist*, February 7, 2008. [online]. [cited August 30, 2011]. grist.org/article/the-butz-stops-here.
11. Ruth Stout. *How to Have a Green Thumb without an Aching Back: A New Method of Mulch Gardening*. Galahad, 1974; *Gardening Without Work: For the Aging, the Busy, and the Indolent*. Devin-Adair, 1961.
12. Rachel Carson. *Silent Spring*. Houghton Mifflin, 1962.
13. Cary Fowler and Pat Mooney. *Shattering: Food, Politics, and the Loss of Genetic Diversity*. University of Arizona, 1990.

14. Dan Morgan. *Merchants of Grain: The Power and Profits of the Five Giant Companies at the Center of the World's Food Supply.* Viking, 1979.

15. Vandana Shiva. "The Politics of Diversity." *Permaculture Activist* #23 (1991).

16. *Mother Earth News.* [online]. [cited August 30, 2011]. motherearthnews.com/; James Adams. "Harrowsmith Country Life Cuts Staff, Ceases Publication." *The Globe and Mail*, August 9, 2011. [online]. [cited November 8, 2011]. theglobeandmail.com/news/arts/books/harrowsmith-country-life-cuts-staff-ceases-publication/article2123885/; *Acres USA.* [online]. [cited August 30, 2011]. acresusa.com/magazines/magazine.htm.

17. Wendell Berry. *The Unsettling of America: Culture and Agriculture.* Sierra Club, 1977.

18. Wes Jackson. *New Roots for Agriculture.* University of Nebraska, 1980; The Land Institute [online]. [cited January 13, 2012]. landinstitute.org/.

19. Robyn Van En. *Basic Formula to Create Community Supported Agriculture.* 1988. This work was expanded in Trauger M. Groh and Steven S. H. McFadden. *Farms of Tomorrow: Communities Supporting Farms, Farms Supporting Communities.* Steiner, 1990.

20. Bob and Bonnie Gregson. *Rebirth of the Small Family Farm: A Handbook for Starting a Successful Organic Farm Based on the Community Supported Agriculture Concept.* IMF, 1996.

21. Andre Voisin. *Grass Productivity.* Crosby Lockwood & Son, 1959.

22. Joel Salatin. *Salad Bar Beef.* Polyface, 1995. Further documentation of Salatin's animal polycultures may be found in the pages of *Acres USA,* to which he is a frequent contributor. My own article on Polyface Farms appears in *Permaculture Activist* #32 (1995).

23. Gene Logsdon. *All Flesh Is Grass: The Pleasures and Promises Of Pasture Farming.* Swallow, 2004.

24. See Richard Heinberg. *The Party's Over: Oil, War and the Fate of Industrial Societies.* New Society, 2005 and subsequent writings. Underneath these more immediate issues lurks the question of complexity — or in Odum's terms EMERGY — eloquently posed by Joseph Tainter. *The Collapse of Complex Societies.* Cambridge, 1990.

25. Jane Jacobs. *The Death and Life of Great American Cities.* Random House, 1961. Chapter 22 lays out a methodology for holistic understanding of urban and biological complexity that is highly relevant to permaculture design.

26. David Pimentel et al. "Environmental and Economic Costs of Pesticide Use." *BioScience* Vol. 42#10 (November 1992), pp. 750–760. [online]. [cited August 31, 2011]. savepanjab.org/images/pdf/Environmental and Economic Costs of Pesticide Use.pdf.

27. Environmental Working Group. "Fact and fiction." Chemical Industry Archives. [online]. [cited August 29, 2011]. chemicalindustryarchives.org/factfiction/facts/4.asp.

28. I. K. Hosein et al. "Clinical Significance of the Emergence of Bacterial Resistance in the Hospital Environment." *Journal of Applied Microbiology*, Symposium Supplement Vol. 92 (2002), pp. 90S–97S. [online]. [cited August 29, 2011]. onlinelibrary.wiley.com/doi/10.1046/j.1365-2672.92.5s1.1.x/pdf.

29. Limits to Growth website. [online]. [cited August 29, 2011]. limitstogrowth.net/index.htm.

30. John Russell Smith. *Tree Crops: A Permanent Agriculture* (1929) reprinted Island, 1987.

31. David Holmgren and Bill Mollison. *Permaculture I: A Perennial Agricultural System for Human Settlements.* Tagari, 1978.

32. Howard T. Odum. *Environment, Power, and Society.* Wiley, 1971 and *The Energy Basis for Man and Nature*, 2nd ed. McGraw Hill, 1981.

33. Bill Mollison. *Permaculture II: Practical Design for Town and Country in Permanent Agriculture.* Tagari, 1979.

**Chapter 2: Who Am I to Farm?**

1. US Department of Agriculture. *2007 Census of Agriculture – Demographics.* [online]. [cited November 9, 2011]. agcensus.usda.gov/Publications/2007/Online_Highlights/Fact_Sheets/demographics.pdf; Statistics Canada; 2011 Census of Agriculture. [online]. [cited November 9, 2011]. statcan.gc.ca/ca-ra2011/index-eng.htm.

2. M. G. Kains. *Five Acres and Independence,* 2nd. ed. rev. Garden City, 1940.

3. Helen and Scott Nearing. *Living the Good Life: How to Live Sanely & Simply in a Troubled World.* Schocken, 1954. The Good Life Center at Forest Farm perpetuates the legacy of the Nearings: goodlife.org.

4. US Census 2000 Special Report. *Demographic Trends of the 20th Century*. November 2002. [online]. [cited August 29, 2011]. census.gov /prod/2002pubs/censr-4.pdf.

5. New York Times. "Farmer Suicide Rate Swells in 1980s, Study Says." October 14, 1991. [online] [cited September 6, 2011]. www.nytimes.com /1991/10/14/us/farmer-suicide-rate-swells-in -1980-s-study-says.html.

6. James Howard Kunstler. "Remarks by James Howard Kunstler at the meeting of The Second Vermont Republic — October 28, 2005." [online]. [cited November 9, 2011]. kunstler.com /spch_Vermont%20Oct%2005.htm.

7. R. P. Dore. "Land Reform and Japan's Economic Development." [online] [cited September 6, 2011] ide.go.jp/English/Publish/Period icals/De/pdf/65_04_06.pdf; "How land reform can contribute to economic growth and poverty reduction: Empirical evidence from International and Zimbabwean experience." [online]. [cited September 6, 2011]. siteresources .worldbank.org/INTARD/825826-11111486068 50/20431879/Zimbabwe.pdf.

8. Bruce Horovitz. "Recession Grows Interest in Seeds, Vegetable Gardening." *USA Today*, February 19, 2009. [online] [cited September 7, 2011]. usatoday.com/money/industries/food /2009-02-19-recession-vegetable-seeds_N.htm.

**Chapter 3: Gardening the Planet**

1. Jack Harich. *The Dueling Loops of the Political Powerplace: Why Progressives Are Stymied and How They Can Find Their Way Again*. Lulu, 2007.

2. Peter Bane. "Keystones and Cops." *Permaculture Activist* #50 (2003). I argue that the loss of two keystone species from eastern US forests within a generation — the American chestnut and the passenger pigeon — is undoubtedly linked to their dramatic interdependence.

3. Jared Diamond. *Collapse: How Societies Choose to Fail or Succeed*, rev. ed. Penguin, 2011.

4. Global Footprint Network. [online]. [cited September 9, 2011]. footprintnetwork.org.

5. Richard Meyer. "The Potential of Solar Energy for Replacing Fossil Fuels." 7th annual conference of the Association for the Study of Peak Oil (ASPO), Barcelona, Spain. 2008. [online]. [cited August 30, 2011]. aspo-spain.org/aspo7 /presentations/Meyer-CSP-ASPO7.pdf.

6. "Squirrels Cannot Live by Truffles Alone: A Closer Look at a Northwest Keystone Complex." *PNW Science Findings*, January 2004. [online]. [cited August 30, 2011]. fs.fed.us/pnw /sciencef/scifi60.pdf; D.A. Perry et al. "Bootstrapping in Ecosystems." *Bioscience* Vol. 39#4 (April 1989). [online]. [cited August 30, 2011]. esf.edu/efb/horton/Perry%20etal%201989.pdf; Keith B. Aubrey et al. "The Ecological Role of Tree-dwelling Mammals in Western Coniferous Forests." [online]. [cited August 30, 2011]. plexusowls.com/PDFs/ecological_role_tree dwelling_mammals.pdf.

7. The Permaculture Credit Union. [online]. [cited September 9, 2011]. pcuonline.org. It operates from offices in Santa Fe, NM but has members and lends in all 50 US states.

8. Gaia University. [online]. [cited September 9, 2011]. GaiaUniversity.org.

9. The Global Ecovillage Network. [online]. [cited September 9, 2011]. gen.ecovillage.org/.

**Chapter 4: Permaculture Principles**

1. David Holmgren. *Permaculture: Principles and Pathways Beyond Sustainability*. Holmgren, 2002.

2. "Sacrifice of First Fruits — Demeter Cult 1." *Theoi Greek Mythology*. [online]. [cited August 31, 2011]. theoi.com/Cult/DemeterCult.html.

3. Tyler Volk. *Gaia's Body: Toward a Physiology of Earth*. MIT, 2004.

4. Helena Norberg-Hodge. *Ancient Futures: Learning from Ladakh*. Random House, 1991.

5. Christopher Alexander et al. *A Pattern Language: Towns, Buildings, Construction*. Oxford, 1977; Christopher Alexander. *The Timeless Way of Building*. Oxford, 1979.

6. Andrew Nikiforuk. "You and Your Slaves." *Energy Bulletin*, May 2011. [online]. [cited August 31, 2011]. energybulletin.net/stories/2011-05-09 /you-and-your-slaves.

7. Associated Press. "NTSB Blames Texting in Deadly Calif. Rail Crash." January 21, 2010. [online]. [cited September 14, 2011]. msnbc.msn .com/id/34978572/ns/us_news-life/t/ntsb-bla mes-texting-deadly-calif-rail-crash/#.TnFmg BzIu68.

8. Ivan Illich. *Energy and Equity*. Marion Boyars, 1974.

9. UN Food and Agriculture Organization. *Agricultural Production Domain*. [online]. [cited

August 31, 2011]. faostat.fao.org/site/339/default
.aspx.

10. Paul B Hamel, Mary U. Chiltoskey. *Cherokee Plants: Their Uses — A 400 Year History*. 1975.

11. Peter Bane, "Climate Change and Tree Migration." *Permaculture Activist* #68 (May 2008).

12. Alan Kapuler was director of research for Seeds of Change. His writing is directed to serious plant breeders but circulates largely outside the academic world. See: Peace Seeds website. [online]. [cited September 9, 2011]. peaceseeds .cn; Mushroom Blog. [online]. [cited September 9, 2011]. mushroomsblog.blogspot.com /2005_01_01_archive.html.

13. Community Based Food and Farming. "Nikolai Vavilov's Seeds and the Siege of Leningrad." [online]. [cited September 14, 2011]. safs.msu .edu/culturaldiv/seed%20and%20gardening% 20stories.htm.

14. ETC Group. "Oligopoly, Inc. 2005." *Communique* #91 (November/December 2005). [online]. [cited November 12, 2011]. etcgroup.org/up load/publication/44/01/oligopoly2005_16dec .05.pdf.

15. John Michael Greer. *The Long Descent: A User's Guide to the End of the Industrial Age*. New Society, 2008.

### Chapter 5: Learning the Language of Design

1. USDA Natural Resource Conservation Service (formerly Soil Conservation Service). [online]. [cited September 19, 2011]. nrcs.usda.gov; USDA NRCS Soils website. [online]. [cited September 20, 2011]. soils.usda.gov.

2. Agriculture and Agri-Food Canada. Canadian Soil Information Service. [online]. [cited September 19, 2011]. sis2.agr.gc.ca/cansis/intro .html.

3. NNDC Online Climate Data Directory. [online]. [cited September 20, 2011]. lwf.ncdc.noaa .gov/oa/climate/climatedata.html; Environment Canada. *Climate and Historical Weather*. [online]. [cited September 20, 2011]. ec.gc.ca /meteo-weather/default.asp?lang=En&n=17A7 AAB9-1.

4. Ian L. McHarg. *Design with Nature*. Wiley, 1995.

5. Dave Jacke with Eric Toensmeier. *Edible Forest Gardens, Volume 2: Ecological Design And Practice For Temperate-Climate Permaculture*. Chelsea Green, 2005, pp. 174–179.

6. The phrase *pattern language* came to currency

with the publication of Alexander et al's *A Pattern Language*.

7. Reliable Prosperity presents a hyperlinked but unnumbered pattern language for the Pacific Northwest bioregion that is a model for other regions: Reliable Prosperity. [online]. [cited September 19, 2011]. reliableprosperity.net/.

8. Douglas Schuler. *Liberating Voices: A Pattern Language for Communication Revolution*. MIT, 2008. See also: Public Sphere Project. [online]. [cited September 19, 2011]. Publicsphereproject .org.

9. Dave Jacke and Eric Toensmeier. *Edible Forest Gardens*. Chelsea Green, 2005.

### Chapter 6: A Garden Farming Pattern Language

1. Alexander et al. *A Pattern Language*.

2. Rob Hopkins. *The Transition Handbook: From Oil Dependency to Local Resilience*. Chelsea Green, 2008.

3. University of Michigan Center for Sustainable Systems Factsheets. *U.S. Cities*. [online]. [cited September 19, 2011]. css.snre.umich.edu /css_doc/CSS09-06.pdf; Sukkoo Kim. *Changes in the Nature of Urban Spatial Structure in the United States, 1890–2000*. [online]. [cited September 19, 2011]. soks.wustl.edu/density.pdf.

4. Alexander et al. *A Pattern Language*, pp. 81–82; Robert and Diane Gilman. *Eco-Villages and Sustainable Communities*. Context Institute, 1991, p. 7.

5. University of Liverpool. "The Ultimate Brain Teaser." *Research Intelligence* #17 (August 2003). [online]. [cited September 19, 2011]. liv.ac.uk /researchintelligence/issue17/brainteaser.html.

6. Hildur Jackson and Karen Svensson, eds. *Ecovillage Living: Restoring the Earth and Her People*. Green, 2002.

7. Kathryn M. McCamant et al. *Cohousing: A Contemporary Approach to Housing Ourselves*, 2nd ed. Ten Speed, 1994.

8. Diana Leafe Christian. *Creating a Life Together: Practical Tools to Grow Ecovillages and Intentional Communities*. New Society, 2003.

9. The figure of 10 gallons of water use per person per day is based on multiyear monitoring of our own household across seasons, rates of occupancy and activity levels. We employ low-flow shower fixtures and low-volume flush toilets. We reuse dishwater for toilet flushing, and average five loads of laundry each month for

two people. Larger households are able to realize greater efficiencies from shared cooking
and dishwashing.

10. National Hedgelaying Society website. [online].
[cited September 19, 2011]. hedgelaying.org.uk/.

11. Arthur Hollins, a British farmer, pioneered the
renewal of *foggage* on his Fordhall Farm, reducing dependence on purchased inputs and
even on feeding of winter hay. His successors
continue this work. Fordhall Community Land
Initiative. *Foggage Farming*. [online]. [cited
September 19, 2011]. fordhallfarm.com/fordhall
_farm.php?pid=9.

12. John Jeavons. *How to Grow More Vegetables
(Than You Ever Thought Possible on Less Land
Than You Can Imagine)*, 7th ed. Ten Speed, 2006.

### Chapter 7: Land—Scales and Strategies

1. Mark Dowie. *Conservation Refugees: The Hun-
dred-Year Conflict between Global Conservation
and Native Peoples*. MIT, 2009.

2. US Department of Agriculture. "TABLE 3. Estimated Calorie Requirements (in Kilocalories) for Each Gender and Age Group at Three
Levels of Physical Activity." Chapter 2, *Dietary
Guidelines for Americans 2005*. [online]. [cited
September 28, 2011]. health.gov/dietaryguide
lines/dga2005/document/html/chapter2.htm.

3. Pat Murphy. *Plan C: Community Survival Strategies for Peak Oil and Climate Change*. New Society, 2008, Chapter 13.

4. A. N. Duckham and G. B. Masefield. *Farming
Systems of the World*. Praeger, 1969.

5. Dr. Fuhrman. *Nutrient Density*. [online]. [cited
November 28, 2011]. drfuhrman.com/library
/article17.aspx; Adam Drewnowski. "Concept
of a Nutritious Food: Toward a Nutrient Density Score." *American Journal of Clinical Nutrition*, Vol. 82#4 (October 2005), pp. 721–732.
[online]. [cited November 28, 2011]. ajcn.org
/content/82/4/721.full; Virginia Worthington.
"Nutritional Quality of Organic vs. Conventional Fruit, Vegetables, and Grains." *Journal
of Alternative and Complementary Medicine*,
Vol. 7#2 (2001), pp. 161–173. [online]. [cited
November 28, 2011]. liebertonline.com/doi/abs
/10.1089/107555301750164244.

6. David Pimentel and Marcia Pimentel. "Land,
Water, and Energy Versus the Ideal U.S. Population." *NPG Forum*, January 2005. [online].
[cited September 29, 2011]. npg.org/forum
_series/land_water_energy%20_pimentel
_forumpaper.pdf.

7. UN Food and Agriculture Organization.
FAOSTAT. [online]. [cited September 28, 2011].
faostat.fao.org/site/377/DesktopDefault.aspx?
PageID=377#ancor; Ruben N. Lubowski et al.
*Major Uses of Land in the United States, 2002*.
US Department of Agriculture Economic Research Service. [online]. [cited September 28,
2011]. ers.usda.gov/publications/EIB14/eib14a
.pdf.

8. Sue Kirchhoff. "Surplus U.S. Food Supplies
Dry Up." *USA Today*, May 2, 2008. [online].
[cited September 28, 2011]. usatoday.com/
money/industries/food/2008-05-01-usda-food
-supply_N.htm; Gardenserf. *The Strategic
Grain Reserve*. February 5, 2011. [online]. [cited
September 28, 2011]. gardenserf.wordpress.com
/2011/02/05/the-strategic-grain-reserve/.

9. US Department of Agriculture. *Economics, Statistics, and Marketing Information System*. [on-
line]. [cited September 28, 2011]. usda.mannlib
.cornell.edu/MannUsda/homepage.do.

10. R. L. Dalrymple. *Controlled Rotation Grazing
Unit*. Samuel Roberts Noble Foundation. [on-
line]. [cited November 28, 2011]. noble.org/ag
/forage/rotation/forgcrgu.htm.

11. Alan Chadwick. *Performance in the Garden: A
Collection of Talks on Biodynamic French In-
tensive Horticulture*. Logosophia, 2007; John
Jeavons, *How to Grow More Vegetables*.

12. Weston A. Price. *Nutrition and Physical Degen-
eration* originally published in 1938. [online].
[cited September 29, 2011]. gutenberg.net.au
/ebooks02/0200251h.html.

13. Masanobu Fukuoka. *The One-Straw Revo-
lution: An Introduction to Natural Farming*.
Rodale, 1979, reprinted in 2009 by New York
Review of Books.

### Chapter 8: Labor—Can You Lend a Helping Hand?

1. U.S. Census Bureau. "Figure 5-3, Average
Household Size 1900 and 1930 to 2000." *Demographic Trends in the 20th Century*, CENSR-4,
November 2002, p. 143. [online]. [cited November 21, 2011]. census.gov/prod/2002pubs
/censr-4.pdf.

2. Pew Social Trends Staff. *The Return of the
Multi-Generational Family Household*. Pew Re-
search Centre, March 18, 2010. [online]. [cited
October 1, 2011]. pewsocialtrends.org/2010

/03/18/the-return-of-the-multi-generational-family-household.

3. Gregson. *The Rebirth of the Small Family Farm.*

4. IBISWorld. *Fertilizer Manufacturing in the US: Market Research Report.* NAICS 32531, August 2011. [online]. [cited October 1, 2011]. ibisworld.com/industry/default.aspx?indid=480.

5. People-Powered Machines. *Cleaner Air: Gas Mower Pollution Facts.* [online]. [cited October 1, 2011]. peoplepoweredmachines.com/faq-environment.htm.

6. Michael J. Rosenfeld. "American Couples: How Couples Meet, and Whether They Stay Together." April 1, 2007. [online]. [cited October 1, 2011]. stanford.edu/~mrosenfe/concept%20sheet,%20how%20couples%20meet.pdf.

7. WWOOF — World Wide Opportunities on Organic Farms website. [online]. [cited September 12, 2011]. wwoof.org/index.asp.

8. For example (all [online]. [cited October 1, 2011]): NOFA Massachusetts. *2011 Directory of Organic Farming Apprenticeship Programs.* nofamass.org/programs/apprentice/apprentice.php; Ohio Ecological Food and Farm Association. *The Apprenticeship Program.* oeffa.org/app-overview.php; Oregon Tilth. *Farm Job and Internship Resources.* tilth.org/education-research/organic-education-center/agriculture-internship-resources; Carolina Farm Stewardship Assn. *Internship Referral Service.* carolinafarmstewards.org/internshipboard.shtml.

### Chapter 9: Running on Sunshine

1. Biodiversity International, J. G. Hawkes. "Back to Vavilov: Why Were Plants Domesticated in Some Areas and Not in Others?" *Distribution of Agricultural Origins, Part 1 — Centers of Origin of Crop Plants and Agriculture.* [online]. [cited October 7, 2011]. www2.bioversityinternational.org/publications/Web_version/47/ch06.htm#Part%201.%20Centers%20of%20Origins%20of%20Crop%20Plants%20and%20Agriculture.

2. As Copernicus pointed out 500 years ago, in the solar system, the sun is the fixed center, and the Earth and the other planets and asteroids do all the moving, but for purposes of analyzing our solar environment on Earth, I'll follow the common convention of talking about the sun's apparent movement through the sky.

3. For instance: SunEarthTools.com. *Sun Position.*

[online]. [cited October 7, 2011]. sunearthtools.com/dp/tools/pos_sun.php.

4. US Environmental Protection Agency. *Indoor Air Pollution: An Introduction for Health Professionals.* Footnote #1. [online]. [cited October 6, 2011]. epa.gov/iaq/pubs/hpguide.html.

### Chapter 10: Water from Another Time

1. Chorus from: John McCutcheon. "Water from Another Time." *Water from Another Time*, 1985. Elyrics.net. [online]. [cited October 7, 2011]. elyrics.net/read/j/john-mcCutcheon-lyrics/water-from-another-time-lyrics.html.

2. Toby Hemenway. "The Wisdom of the Beaver." *Permaculture Activist* #47 (May 2002).

3. Robin Clarke. *Water: The International Crisis.* MIT, 1991.

4. The Bloomington Peak Oil Taskforce. *Redefining Prosperity: Energy Descent and Community Resilience.* October 2009. [online]. [cited December 8, 2011]. http://bloomington.in.gov/media/media/application/pdf/6239.pdf.

5. Traditional water storage tanks in dry regions of rural India were filled with sand to prevent evaporation. Water stored in the pore space of the sand could be drained through a tap in the bottom of the open-topped tank. The sand surface also prevented both the growth of mosquitos and contamination of the water source. Personal communication, Mike Feingold. See also: Ake Nilsson. *Groundwater Dams for Small-Scale Water Supply.* Practical Action, 1988.

6. David Johnson. "The Nasca Lines: Geoglyphs Reveal an Ancient Water Map." *Permaculture Activist* #51 (February 2004).

7. Dr. Dieter Prinz. "The Role of Water Harvesting in Alleviating Water Scarcity in Arid Areas." Keynote address, International Conference on Water Resources Management in Arid Regions, March 23–27, 2002. [online]. [cited October 7, 2011]. ipcp.org.br/References/Agua/aguaCapta/WaterHarvesting.pdf.

8. Wellowner.org. *The Ground Water Supply and Its Use.* [online]. [cited October 7, 2011]. wellowner2.org/2009/index.php?option=com_content&view=category&id=48&layout=blog&Itemid=46.

9. US Geological Survey. "Ground-Water Depletion Across the Nation." Fact Sheet 103-03, November 2003. [online]. [cited October 7, 2011]. pubs.usgs.gov/fs/fs-103-03/.

10. Brian Turnbaugh. "EPA Finds Secret Fracking Chemical in Drinking Water." OMB Watch, August 31, 2009. [online]. [cited October 7, 2011]. ombwatch.org/node/10353.

11. A search engine for locating industry professionals: American Rainwater Catchment Systems Association. Online Directory. [online]. [cited October 7, 2011]. arcsa.org/AF_Member Directory.asp.

12. S. B. Watt. *Ferrocement Tanks and their Construction. Practical Action*, 1978; Art Ludwig. *Water Storage: Tanks, Cisterns, Aquifers, and Ponds for Domestic Supply, Fire and Emergency Use—Includes How to Make Ferrocement Water Tanks.* Oasis, 2005.

13. Nigel Dunnett and Noël Kingsbury. *Planting Green Roofs and Living Walls*, rev. ed. Timber, 2008.

14. Art Ludwig. *Creating an Oasis with Greywater: Choosing, Building and Using Greywater Systems—Includes Branched Drains*, 5th ed. Oasis, 2006.

15. Carol Steinfeld. *Liquid Gold: The Lore and Logic of Using Urine to Grow Plants.* EcoWaters, 2004.

16. Brad Lancaster. "Watergy: Where Water and Energy Meet." *Permaculture Activist* #78 (November 2010).

17. Paul Stamets. *Mycelium Running: How Mushrooms Can Help Save the World.* Ten Speed, 2005.

18. William O. McLarney. *The Freshwater Aquaculture Book: A Handbook for Small Scale Fish Culture in North America.* Hartley & Marks, 1984.

19. Takao Furuno. *The Power of Duck.* Tagari, 2001.

20. Erik Andrus. "Rice—in Vermont?" *Permaculture Activist* #82 (November 2011).

21. Michael R. Miltner. "The Farmed Wetland Alternative: Crawfish and Baldcypress Co-propagation in the South Delta Region." *The International Permaculture Solutions Journal*, Vol. 1# 1 (1990).

22. McClarney. *The Freshwater Aquaculture Book.*

23. Bill Mollison. *Permaculture: A Designers Manual.* Tagari, 1988, chapter 13.

24. Common carp (*Cyprinus carpio*) are completely naturalized in North America except in Maine, Alaska and peninsular Florida. They are found in Canada in the St. Lawrence River, the Great Lakes and tributaries, in the Lake Winnipeg drainage and in other rivers near the US border.

25. Steven Van Gorder. *Small Scale Aquaculture.* Alternative Aquaculture Association, 2000.

26. McLarney. *The Freshwater Aquaculture Book*, p. 21.

27. Van Gorder. *Small Scale Aquaculture.*

28. Laurence Hutchinson. *Ecological Aquaculture: A Sustainable Solution.* Permanent Publications, 2006.

29. McLarney. *The Freshwater Aquaculture Book.*

**Case Study A**

1. CRMPI. *Natural Controls for Noxious Weeds Part I.* [online]. [cited January 23, 2012]. you tube.com/watch?v=W2BQyfaywuw.

**Chapter 11: Soil – the Real Dirt**

1. Albert K. Bates. *The Biochar Solution: Carbon Farming and Climate Change.* New Society, 2010.

2. Jan D. Elsas, Janet K. Jansson and Jack T. Trevors. *Modern Soil Microbiology*, 2nd ed. CRC Press, 2006, p. 115.

3. Stamets. *Mycelium Running.*

4. John B. Marler and Jeanne R. Wallin, "Human Health, the Nutritional Quality of Harvested Foods, and Sustainable Farming Systems." nutritionsecurity.org/PDF/NSI_White%20 Paper_Web.pdf [online]. [cited December 26, 2011].

5. Alan M. Smith. "Living Soil." *Permaculture International Journal* #7 (March 1981); Lea Harrison. "Soil Fertility." *Permaculture Activist* #26 (May 1992).

6. Ronald Nigh. "Trees, Fire, and Farmers." *Permaculture Activist* #72 (May 2009).

7. Preston Sullivan. "Tillage, Organic Matter, and Plant Productivity." ATTRA *Sustainable Soil Management Soil System Guide*, July, 1999. [online]. [cited October 20, 2011]. soilandhealth.org/01aglibrary/010117attrasoilmanual/010117 attra.html#tillage.

8. Susana Lein. "Growing Corn Among Three Sisters." *Permaculture Activist* #82 (November 2011).

9. William Albrecht. "Soil Reaction (pH) and Balanced Plant Nutrition." December 1967. [online]. [cited October 20, 2011]. soiland health.org/01aglibrary/010143albpap/ph.bal anced%20nutrition/ph.bal.nut.htm. Though he deserves credit for puzzling out the issue

of mineral balance, now widely accepted by agronomists and farmers, Albrecht's understanding of the soil's self-fertilizing processes was arguably incomplete, lacking as it did the insight provided by Alan Smith's research into the oxygen-ethylene cycle in the 1970s.

10. For the US see: Benjamin A. Goldman. *The Truth About Where You Live*. Times, 1991. Regarding Canada: Dr. Ian Fairlie. *Tritium Hazard Report: Pollution and Radiation Risk from Canadian Nuclear Facilities*. Greenpeace, June, 2007. [online]. [cited December 20, 2011]. greenpeace.org/canada/en/campaigns/end-the -nuclear-threat/Resources/Reports/tritium -hazard-report-pollu/.

11. Ag-USA. "Brix Readings and What They Tell Us." [online]. [cited October 19, 2011]. ag-usa .net/brix_test_meaning.htm; Toby Balsom and Graham Lynch. "Monitoring pasture quality using brix measurements." Novel Ways. Hamilton, New Zealand, December 17, 2008. [online]. [cited December 26, 2011]. novel.co.nz/up loads/76545/files/136209/Brix_Measurements .pdf.

12. US Environmental Protection Agency Ag 101. "Environmental Concerns Related to Soil Preparation: Soil Erosion." [online]. [cited October 19, 2011]. epa.gov/agriculture/ag101/cropsoil .html; Patricia S. Muir. "B. EROSION — Erosion from Inappropriate Agricultural Practices on Crop Lands." Oregon State University: BI301 Human Impacts On Ecosystems. [online]. [cited October 19, 2011]. people.oregon state.edu/~muirp/erosion.htm.

13. US EPA. "2006–2007 Pesticide Market Estimates — Historical Data." [online]. [cited October 19, 2011]. epa.gov/opp00001/pestsales /07pestsales/historical_data2007_3.htm#5_2; Agriculture and Agri-Food Canada. "A Review of Agricultural Pesticide Pricing and Availability in Canada." 3.0 Industry Structure, Trends and Overview (Part I). [online]. [cited October 19, 2011]. www4.agr.gc.ca/AAFC-AAC/display -afficher.do?id=1180033775119#a3.2.

14. Norman Thomas Uphoff. *Biological Approaches to Sustainable Soil Systems*. CRC Press, 2006, p. 696.

15. Brad Lancaster. *Rainwater Harvesting for Drylands and Beyond (Vol. 2): Water-Harvesting Earthworks*. Rainsource, 2007.

16. The information for legumes and grasses is based primarily on Edwin McLeod's indispensable *Feed the Soil* (McLeod, 1982). Lys de Bray's *The Wild Garden: An Illustrated Guide to Weeds* (Mayflower, 1978) and F. Newman Turner's *Fertility Pastures and Cover Crops*, 3rd ed. (Bargyla and Gylver Rateaver, 1975) provided the basis for much of the section on forbs, while enriching the whole. For nomenclature and other guidance, I have consulted U. P. Hedrick, *Sturtevant's Edible Plants of the World* (Dover, 1972), Stephen Facciola's *Cornucopia II: A Source Book of Edible Plants* 2nd ed. (Kampong, 1998) and USDA Natural Resources Conservation Service. *Plants Database* [online]. [cited January 2012]. plants.usda.gov/java/.

17. Jeff Lowenfels and Wayne Lewis. *Teaming with Microbes: The Organic Gardener's Guide to the Soil Food Web*, rev. ed. Timber, 2010.

18. Steinfeld. *Liquid Gold*.

19. Stamets. *Mycelium Running*.

20. See: Sunseed Desert Technology. "Making Your Own Mycorrhizal Inoculum." *Permaculture Activist* #61 (August 2006).

## Chapter 12: Plants, Crops and Seeds

1. S. Padulosi et al. "Underutilized Crops: Trends, Challenges, and Opportunities in the 21st Century." Presented at Accra, Ghana, November 16–18, 2006. [online]. [cited October 26, 2011]. moringanews.org/doc/GB/GFU/Document8 .pdf.

2. Facciola. *Cornucopia II*. An extremely useful database for exploring plant diversity is ZipcodeZoo. [online]. [cited January 5, 2012]. zipcodezoo.com.

3. I use a term that expresses relationship, rather than the traditional term *kingdom*.

4. Thomas J. Elpel. *Botany in a Day*, 5th ed. HOPS, 2004.

5. North American Plant Distributions. [online]. [cited January 3, 2012]. hua.huh.harvard.edu/ FNA/Outreach/FNA-lesson_biomes.shtml. Also, Nature Serve: Conservation Issues. [online]. [cited January 3, 2012]. natureserve.org /consIssues/caAtlas.jsp.

6. Cary Fowler and Pat Mooney. *Shattering: Food Politics, and the Loss of Genetic Diversity*. University of Arizona, 1990.

7. Jack Doyle. *Altered Harvest: Agriculture, Genetics, and The Fate of the World's Food Supply*. Penguin, 1986.

8. Ibid.

9. Craig Mackintosh. "Australia's First Legal Attack on Monsanto for GM Contamination of Organically Certified Crops." Permaculture Research Institute, February 1, 2011. [online]. [cited October 26, 2011]. permaculture.org.au/2011/02/01/australias-first-legal-attack-on-monsanto-for-gm-contamination-of-organically-certified-crops/.

10. Peter Costantini. "Haitian Farmers Leery of Monsanto's Largesse." Inter Press News Service via CorpWatch, July 1, 2010. [online]. [cited October 26, 2011]. corpwatch.org/article.php?id=15608.

11. Kirk Makin. "Canada Rules in Favor of Monsanto over Seed Saving Farmer Percy Schmeiser." Globe and Mail Update, May 21, 2004. [online]. [cited October 28, 2011]. organicconsumers.org/ge/schmeiser.cfm.

12. Seeds of Deception. *The Health Risks of GM Foods: Summary and Debate.* [online]. [cited October 28, 2011]. seedsofdeception.com/Public/GeneticRoulette/HealthRisksofGMFoodsSummaryDebate/index.cfm.

13. Toby Hemenway. *Gaia's Garden: A Guide To Home-Scale Permaculture.* Chelsea Green, 2000, p. 149.

14. Carol Deppe. *The Resilient Gardener: Food Production and Self-Reliance in Uncertain Times.* Chelsea Green, 2010.

15. Smith. *Tree Crops.* Re roots of honey mesquite (*Prosopis glandulosa*), see Paul W. Cox and Patty Leslie. *Texas Trees: A Friendly Guide.* Corona, 1991.

16. Carol Deppe. *Breed Your Own Vegetable Varieties: The Gardener's & Farmer's Guide to Plant Breeding & Seed Saving*, 2nd ed. Chelsea Green, 2000; Suzanne Ashworth and Kent Whealy. *Seed to Seed: Seed Saving and Growing Techniques for Vegetable Gardeners*, 2nd ed. Seed Savers, 2002; Nancy Bubel. *The New Seed Starter's Handbook.* Rodale, 1988.

17. David Tracey. *Guerrilla Gardening: A Manualfesto.* New Society, 2007.

**Chapter 13: Setting Plant Priorities**

1. For an excellent overview of pumpkin/squash genetics and varieties, their culinary and keeping qualities, see Deppe's *The Resilient Gardener.*

2. Frances Moore Lappé. *Diet for a Small Planet.* Ballantine, 1991.

3. Gene Logsdon. *Small-Scale Grain Raising*, 2nd ed. Chelsea Green, 2009.

**Case Study C**

1. Logsdon. *Small-Scale Grain Raising.*

**Chapter 14: Animals for the Garden Farm**

1. Kentucky Equine Research. "Changes in the Horse World." *Equine Review* (2007), *Horse World 37.* [online]. [cited November 5, 2011]. ker.com/library/EquineReview/2007/HorseWorld/HW37.pdf.

2. Fernando Funes et al, eds. *Sustainable Agriculture and Resistance.* Food First, 2002.

3. US National Research Council. *Microlivestock: Little-Known Small Animals With a Promising Economic Future.* National Academy Press, 1991.

4. Rare Breeds Survival Trust UK. [online]. [cited November 5, 2011]. rbst.org.uk; American Livestock Breeds Conservancy. [online]. [cited November 5, 2011]. albc-usa.org. Both are membership organizations.

5. Sally Fallon and Mary G. Enig. *Nourishing Traditions: The Cookbook that Challenges Politically Correct Nutrition and the Diet Dictocrats*, 2nd ed. rev. New Trends, 1999.

6. Dave Holderread. *Raising the Home Duck Flock.* Storey, 1980; *The Book of Geese: A Complete Guide to Raising the Home Flock.* Hen House, 1993.

7. Saubine Maubouche. "Carrier Pigeons." *Washington Post*, December 2, 1986, quoted in *Microlivestock*, p. 143.

8. American Rabbit Breeders Association. *ARBA Recognized Breeds.* [online]. [cited November 5, 2011]. arba.net/breeds.htm.

9. Jane Hunnicutt. "Rabbits Love Roses." *Permaculture Activist* #47 (June 2002), p. 47.

10. Marjory Wildcraft. *Food Production Systems for a Backyard or Small Farm.* DVD. Backyard Food Production, 2009.

11. See *Microlivestock* and *Eating Guinea Pigs.* [online]. [cited November 10, 2011]. youtube.com/watch?v=AoBTctUfPvI&feature=related.

12. Simon Fairlie. *Meat: A Benign Extravagance.* Chelsea Green, 2010.

13. Peter Bane. "Animal Polyculture." *Permaculture Activist* #32 (April 1995).

14. R. H. Jongman. *The New Dimensions of the European Landscape.* Springer, 2004, p. 144;

Joachim Radkau. *Nature and Power: A Global History of the Environment.* Cambridge, 2008, p. 61.

15. Jim Corbett. *Goatwalking: A Guide to Wildland Living.* Penguin, 1992.
16. David MacKenzie. *Goat Husbandry*, rev. 5th ed. Faber, 1993.
17. Wildcraft. *Food Production Systems.*

**Chapter 15: Living with Wildlife**
1. Hemenway. *Gaia's Garden.*

**Chapter 16: Trees and Shrubs, Orchards, Woodlands and Forest Gardens**
1. Elizabeth Mygatt. "World's Forests Continue to Shrink." Earth Policy Institute: Eco-Economy Indicators, Forest Cover, April 4, 2006. [online]. [cited November 9, 2011]. earth-policy .org/indicators/C56; World Resources Institute. *State of the World's Forests.* January 8, 2009. [online]. [cited November 9, 2011]. wri.org /map/state-worlds-forests.
2. Bates. *Biochar Solution.*
3. See, for example, Peter Dauvergne. "The Political Economy of Indonesia's 1997 Forest Fires." *Australian Journal of International Affairs* Vol. 52#1 (1998). [online]. [cited November 9, 2011]. politics.ubc.ca/fileadmin/user_upload/poli _sci/Faculty/dauvergne/AustralianJournalofIA .pdf; Susan E. Page et al. "The Amount of Carbon Released from Peat and Forest Fires in Indonesia during 1997." *Nature* 420 (November 7, 2002), pp. 61–65. [online]. [cited November 9, 2011]. nature.com/nature/journal/v420/n6911 /abs/nature01131.html.
4. Ruth Loomis with Merv Wilkinson. *Wildwood: A Forest for the Future.* Reflections, 1990; "Seven Generations of Forestry." *Permaculture Activist* #40 (December 1998).
5. James DeMeo. *Saharasia: The 4,000 BCE Origins of Child Abuse, Sex-Repression, Warfare and Social Violence, In the Deserts of the Old World.* Natural Energy Works, 2006.
6. Charles C. Mann. *1491: New Revelations of the Americas Before Columbus.* Knopf, 2005.
7. William H. McNeill. *Plagues and Peoples.* Doubleday, 1967.
8. Badgersett Research Corporation: Woody Agriculture Research and Development. [online]. [cited November 9, 2011]. badgersett.com.

9. Oikos Tree Crops. [online]. [cited November 9, 2011]. oikostreecrops.com.
10. Bane. "Keystones and Cops."
11. Mollison. *Permaculture Designer's Manual*, p. 64. See Rob Scott and William C. Sullivan. "Black Walnut Polycultures." *Permaculture Activist* #68 (May 2008). I have had success with salad crops and tomatillos under *J. nigra*, even though Solanum genus plants are reputedly the most sensitive to juglone.
12. A lengthy list of nurseries specializing in edible and economic species can be found at: Permaculture Activist "Nursery Sources for Edible & Useful Plants." [online]. [cited October 20, 2011]. permacultureactivist.net/nurseries/Plnt Nursrys.htm.
13. The most comprehensive treatment of the subject to date is: Dave Jacke and Eric Toensmeier. *Edible Forest Gardens.* Chelsea Green, 2005. See also notes 14, 15, 18.
14. *Agroforestry News* is a quarterly journal of temperate agroforestry and forest gardening published by Agroforestry Research Trust, UK and edited by Martin Crawford. The journal is now in its 20th year. Current subscriptions and back issues are available in North America through permacultureactivist.net.
15. Robert A. de J. Hart. *Forest Gardening.* Green, 1991.
16. Dave Jacke and Eric Toensmeier. "A Story of Robert" in *Edible Forest Gardens* Vol. 2, p. 451.
17. Simon Henderson. "Raising the Dragon: Bamboo Agroforestry in Vietnam." *Permaculture Activist* #34 (June 1996); Geoff Lawton. "Establishing a Food Forest." ecofilms.au.com. [online]. [cited November 10, 2011]. youtube.com /watch?gl=AU&hl=en-GB&v=-5ZgzwoQ-ao.
18. Martin Crawford. *Creating a Forest Garden: Working with Nature to Grow Edible Crops.* Green, 2010.
19. Masanobu Fukuoka. *The Natural Way of Farming: The Theory and Practice of Green Philosophy.* Japan Publications, 1985.
20. Stuart B. Hill. "Controlling the Plum Curculio, The Hunchback of the Apple Orchard." Ecological Agriculture Projects, McGill University, 1989. [online]. [cited November 10, 2011]. eap .mcgill.ca/publications/EAP60.htm; M. Sean Clark and Stuart H. Gage. "Effects of Free-range Chickens and Geese on Insect Pests and

Weeds in an Agroecosystem." *American Journal of Alternative Agriculture*, Vol. 11 (1996), pp. 39–47. [online]. [cited January 3, 2012]. journals.cambridge.org/action/displayAbstract?fromPage=online&aid=6362368.

21. Mountain Gardens. [online]. [cited November 10, 2011]. mountaingardensherbs.com.

22. Christopher Alexander. *The Nature of Order: An Essay on the Art of Building and the Nature of the Universe, Book 1 — The Phenomenon of Life*. Center for Environmental Structure, 2001, p. 134.

23. The Wildlands Network. *Wildways*. [online]. [cited November 10, 2011]. twp.org/wildways.

## Chapter 17: Productive Trees and Where to Grow Them

1. Images of 70-year-old living fences in Germany, grown from whitebeam and forsythia; many of the trees are pleached, even woven together in striking patterns: Marc's Projekte Homepage. [online]. [cited November 12, 2011]. d-marc.de/natur/ohrdruf/index.htm.

2. Patrick Whitefield. *The Earth Care Manual: A Permaculture Handbook for Britain & Other Temperate Climates*. Permanent, 2004, p. 276.

3. Andrew M. Gordon and Steven M. Newman, eds. *Temperate Agroforestry Systems*. CABI, 1997.

4. Michael R. Miltner. "The Farmed Wetland Alternative: Crawfish and Baldcypress." *The International Permaculture Solutions Journal*, Vol. 1#1 (1990).

5. Richard Wade, personal communication, May 2005.

6. [all cited November 11, 2011]. *Willow Evapotranspiration Systems*. Backlund. [online]. backlund.dk/prod05.htm; Jan Vymazal and Lenka Kröpfelová. *Wastewater in Constructed Wetlands with Horizontal Subsurface Flow*. Springer, 2008. [online]. scribd.com/doc/51369272/30/Zero-discharge-systems; Ioannis Dimitriou and Pär Aronsson. "Willows for Energy and Phytoremediation in Sweden." FAO Forestry Department, 2008. [online].

7. Stamets. *Mycelium Running*.

8. A cord is a volume measure of stacked wood 8 × 4 × 4 feet. A face-cord is ⅓ of that: 8 × 4 feet × 16 inches, also called a rick. A typical pick-up truck will carry about a rick of wood.

9. The weight of wood varies by species, and also wood loses weight as it dries. A cord of oak weighs about 3,800 pounds, depending on which species is measured. A cord of silver maple, generally thought to be almost a weed tree, weighs about 3,000 pounds. Most hardwoods fall in between these values. Softwoods are somewhat lighter. All wood has much the same heat value per pound.

10. Stamets. *Mycelium Running*.

## Chapter 18: Structures, Energy and Technology

1. Eben V. Fodor. *The Solar Food Dryer: How to Make and Use Your Own Low-Cost, High Performance, Sun-Powered Food Dehydrator*. New Society, 2006.

2. Greg Travis, personal communication; see also The City of Bloomington. *Peak Oil Task Force*. [online]. [cited January 4, 2012]. bloomington.in.gov/peakoil/.

3. Our all-steel wheelbarrows are made by Jackson Professional Tools in Harrisburg, PA (jacksonprofessional.com/products. [online]. [cited November 11, 2011]). The model shown is the 6-cubic-foot, narrow tray contractor model. It appears that Sears carries or can supply these to retail customers.

4. The Original "Sunny John" Solar Moldering Toilet. [online]. [cited October 26, 2011]. sunnyjohn.com/toiletpapers2.htm; Joseph Jenkins. *The Humanure Handbook: A Guide to Composting Human Manure*, 3rd ed. Jenkins, 2005.

5. Ecosystems Design identifies its under-bed greenhouse heating technology as "Subterranean Heating and Cooling System," however, in ordinary conversation, the principals of the firm, Michael Thompson and Jerome Osentowski (who credit John Cruickshank with much of the innovation involved), refer to the system incessantly as "the climate battery." Personal communication, August 2011; EcoSystems Design, Inc. *Greenhouse Designs*. [online]. [cited November 12, 2011]. ecosystems-design.com/Greenhouse Designs.html.

## Case Study D

1. Mollison. *Permaculture*.

2. US Department of Agriculture. *Farm Service Agency*. [online]. [cited November 23, 2011]. fsa.usda.gov.

3. Emilia Hazelip. "The Synergistic Garden." DVD, 1995. Emilia Hazelip was the leading permaculture teacher in France, Spain and Italy until her untimely death in 2003. She studied biointensive methods of gardening with Alan Chadwick in California in the 1960s. This film is the primary testament to her life work championing the natural farming philosophy and no-tillage methods of Masanobu Fukuoka and French grain cultivator Marc Bonfils. Hazelip's legacy also includes the work of Las Encantadas, an association of practitioners for research and teaching permaculture, now located in Brussels.

### Chapter 19: Diet and Food

1. Duckham and Masefield. *Farming Systems of the World.* These calories are really kilocalories in strict scientific jargon.
2. Richard Heinberg and Michael Bomford. *The Food and Farming Transition: Toward a Post Carbon Food System.* Post Carbon Institute, March 30, 2009. [online]. [cited November 15, 2011]. postcarbon.org/report/41306-the-food-and-farming-transition-toward.
3. Luis de Sousa, "What Is a Human Being Worth (in Terms of Energy)?" The Oil Drum: Europe, July 20, 2008. [online]. [cited November 15, 2011]. theoildrum.com/node/4315.
4. USDA Agriculture Factbook. "Chapter 2 — Profiling Food Consumption in America." [online]. [cited November 15, 2011]. usda.gov/factbook/chapter2.pdf.
5. John B. Marler and Jeanne R. Wallin. "Human Health, the Nutritional Quality of Harvested Food, and Sustainable Farming Systems." Nutrition Security Institute, 2006 [online]. [cited November 15, 2011]. nutritionsecurity.org/PDF/NSI_White%20Paper_Web.pdf.
6. Fairlie. *Meat.*
7. Lappé. *Diet for a Small Planet.*
8. Stamets. *Mycelium Running.*

### Chapter 20: Culture and Community

1. *World Footprint.* Global Footprint Network. [online]. [cited November 19, 2011]. footprintnetwork.org/en/index.php/GFN/page/world_footprint/.
2. Darrell Posey. "Kayapo Indians: Experts in Synergy." *The Overstory Agroforestry Ejournal* #34. [online]. [cited November 19, 2011]. agroforestry.net/overstory/overstory34.html.
3. Living Machines — Portfolio. [online]. [cited November 19, 2011]. livingmachines.com/portfolio/.
4. Jenkins. *The Humanure Handbook.*
5. USDA National Agricultural Library. *Community Supported Agriculture.* [online]. nal.usda.gov/afsic/pubs/csa/csa.shtml; Ontario CSA Directory. [online]. csafarms.ca/CSA%20farmers.htm. [both cited November 19, 2011].

### Chapter 21: Markets and Outreach

1. Michigan Department of Agriculture and Rural Development. *Michigan Cottage Foods Information.* [online]. [cited January 8, 2012]. michigan.gov/mdard/0,4610,7-125-50772_45851-240577--,00.html.
2. Eliot Coleman. *The Winter Harvest Handbook: Year Round Vegetable Production Using Deep Organic Techniques and Unheated Greenhouses.* Chelsea Green, 2009.
3. Joel Salatin. *Pastured Poultry Profits.* Polyface, 1993 and *Salad Bar Beef.*
4. Van En. *Basic Formula to Create Community Supported Agriculture*; Groh and McFadden. *Farms of Tomorrow.*
5. Gregson. *Rebirth of the Small Family Farm.*
6. For example [both cited November 21, 2011]: Local Harvest. [online]. localharvest.org; Eat-Well Guide. [online]. eatwellguide.org/i.php?pd=Home.
7. For example [both cited November 21, 2011]: The National Sustainable Agriculture Coalition. [online]. sustainableagriculture.net/about-us/members; Canadian Organic Growers. [online]. cog.ca/about_organics/organic-links/.
8. Christopher Bedford. "The Organic Opportunity." DVD. Center for Economic Security, 2007. This inspiring short film tells the story of how Woodbury County in western Iowa used organic agriculture as an engine of economic development.

### Chapter 22: Making the Change

1. Christian. *Creating a Life Together.*
2. Alexander et al. *A Pattern Language*, Pattern 104.

# Index

Page numbers in *italics* indicate illustrations.

# About the Author

PETER BANE has published *Permaculture Activist* magazine for over 20 years and has taught permaculture design widely in the temperate and tropical Americas.

Peter is a native of the Illinois prairie whose interest in good food and simple living led him at mid-life to become a writer and teacher of permaculture design. Its revolutionary ideas of ecological living exposed him to exotic cultures from the Himalayas to the Norse fjords, and the Caribbean isles to the Argentine pampas. They also drew him into the arcane world of intentional community as fate presented the opportunity to help create and build Earthaven Ecovillage in the southern Appalachian Mountains. There

he discovered his inner architect in the course of building a small off-grid solar cabin and later took on the more prosaic job of rehabilitating a pair of suburban ranch houses in the Midwestern college town of Bloomington, Indiana. That was the first step toward creating a small suburban farmstead where he now lives with his partner and apprentices. A prolific writer in journals and collections on forestry, building and all things sustainable, he consults with universities and municipal governments as well as for private landowners.

[Credit: Keith D. Johnson]

If you have enjoyed *The Permaculture Handbook* you might also enjoy other

# BOOKS TO BUILD A NEW SOCIETY

Our books provide positive solutions for people who want to
make a difference. We specialize in:

**Sustainable Living • Green Building • Peak Oil • Renewable Energy
Environment & Economy • Natural Building & Appropriate Technology
Progressive Leadership • Resistance and Community
Educational & Parenting Resources**

---

## New Society Publishers

### ENVIRONMENTAL BENEFITS STATEMENT

New Society Publishers has chosen to produce this book on recycled paper made with
**100% post consumer waste**, processed chlorine free, and old growth free.

For every 5,000 books printed, New Society saves the following resources:[1]

| | |
|---|---|
| 99 | Trees |
| 8,974 | Pounds of Solid Waste |
| 9,874 | Gallons of Water |
| 12,879 | Kilowatt Hours of Electricity |
| 16,314 | Pounds of Greenhouse Gases |
| 70 | Pounds of HAPs, VOCs, and AOX Combined |
| 25 | Cubic Yards of Landfill Space |

[1]Environmental benefits are calculated based on research done by the Environmental Defense Fund and
other members of the Paper Task Force who study the environmental impacts of the paper industry.

---

*For a full list of NSP's titles, please call* 1-800-567-6772 *or check out our website* at:

**www.newsociety.com**

Deep Green for over 30 years